Acoustical Imaging

Volume 16

Acoustical Imaging

A Continuation Order Plan is available for this series. A continuation order will bring delivery of
each new volume immediately upon publication. Volumes are billed only upon actual shipment.
For further information please contact the publisher.

Acoustical Imaging

Volume 16

Edited by

Lawrence W. Kessler

Sonoscan Inc.
Bensenville, Illinois

PLENUM PRESS · NEW YORK AND LONDON

The Library of Congress cataloged the first volume of this series as follows:

International Symposium on Acoustical Holography.

Acoustical holography; proceedings. v. 1–
New York, Plenum Press, 1967–

v. illus. (part col.), ports. 24 cm.

Editors: 1967– . A. F. Metherell and L. Larmore (1967 with H. M. A. el-Sum)
Symposium for 1967– held at the Douglas Advanced Research Laboratories, Huntington Beach, Calif.

1. Acoustic holography — Congresses — Collected works. I. Metherell. Alexander A., ed. II. Larmore, Lewis, ed. III. el-Sum, Hussein Mohammed Amin, ed. IV. Douglas Advanced Research Laboratories, v. Title.

QC244.5.I.5   69-12533

ISBN 0-306-43011-8

Proceedings of the 16th International Symposium on Acoustical Imaging,
held June 10–12, 1987, in Chicago, Illinois

© 1988 Plenum Press, New York
A Division of Plenum Publishing Corporation
233 Spring Street, New York, N.Y. 10013

Printed in the United States of America

PREFACE

This book contains the technical papers presented at the 16th International Symposium on Acoustical Imaging which was held in Chicago, Illinois USA from June 10-12, 1987. This meeting has long been a leading forum for acoustic imaging scientists and engineers to meet and exchange ideas from a wide range of disciplines. As evidenced by the diversity of topical groups into which the papers are organized, participants at the meeting and readers of this volume can benefit from developments in medical imaging, materials testing, mathematics, microsocopy and seismic exploration.

A common denominator in this field, as its name implies, is the generation, display, manipulation and analysis of images made with mechanical wave energy. Sound waves respond to the elastic properties of the medium through which they propagate, and as such, are capable of characterizing that medium; something that cannot be done by other means. It is astonishing to realize that acoustic wave imaging is commonly performed over about eight decades of frequency, with seismology and microscopy serving as lower and upper bounds, respectively. The physics is the same, but the implementations are quite different and there is much to learn.

The conference chairman and editor wishes to express his appreciation to those who helped run the symposium - namely the **Technical Review Committee** and **Session Chairmen** including Floyd Dunn, Gordon S. Kino, Greg Lapin, Hua Lee, William D. O'Brien, Jr., Michael G. Oravecz, Glen Wade, and the staff at Sonoscan, Inc. who organized the various functions, the IEEE UFFC group for providing publicity and the appropriate mailing lists, and to Sonoscan, Inc. for their financial support.

The editor wishes to express his appreciation to the authors for their manuscripts and also wishes to apologize for the delays in publication of this volume owing to late arriving manuscripts which were considered to be an important part of the proceedings.

Lawrence W. Kessler
Sonoscan, Inc.
Bensenville, Illinois   USA

CONTENTS

TISSUE CHARACTERIZATION

INVERSE SCATTERING

IMAGING TECHNIQUES II

ACOUSTIC MICROSCOPY

SEISMIC, UNDERWATER AND SOURCE LOCATION

REFLEX TRANSMISSION IMAGING

Philip S. Green, Joel F. Jensen, and Zse-Cherng Lin

SRI International
333 Ravenswood Avenue
Menlo Park, California 94025

ABSTRACT

Reflex transmission imaging (RTI) is a method for making orthographic acoustic transmission images without requiring the presence of an acoustic source on the opposite side of the object. It is well-suited to imaging specific planes in objects of moderate scatter cross section, such as human tissue. Each pixel of the image is formed by pulsing a transducer that is well-focused at the image plane, then rectifying and integrating the signals received from a selected range zone beyond the focus. This zone acts as a spatially and temporally incoherent insonification source. The pixel value is diminished twice by the attenuation at that focal point. A complete image is produced by raster scanning or bidirectional sector scanning; a reflection C-scan can be acquired simultaneously. Reflex transmission images can be made with a relatively small transducer. For medical diagnosis, RTI can be readily integrated with a real-time B-scanner, allowing multimodal imaging. The RTI process also can be used to form attenuation B-mode images.

Several factors affecting image quality are evaluated, including statistical fluctuations in the integral of the random backscatter and variations in backscatterer uniformity. Ex-vivo images of various organs illustrate RTI's diagnostic potential.

INTRODUCTION

In contrast to B-scan, transmission imaging requires acoustic coupling to large areas on both sides of the body for most medical diagnostic applications. This constraint has restricted the potential use of transmission imaging to those diagnostic applications in which water immersion is practical. Moreover, the masking effect of ultrasonic opacities (e.g. bone, gas in the lung and gut) further restricts the role of transmission imaging. Both of these impediments can be overcome, to a large degree, by using reflex transmission imaging (RTI).

Reflex transmission imaging permits transmission images to be made with a single B-scan-size transducer probe of special design. RTI is totally compatible with B-mode equipment, and affords the opportunity to generate orthographic transmission images, in a selected plane, that can be directly correlated with the B-scan. Orientation of RTI relative to the B-scan is shown in Fig. 1. A reflection C-scan can be made simultaneously with RTI, further expanding the potential diagnostic capability.

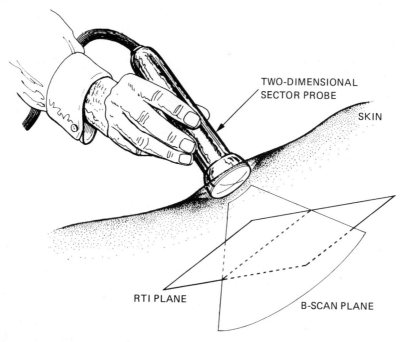

TWO-DIMENSIONAL SECTOR PROBE

SKIN

RTI PLANE

B-SCAN PLANE

Fig. 1. Relationship of RTI and B-scan image planes that would be produced with a two-dimensional, dynamically focused sector-scan probe.

BASIC CONCEPT

The reflex transmission imaging process can be initiated in the same manner as is a B-scan: A brief ultrasonic pulse is launched into the body from a well-focused transducer. However, RTI processing of the received signals is significantly different. With reference to Fig. 2, only reflections from a selected range zone beyond the focal region are utilized. They are amplified, usually with time-gain control, then detected and integrated. The value of this integral is strongly dependent on the attenuation in the transducer's focal region. Of course, it is also dependent on the attenuation in the entire focal cone and on the reflectivity of the tissues within the selected integration zone. The latter factors need not inhibit the formation of an image of high resolution, good dynamic range, and short depth of field, and they can be corrected, as we describe below.

The derived integral value is stored as one pixel of the RTI image. To produce a complete orthographic image, the focal point is scanned in a two-dimensional pattern, using either a rectilinear or dual sector scan, as shown in Fig. 3, or one of several combination scanning patterns.[1] Various mechanical, linear-array, and phased-array scanning methods can be used. The resulting focal plane will be flat in the rectilinear case, but spherical in the dual sector unless dynamic control of the focal range and range-gate onset are used. Only with the

2

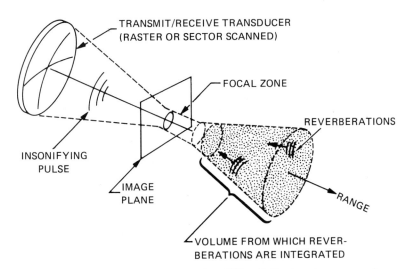

TRANSMIT/RECEIVE TRANSDUCER
(RASTER OR SECTOR SCANNED)

FOCAL ZONE

REVERBERATIONS

INSONIFYING
PULSE

IMAGE
PLANE

RANGE

VOLUME FROM WHICH REVER-
BERATIONS ARE INTEGRATED

Fig. 2.   Geometry of reflex transmission imaging (RTI). Reflections from a zone beyond the transducer's focus serve as an insonification source for transmission imaging.

rectilinear scan does the area imaged remain of constant size at all depths. A two-dimensional sector scan probe could closely resemble the larger annular-array B-scan probes now available commercially.

By reference to previous developments in transmission imaging,[2] we can consider the waves reverberated from beyond the focal zone to be (in effect), a source of incoherent insonification for transmission imaging of the focal plane. Because the transmitted pulse passes through the focal zone, the reverberations are already proportional to the attenuation at the focus. Owing again to the transducer's focal properties, only the reverberated waves returning through the focal zone are received with high sensitivity. Thus, focusing is achieved both on transmission and reception, doubling the sensitivity to focal-zone attenuation as compared with conventional transmission imaging. Because focusing is invoked twice, the resolution also is superior to that of a conventional transmission image made using the same numerical aperture and wavelength. Of course, RTI can be accomplished with focusing only on transmission or on reception. There are potential advantages to both of these variants that could justify the attendant loss of image quality.

Although RTI does benefit from a reasonably uniform body of tissue beyond the focal plane to serve as a source of reverberation, uniformity becomes less important as the integration period and numerical aperture are increased. Moreover, as we will describe in a future publication, there is an easily implemented method for removing artifact arising from non-uniform backscatter. Reflex transmission images of appendicular structures such as the hand and testicle could be made using an ultrasound scattering material (e.g. a particle-laden gel pad or water-filled sponge) behind the structure.

The noise in an RT image, like that of the B-scan, has both additive and multiplicative components. In RTI, additive noise (usually from the preamplifier) is suppressed by the signal-integration process; the longer the useful integration interval, the better the signal-to-noise ratio. Time/gain control can be used within the integration interval to optimize this ratio.

3

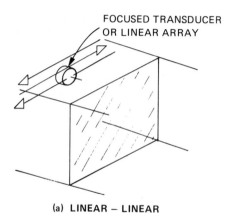

FOCUSED TRANSDUCER
OR LINEAR ARRAY

(a) LINEAR – LINEAR

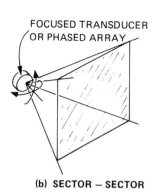

FOCUSED TRANSDUCER
OR PHASED ARRAY

(b) SECTOR – SECTOR

Fig. 3.   Two of the several scan formats suitable for RTI (see Ref. 1). The rectilinear scan (a) was used in a water tank to produce the images in this paper. The dual-sector scan mode (b) could be suitable for most clinical applications. Dynamic control of focus and range gate would be required for a flat image plane.

In the B-scan, multiplicative noise of high modulation index is always present. Referred to as 'speckle,' it arises from the interference of pulse reflections from adjacent scatterers, and has become accepted as an integral part of the image. This pulse-interference phenomenon also produces noise in RTI, but of a substantially lower modulation index. We can consider the reflex insonification source to be composed of concentric scatterer-filled "isochronous" laminae or shells of roughly spherical shape, lying beyond the focal zone.[3] If the shell thickness is taken to be about half the pulse length, we can consider each shell to be a single composite source with a Rayleigh-distributed random amplitude, uncorrelated with the amplitudes of reverberation from all other shells.

To estimate the signal-to-noise ratio, we first recall that the variance of a Rayleigh-distributed variable is approximately one-quarter of the square of its mean. Because the integration process sums the contributions of independent sources, we can apply the central limit theorem to arrive at a normally distributed reflex insonification signal of mean nA and variance $nA^2/4$, where n is the number of statistically independent shells and A is the average backscatter amplitude from each shell (presumed to be set equal by time/gain control). The number of shells is given by $n = t_R/t_P$, where t_P is the pulse duration and t_R the total integration time. Thus, the signal-to-noise ratio for multiplicative noise is

$$SN_mR = 10 \log (4t_R/t_P) \text{ (dB)} \quad . \tag{1}$$

For example, if the pulse duration T_P were 0.5 µs and the duration T_R of the integration zone 25 µs (equivalent to just under 2 cm of soft tissue), the signal-to-noise ratio would be 23 dB. The results of experiments undertaken to confirm this analysis are included in the next section.

EXPERIMENTAL RESULTS

Initial Demonstration

The preliminary experiments on RTI were undertaken using a computer-controlled raster-scan water tank system. A 5-MHz, broadband ceramic transducer was used in conjunction with polystyrene lenses of various focal lengths. The signal-processing circuitry, depicted in Fig. 4, consisted of a B-scan front end, augmented by a range-gated integrator and an A/D interface to the computer. Figure 5(a) is a reflex transmission image of a 2.5-cm-long plastic paper clip suspended on a rubber band, with a uniform-gel scattering phantom behind it. The multiplicative noise in the background is well depicted as a fine, granular pattern, a result of the minimal correlation between the reflex insonification at adjacent pixels. A conventional reflection C-scan, shown in Fig. 5(b), was made with a narrow range gate centered on the focal plane.

Figure 6 shows an image of a lamb kidney, using a 35-mm-thick backscatter zone within a contiguous beef kidney. The depth of focus is shallow; planes a few millimeters apart are differentiable.

Compared to Fig. 5(a), the multiplicative noise in Fig. 6 is much less perceptible, owing to a more moderate contrast setting, a longer integration zone, and a larger field-of-view. The nonuniformity of the backscatter from the inhomogeneous beef kidney is evident only as low-amplitude, low-spatial-frequency modulation of the image; it does not seriously diminish image quality.

Quantitative Evaluation and Methods for Improvement

Signal Processing. We expected the pulse length and the duration of the integration range gate to have a strong effect on SN_mR, as each RTI pixel value is the sum of Rayleigh-distributed random samples, the number of samples equaling the number of statistically independent reflections from the backscatterer zone. Range-gate position was expected to have little effect on SN_mR. To explore these relationships, we positioned a uniformly scattering gel phantom behind the empty focal plane and collected RTI signals over a 6.7×6.7-cm field for many combinations of range-gate duration, range-gate position, and pulse length. The pixel-to-pixel fluctuations were statistically analyzed (the mean and standard deviation computed) and a signal-to-noise ratio derived. Figure 7 shows the results of these experiments.

The dependence on integration range-gate duration conforms generally to that predicted. The experimental values were 1 to 2 dB lower than the theoretical ones, which could be a consequence of the way pulse length was defined. We found essentially no dependence on the range gate's time of onset.

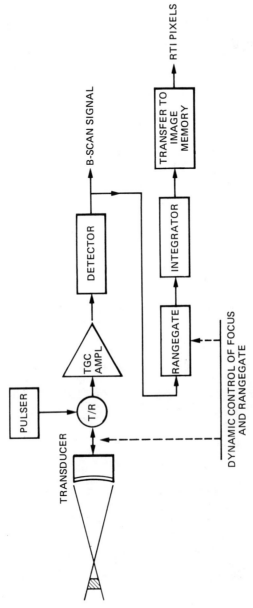

Fig. 4.   Signal-processing configuration, which can be implemented digitally.   Dynamic control of focus and range gate is required to sector-scan a flat image plane.

6

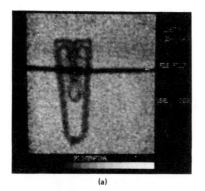
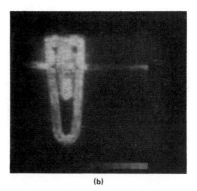

(a)   (b)

Fig. 5.   Reflex transmission (and reflection C-scan) images of simple objects, with a gel phantom backscatterer. Field of view is 3.3 cm^2. (a) RTI of a 1-inch long, thin plastic paper clip; contrast increased electronically to show multiplicative noise. (b) Reflection C-scan of same object.

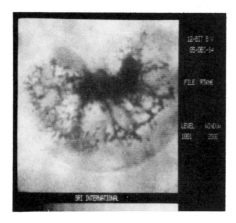

Fig. 6.   Reflex transmission images of a lamb kidney (*in-vitro*) with a contiguous beef kidney as a backscatterer. Field of view 6.7 cm^2.

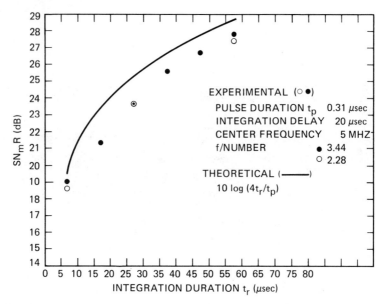

Fig. 7. Signal-to-(multiplicative)-noise ratios. The measured SN_mR follows closely the predicted values.

Images made with SN_mR in the range of 25 to 28 dB look encouraging; however, we would like to be able to discern high-spatial-frequency attenuation differences of, say, 0.2 dB/cm, which would require an SN_mR of about 33 dB. We examined one possible point of SN_mR improvement—the detector/integrator. We had been using a full-wave detector followed by an integrator, which has an output of the form

$$\int_{T_1}^{T_2} |u(t)| \, dt \quad , \tag{2}$$

where u(t) is the detected ultrasonic pressure signal. To conveniently compare this detector to others, we implemented on a computer 200 statistically independent wave trains composed of randomly delayed replicas of a typical broadband ultrasonic pulse. We checked the statistics and found the distribution to be approximately Rayleigh, as expected. We then modeled both the estimator described above and what should be the best estimator, which has the form

$$\left[\int_{T_1}^{T_2} u^2(t) \, dt \right]^{1/2} \quad , \tag{3}$$

and compared the SN_mR of the two. The latter was found to be consistently better, but only by a disappointing 0.15 dB to 0.43 dB, depending on range-gate duration. Nevertheless, because the image insonification is inherently incoherent, square law detection may be preferable for other image-enhancement purposes.[4]

We also investigated the possibility that the spatial sampling density used in our experimental scans was too low, and that sampling more densely would allow us to detect additional statistically independent variations in multiplicative noise. If this were true, it would be possible to improve SN_mR by spatial averaging. However, we discovered that we had undersampled only slightly, and that little improvement could be expected.

Effects of Backscatterer Nonuniformity. To evaluate the effects of nonuniform reflex insonification of the focal plane arising from a grossly inhomogeneous backscatterer, we conducted imaging studies using a specially constructed backscattering phantom that loosely simulates a variety of tissue types. The phantom, shown schematically in Fig. 8(a), is composed of a uniformly scattering gel body with four columnar inclusions. The scatterers are glass beads, 25 to 40 μm in diameter. The bead concentration in the body of the phantom is 11 g/l. The scattering cross section is somewhat higher than that of typical organ tissue, the attenuation coefficient is lower, and the sound speed is about the same. The rectangular columns are characterized as follows:

- Column 1: Styrofoam (highly reflective, simulates gas and bone).
- Column 2: Gel with scatterer concentration of 200 g/l.
- Column 3: Gel with scatterer concentration of 100 g/l.
- Column 4: Gel only (simulates blood vessel, bladder, or cyst).

We positioned the phantom in the water tank behind the focal plane of an f/2.32 transducer, which was left empty (except for water); thus, any departure from uniformity in the image was a result of the backscatterer variation. The field scanned, 10×10 cm, encompassed all four inclusions. The pixel values along each vertical line of the image were averaged to minimize the effect of statistical fluctuations.

We performed three experiments with the range gate (i.e. integration zone) set before, coincident with, and after the inhomogeneities [A, B, and C in Fig. 8(a)]. The results are shown in Fig. 8(b). It is apparent that gross backscatterer inhomogeneities can produce substantial low-spatial-frequency variations in image amplitude, as we had anticipated. In response to this, we have devised methods of compensation that will be reported later.

Further In-Vitro Studies

To demonstrate the potential application of RTI to the kidney (and particularly to evaluate RTI's possible application for monitoring kidney stone extracorporeal lithotripsy), we made multiplane images of ex-vivo lamb kidneys into which human kidney stones had been placed. We used a tissue composite in which the lamb kidney was backed by a beef liver (the backscatterer) and overlaid by beef muscle. One of these images is shown in Fig. 9(a). We also imaged several human cadaver livers, normal and cirrhotic; an example of the latter is shown in Fig. 9(b).

FUTURE INSTRUMENTATION

Clinical RTI Scanners and Multiplane Imaging

We are nearing completion of the installation of an RTI subsystem in a SRI-developed, 10-MHz, dynamically focused peripheral-anatomy scanner.[5] Installation of RTI in a general-purpose commercial scanner also is anticipated in the near future. A hand-held two-dimensional scan probe will be developed for this system. We also intend to install a multipulse scheme, illustrated in Fig. 10, which has been devised to collect several image planes at

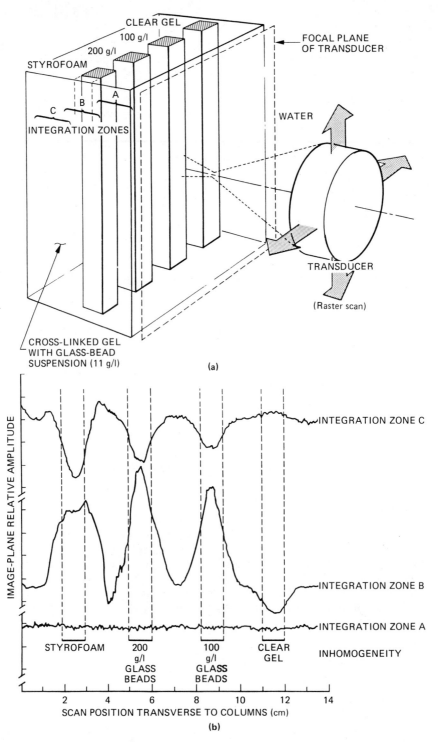

Fig. 8.   Effect of severe backscatterer inhomogeneity on image. (a) is phantom used to determine effect of image on inhomogeneous backscatterer. The transducer was focused in front of the phantom. Three integration zones—A, B, and C—were evaluated. (b) shows apparent focal plane transmission produced by integrating pixel values along each vertical line in the image plane. Image plane (i.e. transducer focus) is located in water in front of the phantom. Integration zones refer to (a). Traces have been shifted vertically for clarity.

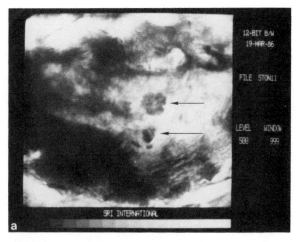

(a) IMAGE OF HUMAN KIDNEY STONES IN A LAMB KIDNEY OVERLAID BY BEEF MUSCLE AND BACKED BY BEEF LIVER TO PROVIDE BACKSCATTER

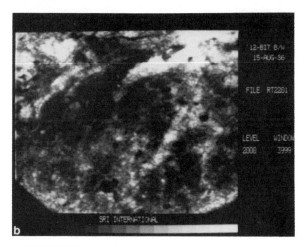

(b) IMAGE OF HUMAN LIVER, BACKSCATTER FROM POSTERIOR OF LIVER

Fig. 9.   Reflex transmission images of ex-vivo organs.

once. This method requires dynamic focus, and relies on the fact that very short pulses—while beneficial—are not essential to RTI. A series of brief bursts of ultrasound of different center frequencies and substantially nonoverlapping spectra would be transmitted. Each burst would be of lower center frequency than the previous one; their reverberations would be associated with increasing range (and thus greater high-frequency attenuation). During transmit, the dynamically focused transducer would be focused for each pulse to the desired depth of image to be formed for that pulse. Similarly, on reception, either the array focus will be

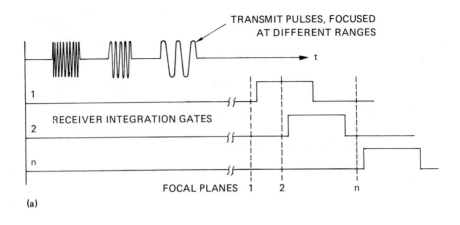

(a)

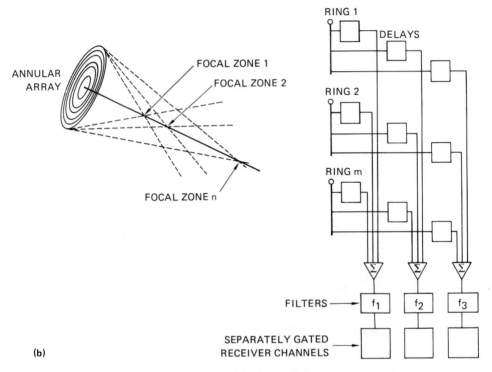

(b)

Fig. 10. Method for simultaneous acquisition of multiplane RTI. (a) Transmit pulse sequence and multiple integration range gates associated with several focal planes. (b) Parallel-processing receiver; each channel focuses the annular array at a separate range and filters the received signals to respond only to the reverberations of the pulse focused at that depth.

switched for each pulse or, if the several integration zones overlap, parallel sets of delay lines would be required, as shown in Fig. 10(b). In all cases, filters would be required that respond only to the reverberations associated with the individual planes.

The Compound RTI B-Scan

In principle, the reflex transmission concept also can be applied to produce attenuation images in the B-scan plane, rather than orthogonal to it as we have demonstrated above. This would be accomplished by displaying, as a function of range, the integrated RT signal from a range zone just beyond each point in range. This will require both dynamic focus and a con-

12

tinuous time-shifting of the zone of integration (a process that could be easily implemented in the digital domain). However, although such an image will have good lateral resolution, its range resolution will be poor compared to the conventional reflection B-scan, because it will be determined by the length of the transducer's focal zone, rather than by the pulse length. This calls to mind the early reflection B-scans, where the use of unfocused transducers resulted in good range resolution but poor lateral resolution. The compound scan was devised, in part, to overcome this limitation. Similarly, RTI B-scans could be compounded and averaged or (more rigorously), back-projected and spatially filtered to provide good spatial resolution in all directions in the plane. RTI B-scans would bear a similarity to ultrasound computed-tomography attenuation images. They would not have the inherent quantitative accuracy, but would be more easily produced using modified B-mode equipment.

DISCUSSION

Transmission images provide information that is complementary to B-scans. Reflex transmission imaging provides the viewing perspective and the sensitivity to attenuation of orthographic transmission imaging and does so using modified B-mode equipment. With RTI both can be made with the same instrument and presented on the same display.

Although we are encouraged by the technical developments and early in-vitro results, the range of clinical applications for RTI remains to be determined. In general, in conjunction with B-scan, RTI should provide a more assured diagnosis than B-scan alone, especially because the images can be precisely correlated using electronic screen cursors. For example, for the breast, RTI could be complementary to B-scan, because transmission imaging is more effective than B-scan for visualizing tumors surrounded by fatty tissue. The potential use of RTI to monitor kidney-stone and gall-stone lithotripsy merits further investigation. Peripheral-anatomy and intraoperative imaging also merit exploration.

ACKNOWLEDGMENT

This work was supported by PHS Grant CA41579.

We gratefully acknowledge the assistance of Mr. Leonard Kelly of the Department of Pathology, Palo Alto Veterans Administration Hospital, who provided tissues for our in-vitro studies.

REFERENCES

1. P.S. Green and M. Arditi, "Ultrasonic Reflex Transmission Imaging," *Ultrasonic Imaging,* 7:201-214 (1985).
2. J.F. Havlice, P.S. Green, J.C. Taenzer, and W.F. Mullen, Removal of spurious detail in acoustic images using spatially and temporally varying insonification, *in:* "Ultrasound in Medicine," 3B, D.N. White and R.E. Brown, eds., pp. 1827-1828, Plenum Press, New York (1977).
3. M.A. Fink and J.F. Cardoso, Diffraction effects in pulse-echo measurement, *IEEE Trans. Sonics and Ultrasonics,* SU-31, 4 (1984).
4. Z-C Lin, P.S. Green, and J.F. Jensen, Multiplane deconvolution in orthographic ultrasonic transmission imaging (this volume) (1986).
5. K.W. Marich, S.D. Ramsey, D.A. Wilson, J.F. Holzemer, D.J. Burch, J.C. Taenzer, and P.S. Green, An Improved Ultrasonic Imaging System for Scanning Peripheral Anatomy, *Ultrasonic Imaging,* 3:309-322 (1981).

IMAGING BY SOURCE CANCELLATION

Jim Burt and Alain Gaudefroy

Physics Department, York University
4700 Keele St., Downsview, Ont. M3J 1P3
Canada

INTRODUCTION

This article describes a new technique which we have developed to find solutions to the Huyghens numbers equations for a simple imaging problem. The imaging problem may be described as detecting two point sources in a two-dimensional domain, or, equivalently, as detecting the position of a horizontal layer illuminated by a point source.

The Huyghens numbers equations were devised by Gaudefroy[1-8] to solve optimum configuration for anti-noise problems. They may also be applied to acoustical imaging. Using the Huyghens numbers method which, importantly, is a rigorous analytic inversion method, the acoustical problem of sources and reflectors is built up from a point source and point reflector description. The method can also deal with continuous curved sources and reflectors but that is not our concern here. The underlying physics of the Huyghens numbers theory is particularly simple. Suppose we wish to locate a single point source in a space. We process the signals obtained at a number of receivers and calculate how the signals would be modified if a fictitious mathematical anti-source is moved through the space. We adjust the magnitude of this anti-source when it arrives at each space point and observe the effect on the receivers. Clearly, if the anti-source, or equivalent source, as we shall name it, is located at the source point and has opposite field strength to that of the source we will detect a null on the receivers. We have thereby imaged (located) the source, and found its magnitude by source cancellation.

The Huyghens numbers equations have the following form:

$$HN(R_1, \ldots R_I, \ldots R_n) = HN(R_1, \ldots F, \ldots R_n) \tag{1}$$

where
$$HN = \frac{\det\left(\dfrac{\partial V_{R1}}{\partial x}, \ldots \dfrac{\partial V_{Rn}}{\partial x}\right)}{\det(V_{R_1}, \ldots V_{R_n})}$$

and

$$V_{R_j} = \begin{pmatrix} (R_1, R_j) \\ \vdots \\ (R_n, R_j) \end{pmatrix} , \quad \frac{\partial V_{R_j}}{\partial x} = \begin{pmatrix} (\frac{\partial R_1}{\partial x}, R_j) \\ \vdots \\ (\frac{\partial R_n}{\partial x}, R_j) \end{pmatrix} .$$

Referring to our simple physical example, above, the F represents the total source field, which, for example, in a tomography problem could be exclusively the reflected, transmitted, and refracted field due to the object we are trying to image. The R_i represent individual unit point equivalent sources. The receivers appear only implicitly since both R_i and F are made known through their received signals, which constitute a vector of received signals whose number of elements is the number of receivers chosen. The solution of the Huyghens numbers equations is guaranteed since there are 3n linear independent equations in the coordinates of the equivalent sources for a three-dimensional problem. But since the equations are nonlinear, their solution is not straightforward. Using approximation theory and the projection theorem, Gaudefroy showed that equations (1) have a solution and are optimum. However, Huyghens numbers theory does not show how to derive practical algorithms to solve them. We have considered 2-D problems only and for the case of 1 single point source, utilising the signals taken from a selected pair out of 3 receivers, 1 equivalent source, 1/r field, we found three techniques which were successful:

(1) We solved the Huyghens numbers equations directly.
(2) We used the receiver energy method which plots the energy received in the 2-D domain as a function of the single equivalent source coordinates.
(3) We used the receiver-pair circle method which is a short, direct geometric construction.

The results obtained for a 1/r field simulate acoustical imaging problems since they represent magnitude of the far-field. We verified that the theoretical results agreed with experiment by using an electrostatic sheet rather than an acoustic tank. This study was reported at the Canadian Society of Exploration Geophysicists meeting in Calgary, May 1987, and will be the subject of a printed communication. Then, for a true acoustic field, as described by a Helmholtz equation, we verified that our developed method worked as well as for the 1/r-field case. This we explain below.

TWO-EQUIVALENT-SOURCE IMAGING PROBLEM

In fig. 1 we show the imaging problem of the two sources. Note that Source S2 could just as well be the image source arising from a reflecting horizontal layer half way between S2 and S1. The field at each of the 5 receivers consists of direct rays from the sources S1 and S2 and the equivalent sources M1 and M2. The problem is to locate S1 and S2 and their intensities from the signals of the form $\frac{\cos kr}{r}$ where r is the source receiver distance and k = const. is the wave number in a Helmholtz equation. To do this we let M1 and M2 vary over the search domain indicated in fig. 1. The Huyghens numbers equations are effectively solved by solving the receiver energy equations:

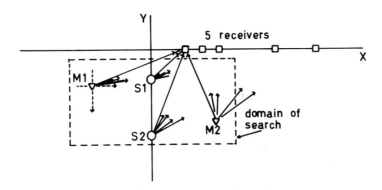

Fig. 1. Imaging two sources, S1, S2, by two equivalent sources, M1, M2. Sinusoidal excitation.

$$(R_1, R_1)Z_1 + (R_1, R_2)Z_2 = (R_1, F)$$

$$(R_2, R_1)Z_1 + (R_2, R_2)Z_2 = (R_2, F)$$

$$(\frac{\partial R_1}{\partial x}, R_1)\ Z_{x1} + (\frac{\partial R_1}{\partial x}, R_2)Z_{x2} = (\frac{\partial R_1}{\partial x}, F)$$

$$(\frac{\partial R_2}{\partial x}, R_1)Z_{x1} + (\frac{\partial R_2}{\partial x}, R_2)Z_{x2} = (\frac{\partial R_2}{\partial x}, F) \qquad (2)$$

$$(\frac{\partial R_1}{\partial y}, R_1)Z_{y1} + (\frac{\partial R_1}{\partial y}, R_2)Z_{y2} = (\frac{\partial R_1}{\partial y}, F)$$

$$(\frac{\partial R_2}{\partial y}, R_1)Z_{y1} + (\frac{\partial R_2}{\partial y}, R_2)\ Z_{y2} = (\frac{\partial R_2}{\partial y}, F)$$

Here F is the field at the receivers due to S1 and S2, while the field from M1 is indicated as R_1, and from M2 as R2. The inner products are direct products over the 5 receivers. The variables, Z_1, Z_{x1}, and Z_{y1}, are different estimates for the intensity of M1. Similarly for M2.

It can be shown that when the Huyghens numbers equations are satisfied, then $Z_1 = Z_{x1} = Z_{y1}$ and $Z_2 = Z_{x2} = Z_{y2}$. To determine this condition numerically we calculate, for a chosen position of M2, the value of the squared differences of the Z variables taken in pairs (this is proportional to signal energy and so our solution represents a minimum received energy condition). We make this calculation for a grid of M1 coordinates over the search domain. We look for the absolute minimum in the grid of energy values. We then repeat for an other position of M2 and compare with the previously obtained minimum. When the absolute minimum of all the searches is reached, M2 will be located at S1 and M1 at S2.

17

We made our calculation for a frequency corresponding to a wave-length of one-tenth of the distance from S1 to the first receiver. We found that the receiver energy method produces spurious minima. For the search domain indicated in fig. 1, two energy minima were found for each grid of M1 coordinates corresponding to a chosen M2 position. Neither was exactly the location of S1 or S2. However, a reduced search domain around one of the minima did include S1. At this stage in our develop-ment of algorithms to search for solutions of the Huyghens equations, we conclude that a preliminary search reduces the overall area which must subsequently be searched on a finer grid of M1 coordinates to find the absolute minimum. However, in spite of this implied improved search efficiency there remain important questions:

(1) How many spurious minima are produced when a larger search domain is taken to begin with?
(2) What is the radius of convergence of the search where, using the minimum found as coordinates for M2 in the next search, we will successively bring M2 to S1?

A theorem developed by Gaudefroy shows how many receivers we must take in the group for the vectors in the inner products of the equa-tions (2) in order to effect an efficient search. For a single equi-valent source the number is 2 while for two equivalent sources, 5 receivers are required. If fewer are used then the number of spurious minima proliferates. We verified that the method works both for a 1/r field, and for an acoustic Helmholtz equation field. In both cases, these are the fields for a planar, infinite domain, however, no parti-cular difficulty arises in incorporating boundary conditions provided the Green's function is known for the chosen boundary configuration.

CONCLUSIONS

In developing techniques to solve the Huyghens numbers equations, we have found the receiver energy method to be a reliable though lengthy procedure. The method is valid for the acoustic field described by a Helmholtz equation but it works as well for a 1/r field. Its disadvan-tage is that, to improve image definition, additional equivalent point sources must be added. At the present, this entails more receiver stations and greatly increased computation time. On the other hand, the Huyghens number equations present a true, rigorous solution to the imaging inversion problem and one which is novel and therefore merits investiga-tion.

REFERENCES

1. A. Gaudefroy, C.R. Acad. Sc. Paris 296B, 1139 (1983).
2. A. Gaudefroy, Acoustics Letters 4, 145 (1981).
3. A. Gaudefroy, Acoustics Letters 4, 136 (1981).
4. A. Gaudefroy, C.R. Acad. Sc. Paris 290A, 67 (1980).
5. A. Gaudefroy, C.R. Acad. Sc. Paris 290B, 405 (1980).
6. A. Gaudefroy, C.R. Acad. Sc. Paris 290B, 187 (1980).
7. A. Gaudefroy, C.R. Acad. Sc. Paris 288B, 379 (1979).
8. A. Gaudefroy, C.R. Acad. Sc. Paris 288B, 405 (1979).

SPARSELY-SAMPLED PHASE-INSENSITIVE TWO-DIMENSIONAL ARRAYS:

SPATIAL INTERPOLATION AND SIGNAL-DEPENDENT APERTURE

Patrick H. Johnston

NASA Langley Research Center
Mail Stop 231
Hampton, Virginia 23665-5225

INTRODUCTION

Phase-cancellation at a piezoelectric receiving transducer is an instrumental effect which arises because the voltage generated by a piezoelectric receiving element is proportional to the integral of the pressure over its aperture. Although this is the same property which yields the desirable directional characteristics of piezoelectric transducers, under some experimental conditions phase-cancellation results in quantitative errors in estimates of the energy in an ultrasonic field. Phase-cancellation has been shown to result in artifacts both in transmission measurements[1] and in scattering measurements[2]. A number of phase-insensitive detection schemes have been developed to reduce or eliminate this effect. These methods include true power or energy detection via radiation force measurements[3], the transient thermoelectric effect[4], and the acoustoelectric effect in piezoelectric semiconductors[5]. Other approaches have been based on linear and phased arrays[6,7]. Shoup, et. al.[8], proposed the use of sparsely-sampled two-dimensional arrays a number of years ago, and the research group at the University of Michigan[9] currently builds and uses actual sparsely-sampled two-dimensional arrays for the phase-insensitive detection of transmitted ultrasonic beams and scattered ultrasound.

The purpose of this paper is to outline a number of methods which we are investigating for interpreting the output from a sparsely-sampled two-dimensional receiving array used in transmission experiments. We describe three basic methods: description of a sampled beam in terms of the first few lower-order spatial moments of the sampled distribution of energy; the use of a signal-dependent cutoff to limit the extent of the effective receiver aperture; and the use of spatial interpolation to increase the sampling density during computation.

PHASE-INSENSITIVE DETECTION USING SPARSELY-SAMPLED ARRAYS

Phase-cancellation arises from the integration of a spatially-varying pressure over the face of a receiving piezoelectric receiving transducer. As a result, the energy content of the voltage generated by the transducer is not representative of the energy content of the incident ultrasonic field. One approach to reducing phase-cancellation is to employ a smaller receiving aperture[1]. This approach reduces the overall signal level, however, and is susceptible to refraction errors. An approach to phase-insensitive detection using arrays is to compute the energy content of the RF signal from each small-aperture element prior to combining them to form the output of the array. Such an approach results in a spatial sampling of the distribution of energy in the beam. Because the energy in a transmitted beam varies more slowly with lateral position than (potentially) does the pressure, a larger element spacing may be used. This reduces the number of elements per unit area of total aperture, easing somewhat the complexity of the mechanical and electronic design of a two-dimensional array.

SPATIAL MOMENTS

Most of the energy in a typical transmitted ultrasonic field is concentrated within a main lobe, surrounded by smaller-amplitude sidelobes. The essential features of a transmitted beam of ultrasound may thus be described by the two-dimensional moments of the spatial distribution of energy sampled across a receiving aperture. The method of moments as applied to two-dimensional arrays was initially described by Shoup, et. al.[8], and continued by the present author[10]. Two-dimensional spatial moments are defined in a discrete form by the equation

$$ m_{pq} = \sum_i E_i (x_i, y_i) \; x_i^{\,p} \; y_i^{\,q} \; \Delta x \; \Delta y \quad . $$

Here $E_i(x_i, y_i)$ is the energy measured at the ith receiving element located at position (x_i, y_i) on the plane of the receiving array. An estimate of the total energy I_0 expressed on a decibel scale is obtained from the zeroth-order moment m_{00} according to

$$ I_0 = 10 \; \log_{10} (\; m_{00} \;) \quad . $$

The centroid (C_x, C_y) of the energy distribution is determined from the first- and zeroth-order moments according to

$$ C_x = m_{10} \; / \; m_{00} \quad , \quad C_y = m_{01} \; / \; m_{00} \quad . $$

A measure of the beam half-width W is obtained from the second- and lower-order moments as

$$ W^2 = (\; m_{20} + m_{02} \;) \; / \; m_{00}^2 - C_x^2 - C_y^2 \quad . $$

These moments-based parameters provide additional information about the beam that would be lost to a single-element receiver. Further, because the energy from each element is used in the calculation, the moments are inherently phase-insensitive.

SIGNAL-DEPENDENT CUT-OFF

The formula for m_{pq} contains a factor $x_i^{\,p} \; y_i^{\,q}$ which applies a heavier weight in the computation to noise which lies away from the interesting main lobe of the transmitted beam. Moments computed under these conditions may be difficult to interpret due to the influence of the noise. It may be desirable to limit the computation to the main lobe of the beam. One approach is to use an effective aperture of fixed geometry smaller than the total sampled aperture, and center it about the element having the maximum value[11]. Our approach is to allow the main lobe of the beam to determine the shape, size, and location of the effective aperture used in the computation of moments. We do this by computing moments using only those values which lie above a specified fraction of the peak value measured within the array aperture.

In Fig. 1 is shown a cross-section of a sampled ultrasonic beam from a circular disk transducer (open boxes) along with the expected $2J_1(x)/x$ function (solid curve). Horizontal lines indicate cutoff levels of 3 dB, 6 dB, 9 dB, etc. below the peak value. For example, when computing spatial moments using a 15 dB cutoff level, all parts of the sampled beam falling 15 dB or more below the peak are excluded from the summations. This excludes in a consistent manner lower-level signals in favor of the higher signals present in the main lobe.

In Fig. 2 is presented the beam width W determined from moments as a function of the cutoff level employed in the moments calculation. The lower, solid curve is the result for a noise-free simulation. Note the smooth increase of width to a plateau as the cutoff magnitude increases from 1 dB to 17 dB, where an abrupt increase in the width occurs as the moments calculations begin to include the peak of the first sidelobe. A second abrupt increase occurs at about 23 dB as a result of the peak of the second sidelobe. The data shown as filled boxes were measured with a system in which the average noise level occurred at approximately 29 dB down from the peak signal. The estimated beam width increases between 12 dB and 18 dB as expected when the computation began to include the first sidelobe. For cutoff levels of greater magnitude, the width continues to increase as the sidelobes and the noise level begin to dominate the calculation. The effects of noise are even more evident in the data plotted as open boxes, where the average noise level occurred at approximately 16 dB. In this case, the estimated beam width rises sharply between 15 dB and 18 dB to a maximum value which is approximately what one would obtain for a constant signal over the aperture. In each of the cases shown in the figure, the beam width estimated using cutoff levels of 12 dB or smaller in magnitude yielded results in line with the theoretical expectations. The criteria for setting cutoff level were determined to be that the cutoff level should be above 17 dB to eliminate sidelobes from the moments calculations and at least 6 dB above the average noise level to virtually eliminate the effects of noise. In the remainder of this paper, a cutoff value of 12 dB was employed.

SPATIAL INTERPOLATION

In Fig. 3, panel A is a grayscale presentation of a (simulated) beam sampled over a 21 x 21 element array. (In this and all other grayscale figures in this figure, the data are presented on a logarithmic scale, with white corresponding to the peak value (0 dB) and with black corresponding to -32 dB). Note the blocky appearance of the figure resulting from sparse sampling. Panel B shows the same data with a 12 dB cutoff applied, i.e. with all values less than 12 dB below the peak set to zero and shown here as black. The blocky character of the effective aperture being imposed is evident. Note also that the aperture is not symmetric about its center, although the beam is perfectly circular in cross-section. These results lead us to ask whether the sampled energy distribution could be

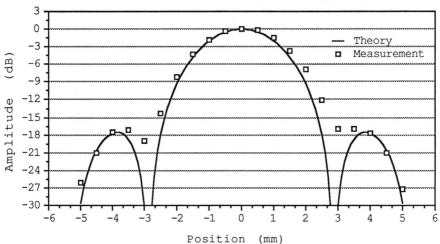

Fig. 1.   Measured scan across a transmitted beam (squares) and theoretical
beam function. Horizontal lines indicate cutoff levels. For example, a 12 dB cutoff implies that everything below -12 dB is ignored.

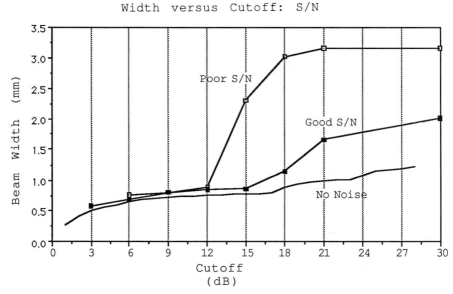

Fig. 2.   Computed beam width as a function of cutoff level applied. Lower
curve results from a noise-free simulation. Middle curve (filled
boxes) results from a measured beam with noise level approximately 29 dB below the peak. Upper curve (open boxes) results from a
measured beam with noise level approximately 16 dB below the peak.

interpolated in order to improve the geometry of the effective aperture and the numerical accuracy of the moments calculations.

According to sampling theory, if a bandlimited function is sampled at a rate faster than twice the highest frequency component present in the function, the function can be reconstructed by interpolation. The far field of a transmitting transducer has a lateral beam shape which is the Fourier transform of the aperture. For a spatially limited transmitter, this field is thus spatially bandlimited. Thus, with adequate spatial sampling, the full field distribution may be reconstructed.

In panel C of Fig. 3 is shown a grayscale representation of the power spectrum of a simulated sampled beam. The center of the image corresponds to zero spatial frequency. We note that the power spectrum falls to very small values beyond some radius in the spatial frequency domain. Although the square shape of the sampling aperture is evidenced by the plus-shaped distortion of the spatial power spectrum, the result is effectively band-limited, and yields to interpolation. Our approach to interpolation is to extend the spatial frequency domain with zeroes to the desired frequency, followed by inverse transformation to yield an interpolated beam. This approach is equivalent to interpolating the beam in the spatial domain with an appropriate sinc(x)sinc(y) function.

In Fig. 3 panel D is presented the result of interpolating the sampled beam in panel A from its original 21 x 21 dimension to 84 x 84. The result is a much smoother representation of the actual distribution of energy in the transmitted beam. Note that the interpolated beam with a 12 dB cutoff applied (panel E) more closely matches the known circular symmetry of the transmitted beam than the original sampled beam with cutoff applied (panel B). The beam itself determines the size and shape of the effective aperture, with the symmetry of the image of the beam determined by the degree of interpolation.

EXAMPLES

As an example of the improvement in performance of a sparsely-sampled array using these methods, we computed the centroids of simulated beams sampled by a 21 x 21 element array with 1-mm element spacing as the beam was displaced with respect to the center of the receiving aperture. The coordinates of eight such target points are given in Table 1.

Table 1: Target positions for centroid computations.

Point Number	X value (mm)	Y value (mm)
1	0.206	0.103
2	0.413	0.206
3	0.619	0.310
4	0.826	0.413
5	1.032	0.516
6	2.064	1.032
7	3.096	1.548
8	4.128	2.064

A 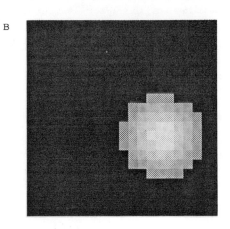 B

C

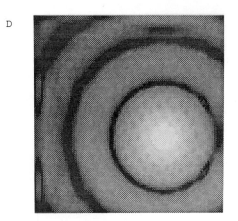 D  E

Fig. 3. Interpolation of sampled beam. A) 21 x 21 sampled beam. B) 21 x 21 beam with 12 dB cutoff applied. C) Power spectrum of 21 x 21 sampled beam. D) Sampled beam interpolated to 84 x 84 points. E) Interpolated beam with 12 dB cutoff applied.

The error in centroid position calculated by moments was determined according to

$$Error = \left((C_x - C_x^{target})^2 + (C_y - C_y^{target})^2 \right)^{1/2} .$$

The results are plotted in Fig. 4. The upper curve (open boxes) is the error obtained when computing the moments directly from the total sampled aperture. Errors from 0.08 to 2.0 mm were found. The results obtained when a 12 dB cutoff was applied prior to computation of moments are plotted as the middle curve (filled boxes). We note a 1 to 2 order of magnitude decrease in the error resulting from the application of cutoff. The lower curve (filled triangles) represents the results obtained when the sampled beam was interpolated from 21 x 21 to 84 x 84 prior to application of a 12 dB cutoff and moments computation. The result was roughly another order of magnitude decrease in error.

As another example, we consider the beam width W. From theoretical considerations, we know that the beam generated by a circular disk transducer diverges in the far field, i.e. that the beam width increases linearly with axial range. In Fig. 5 we present the results of a simulated beam from a 1/4" diameter 5 MHz disk transmitter, sampled by a 21 x 21 element array with 1 mm spacing at axial ranges spanning the distance 100 to 200 mm from the transmitter. The results obtained from direct computation of the width are plotted as the upper curve. These results exhibit a deviation from the expected linear increase with range. After interpolation to 84 x 84 elements and application of a 12 dB cutoff, the beam width exhibits the expected linear behavior (lower curve).

Another characteristic of the beam in the far field of a disk transmitter is that the amplitude falls off inversely with axial range. Thus, we would expect that the zeroth moment, being roughly the product of the beam area ($\sim range^2$) times the beam intensity ($\sim range^{-2}$), would exhibit a constant value as a function of range. In Fig. 6 we present the zeroth moment as a function of range for the same simulation as the preceding example. Note in panel A that the zeroth moment computed with no cutoff decreases monotonically over the range, whereas the zeroth moment computed using a 12 dB cutoff oscillates about a constant value. The oscillations result as a consequence of the coarse sampling of the beam, and are reduced through interpolation of the beam prior to applying a cutoff. The reduction in oscillation obtained by interpolating is shown in the panel B of Fig. 6.

In Fig. 7 we present measured values for zeroth moment versus axial range at three frequencies. The measured beam exhibits the same behavior as the simulated beam, having a constant zeroth moment with range. The three upper curves (filled boxes) represent data taken in a water path, while the three lower curves (open boxes) are results obtained with a 10 mm thick plate of plexiglas placed into the beam. The range-independence of the zeroth moment is maintained with a sample in place, suggesting that the beam-dependent cutoff approach offers automatic correction for diffraction effects in measurements made in the far field of the transmitter.

CONCLUSION

We have found that a sparsely-sampled array of small-diameter transducer elements can be employed to characterize the main features of a transmitted ultrasonic beam. The first few lower-order moments of the

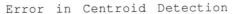

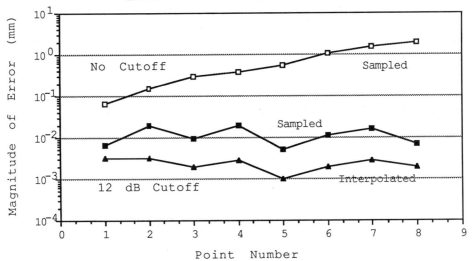

Fig. 4. Magnitude of deviations from known values of computed centroids
of sampled beams in noise-free simulation. Upper curve (open
boxes) results from 21 x 21 sampled beam. Middle curve (filled
boxes) results from 21 x 21 sampled beam with 12 dB cutoff applied. Lower curve (filled triangles) results from sampled beam
interpolated to 84 x 84 and subject to 12 dB cutoff.

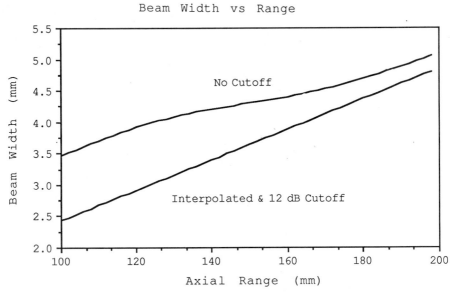

Fig. 5. Beam width computed from simulated sampled beam. Upper curve results from 21 x 21 sampled beam. Lower curve results from sampled beam interpolated to 84 x 84 and subject to a 12 dB cutoff.
Expected behavior is a linearly increasing curve.

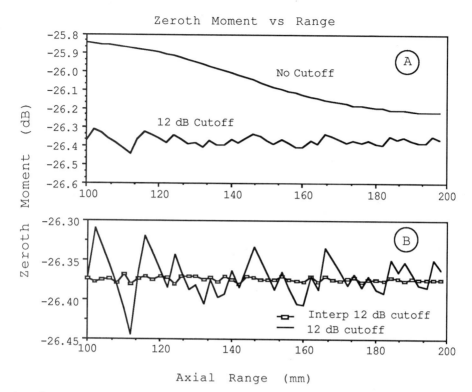

Fig. 6. Zeroth-order moment versus axial range. A) Upper curve results from 21 x 21 sampled beam. Lower curve results from sampled beam with 12 dB cutoff applied. Jagged character is due to the coarse sampling. B) Solid curve is same as lower curve in A). Results from beam interpolated to 84 x 84 and subject to 12 dB cutoff (open boxes) is much smoother.

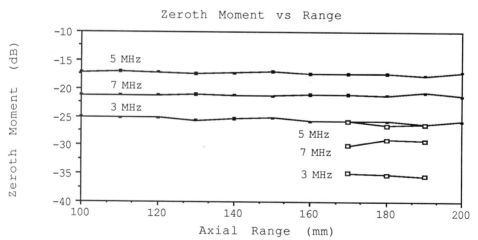

Fig. 7. Zeroth moments from measured beams as function of axial range at three frequencies, with 12 dB cutoffs applied. Zeroth moment remains constant in water only (filled boxes) and with 10 mm thick sample of plexiglas in path (open boxes).

distribution of energy sampled by such an array yield a phase-insensitive description of the width, the centroid, and the total energy content of the main lobe. These characteristic parameters can be computed with improved accuracy by employing spatial interpolation to increase the effective sampling rate of the data, and by then applying a signal-dependent cutoff in which only values above a given fraction of the peak value are used in computing the moments. With a cutoff chosen above the level of any sidelobes and sufficiently above the noise level, this approach provides automatic correction for far-field diffraction effects. Interpolation provides a possible means for detecting very small shifts of the beam or very small changes in the beam width.

ACKNOWLEDGEMENTS

    The author gratefully acknowledges the contribution to this work of Professor James G. Miller and others in his research group at the Department of Physics of Washington University in St. Louis. This work was performed in part at Washington University under grant NIH RR01362.

REFERENCES

1. J.R. Klepper, G.H. Brandenburger, L.J. Busse, and J.G. Miller, Phase Cancellation, Reflection, and Refraction Effects in Quantitative Ultrasonic Attenuation Tomography, Proc. 1977 IEEE Ultrasonics Symposium, 182-188 (1977). (IEEE Catalog No. 77CH1264-1SU).
2. Patrick H. Johnston and J.G. Miller, Phase-Insensitive Detection for Measurement of Backscattered Ultrasound, IEEE Trans, Ultrasonics, Ferro. and Freq. Cont. UFFC-33:713-721 (1986).
3. F. Dunn, A.J. Averbuch, and W.D. O'Brein, Jr., A Primary Method for the Determination of Ultrasonic Intensity with the Elastic Sphere Radiometer, Acustica 38:58-61 (1977).
4. D.W. Duback, L.A. Frizzell, and W.D. O'Brien, Jr., An Automated System for Measurement of Absorption Coefficients Using the Transient Thermoelectric Technique, Proc. 1979 IEEE Ultrasonics Symposium, 388-391 (1979). (IEEE Catalog No. 79CH1482-9).
5. P.D. Southgate, Use of a Power-Sensitive Detector in Pulse-Attenuation Measurements, J. Acoust. Soc. Am. 39: 480-483 (1966).
6. L.J. Busse and J.G. Miller, A Comparison of Finite Aperture Phase Sensitive and Phase Insensitive Detection in the Near Field of Inhomogeneous Material, Proc. 1981 IEEE Ultrasonics Symposium, 617-626 (1981). (IEEE Cat. No. 81 CH 1689-9).
7. M. O'Donnell, Phase-Insensitive Pulse-Echo Imaging, Ultrasonic Imaging 4:321-335 (1982)
8. Thomas A. Shoup, Gary Brandenburger, and J.G. Miller, Spatial Moments of the Ultrasonic Intensity Distribution for the Purpose of Quantitative Imaging in Inhomogeneous Media, Proc. 1980 IEEE Ultrasonics Symposium, 973-978 (1980). (IEEE order no. 80 CH 1602-2).
9. D. Fitting, P. Carson, J. Giesey and P. Grounds, A Two-Dimensional Array Receiver for Reducing Refraction Artifacts in Ultrasound Computed Tomography of Attenuation, IEEE Trans, Ultrasonics, Ferro. and Freq. Cont. UFFC-34:346-356 (1987).
10. Patrick H. Johnston, PhD Thesis, Washington University, St.Louis, MO, August, 1985.
11. D.W. Fitting, J. Giesey, and P.L. Carson, Adaptive Processing of Signals from a Two-Dimensional Receiving Array in Attenuation Tomography, Ultrasonic Imaging 8:48 (1986). Abstract.

UNBIASED CALIBRATION OF ANNULAR ARRAY PROBES

M. Sigwalt and M. Fink

Groupe de Physique des Solides de l'E.N.S., Université Paris
VII - Tour 23, 2 Place Jussieu
75251 Paris Cedex 05 - France

INTRODUCTION

Axial resolution and sensitivity of annular arrays scanners are linked to the acousto-electric response of each element of the annular array probe. The classical experimental measurement of this response, using the pulse-echo from a flat mirror, is strongly affected both in shape and amplitude by diffraction effects.

In order to calibrate plane disk transducers, the mirror is usually located in the near field of the probe to minimize diffraction losses. In the case of annular transducers, we have observed strong perturbations of the pulse-echo waveform and very different sensitivities between adjacent transducer elements. It will be shown that the radiation coupling between an annulus and a flat mirror leads to these effects. The annulus pulse echo signal contains, added to the true acoustoelectric response of the element, a shifted low pass filtered replica of this response. An analytical calculation of the radiation coupling response is given and explains the observed results.

Moreover, from this formula, a deconvolution technique is developed allowing to compensate the diffraction filter for all the annuli. It gives an absolute measurement of the element sensitivity and it allows the knowledge of the annulus spectral response.

I - EXPERIMENTAL RESULTS

A plane mirror is usually used in order to measure the acousto-electric response of an ultrasonic probe in pulse-echo mode. For a plane transducer, the radiation coupling observed in such an experiment acts as a filter which varies with the mirror position. Rhyne [1] has shown that as long as the mirror is located in the near field, the radiation coupling can be treated as a pure delay. The following inequality must be satisfied

$$z < n \frac{d^2}{8\lambda} , \ n \ \epsilon [1,3] \qquad (1)$$

where n is of the order of a few units, z represents the distance between the probe and the mirror, d is the smallest dimension of the aperture and λ the wavelength.

In the case of narrow annular transducer this condition is difficult to implement experimentally, since the echo coming from the mirror would not be separated in time from the trail of the excitation signal. By locating the mirror far enough to separate these contributions, we have observed that the waveform becomes strongly dependent on the annulus geometry. In the case of broadband transducers, these effects are predominant and the observed pulse-echo responses present anomalous long trail. Furthermore, for long radius annuli, it appears that this trail has the appearance of a low pass filtered replica of the primary response. Depending on the annulus geometry this replica may overlap with the primary response. Such behavior would indicate unacceptable performance for the - 20 db axial resolution. Annular elements also present very different apparent sensitivities. For equiarea annuli the echographic responses decreases as the radius increases.

Different annular probes have been realized, working between 2.5 MHz and 3 MHz. The external diameter of the probes is 30 mm. Each probe is subdivided in 6 elements of identical area. Thus the external radius of these elements are given by : $a_n = 15 \sqrt{n/6}$ mm.

Figure 1 shows, for a probe made in lead metaniobate, the 6 different pulse-echo responses from a steel flat mirror located at a depth $z = 50$ mm. We may noticed the different behaviors of the elements. The - 20 db axial resolution varies strongly between adjacent annuli : 4λ ; 4.5λ ; 5.3λ ; 6.7λ and 10.6λ. The insertion loss for the corresponding annuli are : - 27 db ; - 35.4 db ; - 37.7 db ; - 38.5 db ; - 41.1 db and - 49 db.

Figure 2 shows the power spectra corresponding to these echographic responses. It must be noticed that as the external radius of the element increases the power spectrum presents oscillations of increasing frequency. Such a behavior leads to biased bandwidth estimations. We found for the - 6 db bandwidth : 89.2% ; 85.5 % ; 72.9% ; 96% ; 42% and 2%. This result seems to be inconsistent and is related to the radiation coupling between the annular transducers and the flat mirror.

The bandwith and the pulse shape of the different annuli are strongly altered by the radiation coupling operator. The radiation coupling transfer function cannot be treated as a pure delay operator. The radiated ultrasonic field is quite different from the one of a plane wave, and the 1D propagation model is not valid.
For the central disk plane transducer, the radiated beam is well collimated till the Fresnel distance $a^2{}_1/\lambda$, and practically the whole of the reflected field is received by the transducer in the receive mode. However, as the difference $a_{i+1} - a_i$ decreases, i.e the annuli are thiner, the radiated field spreads out and the part of the reflected field received by the transducer becomes smaller.
The radiation coupling transfer function acts as a complex filter which modifies differently each frequency component of the acoustoelectric spectrum.
Thus the evaluation of the exact radiation coupling impulse response (and transfer function) is needed in order to understand and compensate these effects.

II - THE ANNULUS RADIATION COUPLING IMPULSE RESPONSE

The radiation coupling response between a transducer working in transmit-receive mode and a plane mirror is defined as the acoustic force perceived by the receiving area S_R corresponding to an impulsive acoustic force experienced by the emitting area S_E (Fig. 3).

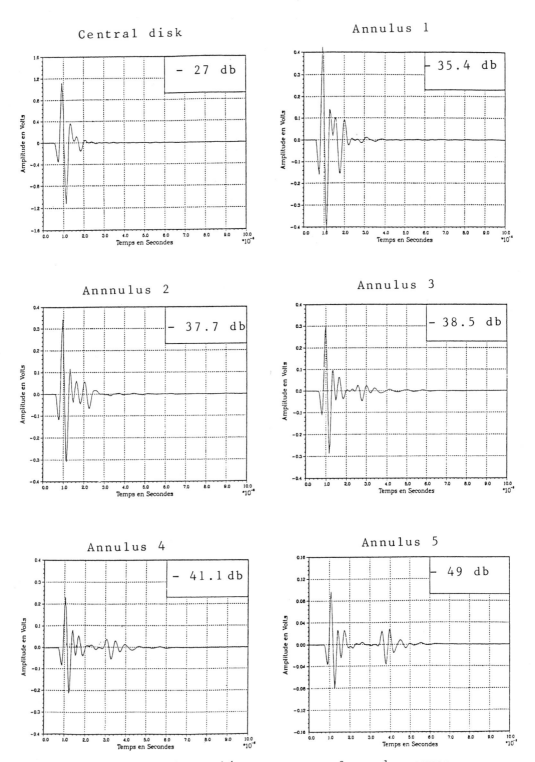

Figure 1   Echographic responses of annular array

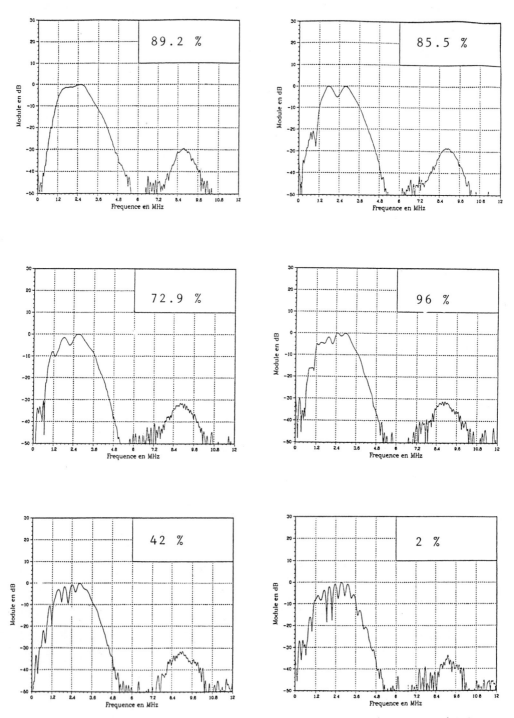

Figure 2   Spectral responses and 6 db Relative bandwidth

The Huygens-Green superposition integral must be carefully used; taking into account the fact that the normal velocity field radiated must vanish on the mirror plane (infinite acoustic impedance). The Green function, i.e the field radiated by a point source, must also vanish on the mirror plane. For a mirror located at a depth z from the emitting area, the Green function is then the linear superposition of 2 spherical waves, one coming from the point source located on z = 0, and the other one coming from the point source image located symmetrically at a depth 2z. Thus the configuration of a transducer working in transmit-receive mode located a depth z = d/2 from a flat mirror is equivalent to the configuration of two transducers separated by a distance (2z = d), one working as a transmiter and the other one working in receive mode.

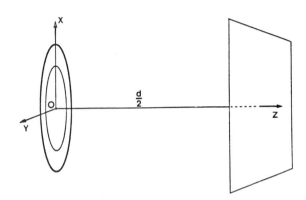

Figure 3

For a velocity impulse excitation of the surface S, the total force F_E on this surface is expressed by :

$$F_E \ \delta(t) = S_E \ \rho c \ \delta(t) \qquad (2)$$

On the other hand the diffraction impulse response defined by Stepanischen [2] makes it possible to calculate the pessure field at any point M in the space from the relation.

$$p(S_E| \vec{M},t) = \rho_o \frac{\partial}{\partial t} \ h(S_E| \vec{M},t)$$

with

$$h(S_E| \vec{M},t) = \iint_{S_E} \frac{\delta(t - ||\vec{M} - M_e||/c.}{2\pi.||\vec{M} - M_e||} \ dS_E \qquad (3)$$

Integration of he pressure over the receiver surface S_R leads naturally to the total acoustic

$$F(S_E|S_R,t) = \frac{\partial}{\partial t} \ \iint_{S_R} h(S_E|M,t).dS_R \qquad (4)$$

The definition of a potential radiation coupling response is then judicious and gives

$$H(S_E|S_R,t) = \iint_{S_R} h(S_e|\vec{M}, t) \, dS_R \qquad (5)$$

For an annulus of internal radius a_i and external radius a_e, the diffraction impulse response can be derived from the one of a disk plane piston transduce by

$$h(a_e ; a_i|\vec{M},t) = h(a_e|\vec{M},t) - h(a_i|\vec{M},t) \qquad (6)$$

where $h(a|\vec{M},t)$ is the classical Stephanishen impulse response of a disk plane of radius a.
Thus the potential radiation coupling Impulse response is

$$H(a_e ; a_i|a_e ; a_i,t) = H(a_e|a_e,t) + H(a_i|a_i,t)$$
$$- H(a_e|a_i,t) - H(a_i|a_e,t)$$

By using the commutativity of H

$$H(S_E|S_R,t) = \iint_{S_R} dS_R \iint_{S_E} \frac{\delta(t - ||\vec{M}_r - \vec{M}_e||/c)}{2\pi ||\vec{M}_r - \vec{M}_e||} \, dS_E$$

$$= \iint_{S_E} dS_E \iint_{S_R} \frac{\delta(t - ||\vec{M}_e - \vec{M}_R||/c)}{2\pi . ||\vec{M}_r - \vec{M}_e||} \, dS_R$$

$$= H(S_R | S_E,t) \qquad (8)$$

We obtained   (9)

$$H(a_e ; a_i|a_e ; a_i,t) = H(a_e|a_e,t) + H(a_i|a_i,t) - 2 H(a_e|a_i,t)$$

The expression $H(a_e|a_e,t)$ and $H(a_i|a_i,t)$ are the potential radiation coupling for disk plane transducers, and have been derived by Th. L. Rhyne [1].
We must evaluated the cross term $H(a_e|a_i,t)$. This expression represents the radiation coupling between two coaxial disks of radius a_e and a_i. By using the Rhyne analytical formulation of the velocity potential field integral

$$I(x,y) = 2 \int_x^y r \cos(u) dr \qquad (10)$$

which is equal to

$$I(x,y) = [r^2 \cos^{-1}(u) - a^2 \cos^{-1}(v) - rT(1-u^2)]^{1/2} \Big|_x^y \quad (11)$$

with

$$T = [(ct)^2 - d^2]^{1/2}$$

$$u = \frac{T^2 + r^2 - a^2}{2rT}$$

$$v = \frac{-T^2 + r^2 - a^2}{2aT}$$

We may found $H(a|b,t)$ and its time derivative $F(a|b,t)$ where we replaced a_e by a and a_i by b [Fig. 4] $F(a|b,t)$ may be written as [3]

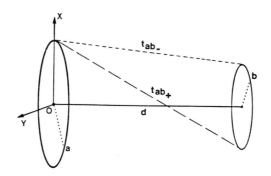

Figure 4

t	F(a ; b| a ; b, t)
$t < t_{ab-}$	$c.\pi.\ [Min(a,b)]^2.\delta(t-d/c)$
$t_{ab-} < t < t_{ab+}$	$-\dfrac{c^4\ t}{T^2}\ \{[(t^2 - (d/c)^2)\ (t^2_{aa} - t^2)]^{1/2}$ $+ [(t^2 - (d/c)^2)\ (t^2_{bb} - t^2)]^{1/2}$ $- 2.[(t^2 - t^2_{ab-})\ (t^2_{ab+} - t^2)]^{1/2}\ \}$
$t_{ab+} < t$	0

$$(12)$$

where $d = 2z$ is the distance between the two coaxial disk of radius a and b and where $t_0 = d/c$; $t_{ab-} = [(a-b)^2 + d^2]^{\frac{1}{2}}/C$ and $t_{ab+} = [(a+b)^2 + d^2]^{\frac{1}{2}}/C$.

It must be noticed that the term $\delta(t-d/c)$ corresponds to a pure delay related to a plane wave propagation. However extra terms appear and correspond to the different edge waves originated from the annulus. Finally the annulus force radiation coupling impulse response can be evaluated as

$$F(a;b|a;b,t) = F(a|a,t) + F(b|b,t) - 2\ F(a|b,t) \qquad (13)$$

and figure 5 shows, for the 6 element the corresponding force impulse responses.
We must noticed on these curves that after the geometric plane wave term the different contributions give a resulting field which presents a secondary maximum at a time t_{max} included in the interval (t_{ab-}, t_{ab+}). The time of arrival of this secondary maximum increases as the annulus number increases. Note also that the bump corresponding to this secondary maximum becomes smoother as the annulus number increases.

The complete echographic response, which is obtained by a convolution product with the acoustoelectric responses, is made of two waveforms which overlap for the first annuli, and which may be well separated for the last annuli. Furthermore, the effect of the smoother bump in the convolution process is to give a smoother replica of the principal waveform. This replica is then a low pass filtered version of the initial waveform. A short time Fourier analysis of the pulse echo signal would shown a lower frequency content of the replica signal. For narrow band transducers the frequency shift of the replica signal may be superior to the transducer bandwidth and then these replica dissapear on the echographic signal. We have observed such a behavior with undamped annular transducers of equivalent geometry.

Finally the behavior of force impulse responses, when increasing the annulus number, explains the power spectrum properties observed on figure 2. The spectral oscillations whose frequency increases with the annulus number (Note that this frequency as the dimension of a time) are related to the impulse response shapes. By considering an approximation of these responses by a couple of two dirac pulses located at time t_o and t_{max}, we notice that the associated transfer function (Fourier spectrum) is an oscillatory function, whose frequency increases with $(t_{max} - t_o)$.

III - CALIBRATION OF THE ACOUSTOELECTRIC RESPONSE AND DECONVOLUTION PROCEDURE

In order to calibrate the acoustoelectric response of annular transducers and to verify the validity of the annulus radiation coupling response, we have developed a complete calibration procedure.

1 - In the first step, sampled values of the measured echographic reponses were taken at 32 MHz/8 bits sampling rate and entered into a computer. For an annulus of area S_i we experimentally measured the response from a flat mirror located at a depth $z = 35$ mm. Fig. 6a shows the response of the second annulus of a probe made with PT Ceramic.

2 - In the second step, we deconvolute the pulse echo signal by the theoretical annulus radiation coupling response $F(Si|Si,t)$. It gives an absolute measurement of the Acoustoelectric response $AE_i(t)$ in the transmit-receive mode.

$$Echo_i(t) = AE_i(t) \ @ \ F(Si|Si,t)$$

$$AE_i(t) = Echo_i(t) \ @ \ F^{-1}(Si|Si,t)$$

where $F^{-1}(Si|Si,t)$ is the inverse of F for deconvolution product.
This deconvolution is obtained by Fourier transformation. It must be noted that, due to the fact that $H(Si|Si,t)$ can be more easily sampled than $F(Si|Si,t)$, it is easier to calculate the time derivative of $AE_i(t)$.

$$AE'_i(t) = Echo_i(t) \ @ \ H^{-1}(Si|Si,t)$$

where H^{-1} is the inverse of H for convolution.
Fig. 6b shows the computed value of $AE'_i(t)$ for the second annulus.

3 - In the third step, once $AE'_i(t)$ is computed we may predicted the theoretical echographic response from a mirror located at a different depth from the annulus. This is done by a convolution product between

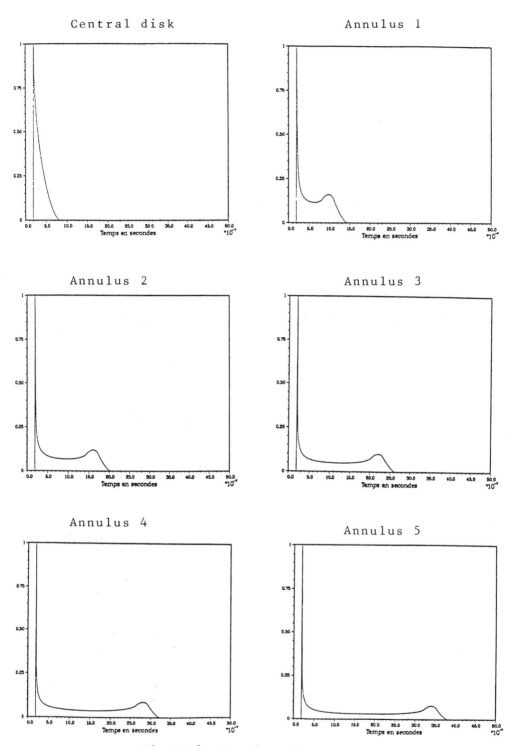

Figure 5   Impulse responses

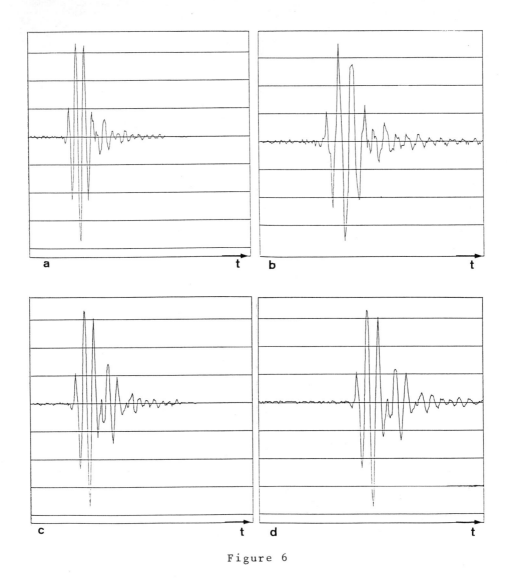

Figure 6

AE'$_i$(t) and the theoretical annulus response corresponding to a new
location of the mirror. In figure 6c we presented the predicted response
for a mirror located at z = 50 mm.

4 - Then in the last step we compared this prediction with the experimental response corresponding to a mirror located at z = 50 mm (Fig. 6d).

We note a very good agreement between the two waveforms of Fig. 6c
and Fig. 6d. The sensitivity difference between z = 35 mm and z = 50 mm
is also very well predicted (the error is less than .2 db).

CONCLUSION

Such a good comparaison between the theoretical model and the experience means that the annulus response model is a good assumption. If we note that this model is based on the piston mode assumption, this means that this annulus vibrates in a piston mode. Development of tracking focusing techniques with annular arrays needs the knowledge of the elements vibration states. Therefore the described 4 steps procedure allows a complete transducer calibration.

REFERENCES

[1] T.L. Rhyne, Radiation coupling to a disk to a plane and back or a disk to a disk : an exact solution, J.A.S.A. 61 (1977), 318-324.
[2] P.R. Stepanishen, Transient radiation from pistons in an infinite planar baffle, J.A.S.A. 49 (1971), 1629-1638.
[3] M. Sigwalt, Thèse, Université Paris VII, Mai 1987.

THE BOREHOLE TELEVIEWER DIGITAL IMAGE ANALYSIS AND PROCESSING

Xie Huchen[1,3], Cao Yimei[2], Zhang Yao[1]
Fu-Pei Hu[4], Zidu Ma[4], and Lu-Lin Hua[4]

[1]University of Science and Technology of China, Anhui
China
[2]Institute of Acoustics, Nanjing University, Nanjing
China
[3]Current address: Instrumentation Research Laboratory
USDA/ARS/SCSI, Beltsville, MD 20705 USA
[4]ADA R&D Group. PO Box 175, Beltsville, MD 20705

ABSTRACT

An ultrasonic borehole televiewer image is the log of the signal reflected from borehole, including the amplitude (log AL) and the transit time (log TTL). This paper presents the effective methods used in AL and TTL processing, and the way to obtain the reflectivity (RI) and distance (DI) images. The processed images can be displayed visually by cross-section pictures (CP) and stereo pictures (SP). The paper also gives some methods for calculating parameters of the borehole. The kernel of the whole system is called the Transportable Image-Processing Software (TIPS), written in the programming language C.

KEYWORDS: Borehole televiewer, image processing, stereo display.

INTRODUCTION

Concepts

An ultrasonic borehole televiewer is used for getting information about an oil well. The logging sonde is centered in a round borehole. The sound is emitted by the sonde, reflected from the wall of the borehole, and collected by the sonde.

Ordinarily the sonde is centered in the borehole and rotates continuously (the typical speed is 3-5 rps), emitting and receiving 512 pulses per revolution. In the meantime, the sonde rises up at 5 ft/min. (2.5 cm/s), so the scanning curve is a spiral on the wall of the borehole.

The amplitude (log AL) and the transit time (log TTL) of the reflected waveform are recorded for each point of the scan line. The two measurements provide complementary information. Other signals such as synchronism, temperature, and position signals are transmitted to the recorder at the same time.

Outline of the System

The system is composed of the sonde, which emits and receives the ultrasonic signal, an A/D converter, and a set of C programs for computer processing of the data. Its features are the following: AL and TTL information is used in producing a composite image, digital computer-image processing methods are widely used, a series of processing procedures are developed accordingly, and the cross-section (CP) and stereo (SP) pictures are visually displayed.

The paper will emphasize computer processing methods which we have implemented, including restore and rectify, obtaining the reflectivity (RI) and distance (DI) images, producing the CP and SP pictures, and calculating other parameters.

PROCESSING

Restore and Rectify

Direct oriented restoring. Some data points are lost due to the selection of the detecting threshold and/or error in transmitting. A lost point is detected by comparing each point with the average of surrounding points. Different from the concept of smoothing, restoring is taken only at these lost points. Suppose the restoring point P's neighborhood N has m non-lost points, the value of restoring is:

$\overline{P}$ = average of non-lost point data in N.

Because of the difference of the horizontal scanning and vertical rising, however, the data will take weight factor in the sum. In televiewer data processing, the weight we use is proportional to the distance.

Point-noise filtering in TTL. Some points with very high/low values scatter over some TTL, as we have observed. The analysis indicates that because the measurement of time delay is based on detecting the reflected waveform, much like the principle of radar, the false alarm/miss probability exists. Point-noise filtering uses a median or mean filter, where acceptable data are within the interval $[t_i, t_h]$ and the two thresholds are closely related to the circuits. (See Fig. 1a&b.)

Line-noise filtering in AL. It is possible that a line is lost in AL due to the loss of scanning the synchronous signal. This missing line restoring uses the method above by means of neighbor lines.

Position rectify. Because the borehole instrument revolves in position while rising up, the images we obtain twist vertically. The magnetometer records the angle between initial scanning pulse and the North Pole. As the sonde rising up, the angle is the function of depth. Through this function, the procedure (line shift is circulatively which is proportional to the angle) makes the beginning of the images in the direction of the North Pole.

42

Obtaining RI and DI

Model. Suppose that the decline coefficient of incident wave from point a is f(d) in the medium, and with angle of incidence θ. Hence, the main diffuse direction follows the reflection theorem; the reflecting angle is also θ. The diffuse wave will spread out symmetrically from this direction, the diffuse coefficient g(a), (f(0)=1; g(0)=1). The diffuse wave amplitude at point b is

$$A_b = A_a \, f(d_1) \, k \, g(a) \, f(d_2) \; ,$$

where k is reflectance. The transit time is given by $T=(d_1+d_2)/v$, where v is the transit speed.

Application. In borehole wall reflection, the reflecting plane is the tangent plane at the reflecting point. The same sonde is used for emitting/receiving, so $d_1=d_2=d$ and a=2θ where 0 incident angle. Hence, the amplitude of received signal is

$$A = A_a \, f^2(d) \, g(2\theta) \, k,$$

and the time delay is T=2d/v.

The v is considered to be constant because of its determination by the transit medium. We can obtain the DI directly from d=(v/2)T.

The RI calculation is more complicated. Let's simplify it by making the following assumptions:

(1) The shape of the borehole wall will not change;

(2) The position of the sonde relative to the wall is fixed.

So, the function $f^2(d) \, g(2\theta)$ is periodic with a period of one revolution time (note that d=d(t), $\theta=\theta(t)$). By using A(t) which comes from every scanning, we get

$$k(t) = A(t) \, / \, [f^2(d(t)) \, g(2 \, \theta(t))],$$

giving the RI.

It can be demonstrated that

$$\tan(\theta) = (1/d) \, (\Delta d/\Delta \alpha)$$
$$= (1/d) \, \Delta d/(2\tilde{n}/N)$$
$$= (N/2\tilde{n}) \, (\Delta d/d).$$

This relation can be used to rectify RI through the function f, g by CI. With this method, we need not make the assumption that the wall shape or sonde position is fixed, but f and g must be given. We did the experiment using this idea (thinking that f(d) is fixed and g is reasonably approximated by a normally distributed function).

Following the assumption in Ref. (1) and (2), if we suppose further that $\underline{k}$ does not change with depth in a specified segment, then

$$\frac{1}{M} \sum_m A(m,n) = A_a f^2(d) \; g(2\theta) \frac{1}{M} \sum_m k(m,n)$$

$$= A_a f \; (d(n)) \; g(2\;\theta(n)) \; \bar{k}(n)$$

and we can obtain RI from this relation. Note the average is along each scan line.

CP and SP

Parallel projection to certain plane. Parallel projection is somewhat like observing objects at infinity in a certain direction. It is appropriate to use parallel projection in stereo display of televiewer images. If observing direction is (u_1,u_2,u_3), $(u_1^2+u_2^2+u_3^2=1)$, the projective plane intersects with plane $z=0$ at line $\theta=\theta_o$ $(\tan\theta_o=-u_1/u_2)$. Letting one direction be the x axis of the projective coordinate system with origin fixed, the y axis is

$$(u_1 \; u_2 \; u_3) \; x \; (\cos\theta_o \; \sin\theta_o \; 0) =$$
$$(-u_3\sin\theta_o \quad u_3\cos\theta_o \quad u_1\sin\theta_o-u_2\cos\theta_o)$$

The projective coordinate of point $P(r\cos\theta, r\sin\theta, z)$ in space is

$$(r\cos\theta\cos\theta_o+r\sin\theta\sin\theta_o$$
$$-ru_3\cos\theta\sin\theta_o+ru_3\sin\theta\cos\theta_o \quad +zu_1\sin\theta_o-zu_2\cos\theta_o)$$
$$= \; (r\cos(\theta-\theta_o) \; ru_3\sin(\theta-\theta_o) + zu_1/\sin\theta_o)$$

The televiewer data, θ_o, is determined by a certain position (e.g., observing direction), where $\underline{z}$ is the depth. Note that the depth has an effect on the $\underline{y}$ coordinate of the projective image. If the sonde rises along the $\underline{z}$ axis, the projection of the scan line can be obtained by first calculating the projective value of sonde $(0,zu_1/\sin\theta_o)$, and then adding to it the projective value of the wall vector relative to the sonde. (See Fig. 1c&d.)

Piling up stereo picture (hidden-line process). Due to the similarity to the circle of a borehole wall, it is simpler to do hidden-line processing. We start drawing at $\theta-\theta_o = \hat{\pi}$ and add vector $(r\cos \triangle \theta, \; r\sin \triangle \theta \cos \Upsilon)$ to $(0,zu_1/\sin\theta_o)$, making points whose y value is greater than $r\cos \Upsilon \sin \triangle\theta$ equal to zero and repeating the process from $\hat{\pi}$ to $2\hat{\pi}$, then 0 to $\hat{\pi}$. (Υ is the angle to the horizontal plane.) We obtained a piled up stereo picture with increasing $\underline{z}$. (See Fig. 2a&b.)

Calculating Parameters of Borehole Well

Well diameter. TTL reflected directly the distance between the sonde and the wall. Because the transit time is quantized, it needs rescaling to get distance. The scaling parameters are related to quantization voltage and well scaling measures.

Region porosity. The region porosity (η) is defined as:

$$\eta = \text{(area of holes)}/\text{(area of the background)}.$$

The calculation of region porosity is very important to the estimation of oil storage of the stratum, the degree of casing damage, and amount of wall crack. We use a threshold technique to distinguish holes and background. If background points are 1's and hole points are 0's after threshold, then the region porosity is estimated by

$$\eta = \text{(number of 1's)/(number of 0's)}.$$

Stratigraphic visual dip angle. The slope and run of stratum can be obtained by observing the crack pattern of the logs which is something like a sine-shaped curve in unfold images. The visual dip angle is determined by the slope of stratum and slant of the well, and can be easily calculated. The visual dip angle (α) is given by

$$\alpha = \arctan(h/D),$$

where h is the height of the crack in unfold images, and D the diameter. The recognition of cracks and their run can be obtained interactively.

RUNNING ENVIRONMENT

The whole processing programs have been implemented on VAX-11/750 with Model 75 as a monitor at IPL, USTC. The programs are written in language C.

Due to the transportability of C, we have been able to implement the software successfully to microcomputers such as Cromemco, IBM-PC, etc. This Transportable Image Processing Software (TIPS) is a powerful tool to the study or application of image processing and analysis.

CONCLUSION

The paper has presented methods of processing ultrasonic borehole televiewer images. In the part of preprocessing, restore and rectify are used to obtain images with high quality and to remove from them the detecting error and position shifting, etc. We have then emphasized the principle and methods of obtaining RI and/or DI from AL and/or TTL which are much more convenient for engineers. Several parameter calculation methods are given. The experiments show that a borehole televiewer log, after processing by the computer to form RI and DI, is comparable to a picture of a continuous core and may yield even more information.

ACKNOWLEDGMENT

We are thankful to Dr. William R. Hruschka and Ms. Lois Harris for their help in the preparation of this paper.

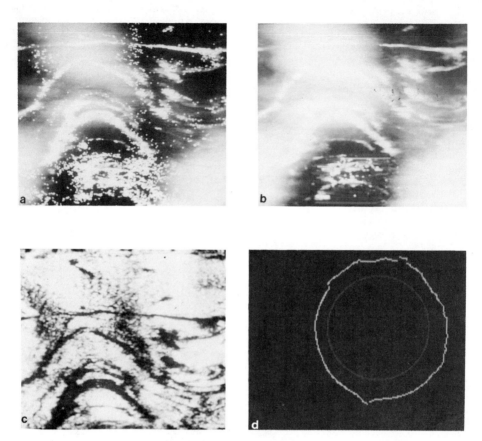

Fig 1. (a) and (b) are the TTL's before and after filtering. (c) AL after processing. (d) CP produced. All of them are taken at XI-12 uncased borehole. Approximate depth 1700m.

Fig. 2 (a) The SP of XI-12 uncased borehole, approximate depth 1700m. (b) The SP of REN-96 cased borehole, 1300m. Note the joint section.

REFERENCES

1. R. A. Broding, Volumetric scanning allow 3-D viewing of the borehole, World Oil, pp. 190-196 (June 1982).
2. T. J. Taylor, Interpretation and application of borehole televiewer surveys, SPWLA 24th Annual Logging Symposium (June 1983).

OPTIMUM FOCUSING IN THE ULTRASONIC ANNULAR ARRAY

IMAGING SYSTEM

T. K. Song, J. I. Koo and S. B. Park

Department of Electrical Engineering
Korea Advanced Institute of Science and Technology
P. O. Box 150, Cheongyangni, Seoul, Korea

ABSTRACT

This paper is concerned with optimum focusing in the ultrasonic annular array pulse echo imaging system– optimum in the sense that the lateral resolution is maximized with a specified sidelobe suppression over the range of interest. First, an effective technique for calculating the impulse response of an annular in the whole image plane is developed and the overall beam pattern is obtained by convolving the transmit field pattern with the receive impulse response. An iterative technique is then applied to minimize the beam widths at two depths of interest under a specified sidelobe suppression. In the receiving mode, a continuous dynamic focusing is carried out at all pixel points along the depth of view with a new digital focusing system.

INTRODUCTION

Improvement of the lateral resolution is most important in enhancing the quality of ultrasonic B-mode images. While the lateral resolution can be improved at the expense of the focused range, the entire situation can be improved by a dynamic focusing in the receive mode. At the same time, the transmit focusing, being flexible with the phased array, plays an important role in the design of the overall beam pattern.

The purpose of this paper is to optimize the delay profile of array elements in the annular array system for the best lateral resolution under a specified sidelobe suppression over the range of interest. First, we develop an effective beam simulation technique appropriate for the annular array. Secondly, an iterative technique is applied to solve the stated optimization problem and the results are shown. Finally, for real time application a new digital focusing system is proposed which enables continuous dynamic focusing in the receive mode.

Many of the ultrasonic pulse echo imaging systems currently in use employ very short pulses, and there have been many works in the study of the wideband transient field pattern[1, 2, 4, 5]. By the classical local baffled piston radiator, Linzer et al . calculated the wideband response of an annular array using the linear system approach for the acoustic lens suggested by Norton and Macovski[11] in 1976. Recently, the transient field response of a certain radially symmetric radiator was analyzed by Foster et al . and Fink[10] , using the instantaneous approach. All of these were concerned primarily with the response of a continuous aperture. Hence, when applied 'to the phased annular array, they had to deal with complicated time intervals resulting from the inner or outer radius of the annulus. We derive a simple analytic expression for the impulse field response, in the whole image plane, of a planar aperture whether it be an annulus or a disk.

The impulse field response by the annular transducer is the sum of contributions from individual array elements with independent delays introduced to them. And the overall beam pattern is obtained by convolving the transmit field response due to some excitation of the transducer (a gaussian excitation is assumed in our case) with the receive impulse response.

In order to carry out continuous dynamic focusing in the receive mode, a new digital focusing system is proposed which completely eliminates the use of conventional analog L-C delay lines - nonideal, bulky, and costly elements. This digital focusing is realized by adopting a pipelining architecture in the Sampled Delay Focusing(SDF) scheme[8] which was proposed recently by the authors . In the SDF scheme, the echoes at array elements are sampled by variable sampling clocks to take into account the round-trip time differences, and the resulting delayed samples are summed to obtain a focused signal. In this case the maximum delay obtainable is limited by the Nyquist sampling rate, which is violated in the near field focusing. This constraint is overcome in this paper by introducing the FIFO memories as pipeline buffers between the sampling and the adder stages in the SDF.

FIELD RESPONSE OF THE CIRCULAR APERTURE

Although either the time domain or the frequency domain approach can be used in studying the diffraction problem in the acoustic field, the former seems preferable in the wide band case and hence will be adopted in this paper. From the classical theory of sound in fluid, the impulse response of velocity potential in the ultrasonic field can be expressed as[10]

$$h_z(\rho, t) = \frac{1}{2\pi} \int_{S_0} \frac{P(r)\delta(t - R/c)}{R} \, dS_0$$

$$= \frac{1}{2\pi} \int_{S_0} \frac{P(r)\delta(t - R/c)}{R} \, r dr d\theta \qquad (1)$$

where $P(r)$ is the pupil function, c is the velocity of ultrasound, S_0 is the source plane, and R denotes the distance from the source point on an aperture to the observation point at (z, ρ).

$$R = [z^2 + \rho^2 + r^2 - 2\rho r\cos\theta]^{1/2} \qquad (2)$$

Refer to Fig. 1 for the coordinate systems employed. In Eq. (1), we first calculate the integral with respect to θ.

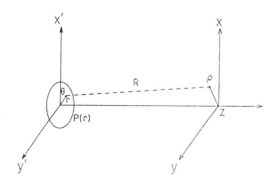

Fig. 1 Geometry of Eq. 1

50

$$h_\theta(\rho, t) = \int_0^{2\pi} \delta(t - R/c) \, d\theta/R \tag{3}$$

By a change of variable, $y = t - R/c$, Eq. (3) can be rewritten as

$$h_\theta(\rho, t) = -\frac{c}{\rho r} \int \frac{\delta(y)}{\sin\theta} \, dy \tag{4}$$

where

$$d\theta = -\frac{cRdy}{\rho r \sin\theta} \quad,$$

$$\sin\theta = \frac{[(2\rho r)^2 - (z^2+\rho^2+r^2-c^2(t-y)^2)^2]^{1/2}}{2\rho r}$$

For the integral of Eq. (4) to exist, there must be a time interval where y equals to zero. That is, for $\tau = ct = R$ we obtain the following inequality by requiring $|\cos\theta| \leq 1$ in Eq. (2).

$$|z^2 + \rho^2 + r^2 - \tau^2| \leq 2\rho r \tag{5}$$

From Eq. (5), the domain of τ in Eq. (3) is defined as

$$(z^2+(\rho-r)^2)^{1/2} \leq \tau \leq (z^2+(\rho+z)^2)^{1/2} \tag{6}$$

And r is bounded by

$$r_0 \leq r \leq r_1 \tag{7}$$

where

$$r_0 = |\rho - (\tau^2 - z^2)^{1/2}| \quad, \quad r_1 = \rho + (\tau^2 - z^2)^{1/2}$$

Then $h_\theta(\rho, t)$ becomes

$$h_\theta(\rho, t) = \frac{-4c}{(4\rho^2 r^2 - (z^2+\rho^2+r^2-\tau^2)^2)^{1/2}} \tag{8}$$

As a result, the impulse response is described by

$$h_z(\rho, t) = \frac{-2c}{\pi} \int_{r_0}^{r_1} \frac{P(r)rdr}{(4\rho^2 r^2 - (z^2+\rho^2+r^2-\tau^2)^2)^{1/2}} \tag{9}$$

Let r_{max} and r_{min} denote boundaries of $H(r-r_0)H(r_1-r)P(r)$ in Eq. (9) where $H(r)$ is the unit step function. Then the impulse response is expressed in a compact form as

$$h_z(\rho, t) = \frac{c}{\pi} \cos^{-1} \frac{\tau^2+\rho^2-z^2-r^2}{2\rho(\tau^2-z^2)^{1/2}} \Big|_{r_{min}}^{r_{max}} \tag{10}$$

The field response due to an arbitrary excitation f(t) is then obtained by

$$U_z(\rho, t) = \frac{d}{dt} f(t) * h_z(\rho, t) \tag{11}$$

where * denotes convolution.

For the annulus, the diffraction impulse can be determined analytically. In this case, the aperture is defined by

$$P(r) = \begin{pmatrix} 1 & , & R_0 \leqslant r \leqslant R_1 \\ 0 & , & \text{elsewhere .} \end{pmatrix} \tag{12}$$

and τ intervals of Eq. (5) can be reexpressed in three regions, as follows :

$$\rho > R_1 \quad : \quad (z^2+(\rho-R_1)^2)^{1/2} \leqslant \tau \leqslant (z^2+(\rho+R_1)^2)^{1/2} \, ,$$

$$R_0 \leqslant \rho \leqslant R_1 \quad : \quad z \leqslant \tau \leqslant (z^2+(\rho+R_1)^2)^{1/2} \, , \tag{13}$$

$$0 < \rho < R_0 \quad : \quad (z^2+(\rho-R_0)^2)^{1/2} \leqslant \tau \leqslant (z^2+(\rho+R_1)^2)^{1/2} \, .$$

Now, the impulse velocity potential response of the disk transducer is derived from the above results as a special case $R_0 \to 0$. On the axis, the impulse response has a constant value c for $z \leqslant \tau \leqslant (z^2+R_1^2)^{1/2}$, and zero elsewhere. Off the axis, the impulse response inside the geometrical projection of the source is expressed as

$$h_z(\rho, t) = c \qquad\qquad\qquad\qquad , \quad z \leqslant \tau \leqslant \tau_1$$

$$= \frac{c}{\pi} \cos^{-1} \frac{\rho^2+\tau^2-z^2-R_1^2}{2\rho(\tau^2-z^2)^{1/2}} \, , \quad \tau_1 \leqslant \tau \leqslant \tau_2 \tag{14}$$

$$= 0 \qquad\qquad\qquad\qquad , \quad \text{elsewhere}$$

where

$$\tau_1 = (z^2+(\rho-R_1)^2)^{1/2} \tag{15}$$

$$\tau_2 = (z^2+(\rho+R_1)^2)^{1/2}$$

Outside the geometrical projection, the impulse response is calculated from Eq. (10). These results are identical, as it should be, to the well-known results[1, 10] derived by the classical approach .

The advantages of the present approach to the field response are
(1) it facilitiates computer programming since the solution contains the annular geometry in a direct way,

(2) it provides an exact impulse response[12] for the case of apodizing with a polynomial [P(r) in Eq. (9)],

(3) it provides an impulse response[12], under para-axial assumption, when the disk transducer is covered with a lens.

BEAM SIMULATION

In this part, we perform computer simulations using the expressions previously derived to investigate the focusing characteristics of the annular array system and the optimum transmit delay profile to produce a large focal depth. We assume a gaussian excitation f(t) with 50 % BW:

$$f(t) = e^{-(t/t_0)^2} \sin(2\pi f_0 t) \tag{16}$$

where f_0 is the carrier center frequency. To obtain the round trip field response, $U(z, \rho, t)$, the transmit field response must be convolved with the receive impulse response:

$$U(z, \rho, t) = \frac{d}{dt}f(t) * \sum h_z^t(\rho, t) * \sum h_z^r(\rho, t) \tag{17}$$

where $h_z^t(.)$ and $h_z^r(.)$ are the transmit and the receive impulse responses, respectively. The value of $f(t)$ under - 40 dB was discarded to reduce the computing time and the impulse response of each annulus was sampled at the rate of $30/\lambda$, when λ is wavelength. In the receive mode, dynamic focusing is carried out continuously along the depth to achieve the optimum resolution.

In order to obtain a line focusing over a large depth of field, several authors have proposed the use of nonspherical transmitters, such as a cone and a horn. The cone transmitter x produces line focus to a depth equal to the radius of the aperture divided by the sine of the cone angle. However, the beam width, which does not vary much over the focusing depth, is inversely proportional to the tangent of the cone angle. To obviate

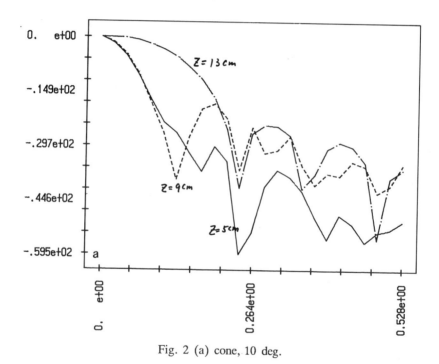

Fig. 2 (a) cone, 10 deg.

this conflict, a horn transmitter was proposed. The horn transmitter can be obtained by slowly tapering the cone angle. Fig. 2 shows the round trip beam patterns along the lateral direction. In the fixed transmit focusing case, the field pattern at the focal plane is much superior to other cases. However, off the focal plane, it shows poorer results. In general, the horn transmitter produces better lateral resolution at far field than the cone transmitter. As expected, nonspherical transmitters produce better lateral resolution, in general, than a fixed focused transmitter.

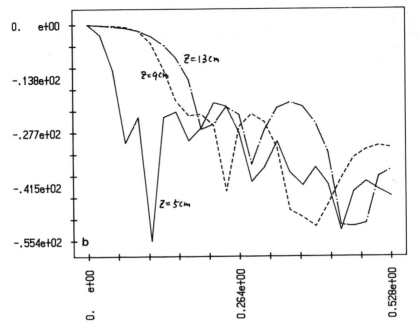

Fig. 2 (b) horn, initial angle=10 deg. final angle=3 deg.

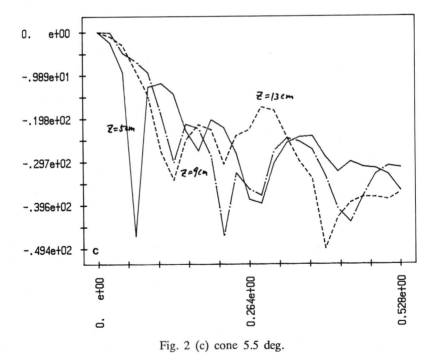

Fig. 2 (c) cone 5.5 deg.

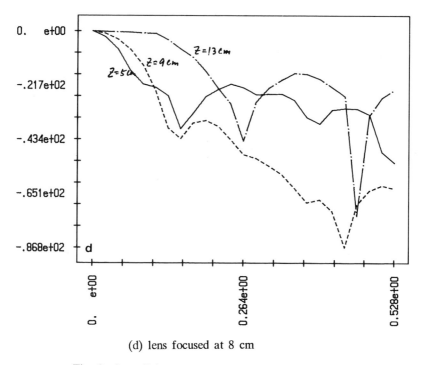

(d) lens focused at 8 cm

Fig. 2. Overall beam patterns for various trnasmeters

OPTIMIZATION OF BEAM PATTERN

In the most general terms the optimization problem can be stated as follows:

minimize a scalar function f(x)

subject to $g_i(x) > 0$, i = 1, 2, 3, ... m   (18)

where x is a set of variables x_1, x_2, ..., x_n to be adjusted. The objective function f(x) and the constraint functions $g_i(x)$ may be specified in many different ways for a specific problem. In our case, one attractive way of formulation of the problem will be as follows:

minimize the sum of $(BW)^N$ at two depths z = z_1 and z_2 (> z_1)   (19)

subject to all sidelobe levels below a specified value L_{max},

where BW denotes the overall beamwidth, N is an even integer > 2, and the variables are the delays ($d_1, d_2, .. d_n$) to be introduced to the array elements. Basically we are interested in the beam patterns in the range $z_1 < z < z_2$, but we consider only two end points for an obvious reason - to save computation time. It is expected that when N is large the larger BW at the two ends tends to be minimized and the beam pattern for $z_1 < z < z_2$ hopefully does not violate the constraints significantly.

Furthermore, in order to facilitate the solution of the problem we convert the constrained optimization problem into an unconstrained one as follows:

$$\text{minimize } f(d) = W_1(BW)^N_{z=z_1} + W_2(BW)^N_{z=z_2} + \sum_{i=n_s}^{n_f} \exp(L_i - L_{max}) \qquad (20)$$

where W_1 and W_2 are weighting factors and L_i's are the side-lobe levels computed at discrete points $i = n_s, .. , n_f$ in the side lobe region (see Fig. 3). BW and L_i are functions of delays $d = \{ d_1, d_2, .. , d_8 \}$. The 3.5MHz annular array has a diameter of 3 cm and consists of eight elements having the same area. Specifically, we chose $W_1 = W_2 = 10^4$, $n_s = 5$, $n_f = 25$, (n_f corresponds to 0.528 cm) N=4, L_{BW} (the level at which the bandwidth is calculated) = -12 dB, L_{max} = -25dB.

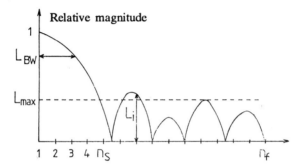

Fig. 3. Specifications for optimization.

For the solution of the optimization problem formulated in the above, the well-known Fletcher-Powell iterative technique[13], was used. The initial value of d_i's were chosen from Fig. 2 (c). After 3 iterations we obtained the beam profiles at the three depths of interest as shown in Fig. 4 where the initial patterns are also shown for comparison. The numerical values of interest are summarized in Table 1.

Table 1. Summary of optimization

	-12dB Bandwidth			Max sidelobe level		
Depth	5 cm	9 cm	13 cm	5 cm	9 cm	13 cm
Initial BW Final BW	1.10 1.460	1.560 3.020	1.860 (mm) 2.460 (mm)	-12dB -27dB	-17dB -29dB	-24dB -31dB
	Initial			Final(3-iteration)		
Transmit delay profile	d_1= 16.623 d_2= 40.26 d_3= 52.468 d_4= 62.273 d_5= 70.65 d_6= 78.117 d_7= 85.0 d_8= 91.30			16.688 (µsec) 29.61 47.922 62.338 71.3 77.338 81.429 91.753		

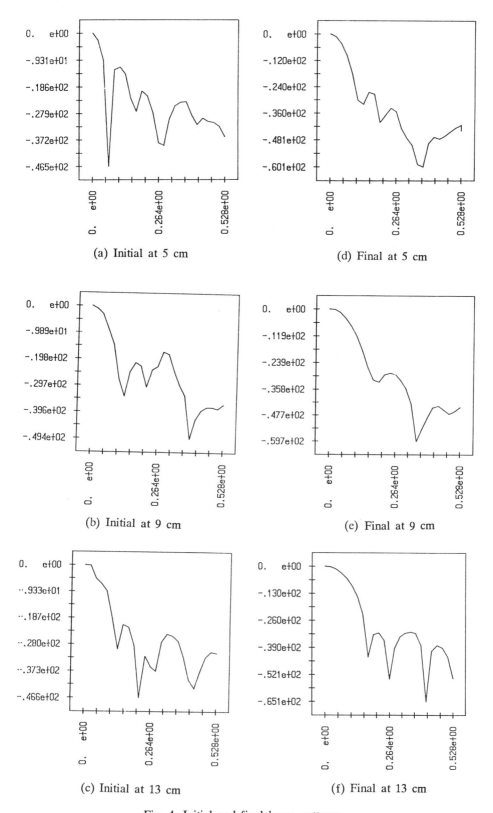

(a) Initial at 5 cm   (d) Final at 5 cm

(b) Initial at 9 cm   (e) Final at 9 cm

(c) Initial at 13 cm   (f) Final at 13 cm

Fig. 4. Initial and final beam patterns,

In order to satisfy the side lobe constraints all along the depth from $z = z_1$ to $z = z_2$, we could, of course, take a few more points to see the beam pattern there, if we do not concern much about the computer time.

A NEW DIGITAL FOCUSING SYSTEM FOR ANNULAR ARRAY

For a dynamic focusing, the delay taps and multiplexing circuitry are indeed complex in a conventional analog focusing system. To completely eliminate the use of delay lines which are nonideal and costly, an all-digital Sampled Delay Focusing(SDF) scheme has been proposed recently by the authors, in which the maximum delay obtainable is , however, limited by the Nyquist sampling rate. This constraint is violated in the near field focusing. In this paper, we propose a new focusing technique which achieves dynamic focusing with a relatively simple circuitry and without the above- mentioned constraint.

In a conventional delay-sum-sampling focusing process, the received echoes for transducer array elements are delayed appropriately and summed to compose a focused signal f(t).

$$f(t) = \sum_{i=0}^{N} X_n(t - \tau_n, t). \tag{21}$$

where N+1 is total number of elements, and X_n and τ_n are respectively the received signal and the focusing delay, of the n-th element. The corresponding sampled version can be expressed as

$$f(kT) = \sum_{n=1}^{N} X_n(t)\delta(t - kT)$$
$$= \sum_{n=1}^{N} X_n(t)\delta[t - (kT - \tau_n, t)]. \tag{22}$$

where T is the sampling interval. From Eq. (22), we see that the focused signal can be obtained by summing the echoes of array elements sampled by the following delayed sampling clock:

$$s(t) = \delta[t - (kT - \tau_n, t)]. \tag{23}$$

This is the key idea of Sampled-Delay-Focusing(SDF). However, since this SDF has no memory buffers between the sampling stage and the adder stage, the maximum delay obtainable is limited by

$$\tau_{max} < T = 1/f_s \tag{24}$$

When the maximum time delay, τ_{max}, is greater than the sampling interval T, we cannot hold the sample value as long as τ_{max}, because the next sample/hold (S/H) pulse must not overlap the previous one. To eliminate overlapping of two consecutive S/H pulses, we must reduce the S/H pulse duration sufficiently. In this case, to add the sampled echo signals to the echo signals sampled later than the sampling interval T, we need a pipelined buffer stage at each array channel to store the previous sample values, as shown in Fig. 5.

The most important factors to be considered in constructing the pipelined buffer are the number of buffer stages and the buffer control mechanism. To determine these, let us consider the focusing step from the point of view of the buffer control. For the transducer array channel indexed as i, where $i = -N/2, \cdots, N/2$, the number of buffer stages, M_i, is determined by the sampling frequency and time delay of the i-th array channel, τ_i, as

$$M_i = f_s \cdot \tau_i . \tag{25}$$

The number of buffer stages required for each transducer array channel changes with the steering direction and the focal depth from zero to a maximum value M_{max} given by

$$M_{max} = \max[M_i] = f_s \cdot \tau_{max}, \qquad \text{for all } i. \tag{26}$$

Therefore, to focus all the points in the scanning area of interest, we need at least M_{max} buffer stages for each array channel and a buffer control mechanism similar to the focusing and steering address generation in the synthetic focusing technique.

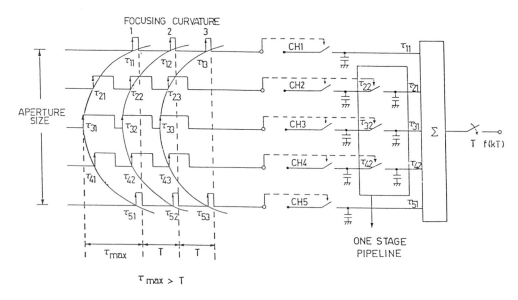

Fig. 5 Sampled delay focusing system

To realize these pipelined buffer stages in a simple fashion, we use an array of analog-to-digital converters (ADC) and first-in-first-out (FIFO) buffers, as shown in Fig. 6, which operates in the following way. When we insert the sequence of sampled data into a FIFO buffer, the data are immediately shifted to the output side of the buffer and ready to be output on a first-in-first-out basis, and the data input/output (I/O) operations can be controlled independently. As a result of the FIFO array, the echo signals sampled at different times to compensate for different propagation path delays will be aligned in the output stages of each FIFO and hence we can easily sum the outputs of the FIFO array to obtain the focused signal. In other words, the FIFO array effectively transforms the input spherical wave front in an arbitrary direction to a normal plane wave front. Note that the above control mechanism can be performed by distributing the address control to the delayed sampling process and the FIFO array. This new scheme is named pipelined sampled-delay focusing (PSDF), and this scheme has no sampling rate constraint except for system bandwidth and requires neither analog L-C delay lines nor complex multiplexers.

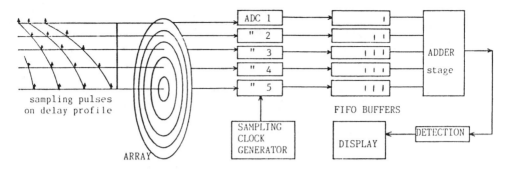

Fig. 6. Digital continuous focusing system using five element annular array.

CONCLUSION

The problem of optimum focusing in the annular array system has been solved by finding optimum delay profiles to be introduced to the array elements in the transmit mode and by using a continuous dynamic focusing in the receive mode which can be implemented in a digital fashion by the sampled-delay-focusing scheme.

Real imaging with the proposed scheme will be reported in the near future.

REFERENCES

1. Peter R. Stepanishen: "Transient radiation from pistons in an infinite planar baffle," J. Acoust. Soc. Am., 49, 1971, pp1629-1638

2. J. C. Lockwood, J. G. Willete: "High speed method for computing the exact solution for the pressure variation in the near field of a baffled piston," J. Acoust. Soc. Am., 53, 1973, pp735-741

3. J. W. Goodman: "Introduction to Fourier Optics," Mcgraw-Hill, New York, 1968, ch.4

4. D. R. Dietz et al.: "Wideband annular response, "IEEE. cat. #78CH., 1344-1 SU, 1978, pp206-211

5. M. Arditi, F. S. Foster: "Transient fields of concave annular arrays," Ultrasonic imaging, 3, 1981, pp37-61

6. M. O'Donnell: "A proposed annular array imaging system for contact B-scan applications," IEEE, Trans. on SU, su-29, 6, Nov. 1982, pp331-338

7. D. R. Dietz: "Apodized conical focusing for ultrasound imaging," IEEE Trans. on SU, su-29, No. 3, May. 1982, pp128-138

8. M. H. Lee, S. B. Park: "New continuous dynamic focusing technique in ultrasound imaging, " Electronics letter, vol.21, No.17, 1985, pp49-751

9. J. H. Kim, T. K. Song, S. B. Park: "A pipelined sampled delay focusing in ultrasound imaging systems, "submitted to Ultrasonic imaging.

10. M. Fink, J. F. Cardoso, Pascal Laugier: "Diffraction effect analysis in medical echography," Acta Electronica, 26, 1-2, 1984, pp59-80

11. S. J. Norton, A. Macovski: "Broadband through a lens: A linear system approach," J. Acoust. Soc. Am. 64, Oct, 1978, pp1059-1063

12. T. K. Song and S. B. Park: "Analysis of broad band ultrasonic field response and its application to the design of focused annular array imaging systems," Ultrasonic technology 1987, Toyohashi International Conference on Ultrasonic Technology, Toyohashi,japan, Edited by Kohji Toda MYU RESEARCH, Tokyo,1987- Printed in Japan

13. R. Fletcher and M. J. D. Powell: "A rapidly convergent descent method for minimization," Computer J., 6, 163-168 (June, 1963).

QUASI-REAL TIME IMAGING OF PULSED ACOUSTIC FIELDS

M.S.S. Bolorforosh and C.W. Turner

Department of Electronic and Electrical Engineering
King's College, London, England

ABSTRACT

Experimental results obtained from optically-scanned imaging of 3MHz
pulsed acoustic fields are presented and discussed in the context of ultrasonic transducers used for B-scan medical imaging and industrial NDE.

The system described is shown to be capable of producing quasi-real
time images of wideband pulses radiated by water-immersed transducers, for
pulse durations down to 1μs in length.   The results obtained are compared
with those measured under C.W. excitation of the same test transducer.
The physical limitations of the system are discussed in respect of the
optical scanning process, the materials parameters and the signal processing
requirements.

INTRODUCTION

The C.W. operation of the optically scanned transducer described in
this paper has been reported previously.[1]   It has been shown to be capable
of producing high resolution, two-dimensional amplitude and phase images of
radiation fields from water-immersed transducers at frame rates of several
Hz, typically for 10^4 pixels.   The principal application has been for
transducer evaluation[2] where the comparatively low cost and ease of operation
provide clear advantages over alternative methods available in the low megahertz frequency range.   The widespread use of pulsed acoustic fields in
industrial NDE and medical diagnostics (e.g. B-scan systems) underlines
the need for instruments capable of rapid assessment of wideband pulsed
transducers or arrays of transducers, particularly in clinical practice.[3]

In this paper we present experimental results obtained for the
optically-scanned transducer imaging 3MHz pulsed acoustic waves and compare
these with the imaging performance under C.W. conditions.   The optically-
scanned transducer is shown to be capable of imaging wideband pulses
without major modifications of the C.W. version, and to provide the basis
for a versatile instrument for transducer evaluation under a wide variety
of excitation conditions.   The performance under pulsed excitation is
assessed in the context of industrial and medical applications of ultrasound and the ultimate limits of the optically-scanned technique are
discussed in relation to the choice of the materials used in the optically-
scanned transducer.

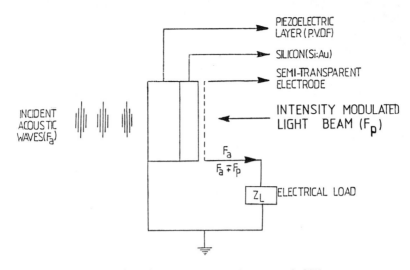

Fig.1   Cross-sectional view of OST

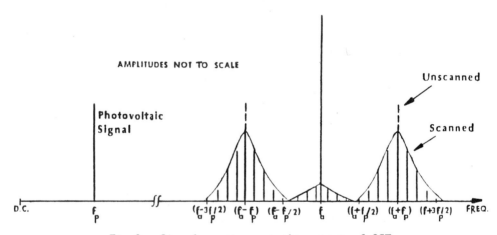

Fig.2   Signal spectrum at the output of OST

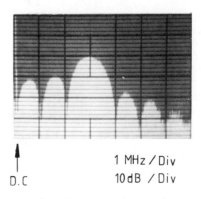

Fig.3   Spectrum of a three cycles pulse centred at 3MHz

BACKGROUND

The cross-section view shown in Fig.1 gives the essential components of the optically-scanned transducer (OST). The acoustic waves at frequency f_a incident upon the PVDF piezoelectric layer induce an electric field distribution which couples to the photoconductive silicon wafer. The scanning optical beam, intensity modulated at a frequency f_p, produces mobile photo-carriers in the photoconductor. The output spectrum from the composite transducer Fig.2 therefore includes sidebands at frequencies $f_a \pm f_p$ containing amplitude and phase information derived from the acoustic field at the location under illumination. A sequentially scanned image of the acoustic field distribution over the surface of the PVDF layer is then displayed on a suitable monitor.

Under C.W. excitation the image information is extracted in the frequency domain, for example at the lower sideband frequency. The wide spectral spread of a pulsed acoustic signal, such as that shown in Fig.3 for a three cycle r.f. pulse centred on 3·5MHz, introduces a number of problems for this approach. Firstly, the OST package is required to have a broadband frequency response in respect of both its acousto-electric behaviour at frequency f_a and the delivery of the output at the lower sideband frequency $f_a - f_p$ to the matching network and signal processing circuits.

The selection of PVDF as the piezoelectric element gives a relatively broadband acoustoelectric response because of its good acoustic impedance and velocity match to water and its low acoustic and electrical Q value.[4] The main problem encountered arises from the frequency dependence of the overall reflection coefficient of the OST caused by the acoustic mismatch at the interface between the PVDF and the silicon wafer which is air-backed. In Figs 4 and 5, reflection measurements obtained using an acoustically transmissive hydrophone show that the incorporation of a thin layer of silicone grease between the PVDF and the silicon wafer significantly improves the frequency response. The grease, which has a relative permittivity of 2·5 , is electrically insulating and does not affect the coupling of the piezoelectric fields from the PVDF to the photocarriers in the silicon.

DESIGN OF THE OST FOR PULSED OPERATION

A satisfactory separation of the lower sideband imaging spectrum from the acoustic carrier spectrum requires the optical pump frequency f_p to be greater the 1MHz. This in turn implies that the photocarrier lifetime in the silicon must be reduced to fractional microsecond levels by gold doping or similar means. Previous work[4] has shown that for a pump frequency of 1MHz a lifetime in region of 0·15μs is required for maximum output. Lifetimes an order of magnitude greater than the optimum value cause more than 30dB reduction in the sideband amplitude. The technique of gold-doping becomes less attractive for lifetimes below 1μs because of the inevitable reduction in the photocarrier mobility and alternative methods such as high energy bombardment by electrons or neutrons or by irradiation must be used. The results reported here were obtained using gold-doped silicon with a lifetime of about 1μs.

These modifications to the OST have enabled the basic design to be adapted for imaging pulsed acoustic waves. A wideband impedance match is used to load the transducer followed by a low noise wideband pulse

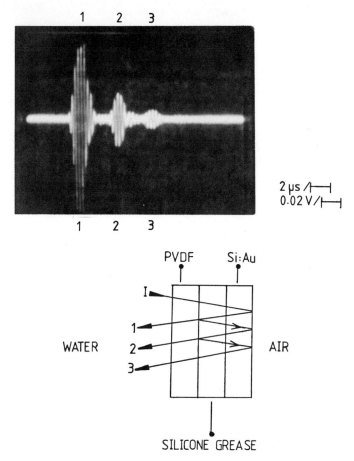

Fig.4   Reflections within OST in the time domain

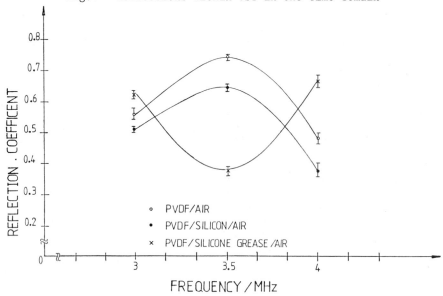

Fig. 5   Reflection coefficient versus frequency for OST
for different acoustic loading conditions

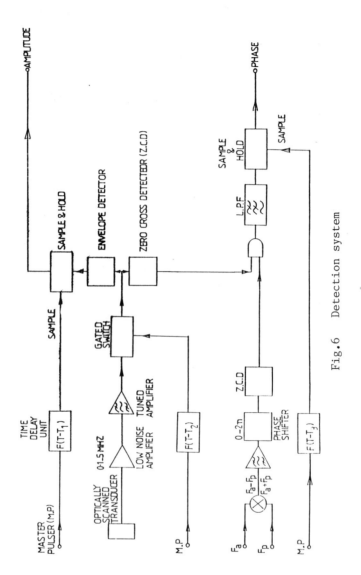

Fig.6   Detection system

67

amplifier centred on the lower sideband frequency of 2MHz. This also suppresses the unwanted acoustic carrier output centred at f_a which contains no image information. A gated amplifier is used to select the arriving pulse at the OST and to reject any reverberations either within the transducer package or in the water tank. Amplitude or phase information is selected by appropriate switching to either an envelope detector with sample and hold or a zero-crossing detector respectively, as shown schematically in Fig.6.

The optical scanning, achieved by a galvanometer mirror system deflecting a collimated 20mW semiconductor laser beam, is synchronized with the acoustic pulse generation to maximize the efficiency of the sampling process. This enables the laser beam to remain stationary at one location during the period in which the complex acoustic field value is sampled. The fly-back step incorporated in conventional raster scanning is eliminated to speed up the process. Means are also provided to control both the time within the pulse cycle at which the sampling is performed and the number of pulses sampled at each location or pixel.

RESULTS

A series of experimental tests was carried out to compare the pulsed performance of the OST with that obtained under C.W. conditions. The test transmitter chosen was a 3·5MHz focussed transducer with a focal length specified to be 9cm. Fig.7 shows the C.W. amplitude distribution with the transmitter at different distances from the OST. The transmitter was then driven under pulsed conditions, with pulse duration of 8µs, 4µs and 2µs, and the acoustic field distribution measured over the scanned area of the OST. The received pulses were also measured in the time domain at various points in the transverse plane. These results are presented in Figs 8, 9 and 10 and show that the field distribution depends on the pulse length and that a better focussing action is obtained for the shorter pulse length with the particular transmitter used. Similar results were obtained for phase images.

The optical scanning arrangement used is extremely versatile. Typically, in the experiments reported here, about 200 pixels were addressed at frame rates of about 1Hz. A maximum of about 10^4 pixels could be used with the present scanning scheme, limited primarily by the loss of signal-to-noise ratio as the number of acoustic pulses received per pixel is reduced. If slower scan rates are acceptable the number of pixels can be increased accordingly. The effect of increasing the scanning speed, or, alternatively, the number of pixels addressed per unit time is shown in Fig.11 where in the limit only one acoustic pulse is received per pixel.

CONCLUSIONS

Reviewing the C.W. and pulsed results it is evident that the OST is capable of producing consistent results that allow ready comparison of transducers under different excitation conditions. The advantages previously demonstrated for the OST in ultrasound transducer evaluation for C.W. excitation have been shown also to apply under pulsed operation. In particular the performance of a given transducer can be examined in a comprehensive way, using amplitude and phase imaging, in respect of beam profile, lateral and axial resolution and behaviour in the time domain as a function of driving frequency and pulse length. Each of these factors is of considerable significance to the assessment of single transducers and arrays used in NDE and medical ultrasonics.

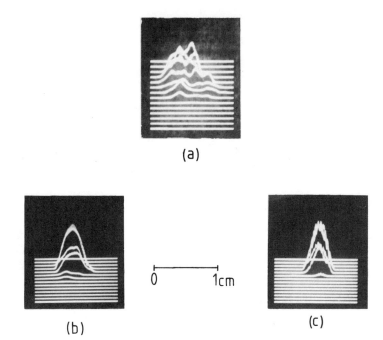

(a)

(b)   (c)

0 ⊢——————⊣ 1cm

Fig.7   Amplitude image of radiation pattern of focused transducer
         (19mm, 3·5MHz, 9cm focus) under C.W. excitation at
               (a) 3cm, (b) 9cm, (c) 15cm from OST

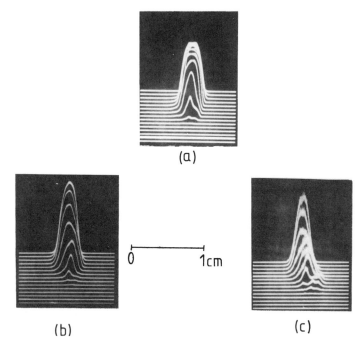

(a)

(b)   (c)

0 ⊢——————⊣ 1cm

Fig.8   Same transducer as in Fig.7 under 8μs pulse excitation
               (a) 3cm, (b) 9cm, (c) 15cm from OST

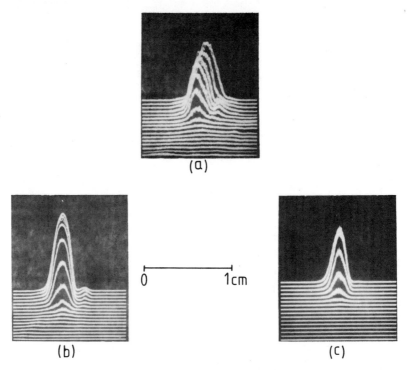

Fig.9   Same transducer as in Fig.7 under 4 μs pulse excitation
(a) 3cm, (b) 9cm, (c) 15cm from OST

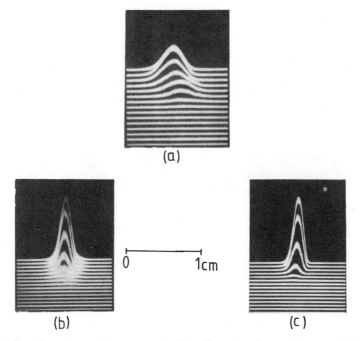

Fig.10   Same transducer as in Fig.7 under 2μs pulse excitation
(a) 3cm, (b) 9cm, (c) 15cm from OST

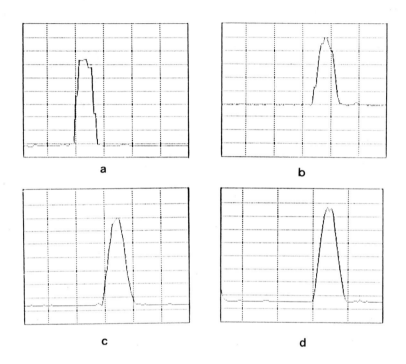

Fig. 11   Line scan through the same focused transducer
output at different scanning speeds

(a)   One pulse per pixel; 0·2 v/div; 0·5 ms/div
(b)   Two pulses per pixel; 0·5 v/div; 1 ms/div
(c)   Three pulses per pixel; 0·5 v/div; 2 ms/div
(d)   Four pulses per pixel; 0.5 v/div; 5 ms/div

Although the shortest pulse lengths used in the experiments reported here were 2μs, preliminary tests with shorter pulses have indicated that the OST can be adapted further to allow imaging of sub-microsecond pulses. At this limit the intensity modulation of the scanning laser beam would be replaced by a stroboscopic pulse synchronized with the acoustic pulse and subsequent correlation in a gated receiver. Re-design of the OST package will also be necessary to achieve sub-microsecond performance. The silicon photoconductor would require a further reduction in photocarrier lifetime to below 1μs, or possibly the use of an alternative semiconductor material such as gallium arsenide. The introduction of a transparent backing material to give a broadband matched acoustic load would also improve the response to very short pulses. These modifications should enable the OST to provide acceptable images of acoustic pulses in the region of 0·5μs in length.

REFERENCES

1. C.W. Turner and S.O. Ishrak, "Two-dimensional Imaging with a High Resolution PVF$_2$/Si Optically-scanned Receiving Transducer", Acoustical Imaging, Vol.11, J.P. Powers, Ed., Plenum Press, New York, 1981, pp77-93.

2. M. Salahi and C.W. Turner, "Improved Evaluation of Acoustic Transducers Using Digital Processing of Radiation Field Images", IEEE Trans. Sonics Ultrason., Vol SU31, pp307-312, July 1984.

3. R.C. Preston et al, "PVDF Membrane Hydrophone Performance, etc.," J. Phys. E. Sci. Instrum., Vol.16. 1983, pp786-796.

4. C.W. Turner and S.O. Ishrak, "Comparisons of Different Piezoelectric Transducer Materials for Optically-scanned Acoustical Imaging". Acoutical Imaging, Vol.10, P. Alais and A. Metherell, Eds., Plenum Press, New York, pp761-778, 1980.

ANALYSIS OF OBLIQUE ANGLE SCANNING IN THE

IMAGING OF MULTILAYERED TARGETS*

Daniel T. Nagle and Jafar Saniie

Department of Electrical and Computer Engineering
Illinois Institute of Technology
Chicago, IL 60616

INTRODUCTION

The ultrasonic scanning of multilayered targets, using pulse-echo detection, results in reverberations which complicate the direct characterization of the boundaries. In the nondestructive material evaluation, the problem of reverberation arises frequently. In fact some structures by their very nature are so reverberant that the reverberations comprise the entire received signal. A practical example of such a problem occurs in connection with the inspection of steam generator tubing. Steam generators currently in use contain inconel tubing which is held loosely within a steel support structure. The integrity of the tube/support structure can be evaluated by sending ultrasonic pulses through the tube wall which are highly reverberant. The multilayered model of the tube/support structure is shown in Figure 1, where Region I is inside the tube, Region II is the tube wall, Region III is the water gap, and Region IV is the support plate. Through earlier investigations [1,2], an ultrasonic classification technique using a normal incident angle scheme has been developed that facilitates the imaging of targets (Region IV) hidden by highly reverberant thin layers (Region II).

Our current studies involve the imaging of multilayered targets using oblique angle scanning. This scanning technique provides an attractive feature of automatic rejection of the first set of reverberant echoes from the front layer while preserving information in the signal of the subsequent layers. When examining elastic mediums, the problem becomes more complex since two modes of propagation (i.e. longitudinal and shear) are supported. The mathematical model of reverberations and the effect of mode conversion for oblique angle scanning will be presented. Furthermore, experimental results that are in close agreement with theoretical analysis using planar and tubular multilayered targets will be presented.

---

* This project is supported by EPRI Grant RP2673-5

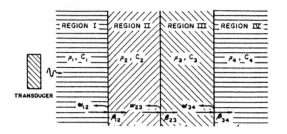

Figure 1. Multilayered structure consisting of four different regions.

NORMAL SCANNING ECHO CLASSIFICATION

Consider the planar model of multiple thin layers as shown in Figure 1. The received signal is comprised of multiple echoes detected after traveling k times in Region II and l times in Region III:

$$r(t) = \sum_{k=0}^{\infty} \sum_{l=0}^{\infty} \gamma_{kl} u(t - 2kT_2 - 2lT_3) \tag{1}$$

where $u(t)$ represents the measuring system impulse response, T_2 represents the traveling time in Region II, T_3 represents the traveling time in Region III, and the term γ_{kl} is the received echo amplitude related to the reflection coeficients, α_{ij}, or the transmission coefficient, β_{ij} (Note: i and j indicate which regions constitute the interface).

To facilitate the characterization of multiple layers, Equation (1) can be reorganized as follows:

$$r(t) = \alpha_{12}u(t) + \sum_{k=1}^{\infty} a_k u(t - 2kT_2) + \sum_{k=1}^{\infty} b_k u(t - 2T_3 - 2kT_2) + \tag{2}$$

$$\sum_{k=1}^{\infty} c_k u(t - 4T_3 - 2kT_2) + \sum_{k=1}^{\infty} d_k u(t - 6T_3 - 2kT_2) + \ldots\ldots$$

where a_k is the amplitude of the class 'a' echoes, which are characterized by the fact that they reverberate in Region II only; b_k is the amplitude of the class 'b' echoes which reverberate continually in Region II, and once in Region III; c_k is the amplitude of the class 'c' echoes which reverberate continually in Region II, and twice in Region III; d_k is the amplitude of the 'd' echoes which reverberate continually in Region II and three times in Region III; etc...

The amplitudes of these classes are given [3]:

$$a_k = (\frac{\beta_{12}\beta_{21}}{\alpha_{21}})A_0{}^k; \quad for \quad k \geq 1 \tag{3}$$

$$b_k = k(\frac{\beta_{12}\beta_{21}}{\alpha_{21}})A_1 A_0{}^{k-1}; \quad for \quad k \geq 1 \tag{4}$$

$$c_1 = \frac{\beta_{12}\beta_{21}}{\alpha_{21}}A_2 \tag{5}$$

$$c_k = (\frac{\beta_{12}\beta_{21}}{\alpha_{21}})k[A_2 A_0{}^{k-1} + \frac{k-1}{2}A_1{}^2 A_0{}^{k-2}] ; \quad for \quad k > 1 \tag{6}$$

$$d_1 = \left(\frac{\beta_{12}\beta_{21}}{\alpha_{21}}\right)A_3 \tag{7}$$

$$d_2 = \left(\frac{\beta_{12}\beta_{21}}{\alpha_{21}}\right)[A_3 A_0 + A_2 A_1] \tag{8}$$

$$d_k = \left(\frac{\beta_{12}\beta_{21}}{\alpha_{21}}\right)[A_3 A_0{}^{k-1} + (k-1)A_2 A_1 A_0{}^{k-2} + \frac{(k-1)(k-2)}{6}A_1{}^3 A_0{}^{k-3}]; \quad for \ k > 2 \tag{9}$$

where

$$A_n = \beta_{23}\beta_{32}\alpha_{34}^n \alpha_{32}^{n-1}\alpha_{21} \tag{10}$$

Class 'a' and 'b' echoes are the most dominant classes of echoes in the received signal. Using Equations (3) and (4), the envelopes of class 'a' and 'b' echoes for highly reverberant multilayered structures are displayed in Figure 2. The envelopes of these two classes differ significantly, where the class 'a' echoes are monotonically decreasing as a function of time, whereas the 'b' echoes increase first and then taper off later on. The envelope of subsequent classes (i.e. 'c', 'd',...) are of similar nature to that of the 'b' echoes. These changes can be seen from experimental measurements shown in Figure 3, where 'a', 'b', 'c', and 'd' are clearly present for several reverberations. The tube thickness (Region II) and gap distance (Region III) can also be determined from this figure. The tube thickness corresponds to delay between the peaks of the echoes within each class, and the gap distance is given by time delay between the sequential classes of echoes ('a' and 'b' or 'b' and 'c', etc.).

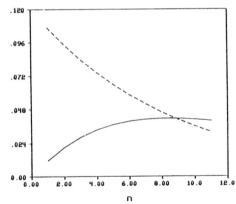

Figure 2. Comparison of the envelope of the class 'a' echoes (dashed line) with the envelope of the class 'b' echoes (solid line).

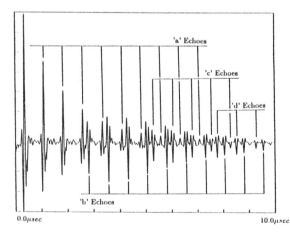

Figure 3. Experimental results depicting the reverberation process of the multilayered model.

OBLIQUE ANGLE SCANNING

Oblique angle scanning (OAS) provides the automatic rejection of the 'a' echoes (tube echoes) while still preserving the information-bearing 'b' echoes (support plate echoes) in the backscattered signal [4]. In general, as the scanning angle varies, the degree of refraction and reflection and energy transfer that occurs at each interface changes dramatically and complicates the evaluation of detected multiple echoes. Thus, for this technique to be most effective the choice of the oblique angle must be conducive towards the optimality criterion, which means the scanning angle must be chosen to reject a sufficient amount of 'a' echoes and maximize the energy of the 'b' echoes in the received signal. Besides satisfying the optimality criterion, the scanning angle is confined by system constraints such as the range of detection of the transducer or the phsysical dimensions of the tube/support structure. In this study, the relationship between the scanning angle and the beam interactions at each boundary are given explicitly and analyzed to satisfy the optimality criterion.

The effect of refraction and reflection in the oblique angle scanning of a simplified planar tube/support structure is illustrated in Figure 4. Mode conversion is evident at the liquid/solid interface, where the incident beam penetrates an elastic medium (solid) and creates two transmitted waves, namely one longitudinal and the other shear. The propagation of these two waves differ in intensity, direction, and velocity. However, liquids do not support the propagation of shear waves since they are inelastic by nature. The effect of mode conversion at the solid interface increases the number of reverberant echoes tremendously.

The redirection of the incident energy is governed by Snell's Law:

$$\frac{sin\theta_1}{c_1} = \frac{sin\theta_2}{c_2} = \frac{sin\gamma_2}{b_2} \tag{11}$$

where θ_i and γ_i are the angles (measured with respect to the normal of the boundary) and c_i and b_i are the velocities of the longitudinal and shear waves, respectively in the given medium i (where medium 1 is liquid and medium 2 is solid). As can be

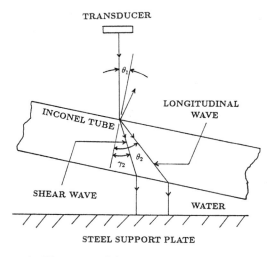

Figure 4. Planar model of oblique angle scanning of tube/support structure.

seen from the above equation, the velocities of the waves before and after impinging the boundary determine the degree of refraction or reflection. The velocity of the longitudinal wave is approximately twice that of the shear wave in most materials which infers that the longitudinal wave will be refracted greater than the shear wave for oblique angles.

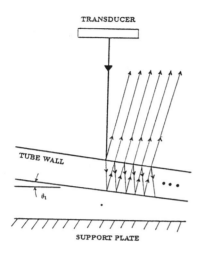

Figure 5. Progression of reverberations of 'a' echoes which
are rejected using OAS.

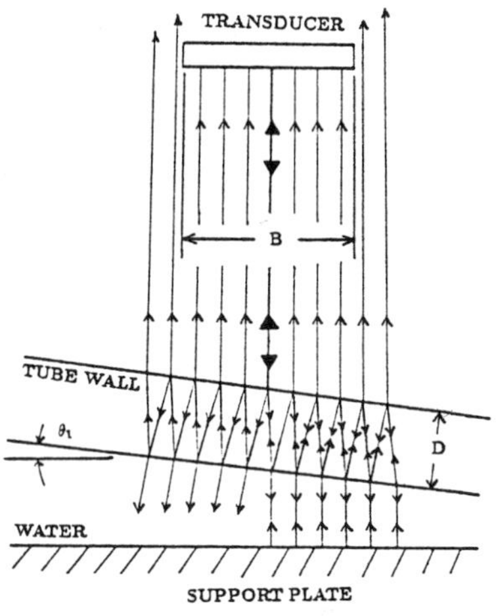

Figure 6. Progression of reverberations of 'b' echoes us-
ing OAS within the Tube/Support
Plate structure.

The reflections and refractions that take place in the tube structure eliminate the detection of 'a' echoes by directing the backscattered echoes away from the transducer. This can be seen in Figure 5 (only the longitudinal class 'a' echoes are shown for visual clarity). If the first reflected wave is out of the detection range, then all subsequent reverberations will not be detected since the returning echoes are shifted laterally as the reverberations progress. The composite effect of this shifting will result in a greater rate of decay of the 'a' echoes. It can be shown from Equation (11) that as the scanning angle is increased, the further the beams are reflected away from the transducer. Thus, a lower bound on the scanning angle can be constructed to eliminate a significant amount of 'a' echoes. This lower bound, $\theta_{1\,min}$, can be derived from the geometry of Figure 5,

$$\theta_{1\,min} = \frac{1}{2}tan^{-1}(\frac{B}{2X_f}) \qquad (12)$$

where X_f is the path distance from the transducer to target and B is the beam field for class 'a' rejection criteria. For example, the 20 MHz transducer used to obtain experimental data has a detection beam field of approximately 2 mm and a target distance of 1.5 cm which corresponds to a minimum angle of $1.9°$.

Unlike the 'a' echoes, the 'b' echoes are preserved since the angle of incidence of the returning echoes is equivalent to the initial incident beam. Figure 6 illustrates the oblique angle scanning reverberation process for the 'b' echoes, not showing the shear waves or the 'a' echo patterns to simplify the figure. Although the returning 'b' echoes have been unaltered in terms of direction in the refraction process, the 'b' echoes have been shifted laterally similar to the 'a' echoes as previously discussed.

As the scanning angle increases, the refracted angles increase and the lateral shifting becomes more dominant. Thus, to capture a mininum number of reverberations, an upper bound on the scanning angle, $\theta_{1\,max}$, may be found by examining the geometry of Figure 6 which gives the following equation:

$$cos\theta_1 = \frac{B}{DNtan\theta_2} \tag{13}$$

where θ_1 is the incident angle, θ_2 is the longitudinal refracted angle, D is the tube thickness, N is the minimum number of returning echo paths detected, and B is the given beam field. Using Snell's Law, Equation (13) can be solved to find $\theta_{1\,max}$,

$$\theta_{1\,max} = sin^{-1}\sqrt{\frac{1}{2}[(1+C^2) - \sqrt{(1+C^2)^2 - 4C^2k^2}]} \qquad \text{for } k < 1 \tag{14}$$

where

$$k = \frac{c_1}{c_2} \qquad and \qquad C = \frac{B}{2ND}$$

The above equation can be used to examine the number of possible 'b' echoes that can be detected. For example, using a tube thickness of 1 mm and requiring five reverberations to be within the detection field forces an upper bound on incident angle, $\theta_{1\,max} = 2.7^o$. In practice, the incident angle θ_1, must be bounded by $\theta_{1\,min}$ and $\theta_{1\,max}$, i.e.,

$$\theta_{1\,min} \le \theta_1 \le \theta_{1\,max} \tag{15}$$

In general, the above condition may not be satisfied if the value of N exceeds the physical limits of the system. If the above example is modified, so that there are ten required reverberations, the result is $\theta_{1\,max} = 1.4^o$, which is contradictory to the optimal criterion. Equation (15) simplifies the analysis of the energy transfer functions of the reverberant tube/support plate structure considerably.

MODE CONVERSION

During the reverberation process of oblique angle scanning, there are three types of mode-boundary wave interactions that occur which are circled in Figure 7 and labeled case 1,2, and 3. All three have different energy transfer characteristics and contribute to the development of the 'b' echoes in some way. The longitudinal wave incident on a liquid/solid boundary is referred to as case 1 (as shown in Figure 7). Case 1 interface determines how much energy is transmitted into the tube wall which should be maximized to reduce the amount of energy that is lost on initial penetration of the tube wall. Case 2 corresponds to the situation where a longitudinal beam is incident on a solid/liquid boundary in which its energy transfer characteristics describes how much energy leaks out of the tube wall and the degree of mode conversion. In order to increase the energy of the 'b' echoes, transmission into the water gap should be maximized. Case 3 refers to a shear wave incident on a solid/liquid boundary and has optimal requirements similar to case 2. The explicit solutions for the energy transfer functions (i.e. shear and longitudinal) of these cases are known [5,6], assuming that the incident beam is a plane harmonic wave and that the boundary is homogeneous (which means that energy is conserved).

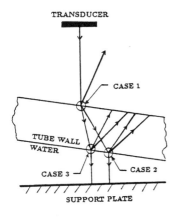

TRANSDUCER

CASE 1

TUBE WALL
WATER

CASE 3

CASE 2

SUPPORT PLATE

Figure 7. The three cases of refraction and reflection
scenarios encountered
during the reverberation
process within the tube/
support plate structure.

Wave generation for case 1 consists of one reflected wave and two transmitted waves in which the energy relationships with respect to the incident beam are given below:

$$\left(\frac{\varphi_1'}{\varphi_1}\right)^2 = \left|\frac{-Z_1 G_{2l} G_{2t} + G_{2t} cos^2 2\gamma_2 + G_{2l} sin^2 2\gamma_2}{Z_1 G_{2l} G_{2t} + G_{2t} cos^2 2\gamma_2 + G_{2l} sin^2 2\gamma_2}\right|^2 \tag{16}$$

$$\left(\frac{\varphi_2}{\varphi_1}\right)^2 = \frac{4\rho_1 tan\theta_1}{\rho_2 tan\theta_2}\left|\frac{G_{2t} cos 2\gamma_2}{Z_1 G_{2l} G_{2t} + G_{2t} cos^2 2\gamma_2 + G_{2l} sin^2 2\gamma_2}\right|^2 \tag{17}$$

$$\left(\frac{\psi_2}{\varphi_1}\right)^2 = \frac{4\rho_1 tan\theta_1}{\rho_2 tan\gamma_2}\left|\frac{G_{2l} sin 2\gamma_1}{Z_1 G_{2l} G_{2t} + G_{2t} cos^2 2\gamma_2 + G_{2l} sin^2 2\gamma_2}\right|^2 \tag{18}$$

where subscript 1 indicates the propagating medium is water, and subscript 2 indicates the propagating medium is inconel (hardened steel). The variables used in all three cases represent:

φ_i: The intensity of the longitudinal wave in region i
φ_i': The intensity of the reflected longitudinal wave in region i
ψ_i: The intensity of the shear wave in region i
ψ_i': The intensity of the reflected shear wave in region i
θ_i: The angle of the longitudinal wave in region i
γ_i: The angle of the shear wave in region i
ρ_i: The density of the region i
c_i: The velocity of the longitudinal wave in region i
b_i: The velocity of the shear wave in region i

Also, in order to simplify the energy relationships, the acoustic impedance and admittance will be defined as follows:

$$Z_1 = \frac{\rho_1 c_1}{cos\theta_1} \qquad Z_{2l} = \frac{\rho_2 c_2}{cos\theta_2} \qquad Z_{2t} = \frac{\rho_2 b_2}{cos\gamma_2} \tag{19}$$

$$G_{2l} = \frac{1}{Z_{2l}} \qquad G_{2t} = \frac{1}{Z_{2t}} \tag{20}$$

where subscripts l and t refer to the longitudinal and shear waves, respectively. The energy conversion characteristics of the reflected and refracted waves are examined using computer simulation, and the value of the parameters used in this simulation are;

79

$$\rho_1 = 1.00(g/cm^3) \qquad c_1 = 1.483(km/sec) \qquad b_1 = 0.000(km/sec)$$
$$\rho_2 = 8.51(g/cm^3) \qquad c_2 = 5.476(km/sec) \qquad b_2 = 3.302(km/sec)$$

The intensity of the normalized waves verses the incident angle, θ_1, are shown in Figure 8. In keeping with the previous constraints of θ_1 being less than approximately 3°, the amount of energy transmitted into the medium is relatively constant in this interval. This can be assessed by observing the reflected energy, $(\varphi_1'/\varphi_1)^2$, in this interval. As can be seen from Figure 8c and d, changing the incident angle would only shift the energy from one mode to the other. It is important to point out that the amount of transmitted shear wave is significantly smaller than the longitudinal wave over the region of interest.

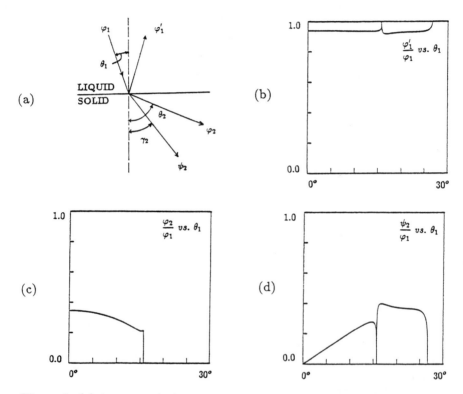

Figure 8. (a) A geometrical representation of refracted and reflected waves for case 1 and (b),(c), and (d) show the intensities of the refracted and reflected as a function of θ_1.

The energy relationships between the transmitted and reflected waves with respect to the incident beam for case 2 are as follows:

$$\left(\frac{\varphi_2'}{\varphi_2}\right)^2 = \left|\frac{Z_1 G_{2l} G_{2t} - G_{2t} cos^2 2\gamma_2 + G_{2l} sin^2 2\gamma_2}{Z_1 G_{2l} G_{2t} + G_{2t} cos^2 2\gamma_2 + G_{2l} sin^2 2\gamma_2}\right|^2 \qquad (21)$$

$$\left(\frac{\varphi_1}{\varphi_2}\right)^2 = \frac{Z_1 G_{2l}}{cos^4 2\gamma_2}\left|1 - \frac{\varphi_2'}{\varphi_2}\right|^2 \qquad (22)$$

$$\left(\frac{\psi_2'}{\varphi_2}\right)^2 = \left(\frac{b_2}{c_2}\right)^4 \frac{tan\theta_2 sin^2 2\theta_2}{tan\gamma_2 cos^2 2\gamma_2}\left|1 - \frac{\varphi_2'}{\varphi_2}\right|^2 \tag{23}$$

The plot of the intensity of the normalized waves against the range of incident angles, θ_1, is shown in Figure 9. Figure 9b reveals that the transmitted energy, $(\varphi_1/\varphi_2)^2$, maximization occurs when the scanning angle is minimized. This minimization also results in making the longitudinal waves more dominant and mode conversion becomes negligible at this interface.

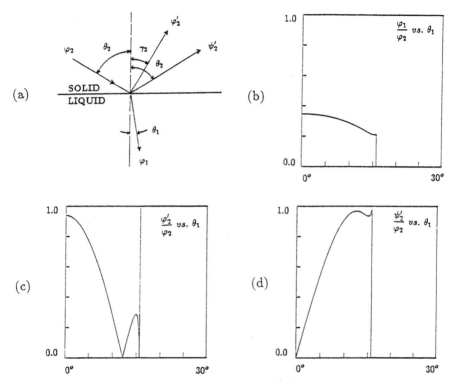

Figure 9. (a) A geometrical representation of refracted and reflected waves for case 2 and (b),(c), and (d) show the intensities of the refracted and reflected as a function of θ_1.

The energy relationship between the transmitted and reflected waves and the incident wave for case 3 are:

$$\left(\frac{\psi_2'}{\psi_2}\right)^2 = \left|\frac{Z_1 G_{2l}G_{2t} + G_{2t}cos^2 2\gamma_2 - G_{2l}sin^2 2\gamma_2}{Z_1 G_{2l}G_{2t} + G_{2t}cos^2 2\gamma_2 + G_{2l}sin^2 2\gamma_2}\right|^2 \tag{24}$$

$$\left(\frac{\varphi_2'}{\psi_2}\right)^2 = \left(\frac{c_2}{b_2}\right)^4 \frac{tan\gamma_2 cos^2 2\gamma_2}{tan\theta_2 sin^2 2\theta_2}\left|1 - \frac{\psi_2'}{\psi_2}\right|^2 \tag{25}$$

$$\left(\frac{\varphi_1}{\psi_2}\right)^2 = \frac{\rho_1 tan\theta_1}{2\rho_2 sin2\gamma_2 sin^2\gamma_2}\left|1 - \frac{\psi_2'}{\psi_2}\right|^2 \tag{26}$$

The above energy relationships versus the incident angle, θ_1, can be seen in Figure 10. Figure 10b indicates that the transmitted energy, $(\varphi_1/\psi_2)^2$ into the water gap will increase by increasing the incident angle. This will also result in increasing mode conversion.

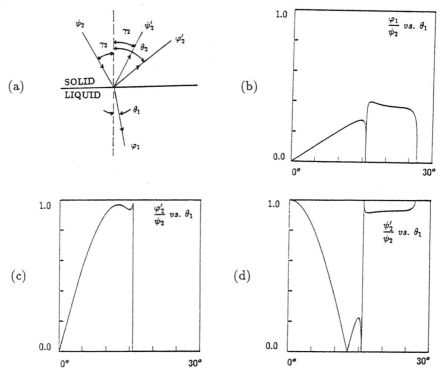

Figure 10. (a) A geometrical representation of refracted and reflected waves for case 3 and (b),(c), and (d) show the intensities of the refracted and reflected as a function of θ_1.

In consolidating the results of the three previous papagraphs, it becomes apparent that minimization of the shear waves and mode conversion would be concurrent with the optimality criteria. This is supported by the fact that cases 1 and 2 are not efficient in producing shear waves. Furthermore, in comparing cases 2 and 3 as far as effectiveness at transmitting energy into the water gap, case 2 transmits considerably more energy. Thus, the smallest scanning angle should be chosen in order to minimize the energy lost in mode conversion and to maximize the intensity of the 'b' echoes.

EXPERIMENTAL RESULTS AND DISCUSSION

Oblique angle measurements were made using planar and tubular models. Relatively small angles were studied and experimental results supported the theoretical analysis. The planar measurements were taken using several angles inside and outside of the bounds described by Equation (15), showing how the characteristics of 'b' echoes are effected as a function of incident angle. Similar results were observed using the tubular model. Both planar and tubular measurements demonstrated the effectiveness of OAS in recovering information from the target hidden by highly reverberant thin layers.

The set-up for the planar OAS measurements is similar to that shown in Figure 4. The tube thickness is 2 mm, a target distance of 5 cm, and the beam field of the transducer used is 7 mm. The constraints on the incident angle are $2^{o} \leq \theta_1 \leq 8.6^{o}$. The three measurements are shown in Figure 11, where the scanning angles are

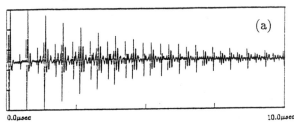

(a)

0.0μsec   10.0μsec

Figure 11. Experimental results from the planar model, where in (a) $\theta_1 = .35^{o}$; (b) $\theta_1 = 1.94^{o}$; and (c) $\theta_1 = 5.1^{o}$.

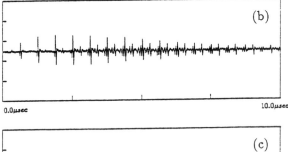

(b)

0.0μsec   10.0μsec

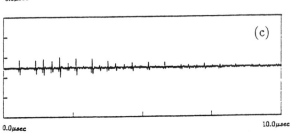

(c)

0.0μsec   10.0μsec

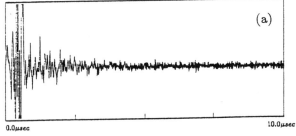

(a)

0.0μsec   10.0μsec

Figure 12. Experimental results from tubular model: (a) tube echoes, and (b) tube/support plate echoes

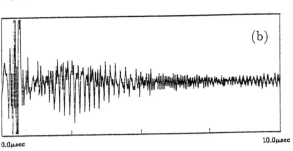

(b)

0.0μsec   10.0μsec

83

$0.35°, 1.94°$, and $5.10°$ respectively. For Figure 11a the 'a' echoes are interfering with the 'b' echoes which makes direct classification difficult. As the angle is increased to $1.94°$ (shown in Figure 11b), the 'a' echoes are completely eliminated from the region of interest. The 'b' echoes are clearly visible bearing minimal distortion. Although, if the angle is increased further to $5.1°$, then the 'b' echoes decrease in magnitude as shown in Figure 11c due to the lateral shifting described earlier. In addition, spurious echoes are also present due to the effect of mode conversion.

The tubular measurements using OAS are shown in Figure 12. These measurements follow the constraints calculated in the theoretical analysis, in which the scanning angle used was about $2°$. Figure 12a shows the results of the OAS of the tube structure without the support plate behind it, showing only 'a' echoes. The rejection of the 'a' echoes is evident, although more noise is observed due to the fact that the measurements are taken in an enclosed space. Figure 12b shows the OAS of the tube/support plate, where the 'b' echoes are clearly visible and resolvable. These results can be beneficial in characterizing the support plate integrity .

In conclusion, ultrasonic oblique angle scanning facilitates the imaging of targets hidden by reverberant thin layers. Rejecting the class 'a' echoes and maintaining the visibility of the 'b' echoes resulted in a range of acceptable incident scanning angles. It must be noted that the overall OAS performance can be limited by the geometrical configuration of the multilayered structure.

REFERENCES

[1] J. Saniie, E.S. Furgason, and V.L. Newhouse, "Ultrasonic Imaging Through Highly Reverberant Thin Layers - Theoretical Considerations",Materials Evaluation, Vol. 40, pp. 115-121 (1982).

[2] J. Saniie, "Identification of Reverberant Layered Targets Through Ultrasonic Wave Classification",Review of Progress in Quantitative Nondestructive Evaluation, Editors: D.O. Thompson and D.E. Chimenti, Plenum Press, pp. 1011-1018 (1984).

[3] J. Saniie, "Ultrasonic Signal Processing: System Identification and Parameter Estimation of Reverberant and Inhomogenous Targets", Ph.D. Thesis, Purdue University (1981).

[4] J. Saniie and D.T. Nagle, "On the Imaging of Tube/Support Structure of Power Plant Steam Generators",Review of Progress in Quantitative NDE, Eds: D.O. Thompson and D.E. Chimenti, Plemnum Press, Vol. 6a, pp. 519-525 (1987).

[5] W.M. Ewing, W.S. Jardetzky, and F. Press, Elastic Waves in Layered Media, McGraw Hill (1957).

[6] L.M. Brekhovskikh, Waves in Layered Media, Translated and Edited by T. Beyer, Academic Press (1980).

REMOVING THE EFFECTS OF SURFACE ROUGHNESS FROM

LOW-FREQUENCY ACOUSTIC IMAGES

P. A. Reinholdtsen and B. T. Khuri-Yakub

Edward L. Ginzton Laboratory
Stanford University
Stanford, California 94305

INTRODUCTION

We have built a low-frequency scanning acoustic microscope (SAM) that measures both amplitude and phase. The majority of SAMs simply measure the amplitude of the reflected signal. Measuring the phase gives a great deal more information. For one thing, the phase is very sensitive to height variations. Measuring the phase also gives us the ability to do signal processing on the resulting images, such as removing the effects of surface features from defocused images of subsurface defects.

THE MICROSCOPE

An efficient broadband focused transducer is excited with a tone burst. The transducer also received the reflected echo from the sample of interest. As shown in Fig. 1, a 3 MHz tone burst is generated from a 192 MHz clock. The 192 MHz clock is also used to generate a 3 MHz cw reference whose phase can be shifted by 1/64 increments of

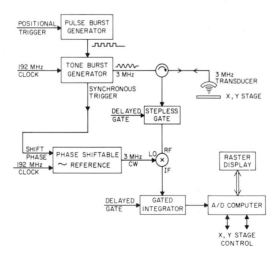

Fig. 1. Schematic of magnitude and phase acoustic microscope.

2π , relative to the transmitted tone burst. The return signal is mixed with the reference signal and the result is integrated and digitized. The digitized result varies sinusoidally with the phase shift of the reference. Taking the first harmonic of the digitized values gives the amplitude and phase of the return signal and removes any DC and higher harmonic effects that may be introduced by the mixer, gated integrator, or digitizer. The amplitude can be measured with 0.1% accuracy and the phase can be measured to better than 0.1° . At 3 MHz , this kind of phase variation corresponds to height variations of less than a micron over large areas (greater than 1 mm across). This is significantly more accurate than the stability of the scanning stage used.

DEFOCUSING

A common technique used with SAMs is defocusing the transducer to obtain enhanced subsurface defect detection. This consists of moving the transducer closer to the sample of interest. This concentrates more acoustic energy below the surface, leading to greater contrast when subsurface features are present. A defocused image contains information from both surface and subsurface features. If there are features on the sample's surface, such as random roughness, they will show up in the defocused image. It would be useful to be able to distinguish between surface and subsurface effects in order to remove the surface effects from a defocused image. With both amplitude and phase information, it is possible to do this to a large degree by also taking an image with the transducer on focus.

By taking an on-focus image, it is possible to numerically defocus the image by an amount equivalent to the defocused image. Because the on-focus image contains surface information, the numerically-defocused image will contain defocused surface features. The difference image between the numerically and experimentally defocused images will contain only subsurface features.

In order to perform this numerical defocusing, we need to do three things: (1) characterize the transducer; (2) determine the effect of the transducer's characteristics on image formation; and (3) determine how these characteristics change as the transducer is defocused.

TRANSDUCER CHARACTERIZATION

The transducer will be characterized by plane wave decomposition and V(z) inversion.[1,2,6] The field produced by a transducer can be decomposed into a superposition of plane waves by taking the two-dimensional Fourier transform of the generated field in a plane perpendicular to the transducer's face.

$$S_{10}(z; \vec{\kappa}) = \mathcal{H}\{s_{10}(z; \vec{X})\} \qquad (1)$$

where S_{10} $(z; \vec{\kappa})$ is the transmitted angular spectrum of the transducer and s_{10} $(z; \vec{X})$ is the acoustic field at a plane z . $\vec{X} = (x,y)$ and $\vec{\kappa} = (k_x, k_y)$ is the transverse wave number.

If a circularly-symmetric transducer is located over a uniform sample, we can compute the transducer's output as a function of its height above the sample, given the sample's reflection coefficient for plane waves as a function of wave number. This is the famous $V(z)$ curve and is given by:

$$V(z) \alpha \int_0^k k_z^2 \; S_{10}^2(k_r) \; R(k_r)e^{2ik_z z} \, dk_r \qquad (2)$$

where $R(k_r)$ is the sample's reflection coefficient as a function of incident angle and

$$k_r = \sqrt{k_x^2 + k_y^2}$$

This assumes the transducer is reciprocal so that its receiving spectrum S_{01} is related to its transmitting spectrum by

$$S_{01}(z; \vec{k}) \; \alpha \; k_z S_{10}(z; \vec{k}) \qquad (3)$$

With a couple of variable changes, this becomes a Fourier relationship

$$V(z) \alpha \; \mathcal{F}^{-1}\{\beta^2 S_{10}^2(\beta)R(\beta)\} \qquad (4)$$

where $\beta = 2k_z$. We can invert this to obtain the product of the sample's reflectance function and the square of the transducer's transmitting spectrum

$$\mathcal{F}\{V(z)\} \; \alpha \; \beta^2 S_{10}^2(\beta) \; R(\beta) \qquad (5)$$

If we know the reflectance function of a reference sample and take $V(z)$ for an unknown transducer using this sample, we can experimentally determine the transducer's spectrum.

Taking the square root in Eq. (4) leads to a 180° phase ambiguity. If we assume the spectrum is continuous, we can flip the sign of the computed square roots in order to minimize the difference between adjacent points in the spectrum.

NUMERICAL DEFOCUSING

Now that we can experimentally characterize the transducer, we need to know what effect the transducer has on image formation. In order to do this, we will make several simplifying assumptions. We assume that the reflected acoustic field at the object's surface is a point-by-point product of the incident field and the object's acoustic field response. This means that the sample is relatively flat compared to the focal depth of the transducer. This is a good approximation if the transducer used has a large F-number (a small aperture). There should also be no mode conversions (such as into pseudo-Rayleigh waves). This is true for many composites or other materials with low acoustic velocities and again if the transducer has a large F-number.

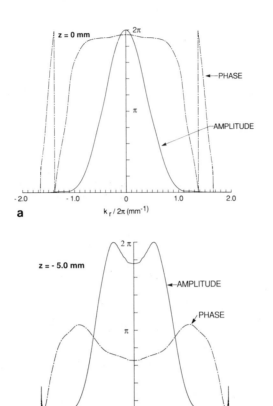

Fig. 2. Round-trip spatial frequency response of F2 transducer operating at 3 MHz. (a) Transducer on focus. (b) Transducer defocused 5 mm.

With these assumptions, the measured image is simply the round-trip impulse response of the transducer convolved with the object's field response

$$i(z; \vec{X}) = i_0(\vec{X}) \ast\ast t(z; \vec{X}) \qquad (6)$$

where the round-trip impulse response is simply the product of the transmitting and receiving impulse responses

$$t(z; \vec{X}) = s_{10}(z; \vec{X}) \, s_{01}(z; \vec{X}) \qquad (7)$$

These, in turn, are simply the Fourier transforms of the transmitting and receiving spectra of the transducer, which we can experimentally determine. If we transform Eq. (6) to the Fourier domain, the convolution becomes a product

$$I(z; \vec{k}) = I_0(\vec{k}) \, T(z; \vec{k}) \qquad (8)$$

By dividing Eq. (8) by itself, for different defocus levels, we can relate images taken at different defocus depths.

$$I(z; \vec{k}) = I(0; \vec{k}) \, \frac{T(z; \vec{k})}{T(0; \vec{k})} \qquad (9)$$

This gives us the ability to take an on-focus image and numerically de-focus it to any depth, if we know the transducer's spectrum, by simply convolving the on-focus image with a defocusing filter.

Figure 2 shows the experimentally-determined two-way spatial fre-quency spectrum of an F2 transducer operating at 3 MHz when the trans-ducer is on focus, and when it is defocused 5 mm . Figure 3 shows the frequency domain filter used to numerically defocus images 5 mm . It is simply the ratio of the two spectra, as given in Eq. (9).

For height variations on a sample made of a single material, the surface acts to a good approximation as a phase object. The reflected field is simply the incident field multiplied by a phase factor that varies with the sample coordinate.

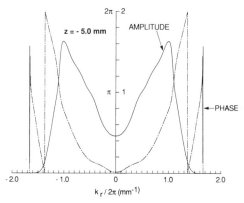

Fig. 3. Defocusing filter in spatial frequency domain for F2 transducer operating at 3 MHz numerically defocuses an image 5 mm by multiplying the image's spatial frequency spectrum. It is simply the ratio of the spectra in Fig. 2.

EXPERIMENTAL RESULTS

In the results that follow, the samples were imaged with an F2 transducer. In picking a transducer to use, there are some tradeoffs: a transducer with a small F-number is good because it has a short depth of field, which leads to better contrast between surface and subsurface defects; if the F-number is too small, it can lead to excitation of sur-face waves because of the larger incident angle of the acoustic waves,

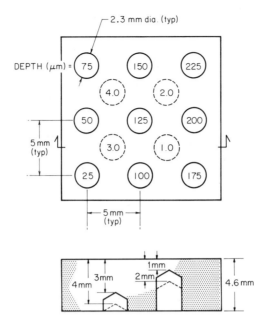

Fig. 4.   Schematic of aluminum plate with surface and subsurface
defects.

which violates the assumptions necessary to numerically defocus images.
The F2 transducer was the smallest F-number transducer available that
did not excite surface waves to a large degree in the samples imaged.

A sample with both surface and subsurface features was imaged with
the F2 transducer operating at 3 MHz . A 0.5 cm thick plate of aluminum had nine depressions 2.3 mm across machined into its surface,
varying in depth from 25-225 μm , as shown in Fig. 4. Four subsurface
defects were made by drilling holes from the back that came to within
1-4 mm of the surface. The sample was imaged at several defocus
depths. The on-focus image was numerically defocused in order to remove
the effects of the surface features on the experimentally-measured,
defocused images.

The results are shown in Fig. 5. In the measured defocused images,
the subsurface features are barely visible. In the processed images,
the contrast of the subsurface features has been dramatically increased
relative to the surface features. The shallower surface defects have
been virtually eliminated from the image, whereas the deeper defects
have only been partially removed. The deeper defects don't act as ideal
phase steps because of large reflections off the sides. This shows that
the effects of small amounts of roughness can be virtually eliminated,
and for larger roughnesses the effects can still be reduced greatly.

The same experimental procedure was tried on another sample to see
if the effects of random surface roughness could be removed. This
sample had nine subsurface holes ranging from 0.5-4.5 mm beneath the
surface of the same aluminum plate used before, as shown in Fig. 6. The
surface was roughened by beating it with a screwdriver and other blunt

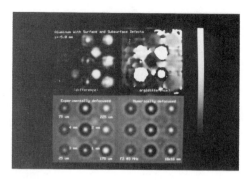

Fig. 5.   Acoustic images of aluminum sample in Fig. 4, defocused
          5 mm . Lower left: magnitude of experimentally-measured,
          defocused image. Lower right: magnitude of numerically-
          defocused image predicted from on-focus image. Upper left:
          magnitude of complex difference between measured and numer-
          ically-defocused images. Upper right: phase of difference.
          Notice increased relative contrast of subsurface defects in
          difference images.

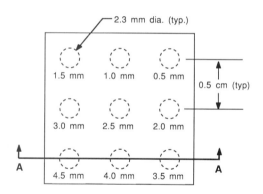

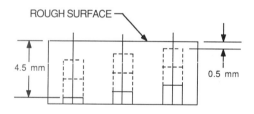

VIEW A - A

Fig. 6.   Schematic of Al plate with random surface roughness and subsur-
          face defects. Surface has 60 μm peak-to-peak height varia-
          tions.

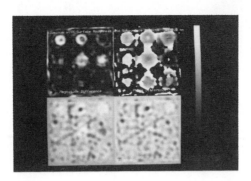

Fig. 7. Acoustic images of Al sample in Fig. 6, defocused 5 mm . Lower left: measured defocused image (magnitude). Lower right: numerically-defocused image. Upper left: magnitude of difference between measured and numerically-defocused images. Upper right: phase of difference. The effects of surface roughness have been virtually eliminated. Subsurface defects are now clearly visible.

instruments. This produced a roughness with peak-to-peak variations of ~60 μm as measurd by a stylus profile measurement system. The acoustic images are shown in Fig. 7. In the unprocessed images, the surface roughness obscures all but the closest subsurface defects. In the processed images, most of the surface roughness is removed and most of the subsurface defects become visible. This is particularly true in the phase images. This shows that we can observe subsurface features that would normally be lost in the signal produced by random surface roughness.

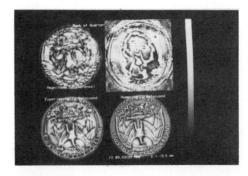

Fig. 8. Imaging front side of quarter through back side. The phase of the difference image (upper right) clearly shows a detailed outline of George Washington's head. Even though the magnitude is highly corrupted, the phase gives a good indication of the height variations of the front side as seen through the back.

For a more graphic demonstration of this technique, we looked at the back side of a quarter in an attempt to image the front (the side with George Washington). This was done with the same F2 transducer operating near its third harmonic at 9.5 MHz . The processed image in Fig. 8 shows the outline of George's head quite well in the phase picture. The surface height variations of the back side of a quarter are greater than an acoustic wavelength (at 9.5 MHz the wavelength in water is 0.156 mm) as determined by the phase variations in the acoustic image. Therefore, this example severely tests the assumption that the sample is planar. Still, the processing does a very good job of pulling out the information of interest.

CONCLUSIONS

With the low-frequency acoustic microscope that measures both amplitude and phase, we can do two-dimensional image processing on images that we could not do with just amplitude images. We can characterize the angular spectrum of transducers by the inversion of V(z) data. This gives the ability to numerically defocus acoustic images and it enables surface features to be removed from experimentally-defocused images, leading to greatly increased relative contrast of subsurface features.

ACKNOWLEDGMENT

This work was supported by the Office of Naval Research under Contract No. N00014-87-K-0269 and by Sandia National Laboratories.

REFERENCES

1. D. H. Kerns, J. Acoust. Soc. America 57 (2) (February 1975).
2. K. K. Liang, G. S. Kino, and B. T. Khuri-Yakub, IEEE Sonics & Ultrasonics SU-32 (2), 213 (March 1985).
3. R. A. Lemons and C. F. Quate, Appl. Phys. Lett. 24, 163 (1974).
4. C. F. Quate, A. Atalar, and H. K. Wickramasinghe, Proc. IEEE 67, 1052 (1979).
5. A. Atalar, J. Appl. Phys. 45, 5130-5139 (October 1979).
6. P. Reinholdtsen and B. T. Khuri-Yakub, Review of Progress in Quantitative Nondestructive Evaluation, D. O. Thompson and D. E. Chimenti, Eds., Plenum Press, 263 (1986).
7. See IEEE Trans. on Sonics and Ultrasonics, Special Issue on Acoustic Microscopy SU-32 (2) (March 1985).

PERFORMANCE EVALUATION OF SPATIAL AND TIME AVERAGING OF ULTRASONIC GRAIN SIGNAL**

Jafar Saniie, Tao Wang and Nihat M. Bilgutay*

Elect. & Comp. Eng. Dept.   * Elect. & Comp. Eng. Dept.
Illinois Institute of Technology   Drexel University
Chicago, IL 60616   Philadelphia, PA 19104

ABSTRACT

The ultrasonic backscattered grain signal consists of interfering multiple echoes with random amplitude and phase corresponding to a highly complex grain structure. Among the various inherent features of the grain signal, attenuation is a measurable feature resulting from the effect of scattering and absorption. The concept of ergodicity of the ultrasonic backscattered grain signal and its usefulness for measuring attenuation is presented in this report. A mathematical model of the grain signal describing both spatial and temporal averaging is developed. Experimental results using steel samples with different grain sizes along with the discussion of reproducbility and sensitivity of spatial and temporal averaging for measuring attenuation is presented.

INTRODUCTION

Common ultrasonic microstructure evaluation is based on a comparison of attenuation measurements of the specimens with unknown grain sizes to specimens with known grain sizes. This is accomplished by wave transmission using either two transducers, or by pulsing a transducer and measuring the amplitude of the echo as it is reflected from the far end surface of the specimen by the same transmitting transducer. Microstructure grain size evaluation using transmission mode testing has some practical limitations: i) flat and parallel surfaces are required; ii) perfect coupling between the transducers and specimen are essential; and iii) more importantly, the measured attenuation coefficient represents an average value over the total sound path where local variation can greatly alter the attenuation coefficient and cannot be effectively evaluated.

** This project is supported by EPRI Grant RP2405-22

An alternative method of evaluating the grain size can be achieved by using the measurements of the backscattered grain signal. This method utilizes the principle that an ultrasonic wave travelling through a medium is subject to scattering and absorption losses. These losses depend on the location of scatterers (i.e. grains), path of propagation and frequency of the ultrasonic transducer. Therefore, the direct characterization of backscattered grain signals yields information pertaining to variation in the scattered energy as a function of depth and, hence, the grain size distribution.

Some of the earlier work in grain size estimation using ultrasonic scattering measurements was performed by Beechman[1] and Fay[2]. Goebbels and Holler[3] used spatial averaging for obtaining attenuation measurements. Saniie and Bilgutay[4] demonstrated various grain size characterization techniques which extract parameters from the backscattered signal related to the frequency dependent attenuation coefficient. In this paper we present a statistical model for ensemble (spatial) and time averaging of the backscattered grain signal. This model is used for obtaining ultrasonic attenuation measurements. The attenuation measurement then serves as the characterization basis for grain size distribution.

In general, when an ultrasonic burst of sound travels through materials its amplitude is attenuated. This attenuation has two major causes,

$$\alpha(f) = \alpha_a(f) + \alpha_s(f) \qquad (1)$$

where the term $\alpha_a(f)$ is a hysteresis loss caused by the anelastic behavior of the materials, and the term $\alpha_s(f)$ is a scattering loss associated with the characteristics of grain and phase boundaries (acoustic impedance discontinuities). Grain scattering loss at ultrasonic frequencies is so large in relation to the hysteresis loss that the latter is negligible. For example, in the Rayleigh region, the scattering coefficient varies with the average volume of the grain and the fourth power of the wave frequency, while the absorption coefficient increases linearly with the frequency[5]. Therefore, the total attenuation coefficient of ultrasound in the Rayleigh region is expressed as:

$$\alpha(f) = a_1 f + a_2 \overline{D}^3 f^4, \qquad \lambda > 2\pi \overline{D} \qquad (2)$$

where $\overline{D}$ is the grain diameter, λ is the sound wavelength, a_1 is the absorption constant, a_2 is the scattering constant and f is the transmitted frequency.

STATISTICAL MODEL OF GRAIN SIGNAL

The measured backscattered echoes corresponding to a given range cell (Figure 1) can be modeled as:

$$r(t) = \sum_{k=1}^{M} A_k u(t - \tau_k) \qquad (3)$$

where M is a random variable and represents the total number of scatterers within the range cell. The term A_k is defined as:

$$A_k = \sigma_{sk} e^{-\alpha \tau_k} \tag{4}$$

where σ_{sk} is a random variable corresponding to the grain scattering cross-section and $e^{-\alpha \tau_k}$ is the effect of attenuation. It is important to point out that both the scattering coefficient σ_{sk} and attenuation coefficient α are functions of frequency which are not included in the above equation in order to simplify mathematical representation. The term $u(t)$ is the basic ultrasonic wavelet and a typical example is shown in Figure 2. Without loss of generality, we make the assumption that

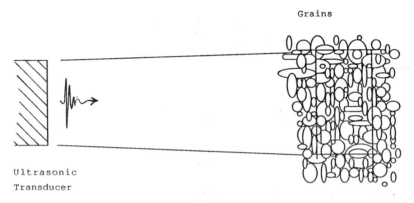

Fig. 1. Range Cell Demonstration.

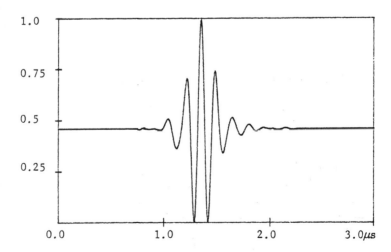

Fig. 2. A Typical Ultrasonic Wavelet.

the wavelet is of Gaussian envelop,

$$u(t) = e^{-\gamma t^2} e^{j\omega t} \tag{5}$$

where ω is the center frequency, γ is a constant representing the timewidth of the wavelet. Let's consider that the range cell represents a small time interval of size 2ϵ,

$$t - \epsilon < \tau_k < t + \epsilon \qquad \forall \quad k \tag{6}$$

Then, the grain signal for a range cell corresponding to a given time t becomes:

$$r(t) = \sum_{k=1}^{M} \sigma_{sk} e^{-\alpha \tau_k} e^{j\omega(t-\tau_k)} e^{-\gamma(t-\tau_k)^2} \tag{7}$$

In the far-field region, $t \gg \epsilon$, Eq. 7 will simplify to,

$$r(t) = e^{-\alpha t} \sum_{k=1}^{M} \sigma_{sk} e^{j\omega \Delta t_k} \tag{8}$$

where $\Delta t_k = t - \tau_k$. Let's define

$$\hat{r} e^{j\theta} = \sum_{k=1}^{M} \sigma_{sk} e^{j\omega \Delta t_k} \tag{9}$$

Then,

$$r(t) = \hat{r} e^{j\theta} e^{-\alpha t} \tag{10}$$

In practice, the number of the scatterers under the illumination of the transducer is very large and they are considered to be independent and identically distributed. Consequently, $\hat{r}$ has a Rayleigh probability density function[6].

$$p_{\hat{r}}(\hat{r}) = \frac{2\hat{r}}{\eta} e^{\frac{-\hat{r}^2}{\eta}} \tag{11}$$

where

$$\eta = E[M] \cdot E[\sigma_s^2] \tag{12}$$

Using Eqs. 10-12, the expected value of the amplitude of the backscattered signal will result in:

$$E[r(t)] = \frac{e^{-\alpha t}}{2} \sqrt{E(M) \cdot E(\sigma_s^2)\pi} = C_1 \cdot e^{-\alpha t} \tag{13}$$

Inspection of Eq. 13 reveals that the expected grain signal is attenuated exponentially as a function of time and the intensity is proportional to the grain scattering cross-section. Based on the above equations, one may obtain the attenuation coefficient by ensemble averaging of the grain signals at a given time. In general, a better estimate of the attenuation coefficient can be obtained by estimating $E[r(t)]$ through ensemble averaging at many different times and applying linear regression after logarithmic transformation. The overall system of estimating the attenuation coefficient is shown in Figure 3. The key to the accuracy of estimating the measurement is the effectiveness of the averaging operation. In the following section, we present the mathmatical analysis of two techniques -spatial and temporal- for obtaining the expected value of the grain signal.

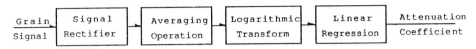

Fig. 3. Overall System for Attenuation Measurement.

SPATIAL AVERAGING

The grain signal consists of interfering multiple echoes with random amplitude and phase from which the attenuation coefficient cannot readily be obtained. Therefore, as was discussed in the previous section, some type of ensemble averaging is required. A practical method is spatial averaging[3]. Spatial averaging is acomplished by scanning the specimen and averaging the rectified backscattered signals. Let's assume that averaging of N measurements at N different position of the specimen is performed,

$$\langle r(t) \rangle = \frac{1}{N} \sum_{i=1}^{N} \hat{r}_i e^{-\alpha t} \tag{14}$$

where $\hat{r}_i e^{-\alpha t}$ is the rectified backscattered signal measured at a given position i. The above estimate is unbiased and as a result,

$$E[\langle r(t) \rangle] = E[\hat{r}] e^{-\alpha t} \tag{15}$$

The signal amplitude is considered to be wide-sense stationary which implies,

$$E[\hat{r}_i] = m \tag{16}$$

and spatial correlation

$$E[\hat{r}_i \hat{r}_j] = R_s(|i - j|) \tag{17}$$

Using the above assumption, the variance of $\overline{r(t)}$ can be determined. After some mathematical simplification, the variance becomes;

$$\sigma^2_{\langle r(t) \rangle} = e^{-2\alpha t} \frac{\sigma^2_{\hat{r}}}{N} + e^{-2\alpha t} \frac{1}{N^2} \sum_{k=1}^{N-1} \{2(N-k)(R_s(k) - m^2)\} \tag{18}$$

If the spacings among all measurements are large, the ensemble measurements are uncorrelated. Hence,

$$R_s(k) = m^2, \qquad k \neq 0 \tag{19}$$

and

$$\sigma^2_{\langle r(t) \rangle} = e^{-2\alpha t} \frac{\sigma^2_{\hat{r}}}{N} \qquad (20)$$

In the above equation, as $N \to \infty$, the term $\sigma^2_{\langle r(t) \rangle} \to 0$. This is the condition for obtaining consistent estimate of $\langle r(t) \rangle$. In practice, the ensemble measurements are not necessarily uncorrelated $(R_s(k) \neq m^2)$ and, consequently, the effective number of averagings is less than the actual value of N.

TIME AVERAGING

Another practical approach to ensemble averaging is time domain averaging. The grain signal is a stochastic process in which randomness is inherent to any single measurement. The temporal fluctuations contain equivalent information to the random spatial fluctuations; therefore, it is appropriate to determine the statistical parameters (e.g., mean and variance) of the process from a single measurement, which is far more practical than using multiple measurements. In general, this approach is valid when using stationary random processes in which time averages are identical to their ensemble averages (i.e., ergodic process)[7].

Time averaging is accomplished by averaging samples of the rectified received signals which are taken at time $t + \Delta t, \ldots, t + N\Delta t$, such that,

$$\overline{r(t)} = \frac{1}{N} \sum_{i=1}^{N} r_i \qquad (21)$$

where r_i is a random variable defined as:

$$r_i = \hat{r}(t + i\Delta t)e^{-\alpha t}e^{-\alpha i \Delta t} \qquad (22)$$

Therefore,

$$\overline{r(t)} = \frac{e^{-\alpha t}}{N} \sum_{i=1}^{N} \hat{r}(t + i\Delta t)e^{-\alpha i \Delta t} \qquad (23)$$

In the above equation, the sum represents averaging of N random variables weighted by $e^{-\alpha i \Delta t}$. The $\hat{r}(t)$ is a stationary random process and the expected value of Eq. 23 becomes:

$$E[\overline{r(t)}] = \frac{e^{-\alpha t}}{N} E[\hat{r}(t)] \sum_{i=1}^{N} e^{-\alpha i \Delta t} \qquad (24)$$

Let's assume the integration period, T, is equal to $N\Delta t$, and the term T remains constant as $N \to \infty$. Then, Eq. 21 is simplified to:

$$E[\overline{r(t)}] = e^{-\alpha t}(E[\hat{r}(t)] \cdot \frac{1 - e^{-\alpha T}}{\alpha T}) \qquad (25)$$

where $E[\hat{r}(t)]$ is a constant due to stationarity, and $\hat{r}(t)$ is the grain signal amplitude in the absence of attenuation. Once T is defined, the term $\frac{1 - e^{-\alpha T}}{\alpha T}$ becomes a known constant. Similar to spatial averaging, time domain integration of the amplitude signal is equal to the attenuation factor, $e^{-\alpha t}$, multiplied by a constant. Therefore, the attenuation coefficient can be obtained by time averaging. The validity of the attenuation is highly dependent on the value of $\sigma^2_{\overline{r(t)}}$

Let's define the temporal autocorrelation of the sampled signal as,

$$R_t(|i - j|) = E[\hat{r}(t + i\Delta t)\dot{\hat{r}}(t + j\Delta t)] \qquad (26)$$

After some algebraic operations, the variance of $\overline{r(t)}$ becomes;

$$\sigma^2_{\overline{r(t)}} = \frac{e^{-2\alpha t}}{N^2}\{\sigma^2_{\hat{r}(t)}\frac{1 - e^{-2\alpha T}}{2\alpha\Delta t} + \frac{2e^{-2\alpha\Delta t}}{1 - e^{-2\alpha\Delta t}}\sum_{j=1}^{N-1}\sigma^2_{co}(j)(e^{-\alpha j\Delta t} - e^{-2\alpha N\Delta t + j\alpha\Delta t})\}$$

$$(27)$$

where

$$\sigma^2_{co}(j) = R_t(j) - (E[\hat{r}(t)])^2 \qquad (28)$$

If the Δt is sufficiently long to ensure the uncorrelation of the sampled signals, then the term σ^2_{co} will be zero for all j. Hence, the $\sigma^2_{\overline{r(t)}}$ will approach zero as $N \to \infty$. Consequently, a consistent estimate for $\overline{r(t)}$ can be obtained.

Similar to spatial averaging, the effective number of time averagings is less than the actual value of N due to the correlation of sampled values of the signal. Furthermore, the accuracy or reliability of time averaging is highly dependent on the choice of T. From an analytical point of view, the larger the value of T, the better the estimate can be. But when T is assigned a large value, signal attenuation becomes significant, causing poor signal-to-noise ratio which must be avoided.

EXPERIMENTAL RESULTS AND DISCUSSION

The object of this work is to evaluate the grain size variation in solids when other physical parameters (e.g., crystal shape, elastic constants, density, and velocites) remain constant. These assumptions allow us to accurately interpret the measurements resulting from the grain size variation. In this study, Type 1018 steel blocks (4×4×10inches) were heat treated for four hours to obtain various grain sizes.

The experimental grain signal measurements were accomplished using a Gamma type transducer manufactured by K-B Aerotech with approximately a 5-MHz center frequency and 3-dB bandwidth of approximately 1.5 MHz. The RF signal is sampled at 100 MHz with 8-bits resolution. The transducer aperture is 0.5 inches and, for obtaining uncorrelated data, significant spacing between all measurements is required. The scanning area is a square shape covered by a 16×16 grid. Figure 4 shows spatial averaging results corresponding to various scanning areas: 0.5×0.5, 0.75×0.75, 1.0×1.0, and 2.0×2.0 inches. All measurements clearly display the effect of attenuation in signal. It can be noted that there is more variation in the rate of decay corresponding to smaller areas of scanning (e.g. 0.5×0.5 inches) than in the larger areas (e.g. 2.0×2.0 inches). This is due to the high degree of correlation from measurement to measurement using a small area of scanning. These results indicate that the estimated attenuation coefficient can vary as much as five percent. Consistent results are obtained in 2.0×2.0 inches scanning area. Comparison of results corresponding to samples of various grain sizes are shown in Figure 5. The attenuation coefficients estimated by using spatial averaging corresponding to samples with different grain sizes are presented in Table I. These results support measuring attenuation as a method of nondestructive grain size characterization.

A similar study has been performed using temporal averaging. As discussed in the theoretical section, the number of samples and sampling intervals are essential in obtaining consistent averaging results. To evaluate temporal averaging

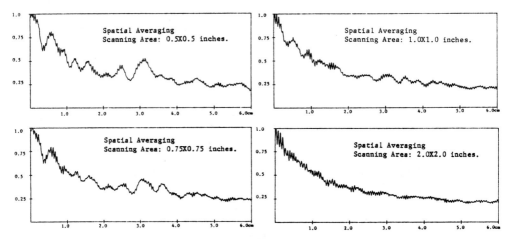

Fig. 4. Comparision of the Backscattered Signals Using Different Scanning Area for Spatial Averaging.

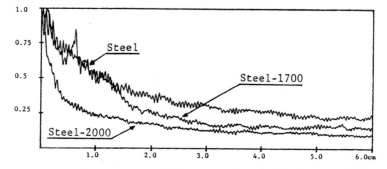

Fig. 5. Spatial Averaging of Ultrasonic Signals Backscattered from the Samples with Different Grain Sizes.

performance, window lengths from 64 up to 512 samples were examined. Experimental results using different window sizes are shown in Figure 6. The estimated attenuation coefficient for different window sizes varied as much as ten percent. Both theory and experiments indicate that similar results can be obtained using temporal and spatial averaging. For example, a comparison of temporal and spatial averaging results is shown in Figure 7.

It should be noted that the single A-scan is more practical and efficient for ultrasonic testing. In fact, in some situations, the geometry of the object interferes with or prohibits the use of multiple measurements. Furthermore, if

Table I. The Estimated Attenuation Coefficient of Samples
with Different Grain Sizes.

Steel	Heat-Treated Temp. (°F).	Estimated Grain Sizes. (μm).	Estimated Atten. Coeff. (dB/cm).
Steel	-----	14	1.8267
Steel-1700	1700	24	2.5350
Steel- 2000	2000	50	2.9011

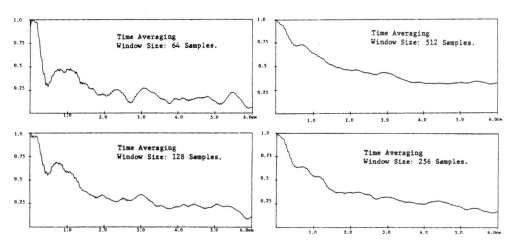

Fig. 6. Temporal Averaging of Ultrasonic Grain Signal Using Different
Window lengths.

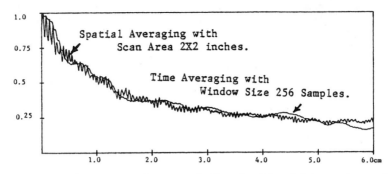

Fig. 7. A Comparison Between the Result of Spatial Averaging and
the Result of Temporal Averaging.

the penetration of the ultrasonic energy is position or orientation dependent,
an assessment of this variation is necessary and must be compensated prior to
averaging. Finally, and most importantly, the use of a single measurement reveals
information confined to a smaller region of the sample relative to the average of
multiple measurements, which displays integrated information pertaining to a
broader region of the sample.

REFERENCES

[1] D. Beecham, "Ultrasonic Scatter in Metals: its Properties and its Application
to Grain Size Determination," Ultrasonics 4, pp. 67-76, (1966).

[2] B. Fay, "Theoretical Consideration of Ultrasonic Backscattering", Acustica 28,
pp. 354-357, (1973).

[3] K. Goebbels and P. Holler, "Quantitative Determination of Grain Size and Detection of Inhomogeneities in Steel by Ultrasonic Backscattering Measurements",
in Ultrasonic Materials Characterization, edited by H. Berger and M. Linzer (National Bureau of Standards, Gaithersburg, MD), Special Publication 596, pp.
67-74, (1980).

[4] J. Saniie and N. M. Bilgutay, "Quantitative Grain Size Evaluation Using Ultrasonic Backscattered Echoes", J. Acoust. Soc. Am. (80), pp. 1816-1824,
(1986).

[5] E. P. Papadakis, "Ultrasonic Attenuation Caused by Scattering in Polycrystalline Metals," J. Acoust. Soc. Am. (37), pp. 711-717, (1965).

[6] J. Saniie, "Ultrasonic Signal Processing: System Identification and Parameter
Estimation of Reverberant and Inhomogeneous Targets", Ph.D. thesis, Purdue
University, (1981).

[7] A. Papoulis, "Probability, Random Variables, and Stochastic Process", Mc-
Graw Hill Book Company, (1984).

HIGH-SPEED AUTOMATED NDT DEVICE FOR NIOBIUM PLATE

USING SCANNING LASER ACOUSTIC MICROSCOPY[*]

M.G. Oravecz, B.Y. Yu,   H. Padamsee
K.L. Riney and L.W. Kessler

Sonoscan, Inc.   Laboratory of Nuclear Studies
530 E. Green Street   Cornell University
Bensenville, IL 60106   Ithaca, NY 14853

ABSTRACT

   This paper presents an NDT device which rapidly and automatically
identifies defects throughout the volume of a 23.4 cm x 23.4 cm x 0.3 cm,
pure niobium plate using Scanning Laser Acoustic Microscopy (SLAM), high-
resolution, 60 MHz, ultrasonic images. A principle advantage of the SLAM
technique is that it combines a video scan rate with a high scan density
(130 lines/mm at 60 MHz). To automate the inspection system we integrated
under computer control the following: the SLAM RS-170/330 video output, a
computerized XY plate scanner, a real-time video digitizer/integrator, a
computer algorithm for defect detection, a digital mass storage device, and
a hardcopy output device. The key element was development of an efficient,
reliable defect detection algorithm using a variance filter with a locally
determined threshold. This algorithm is responsible for recognizing valid
flaws in the midst of 'random' texture. This texture was seen throughout
the acoustic images and was caused by the niobium microstructure. The
images, as analyzed, contained 128 x 120 pixels with 64 grey levels per
pixel. This system allows economical inspection of the large quantities
(eg. 100 tons) of material needed for future particle accelerators based on
microwave superconductivity. Rapid nondestructive inspection of pure
niobium sheet is required because current accelerator performance is largely
limited by the quality of commercially available material. Previous work
documented critical flaws that are detectable by SLAM techniques.

INTRODUCTION

   Superconducting RF (SRF) accelerator technology is beginning to make
contributions in the fields of particle physics, nuclear physics and free
electron lasers <1>. Serious application of this technology is seen in the
attainment of an operational accelerator at HEPL, continuing construction by
the Darmstadt/Wuppertal collaboration in W. Germany, and the choice of SRF
technology in plans by CEBAF in the United States and ALS-II in France.

   Superconducting cavity technology continues to offer attractive
opportunities for further advances in achievable accelerating voltage at
reasonable cost for future accelerators. For niobium, the voltage can be
increased an order of magnitude over current capabilities. Substantial R&D
is in progress aimed towards advancing the technology so that it may
continue to serve in the current and succeeding generation of accelerators.

----
[*] Supported by US DOE, SBIR Contract No. DE-AC02-84ER80180.A003.

Current limitations in the performance of superconducting niobium cavities result from 1) thermal breakdown of superconductivity at localized spots due to heating at defects and 2) field emission at isolated sites <1,2>. An understanding of the size and location of potential defects within the "as manufactured" material is important in choosing an inspection technique. Numerical models of the heat flow due to defects, and subsequent experiments, indicate the defect size at which quenching occurs for a given accelerating gradient. For example if we choose an accelerating gradient of 10 MeV/m, the critical diameter for a normal conducting defect is 72 microns for RRR=40 niobium. For very high purity, RRR=300 niobium, the critical diameter is 480 microns <3>. Niobium with a higher residual resistivity ratio (RRR) is purer and has a higher thermal conductivity. The higher thermal conductivity improves heat dissipation around a normal conducting defect. The size of the field emitter particles on Nb surfaces range from 0.5 to 20 microns, with the most probable size between 0.5 and 1 microns <4,5>.

The use of nondestructive inspection techniques during Nb cavity production is desired due to the high cost of scrapping even a few completed cavities. Large scale cavity production for CEBAF type accelerators is untried and may have less than a 100% yield. It is also desirable to find a low cost alternative to low temperature thermometry diagnostic inspection and repair. Such low temperature procedures are costly; the additional cost is a significant fraction of the production cost of the cavity. An efficient, rapid, high temperature, nondestructive examination technique can be cost effective even though improvements in manufacturing techniques and processing cleanliness have improved the breakdown problem and have allowed attained-field-levels to increase above the current accelerating-field-gradient design specification of 5 MeV/m.

For current accelerator designs the nondestructive technique should be able to identify all defects larger than 50 microns to ensure that the cavities can perform at the design specification. This compares with the optical surface inspection limit of 50 microns diameter defects. To be used for incoming inspection, the technique must detect both surface and interior defects since chemical etching during manufacturing can expose previously buried defects. Also contamination from deep lying defects could contaminate the superconducting surface layer through diffusion during high temperature manufacturing stages. For 100% inspection to be practical, the technique must be rapid and economical since large quantities of material are proposed for future accelerators. 100% inspection is desirable since one defect causes inadequate performance of an entire cavity structure.

Only ultrasonic testing methods can nondestructively detect 50 microns inhomogenieties both at the surface and inside a large area, 0.3 cm thick plate of niobium. Ultrasonic techniques have been in widespread use for many years for the purpose of nondestructive defect detection and materials characterization. In order to find 50 microns defects in niobium, the ultrasound frequency must be in the range of 20 to 100 MHz, due to resolution limits imposed by the ultrasonic wavelength.

The required ultrasonic frequencies are more characteristic of acoustic microscopy than conventional pulsed ultrasonics. Pulsed ultrasonics are not sensitive to near subsurface defects due to a dead zone caused by the comparatively very large echo off the material surface. In addition the mechanical scans used in conventional ultrasonic systems are far too slow to allow timely and cost efficient scanning.

Acoustic microscopy is the only high temperature nondestructive technique that can meet the requirements for testing SRF cavities. There are two completely different acoustic microscopy technologies. In most cases the different technologies are useful in different application areas. The first is a high frequency RF-burst type system specialized to surface analysis. This system uses a tiny focused transducer that is scanned by a mechanical vibrator along one axis at 60 lines/second while the sample is automatically moved in the orthogonal direction. This instrument is known as the Scanning Acoustic Microscope (SAM). It requires about 8 to 10 seconds to produce a standard video image.

The second type of acoustic microscope employs a scanning laser beam as a point detector of ultrasonic waves and is known as the Scanning Laser Acoustic Microscope (SLAM). With laser beam scanning technology, acoustic images are produced very rapidly, that is, at conventional TV rates of 30 images per second. Compared with the SAM, the improvement in speed is about 300. The principles of SLAM operation have been described in the literature <6,7,8,9,10>. In addition to speed of scanning advantages, SLAM can detect defects at the surface, just below the surface, and deep within the bulk. The advantage of SAM is high resolution capability in the surface and near surface (0.005") zone.

Since production testing, especially 100% inspection, is only practical with a SLAM detection system, an automated device and associated methods were developed based on previously existing SLAM detection technology. This paper describes the results of that development. Previous work documented critical flaws that are detectable by SLAM techniques <11>.

INSTRUMENTATION

In performing the research described in this report, niobium was examined with the SLAM using 60 MHz. Higher ultrasonic frequencies are capable of detecting smaller defects and creating images with higher resolution. However, higher frequencies are associated with increased ultrasonic attenuation and visualization of grain structure. Grain structure visualization produces background texture in the image. Also at higher frequencies, each image covers a smaller area. We chose 60 MHz as a compromise that maximizes the image resolution and that minimizes the background texture seen in the acoustic images due to grain structure. This gives the best detectability of small, 50-100 microns diameter defects.

To inspect Nb for defects, the amplitude mode of the SLAM is used with frequency scanning. The amplitude mode produces images on a CRT wherein the grey scale, or brightness of the display, corresponds to the ultrasound transmission level through the sample at the position shown. This mode may be displayed at a single acoustic frequency analogous to monochromatic optical illumination in an optical microscope, or it may be displayed after integration over a limited range of acoustic frequencies (frequency scan) which is analogous to white light optical illumination.

SLAM Modifications

To accomplish the automated scan objectives, several modifications to the SLAM were required: development of 60 MHz capability; sample handling stage and optical system changes; and a flow system for ultrasonic coupling. As mentioned, the 60 MHz ultrasound optimizes the compromise between resolution, penetration, and background texture that can mask the image of the defect. The development of a new frequency like 60 MHz was an extension of the same detection techniques already employed at other frequencies. Several components of the SLAM required redesign: the ultrasonic driver, the preamp, the ultrasonic transducer and the final section of laser scanning optics. The standard stage was replaced by a plate mounting system attached to an X-Y scanner. To allow free space for the large (23.4 cm x 23.4 cm) moving plate, the modular optical system was supported above the new scanning stage, and additional optics were added to extend the detection plane well below the bottom of the optical system. Finally a new dynamic-flow, water coupling system was used instead of an immersion tank. The flow system was chosen since high-speed, high-resolution, mechanical scanning of a large mass of water is impractical (Figure 1).

The Complete System

In this project we established the technical feasibility of making an automated, high-speed NDT device for niobium by developing a working prototype device (Figure 2).

1   2

Figure 1. This picture shows the dynamic water flow system and the plate mounting system. A 23.4 cm x 23.4 cm niobium plate is seen mounted on the X-Y stage. The ultrasonic transducer is held by a 90 degree arm underneath the niobium plate. Notice that the size of the laser-scanned field of view is much smaller than the plate. This makes a second, large scale mechanical scan necessary for full plate coverage.

Figure 2. A working prototype, automated high-speed NDT device for niobium plate. The device is based on a 60 MHz SLAM detection system.

To automate the inspection system we integrated under computer control the following: the SLAM RS-170/330 video output, a computerized XY plate scanner, a real-time video digitizer/integrator, a computer algorithm for defect detection, a digital mass storage device, and a hardcopy output device. The hardware block diagram is shown in Figure 3. The double-line arrows represent the data paths; the single-line arrows represent the control signal paths.

Our host computer was an IBM Personal Computer AT compatible (7.16 MHz 80286, zero wait state RAM, 5 MHz 80287) with a 10 MHz, 32 bit coprocessor board with 4 megabytes of zero wait state RAM. The coprocessor uses a 10 MHz, National Semiconductor 32032 CPU with a NS 32081 Floating Point unit. The coprocessor increases calculation speed by a factor of two or three. Included was a parallel digital I/O interface for controlling the real-time integrator.

An important system design goal was to rapidly scan a 23.4 cm x 23.4 cm plate with a resolution capable of detecting 50 microns flaws. The Scanning Laser Acoustic Microscope uses a small laser spot to detect the sound. At video rates, the scanning laser system combines many spots into a practical field of view. For example, at 60 MHz the laser spot detector is about 30 microns. The scanning laser system creates a 5.00 mm x 3.75 mm field 30 times per second. Thus, for a 23.4 cm x 23.4 cm plate, we need about 61 million laser spots and almost 3000 non-overlapped fields of view. (NOTE: In theory the 60 MHz SLAM can scan a 23.4 cm x 23.4 cm plate in 100 seconds.)

To accomplish the entire plate scan, we chose a two-level scanning system (Figure 4). The scanning laser system rapidly scans the 30 microns laser spot into a 5.00 x 3.75 mm field of view and a mechanical scanning system combines the fields of view to scan the entire 23.4 cm x 23.4 cm plate.

The large-scale mechanical scanning is accomplished by a high resolution (4 micron) X-Y positioner consisting of two linear positioners with DC servo drives and optical encoders. The positioner was under direct control by a numerical controller. The numerical controller and the computer communicate through RS232 serial ports. The exact position of the X-Y stage can be sent to the computer, and the computer can download scanning programs, request position and issue interrupts to stop all motion.

Tests show that an entire plate can be scanned in less than 20 minutes with this system indexing from field-of-view to field-of-view without overlap and dwelling for two frame times (2/30 sec) on each field of view. The time required to move from frame to frame (about 0.3 sec) is much greater than the basic detection time (1/30 sec). The indexed scan was chosen since it can be used with readily available, low-cost frame grabbers. A 'real-time' continuous scan, synchronized with the laser scan, is possible with the same mechanical scanner. However, the continuous scan requires a more expensive, custom designed, real-time line-by-line data acquisition system. This improvement will be implemented as needs require.

The mechanical scanner moves the plate relative to the stationary SLAM detection system and ultrasonic transducer. The plate is positioned between the detection plane (top) and the transducer (bottom). The water flow system couples the transducer/plate and the plate/detection plane interfaces.

The SLAM's analog signal processing circuitry turns the detected signal into an RS-170/330 video signal containing amplitude mode, acoustic images. The video signal is frame averaged, digitized into 6-bits (64 grey levels), and stored in memory that is mapped into the 80286's memory space. For the prototype system, existing equipment was utilized that performed the frame averaging and the mapping into the computer's memory separately. The real-time averager digitized the video and summed each image into 12-bit memory (up to 64 images) at video rates, divided by the number of images, and reconverted to analog video. A frame grabber then redigitized the signal and stored the data into computer memory. The averager used 512 x 480 pixels; the frame grabber used 256 x 240. The frame averager was controlled by the computer through a parallel, digital I/O interface. The frame grabber was tied directly to the computer's bus. In a production version, both functions would be combined into one circuit removing the redundant, second digitization.

Data storage was accomplished using a two-level system, combining 30 Mbytes of hard disk storage with a 60 Mbyte, quarter inch, streaming tape unit. The storage required for each acquired image was minimized by, first, averaging four pixels to one creating a 128 by 120 pixel image and then using an error-free coding algorithm (image compression) to eliminate redundancy. Each reduced image was stored first to disk. The data was then sent to the streaming tape in 30 Mbyte chunks at 5 Mbytes/min. The data for a complete plate could be stored on one 60 Mbyte tape, thus data archival is simple.

Once the data was processed to recognize defects (discussion below) a dot matrix printer was used to print out a map of defective areas on the plate. The map scale was 1:1 so direct comparison to the plate could be done.

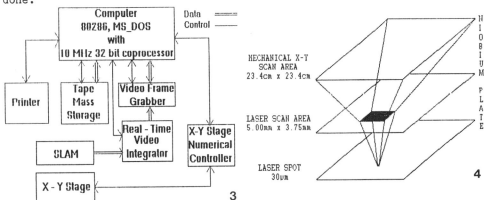

Figure 3. Hardware block diagram. A single line represents a control signal; a double line means a data pipeline.
Figure 4. Illustration of the two-level scan system used to scan an entire 23.4 x 23.4 cm niobium plate with better than 50 microns resolution. The first scan level is a laser scan; the second level is a mechanical scan.

PROCEDURE

This procedure outlines the steps necessary for the complete inspection of CEBAF-type niobium plates using the prototype automatic SLAM.

Upon receipt each plate is scribed with an identification. This number identifies the plate and defines a unique orientation of the plate. The number is scribed on the 3 mm wide edge of the plate. Any initial data such as manufacturer, RRR values, grain size, and plate dimensions are recorded for the plate.

For inspection the plate is cleaned with acetone to remove dust and smudges. The plate is placed in the SLAM mounts noting the orientation of the identification number. The plate mounts are adjusted to raise the plate so that it is in close proximity to the SLAM's detector plane and so that the plate is level during the scan. A small area of the plate is not scanned due to the plate mounts. The mounting areas were chosen so that the central circle, whose diameter equals the plate width, is completely inspected. The central circle is used in CEBAF cavities.

Once the plate is mounted, the SLAM detection system is set-up to obtain an acoustic image. Note that once the SLAM is set-up no adjustments are needed during data acquisition. Set-up includes making standard SLAM adjustments, starting the flow system and adjusting the SLAM's video gain and offset to match the input requirements of the digitizer. Samples with significantly different transmission characteristics require adjustment of these video signal parameters to maximize the accuracy with which transmission variations within the sample are digitized. For sample to sample comparisons, the video adjustments would remain constant.

With set-up complete, control of the experiment is turned over to the host computer. Both data acquisition and data analysis were completely automated. The data analysis could be performed as the data were acquired, or it could be done offline. Since the analysis didn't require the SLAM system, it was more economical to run the analysis overnight with the computer running unattended.

The control program was written in the C programming language and ran on the 32-bit coprocessor which acts as the main computer using the MS-DOS computer as an I/O controller. The program flow chart is shown in Figure 5. This program includes the following steps:

1. The operator inputs the scan set-up parameters as well as plate and test information. The scan set-up parameters include: size of plate, number of frames to average (N), file directory name for storing the digitized images, whether or not to do defect analysis during data acquisition process, and a defect detection threshold value. Plate information includes: identification number, orientation, thickness, manufacturer, known material properties like RRR value and grain size, etc. Test information includes: operator name, date, SLAM settings, etc.

2. The host computer calculates a scanning program for the X-Y stage numerical controller and downloads it to the controller. The scanning program is based on plate size information and standard positions of the plate mounts relative to the plate edges. The program also includes commands allowing control by the computer.

3. The X-Y stage is sent to the initialization position.

4. The computer signals the integrator to average N frames, waits till averaging is finished and then gets X-Y stage coordinates.

5. Frame grabber digitizes the averaged, analog video signal from the real-time integrator and stores it in DOS memory.

6. The 32-bit coprocessor moves the digitized data from DOS memory to its own RAM.

7. If defect analysis is being done on-line, the variance filter, defect detection algorithm looks for defects in current image.

8. If the hard disk is full, dump all the data files in the hard disk to tape backup system and erase them to make space for remaining data.

9. Squeeze 60 kbyte image into 6 to 8 kbyte by averaging four pixels to one and by using an error-free coding algorithm on the resulting 128x120

pixel image. If defect analysis was done, compress variance array also.
10. Save X-Y stage position and coded data to a hard disk file.
11. Computer checks if scan is finished. If finished go to step 13. If not, send a continue signal to X-Y numerical controller so X-Y stage moves to the next position. Note that each new field-of-view overlaps the previous one by 1.1 mm horizontally or 0.6 mm vertically. This overlap ensures 100% coverage and simplifies the detection algorithm.
12. Go back to step 4.
13. If any defect is detected, print out a map showing the apparent location of the defect on the plate. Note that the true location will depend on the defect depth and the ultrasonic propagation angle in the plate, 35x. Defects at or near the surface next to the detection plane are correctly located. Deeper defects appear consistently to one side of the true location with a maximum shift of about 2 mm for a 3 mm thick plate.

When the plate scan is completed, the SLAM is turned off and the plate removed and dried for storage.

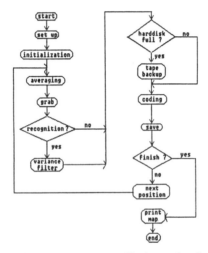

Figure 5. Flow chart for the program that controls data acquisition.

DEFECT RECOGNITION ALGORITHM

Our basic objective was to find a fast algorithm to recognize and locate flaws of different size and shape. Anomalous areas of ultrasonic transmission indicate defects in acoustic images. Thus we first had to understand what anomalous meant in the context of our niobium images.

Defect Images

A typical, 60 MHz acoustic image of RRR=200 niobium is shown in Figure 7a. Throughout the image we can easily see a random texture. This texture is due to ultrasonic scattering caused by the niobium grain structure, and it is seen in all RRR=200 plates from the same manufacturer.

Any elastic inhomogeneity in the material will cause the ultrasound to be scattered as it propagates across the interface(s) between the host material and the inhomogeneity. If the ultrasonic energy in one locale is completely redirected, the acoustic amplitude image in that area will be black. Also, some areas will appear much brighter than expected if the normal, through-transmitted energy is strengthened with additional scattered energy. As one might expect, scattering centers often are identified by a mixture of darker and brighter areas. The ultrasonic energy is redirected from the area in the shadow of an inhomogeneity to the area immediately surrounding the shadow. Thus the inhomogeneity causes unusually high bright-to-dark variations within a small area (Figure 8a).

Decision Parameter

To accomplish our basic objective, the algorithm must have the ability to distinguish between the defect and the random background texture. This is most important.

Extracting a defect from the typical background texture is very difficult because there are many orientations and there is a significant randomness to the texture. We can't use a band pass filter to remove the background, because the filter would be different for each image. Besides, a band pass filter algorithm is very time consuming. If we try a match filter method, then we need different transfer functions and different window sizes for various flaw sizes, orientations and shapes. Note that using a match filter can become complex very quickly. For example with four orientations, two sizes, and two shapes, we already have 16 different values that we need to detect <12>. In addition, we must locate the flaw's position. So, using a match filter is extremely time consuming and impractical.

Obviously, we need some invariants for the background texture so that we can simplify the detection analysis. Fortunately, the variance in a defined domain is invariant in orientation and translation. The variance, V, can be written as:

$$V = S^2 = \langle I^2(X) \rangle - \langle I(X) \rangle^2$$

where S is the standard deviation and I(X) is a two-dimensional digitized function representing the image <13>. $\langle I(X) \rangle$ and $\langle I^2(X) \rangle$ are the expectation values of I(X) and $I^2(X)$ over a chosen region of pixels. It can be easily shown that the variance of I(X) is invariant in translation and rotation.

Our variance filter method of recognizing defects was based on calculating the typical standard deviation of the niobium background texture and comparing this to the standard deviation in the area of a defect. Given the significant bright-to-dark variations in the vicinity of a scattering site, the standard deviation should be larger than normal. Also if there is a large void or other area with little or no bright-to-dark variation, we expect a smaller than normal standard deviation.

Calculation Window

The variance filter method requires a calculation window of a specific size, NxN pixels. If the window is too small compared to the background texture in the image, then the standard deviation calculated from window to window will vary significantly. The window must be roughly equal to at least a single cycle of the lowest, significant, spatial frequency in the background texture. If, however, the window size is too large, then small defects will affect too few pixels to be statistically significant. We found a 16 x 16 pixel window to be optimum for a 60 MHz, RRR=200 niobium image. The image contained 128 x 120 pixels and covered 5.0 mm x 3.75 mm.

The resolution afforded by a 16 x 16 pixel window, 0.6 mm x 0.5 mm, is inadequate for identifying individual defects as small as 50 microns.

We can maintain the necessary window size and increase the defect detection resolution by hopping the window one pixel at a time through the image. We calculate the standard deviation of the first window and map the value to the window's center pixel. The window is then hopped one pixel horizontally. This process repeats by calculating the new standard deviation and mapping it to the new center pixel, adjacent to the last. At the end of each row the window hops down (vertically) one row and proceeds with the horizontal hopping. For a 128 x 120 pixel image and a 16 x 16 pixel window, there are more than 10,000 window positions per image. Note that there is a border around each image that is ignored for reliability of detection and algorithm simplicity. Complete coverage of the plate is assured by overlapping images.

Thus we have performed a special transformation of our acoustic amplitude image into an image that shows the local variation in amplitude as measured by the window standard deviation. This special standard deviation image minimizes the variation due to the background texture.

Define Threshold

Ideally the standard deviation calculated in each window across the plate would be equal. In this simple case any change from this well-defined value would be recognized as a defect.

Due to a degree of randomness in the microstructure of the niobium, there is also some randomness in the acoustic image background texture. Thus, the standard deviation calculated within a window will differ slightly from place to place. This is true even in the absence of defects. Since these differences result from a random variation, the collection of window standard deviation values within a typical acoustic image defines a statistical distribution.

The next simplest means of defect detection compares the absolute difference between the mean value of all of the standard deviations in an image and the standard deviation in a specific window. If the difference is larger than a chosen threshold value, then a defect is said to be detected.

Unfortunately, our experiments showed that the acoustic image contrast varies across a plate due to warpage in the plate itself. The highest contrast and best resolution images occur when the detection plane is in close proximity to the sample surface. If we defined a fixed threshold value using a high-contrast image, the value would be incorrect for a low contrast image. So we needed a method which could adjust for these contrast changes automatically.

Two solutions exist. One solution uses hardware; the other uses software. Additional hardware can be used to maintain a constant distance between the detection plane and the sample surface. This method was originally proposed. However, the degree of warpage in commercial plates is now low enough that the software approach is practical. The software approach was used since it was less costly. In the case of badly warped plates, the hardware solution is still superior.

The software solution recognizes that image contrast can be expressed as the standard deviation of the distribution of all the window standard deviations in an image. We shall call the distribution standard deviation an image standard deviation to distinguish it from the previously discussed window standard deviations. Low contrast images have distributions with a small standard deviation, that is they have a small image standard deviation. High contrast images have a large image standard deviation. The following describes how to determine the standard deviation. From the center-limit theorem of statistics, we can assume that the probability density function (PDF) of the window standard deviation in a specified image is a Gaussian distribution. This is reasonable because there are more than 10,000 window values calculated in a 128 x 120 image. Then using the Gaussian fitting method <14>, we can calculate the image standard deviation of this distribution.

To be practical, the detection criterion must use a soft threshold that floats slowly from image to image as the plate warpage changes the contrast.

The final detection criterion combines a threshold multiplier with the image standard deviation. That is, the threshold value equals the threshold multiplier times the image standard deviation. This threshold multiplier is a number chosen by the operator. To date, the value for this multiplier is chosen based upon experience with typical images. We have found values of 3.5 to 5.5 useful (see below). In principle, a multiplier could be chosen more objectively, depending upon the acoustic frequency and a knowledge of the material's characteristics, such as grain size and orientation, RRR value, plate thickness, etc.

Detection Algorithm Summary

Figure 6 shows the major steps of the defect detection algorithm. A video image from SLAM passes through a temporal averaging filter (32 frames in our experiment) to increase the signal-to-noise ratio. A spatial averaging filter takes four pixels to one to reduce the image size to 128 x 120 pixels. This reduction in image size allows our compressed data to fit within a 60 Mbyte storage space when scanning a 23.4 cm x 23.4 cm plate. Then, the variance filter method discussed above is used to obtain the probability density function of the window standard deviation, S. A Gaussian fitting determines the image standard deviation of the distribution. Finally using the threshold multiplier times the image standard deviation as a threshold value, defects are detected, and their location is indicated in the display.

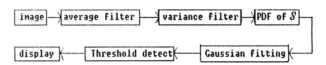

Figure 6. Diagram depicting major steps of the defect detection algorithm.

RESULTS

Figures 7, 8 and 9 illustrate the capability of the defect detection algorithm using experimental data. Each figure is a set of pictures showing important details in the recognition of defects.

Picture (a) in each of the three figures, 7 - 9, is a 60 MHz acoustic amplitude image. These images have already been averaged and processed into 128 x 120 pixel images. Only careful examination reveals the finite pixel size. Picture 7a shows a typical niobium image; Picture 8a shows an image with an easily observed surface defect; and Picture 9a shows an image with a comparatively low contrast internal defect.

These three amplitude images are a small, but representative, sampling of the types of features seen when examining the niobium. Clearly visible is the background texture caused by the scattering of the through-transmitted ultrasonic waves. Measurements of the acoustic velocity in single crystal niobium demonstrate an anisotropy of about 7% along one crystal axis, relative to the other axes <15>. The measured velocities were 5.41 km/s (v_1) and 5.02 km/s (v_2). Since the CEBAF type niobium plates have been heat treated to achieve 100% recrystallization after rolling, the grains are randomly oriented. At grain boundaries the sound often sees changes in the elastic properties from one grain to the next due to their relative orientation. The sound is scattered at such elastic property changes.

Two of the amplitude images, 8a and 9a, have a region that is observed to have anomalous variations in the detected amplitude of sound. With a small amount of experience, an operator will identify these regions as a surface defect, 8a, and an internal defect, 9a. Note that as you become more critical of the images, a greater and greater number of potential defects can be seen, even in the typical image, 7a. For short periods an operator can detect such anomalies reliably. However to inspect one plate, an operator must look at almost 4,000 overlapped, fields of view. (Only 3,000 nonoverlapped fields of view are needed, but this is less reliable.) The monotony of thousands of very similar images quickly causes fatigue, boredom and loss of reliability. We found that one or two weeks are needed for an operator to examine the data from one plate.

Due to the impracticality of operator inspection, a digital computer based algorithm was developed to identify anomalous regions.

The success of the detection procedure is understood by examining the charts showing the distributions of the window standard deviations. These charts show the absolute difference between each window standard deviation and the mean window standard deviation for the current image. The charts in pictures 7b, 8b and 9b were calculated from pictures 7a, 8a and 9a, respectively. The method of calculating the window standard deviations was described above. The most important part of each distribution is the tail-end of high standard deviations, not the bulk of the distribution.

Comparing pictures 8b and 9b with picture 7b, we can easily see they have unusually high window standard deviation values. This special, high window standard deviation information is associated with the regions that a trained operator would identify as defective. Each pixel associated with a variance value more than 3.5 times the image standard deviation are highlighted in pictures 8c and 9c. Such highlighting quickly identifies the defect related image region(s). The typical image in 7a did not contain defective regions.

With an algorithm that can detect defective regions, we can process the data from an entire plate. In one example when we set the threshold to 5.5 standard deviations, 48 locations were recognized as defects by the algorithm. Using the video graphic overlays to highlight these regions, an operator examined each one manually. We found that the variance filter algorithm does detect uncommon regions in the image. These detected zones were due to inner structure in the material, surface defects, and severe streaks. (See <11> for earlier discussion of streaking.)

An important part of our work was to determine how quickly a practical inspection of a CEBAF type niobium plate could be performed. Obviously these numbers reflect the specific hardware chosen for this first working prototype machine. As such these timings represent a good first effort. Significant progress is possible based on what we have learned and using newly available hardware, particularly in the computer field.

Central to the overall data acquisition time is the time it takes to acquire each image. This basic unit of time is repeated many times. We found that only 2.52 seconds is needed to: generate the SLAM video data, time average 32 frames, digitize into computer memory, code the data to reduce storage, and save the data to hard disk. For a 23.4 cm x 23.4 cm plate, almost 4000 images, including overlap, must be processed. Acquiring all 4000 images per plate takes 2.8 hours. An additional overhead is required to perform backups of the hard disk data files to a 60 Mbyte tape. The total backup time is 0.5 hour.

Therefore, we only need 3.3 hours to acquire the 60 million pixels of raw data per plate. The data acquisition time is the most important, because both the automatic SLAM system and a computer system are required during this procedure.

To analyze the data and recognize defect zones takes 14 seconds per image with a 16 x 16 window. The total recognition time per plate is 15.5 hours in the coprocessor enhanced personal computer system. Although this is a significant period of time, only a tape drive and coprocessor enhanced personal computer system are required. The equipment cost for offline data analysis is only about $7,500. Clearly the data analysis time is very economical, and the analysis can run constantly day and night without an operator.

CONCLUSION

We have successfully completed a first, working prototype, high-speed automated NDT device for niobium plate based on 60 MHz SLAM detection technology. The device is capable of detecting defects as small as 50 microns throughout the volume of a 23.4 cm x 23.4 cm x 0.3 cm plate. The digital data acquired during one plate scan can be archived on one 60 Mbyte, quarter inch tape cassette. A hard copy map of detected defects can also be obtained; the map scale is 1:1.

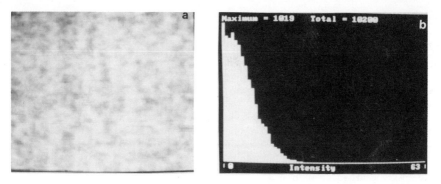

Figure 7. a) Typical 60 MHz acoustic amplitude image of RRR=200 niobium.
b) Distribution of window standard deviations, S, calculated from image (a).

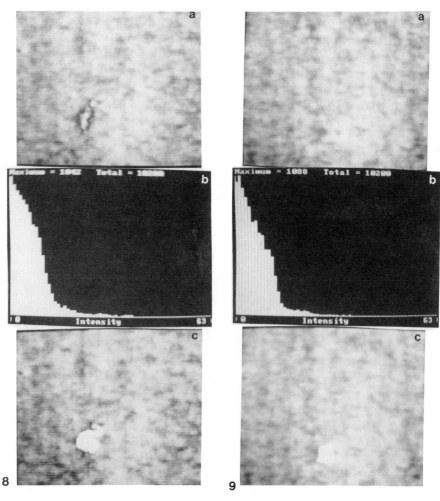

Figure 8. a) A 60 MHz acoustic amplitude image of RRR=200 niobium with a
surface defect. b) Distribution of window standard deviations calculated
from data in (a) with a 16x16 window. c) Picture (a) with a graphic overlay
indicating the defect location. Threshold was 3.5 standard deviations.
Figure 9. a) A 60 MHz acoustic amplitude image of RRR=200 niobium with a
subsurface defect. b) Distribution of window standard deviations calculated
from data in (a) with a 16x16 window. c) Picture (a) with a graphic overlay
indicating the defect location. Threshold was 3.5 standard deviations.

The data presented show that the variance filter algorithm can recognize defects and provide an accurate location of the defect area as it appears in the SLAM image. The recognition depends upon the defect causing variations in the detected acoustic amplitude that are statistically significant. These variations can be either higher or lower than the typical variations observed in the plate. The current algorithm cannot determine the type of flaw. Combining the current data with a scan using the SLAM'S optical reflection imaging ability would easily provide correlative information on surface flaws. A second acoustic scan, with the plate rotated 180 degrees, could provide information about the depth of the detected flaws.

The recognition algorithm is very fast since it consists mainly of multiplications and additions. This speed and simplicity make the algorithm a very good method to process a large number of images with background texture. Commercially available coprocessors, using a digital signal processor integrated circuit, can perform these calculations at video rates. The fast, variance filter processing can extract out those images requiring a more complex pattern recognition method to determine the defect type.

Clearly a major requirement for the long term application of the automatic device is to build experience by inspecting material. Work has already begun in performing correlative work, such as destructive physical analysis, to understand what the detected defects are. Also we hope to scan a significant amount of material that will be used in superconducting RF accelerators. Tracking the performance of accelerators built with screened material will be necessary to establish that 'defect free' material provides higher performance and reliability.

Should the technique prove to be useful, engineering modifications to the detection and sample scanning systems would allow the device to be used on completed cavities. Modifications allowing a manually scanned inspection of the equatorial weld region of a single cell have already been achieved.

REFERENCES

1) H. Padamsee, "Status and Issues of Superconducting RF Technology", CLNS 87/65.
2) G. Mueller and H. Padamsee, "High-Temperature Annealing of Superconducting Cavities Fabricated from High Purity Niobium", 1987 Particle Accelerator Conference, Washington, D.C.
3) Personal communication with Dr. Gunter Mueller, University of Wuppertal, W. Germany.
4) Niederman, Ph. et al., J. Appl. Phys. 59, 892 (1986).
5) Niederman, Ph., Ph.D Thesis No. 2197, U. of Geneva (1986).
6) M. G. Oravecz, "Quantitative Scanning Laser Acoustic Microscopy: Attenuation", Proc. of ICIFUAS-8, Journal de Physique, Colloque C10, supplément au nx12, Tome 46, décembre 1985, pp. 751 - 754.
7) M. G. Oravecz and S. Lees, "Acoustic Spectral Interferometry: A New Method for Sonic Velocity Determination", Acoustical Imaging, Vol. 13, Ed. by Kaveh, Mueller and Greenleaf, Plenum, 1984, pp. 397-408.
8) L. W. Kessler and D. E. Yuhas, "Acoustic Microscopy - 1979", Proc. of IEEE, Invited Manuscript, Vol. 67, No. 4, April, 1979, pp. 526-536.
9) L. W. Kessler and D. E. Yuhas, "Principles and Analytical Capabilities of the Scanning Laser Acoustic Microscope (SLAM)", SEM/1978, Vol. 1, p. 555.
10) L. W. Kessler, "A Review of Progress and Applications in Acoustic Microscopy", J. Acoustic Society America, Vol. 55, pp. 909-918, 1974.
11) M. G. Oravecz, L. W. Kessler and H. Padamsee, "Nondestructive Inspection of Niobium to Improve Superconductivity", Proc. of 1985 IEEE Ultrasonics Symp., Catalog #85CH2209-5, IEEE, pp. 547-552.
12) Woods, J.W., "Image Detection and Estimation", Chapter 3 in Digital Image Processing Techniques, M.P. Ekstrom, Ed., Academic Press, 1984.
13) S. Haykin, Communication Systems 2nd Ed., Wiley 1983, pp. 236-238.
14) Pugh and Winslow, The Analysis of Physical Measurements, Addison-Wesley, pp. 66-68 and pp. 91-95.
15) R. Weber, "Ultrasonic Measurements in Normal and Superconducting Niobium", Phys. Rev., Vol. 133, No. 6A, pp. A1487-A1492, 16 Mar. '64.

EVALUATION OF PLASTIC PACKAGES FOR INTEGRATED CIRCUITS USING SCANNING LASER ACOUSTIC MICROSCOPY (SLAM)

Elizabeth M Tatistcheff

Digital Equipment Corporation
30 Forbes Road
Northboro, MA

INTRODUCTION

In the past several years the electronics industry has been making a concerted effort to improve the quality and reliability of the integrated circuits (ICs) which make up computers. In order to assure this high reliability, a reliable and inexpensive method for the nondestructive evaluation (NDE) of IC packages is needed. Radiographic analysis of plastic packages, primarily used for the location of bond wires and the evaluation of die attach material distribution, does not provide any information about the silicon chip or the internal bond quality and has thus proven inadequate at detecting a number of plastic package defects.

The theory behind the Scanning Laser Acoustic Microscope (Sonomicroscope 2140, of SonoScan Inc, IL) and a number of specific applications have been described in many publications [1-4]. All the acoustic micrographs presented here were taken at an operating frequency of 10 MHz. At this frequency, the theoretical resolution is approximately 250 um or 10 mils [5]. The vertical resolution has been shown to be as low as 6 um in practice [6].

PLASTIC IC PACKAGING TECHNOLOGY

Three typical plastic packages are 1) the Dual In-line Package (DIP), 2) the Small Outline J-bend package (SOJ) which is the surface mountable DIP, and the Plastic Leaded Chip Carrier (PLCC). Plastic packages share a common morphology as shown in Figure 1. The lead frame (flag) is stamped from copper, and the chip (die) is bonded to the lead frame by any of a number of die attach techniques: eutectic bonding, epoxy which may or may not be filled with silver, and in some cases a polyimide die attach material is used. Wire bonds are made to connect the chip to the leads. The encapsulant, typically an epoxy filled with 60-80% SiO2, is injected into a mold containing the chip and lead frame.

During board assembly the package may be brought to temperatures as high as 220-230 degrees Centigrade for several minutes. The stresses on such a part can be separated into three categories: 1) residual molding stresses, 2) thermal stresses due to the different thermal expansion coefficients of silicon, copper, and encapsulant, and 3) stresses caused by heating the excess humidity absorbed by the epoxy. Heating a plastic package well above its softening temperature serves to raise the incidence of package defects by increasing the internal stresses and softening the encapsulant.

Three types of package defects will be discussed here: 1) poor die attach which can be due to uneven distribution of die attach material, a crack in the die attach material, or delamination between die attach material and chip or lead frame; 2) poor package attach which is a delamination between encapsulant and lead frame or chip; and 3) package cracks which typically originate at the corners of the lead frame and propagate through the encapsulant. These defects are often seen in combination with each other, and package cracks are always accompanied with poor package attach.

Radiographic analysis can detect uneven die attach distribution if the die attach material has a metal component but several IC vendors use a simple polymeric material with no metallic filler. An x-ray cannot detect cracks in the encapsulant or die attach material, nor can it evaluate bonding per se.

The results presented here concentrate on the SOJ and PLCC. They are also applicable to the DIP which has a similar internal structure. Figures 2 and 3 are acoustic micrographs of a good SOJ and PLCC, respectively. Note the high acoustic transmission overall and particularly in the die areas (arrowed). Note also the uniform interference fringes in the die areas of the interferograms (Figs 2b and 3b). These parts display none of the defects listed above.

POOR DIE ATTACH

Figure 4 is an acoustic micrograph of an SOJ displaying poor die attach as denoted by the slight shading in the die in Fig 4a and the scrambled interference fringes in the die in Fig 4b. As the die attach used in this package had no metal filler, the package was opened and the lead frame peeled away from the chip, revealing the die attach distribution as in Figure 5. Arrowed areas denote regions void of die attach material which correlate with areas displaying poor acoustic characteristics.

As a metal filled epoxy is a more typical die attach material in these packages, an x-ray is usually sufficient for correlation. Figure 6 is an acoustic micrograph of a PLCC displaying poor die attach. Figure 7 is the x-ray of that part. Note that in this case the x-ray provided much better resolution of die attach voids, whereas SLAM only provided an acoustic signature characteristic in general of voids in the die attach.

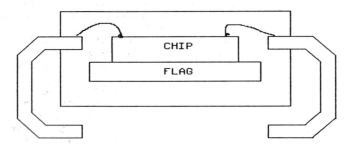

Figure 1.   Internal configuration of a PLCC.

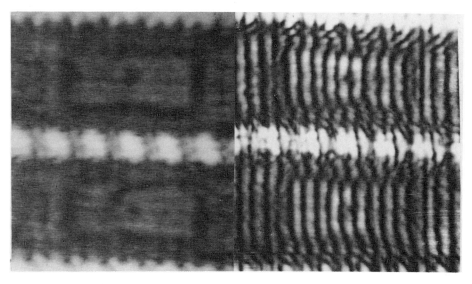

a. Amplitude   b. Interferogram

Figure 2. Two fairly good SOJs. Note the outline of the die area, overall high transmission, and uniform interference fringes.

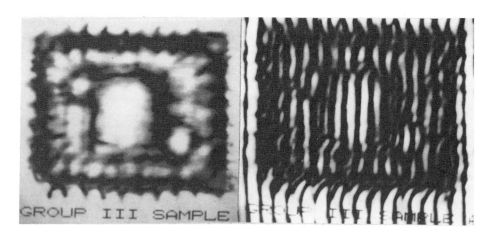

a. Amplitude   b. Interferogram

Figure 3. A good PLCC displaying high acoustic transmission and uniform interference fringes.

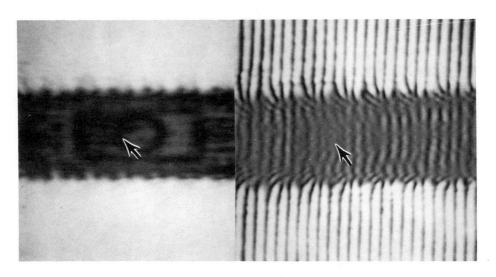

a. Amplitude   b. Interferogram

Figure 4. An SOJ displaying poor die attach as denoted by the lowered acoustic transmission and scrambled interference fringes (arrowed).

Figure 5. The same device with the lead frame peeled away. The light areas on the right are residual die attach material, and the very dark areas are where the encapsulant seeped in between die and flag. In between the bare Silicon can be discerned.

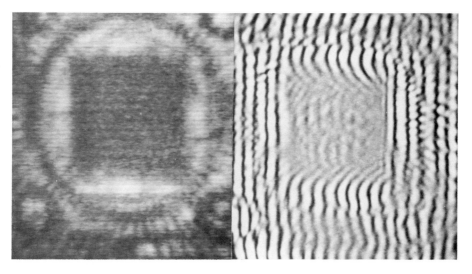

a. Amplitude   b. Interferogram

Figure 6. A PLCC displaying poor die attach as shown by scrambled interference fringes.

Figure 7. An X-ray of that device. Darker areas in the die region are silver-filled die attach epoxy, and lighter areas are large voids.

POOR PACKAGE ATTACH

Figure 8 is an acoustic micrograph of a PLCC with poor package attach. Note the total lack of acoustic transmission in the die area denoted by dark shading in Fig 8a and the obliteration of the interference fringes in Fig 8b. As shown in cross section (Figure 9[6]) the defect is a long delamination, 6 um thick, between lead frame and encapsulant.

Knowledge of the internal specimen configuration is extremely important, especially in this case as the presence of a silicone meniscus on top of the die will cause the SLAM image of such a part to resemble closely Figure 8.

What distinguishes the acoustic signature for poor package attach from that of poor die attach is: 1) the total loss of acoustic transmission as detected by the absence of interference fringes, and 2) the square or rectangular configuration of the area with no acoustic transmission. While this might be confused with a total lack of die attach which might also obliterate the interference fringes in the square or rectangular die shape, this has not proven to be a problem in practice for 3 reasons:

1) the delaminations associated with poor package attach have to date been found to extend over the whole area of either the top of the die or bottom of the lead frame, ie, no partial delaminations have yet been discovered above the die or below the lead frame;

2) the poor die attach associated with excess porosity or uneven application of the die attach material does still provide some bonding, thus giving rise to scrambled interference fringes or obliteration of the interference fringes over a limited portion of the die area;

3) the poor die attach associated with debonding or actual fracture of the die attach material has been found to be more random than with poor package attach, which is to say that as the crack or debonding in this layer has not been found to extend over the entire die area, acoustic examination of this type of defect will reveal interference fringes -- if somewhat scrambled -- in some portion of the die area.

Thus the presence of any form of interference fringes in the die area would eliminate the likelihood of poor package attach as the cause of any defects detected acoustically. While this has been the case to date, it holds no guarantees and destructive physical analysis is recommended on any plastic packages displaying the acoustic signs of partial delamination. The development of a practical method of measuring the depth of such delaminations is expected to eliminate the need for DPA to confirm the distinction between poor package attach and poor die attach.

PACKAGE CRACKS

Figure 10 is an acoustic micrograph of a PLCC with a package crack which propagated through the encapsulant and terminated on the outside of the bottom of the package. Note that in this figure, the acoustic shading is rounded rather than square as in Fig 8. As seen in cross section (Figure 11) the defect is poor package attach (total delamination of encapsulant and lead frame) and package crack.

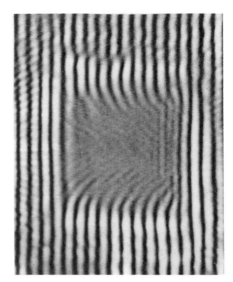

| a. Amplitude | b. Interferogram |

Figure 8. A PLCC displaying poor package attach as shown by lowered acoustic transmission and a total absence of interference fringes in the die area.

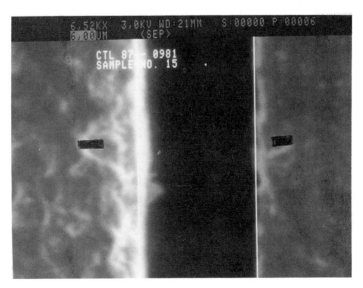

Figure 9. A cross section of that part shows a delamination 6 um thick between flag (right) and encapsulant (left).

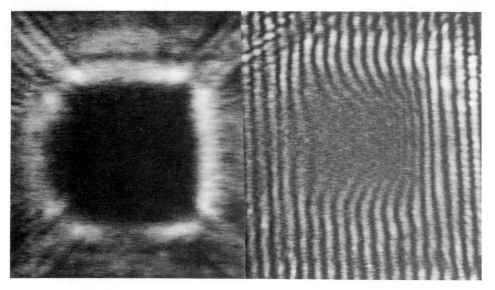

a. Amplitude   b. Interferogram

Figure 10. A PLCC displaying poor package attach as shown by lowered
acoustic transmission and a total absence of interference fringes in the
die area, and a package crack as shown by the irregular shape of the
shaded area.

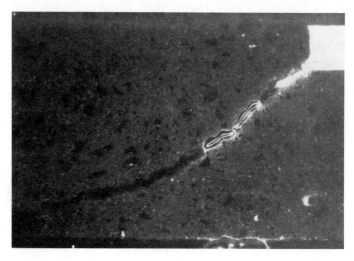

Figure 11. A cross section of that part shows a delamination between
flag and encapsulant, as well as a package crack propagating from the
lower corner of the flag to the bottom of the device.

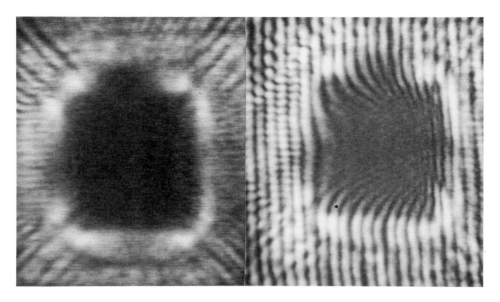

a. Amplitude   b. Interferogram

Figure 12. A PLCC displaying poor package attach as shown by lowered
acoustic transmission and a total absence of interference fringes in the
die area, and two package cracks as shown by the irregular shape of the
shaded area. These cracks are entirely internal.

Figure 13. A cross section of that part shows a delamination between
flag and encapsulant, as well as a package crack propagating from the
lower corner of the flag.

Figure 12 is the acoustic micrograph of a PLCC with entirely internal package cracks as shown in cross section (Figure 13). While the crack in Figures 12 and 13 propagates downward from the lead frame it does not terminate at the bottom of the package, remaining undetected by visual inspection.

The acoustic signature for package cracking is a total loss of acoustic transmission as indicated by the absence of interference fringes in an irregular, often round area larger than the die. As there is no irregularity or roundness to any of the features internal to a plastic package, this acoustic signature unambiguously indicates damage to the encapsulant material, ie, a package crack. As stated earlier, all package cracks detected have been accompanied with poor package attach.

CONCLUSION

Despite the poor lateral resolution at a 10 MHz operating frequency and the difficulty in measuring the precise depth of specific features, some knowledge about the specimens examined (internal configuration and rough geometry of typical defects) has made it possible to identify the specific acoustic signatures as seen in the SLAM which closely correlate to the defects detected by radiographic and destructive physical analyses as presented here. By providing this unique nondestructive capability as well as demonstrating great possibilities for future applications, acoustic microscopy is expected to become an increasingly important tool for the electronics industry as a whole.

REFERENCES

1. L.W. Kessler and D.E. Yuhas, "Principles and Analytical Capabilities of the Scanning Laser Acoustic Microscope (SLAM)", Scanning Electron Microscopy, 1978, Vol. 1, SEM Inc., AMH O'Hare, IL.
2. L.W. Kessler, "Acoustic Commentary: SLAM and SAM, IEEE Transactions on Sonics and Ultrasonics, Vol. SU-32, No. 2, March 1985.
3. L.W. Kessler, J.E. Semmens, and F. Agramonte, "Nondestructive Die Attach Bond Evaluation Comparing Scanning Laser Acoustic Microscopy (SLAM) and X-Radiography", Proceedings of 35th Electronic Components Conference, IEEE, 1985.
4. L.W. Kessler, J.E. Semmens, and F. Agramonte, "Scanning Laser Acoustic Microscopy (SLAM): A New Tool for NDT", 11th World Conference on NDT, Las Vegas, NV, 1985.
5. Notes from the First SLAM Users Advanced Workshop, June 1986, unpublished.
6. R. St. Amand and E.M. Tatistcheff, "An Investigation into PLCC Cracking", DEC Internal Document, June 1987, unpublished.

IMAGE ANALYSIS AS AN AID TO QUANTITATIVE INTERPRETATION OF

ACOUSTIC IMAGES OF DIE ATTACH

J. E. Semmens and L. W. Kessler

Sonoscan, Inc.
530 East Green Street
Bensenville, Illinois   60106

INTRODUCTION

Die attach involves bonding a silicon chip to a substrate or lead frame using one of several methods; eutectic, adhesive material, or solder. Good adhesion of the die is critical in order to provide both mechanical stability of the component and uniform heat dissipation in the circuit. Voids or disbonds at the die attach interface reduce the power dissipation capability which can result in interruption of the proper operation of the component. The SLAM (Scanning Laser Acoustic Microscope) has been used successfully to nondestructively image the die attach interface. The SLAM is very sensitive to laminar flaws such as disbonds or voids in a die attach bond. The SLAM, however, produces a through-transmission image and as such, does not necessarily give the exact depth at which the flaws occur. The C-SAM (C-Mode Scanning Acoustic Microscope) is a reflection mode device and as such, is capable of focusing at different levels within a sample to determine the exact depth of a flaw. Qualitative information is easily obtained from acoustic images as to the presence and location of voids or disbonds in the die attach. In many cases it is neccessary to obtain quantitative information from the qualitative images. The current MIL-SPEC states that only those disbonds which cover a certain percentage of the bond area qualify as failures (Figure 1). Also, quantitative information provides for an objective determination of bond quality, and allows for computer decision of acceptance in automated production applications. In order to derive the quantitative information, image analysis is used. Simple routines can be used to determine the area fractions of the bond and disbond. Other routines can be used to analyze the contrast in the image and to measure the ultrasonic attenuation in the bond area. Examples of how image analysis has been used to aid in the interpretation of acoustic images of die attach bonds will be presented.

TECHNIQUES

Two types of acoustic microscopes were used in this study, the Scanning Laser Acoustic Microscope (SLAM) and the C-Mode Scanning Acoustic Microscope (C-SAM). The SLAM is a through transmission technique, typically operating at frequencies between 10 and 200 MHz. SLAM utilizes a scanning laser detector to obtain real-time (1/30 second

Mil Standard 883C, Method 2030, Ultrasonic Inspection
of Die Attach

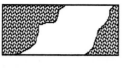 **Reject... void greater than 15%
intended contact area, or
corner void greater than 10%**

 **Accept... no single void
greater than 15% of intended
contact area, corner void
less than 10%**

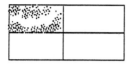 **Reject... quadrant 70%
disbonded**

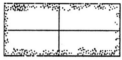 **Accept... all quadrants
less than 70% disbonded**

FIGURE 1 - Accept/reject criteria for die
attach inspection per MIL-STD-883.

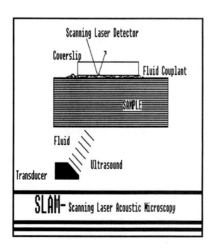

FIGURE 2 - Block diagram of SLAM operation.

ATTENUATION VS. THICKNESS

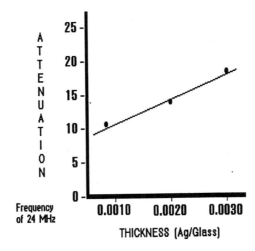

FIGURE 3 - Block diagram of C-SAM operation.

C-Mode Scanning Acoustic Microscope
(C-SAM)

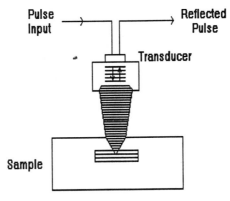

FIGURE 4 - Attenuation characteristic of
AG/Glass die attach material measured at
24 MHz in through-transmission.

per image) ultrasonic images of the internal details. The ultrasound travels through the entire volume of the material, and the scanning laser detects the variations in transmitted ultrasound (Figure 2).

The C-SAM system operates in a reflection mode and uses special high frequency pulsed transducers operating in the 10 - 100 MHz frequency range. Here, the transducer both transmits and receives the ultrasound and it is mechanically scanned in a raster fashion to create the complete image. C-SAM is capable of focusing at specific depths within the material providing a means to determine at which interface the defects may occur (Figure 3). The scanning speed is very high for an instrument of this type. Images are produced in 10-20 seconds.

A Digital Image Analyzer was used to extract the quantitative information from the SLAM and C-SAM acoustic data. The unit interfaces with the RS-170 output from either instrument. The analyzer digitizes the analog video signal at video rates. Each image contains 256 by 240 pixels. Once the image is acquired, software routines for window definition, image enhancement and image analysis are performed.

QUANTIFICATION OF SLAM IMAGES

Attenuation Measurements and Area Fractions

A recent study concerning process variables in the recently developed Ag/glass die attach bonding process gives an example of how area fractions were used to analyze the acoustic data. The die were attached to ceramic substrates and physical defects were readily observed in the acoustic images. In order to establish a relationship between pull strength (destructive) and acoustic information (nondestructive), the images had to be quantified. First, the overall average intensity level was measured in a defined area to determine the total ultrasonic energy lost. Although average attenuation measurements provided correlation with pull tests in most instances, there were some cases where the attenuation readings were misleading. Specifically, samples had been prepared using varying thicknesses of the Ag/glass material. The Ag/glass material itself has significant attenuation with increasing thickness. This was determined by measuring known thicknesses of the bonding material alone (Figure 4). Because of this effect, the average attenuation of the overall die bond could be similar for a uniformly bonded die with a thick layer of Ag/glass and a partially disbonded die having a thin layer of material. In some other die attach methods, for example eutectic, the bond layer is not subject to this type of variation and would not be a problem for average attenuation data.

In order to analyze the area of the die attach which is disbonded, the grey scale image is converted to a binary image and a threshold for the defect level is established using the computer. The histogram routine is then used to provide the percentage of the area which is bonded and the percent disbond.

Figure 5 is an example of an area fraction on a 30 MHz SLAM image of a eutectic die attach. The grey scale image (a) shows dark areas within the outline of the die bond which correspond to disbonds at the interface. In order to obtain quantitative information on the amount of disbond, the specific area of interest (the die area) is defined and is converted to a binary image (b). The computer then provides the percent of bond versus disbond using the histogram routine. This die was found to be 30% disbonded.

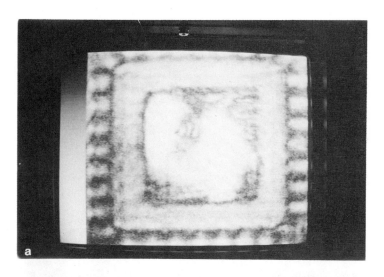

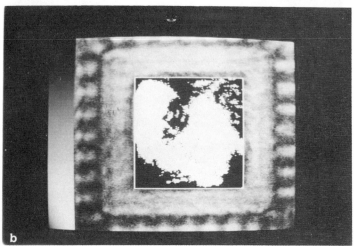

FIGURE 5 - (a) Acoustic image of a silicon die,
eutectically bonded to a ceramic
substrate. Dark areas within the die
area indicate disbonds.

(b) Binary image of the die attach region.

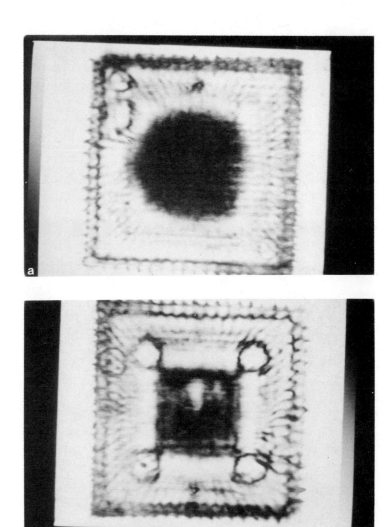

FIGURE 6 – Acoustic images of defective plastic
leaded chip carriers (PLCC's) at 10 MHz.

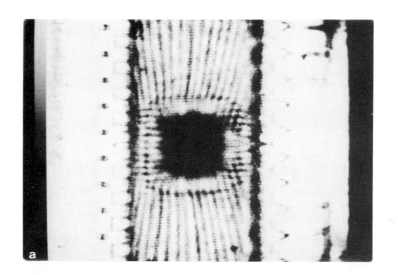

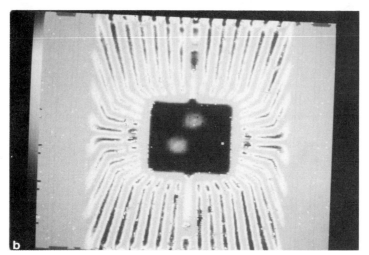

FIGURE 7 - (a) 10 MHz SLAM transmission mode image of a
plastic encapsulated integrated circuit
showing a defective die attach.

(b) 15 MHz C-SAM reflection mode image
showing defective die attach plus two
voids in the plastic.

ANALYSIS OF PLASTIC ENCAPSULATED DEVICES WITH SLAM AND C-SAM

Acoustic microscopy can also be used to evaluate the die attach bond within plastic encapsulated devices. The SLAM acoustic images clearly show defects in the parts and from the size and location of the defects together with knowledge of the part's construction, a determination can be made as to the type of anomaly that exists.

The two SLAM images in Figure 6 show defects in two plastic leaded chip carriers at 10 MHz. Figure 6a shows a large, dark circular feature extending beyond the outline of the die area. This corresponds to a horizontal crack in the encapsulant. In Figure 6b the defect is confined to the die area and could either be related to a die attach problem or a separation of the encapsulant from the die. To establish further information, the C-SAM can be used to find the specific level at which the defect occurs.

Figure 7 shows a 10 MHz SLAM image (a) and a 15 MHz C-SAM image (b) of the same device, a plastic encapsulated, 40 pin, dual-in-line package. The SLAM acoustic image clearly shows a defective area in the die region. The C-SAM acoustic image, focused at the die attach interface, shows that the die is disbonded as well as existence of two voids in the encapsulant at a level above the die.

SUMMARY

Acoustic microscopy provides useful, qualitative information on the quality of die attach bonds and the encapsulation integrity of micro-electronic devices. The SLAM is the most rapid way of detecting an anomaly in a component due to the real-time imaging capabilities. In cases where specific layers of material are of interest, the C-SAM is capable of discerning the exact interface in question. Further, quantitative analysis of die attach bonds can be performed on both SLAM and C-SAM acoustic images using straightforward computer image analysis routines. The quantitative information is needed in order to conform to MIL-SPEC accept/reject criteria. Also the quantitative capabilities would eventually allow for computer accept/reject decisions for automated production applications.

CHARACTERIZATION OF DEFECTS IN Mn-Zn FERRITES BY

SCANNING LASER ACOUSTIC MICROSCOPY (SLAM)

C. W. Boehning, and W. D. Tuohig

Allied Corporation, Bendix Kansas City Division*

P. O. Box 419159 Kansas City, MO. 64141-6159

INTRODUCTION

A scanning laser acoustic microscope (SLAM) has been used to
evaluate the integrity of Mn-Zn ferrite ceramic components which
comprise part of the magnetic circuit in an electromechanical code
interrogation device. Cracking of the ferrites during processing and
assembly emerged as a significant manufacturing problem. Operations
such as grinding, metallization, joining, and welding were suspected
of causing damage, and acoustic microscopy was used to monitor these
processes. Parts which produced suspicious acoustic images were
dismantled and destructively sectioned to identify specific physical
defects. Correlations between the defects and their acoustic
signatures were established. This procedure has provided the basis for
several process modifications and improvements which have resulted in
acceptable production yields. The SLAM is used as an engineering tool
for the detection and characterization of defects and is presently
being used routinely to inspect production ferrite components.

BACKGROUND

Ferrites are magnetic oxides (ceramics) which because of their
efficiency at high frequency have found applications in a variety of
electrical and electronic devices.[1,2] One such device is an
electromechanical code-interrogation switch manufactured by the Bendix
Kansas City Division of Allied Corporation. However, a high incidence
of cracking in the Mn-Zn ferrite components emerged as a concern
during the manufacturing development phase. Ferrite ceramic inserts
are metallized and joined to a stainless steel header body with a
lead-indium solder alloy, Figure 1. Completed header assemblies are
then electron beam welded into the device. The sequence of steps used
to fabricate the header and incorporate it into the next assembly are
outlined in Figure 2.

Cracking of the ceramic was originally thought to be due to
improper handling procedures, but improvements in handling did not

---
*Prepared for the United States Department of Energy Under Contract
Number DE-AC04-76-DP00613

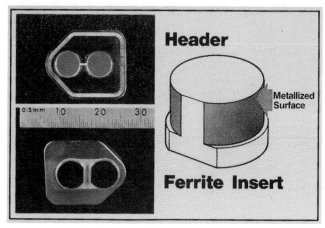

Figure 1. Header Assembly (left) and Detail of Metallized Ferrite Insert (right)

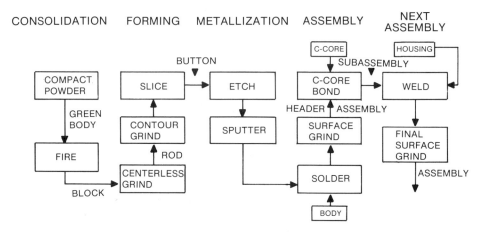

Figure 2. Processing Sequence for Manufacture and Assembly of Header

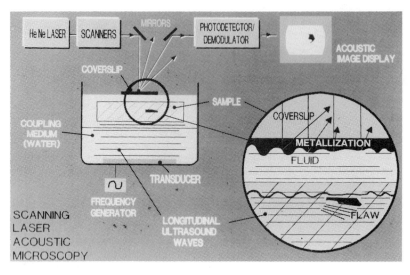

Figure 3. Schematic of SLAM Pattern Generation and Detection

significantly improve yields. Consequently, manufacturing processes, at both the piece part and next assembly levels, were scrutinized to identify the cause of damage to the ferrites.

Scanning laser acoustic microscopy was proposed as a technique to investigate grinding, metallization, joining, and welding operations because it is both rapid and non-destructive. Ferrites which produced suspicious acoustic images but contained no visible flaws were destructively sectioned to permit correlation of those images with internal defects. Feasibility of the technique was first established through correlation of acoustic image patterns with visible defects. This paper presents an overview of results of a SLAM characterization of defects found in the Mn-Zn ferrites.

THE SCANNING LASER ACOUSTIC MICROSCOPE

The SLAM* is a relatively new instrument which uses local variations in acoustic properties to produce magnified, ultrasonic images of the sample. Internal defects such as cracks, delaminations, voids and inclusions can be detected, identified, measured, and imaged directly in real time. Figure 3 illustrates the principles of acoustic pattern generation and detection employed in the SLAM.[3,4] An ultrasonic transducer (30 or 100 MHz) generates a plane-wave signal which is transmitted through the sample to a specially designed coverslip via a coupling medium, usually distilled water. The pattern produced on the metallized coverslip is scanned at a rate of 30 times per second by a helium-neon laser and a photodiode positioned above the stage detects the reflected laser light and converts it to a real-time CRT image. High-resolution (75 microns at 30 MHz) acoustic images can be obtained from materials of variable geometry, thickness, and density with little or no sample preparation.

The presence of an internal defect such as a crack, delamination, or void, alters the amplitude of the transmitted ultrasound wave in the specimen, producing an acoustic "shadow" on the sample surface and on the coverslip. As the laser beam is rastered over the gold surface of the coverslip, the beam is angularly deflected according to the slope of the distortion pattern on the coverslip. In the region of the acoustic shadow the gold film remains relatively flat, and the photo-detector (with a knife edge blocking half of the beam) will detect the variation in intensity in this region of the coverslip. Material which is homogeneous and contains no defects produces an acoustic image which is uniformly bright. If the sample contains an internal defect, the projected image of the flaw will appear on the CRT as a dark area at a location corresponding to the defect in the part.

DEFECT CHARACTERIZATION

In order to establish the capability of the SLAM to detect ferrite defects, header assemblies with visually detectable cracks were examined. Figure 4 compares an optical image with the 30 MHz

*Sonoscan, Inc., Sonomicroscope-System 130, Bensenville, Il.

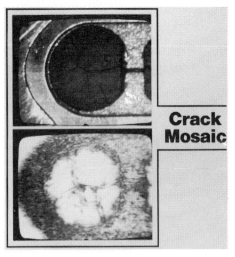

Figure 4. Comparison of Optical and Acoustic Micrographs of a Cracked
Ferrite Insert

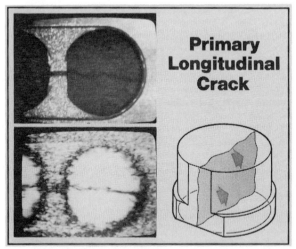

Figure 5. Optical and Acoustic Micrographs of Extended Longitudinal Cracks

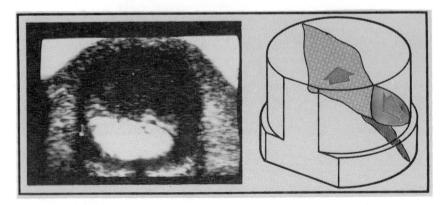

Figure 6. Acoustic Micrograph of a Lateral Crack Originating at the
Step Between the Two Dimeters of the Insert. Arrows
Indicate the Direction of Crack Propagation

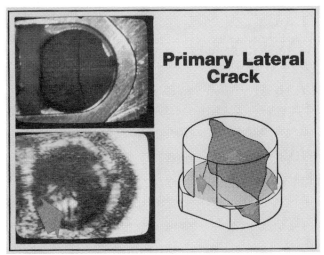

Figure 7. An Optical and Acoustic Micrograph of Another Lateral Crack
Originating at the Step. A Distinct Secondary Crack
Originated on the Primary Crack Surface and Extended to the
Surface

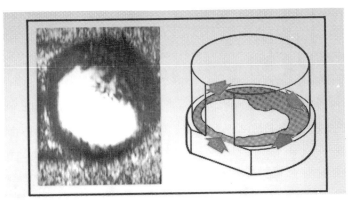

Figure 8. Subsurface Lateral Crack

Figure 9. Initiation and Propagation of a Crack by Environmental
Stresses

acoustic image of a damaged ferrite. The figure shows attenuation of the ultrasound in the sample by fracture interfaces. The shadow mosaic in the acoustic image can be directly correlated with the actual surface crack geometry.

A second example, Figure 5, shows extended longitudinal cracks across both ferrite inserts. Finite element analysis of the header assembly had identified the area near the slot located in the stainless steel housing as a region of high residual stress, and, therefore, as a potential site for crack initiation. Following examination in the SLAM the inserts were removed from the assembly. The drawing to the right in Figure 5 shows the actual location of the crack.

In Figure 6, the acoustic image shows a crack that originated at the step between the two diameters of the insert. The crack initially extended laterally before turning toward the longitudinal direction and intersecting the surface of the part. A second example, Figure 7, is another crack originating at the step. A distinct secondary crack originated on the primary crack surface and extended to the surface. The direction of propagation is indicated by arrows on the drawing.

A lateral crack which is entirely subsurface, Figure 8, may grow in a manner similiar to the two previous examples as a result of environmentally imposed stresses. An experiment was devised to test whether defects could be initiated and propagated by prototypic environmental stresses. Header ferrites were examined by SLAM both before and after the following tests:

1. Three cycles of thermal shock between −65 and 165 degrees C.

2. Random vibration in three test planes.

3. Mechanical shock testing (haversine pulses) in the axial plane of the headers at four levels of intensity.

The majority of the test samples retained their structual integrity. However, a few initially defect-free ferrites, did experience crack initiation and growth during environmental testing. An example is shown in Figure 9. The initial acoustic image in the upper left shows the insert to be free of detectable defects. After three cycles of thermal shock, evidence of a possible crack was observed. The acoustic image in the lower left of Figure 9 shows the suspect crack as a spur at the four o'clock position. Following the mechanical shock tests, the defect had grown to the distinct longitudinal crack shown in the third acoustic image.

The initial SLAM results cited in the previous examples provided impetus for a design modification of the ferrite insert. The step between the two diameters was given a larger radius to reduce stress concentration. This change significantly improved production yields by reducing the incidence of lateral cracks.

Individual manufacturing operations were systematically monitored by examining several serialized parts before and after process steps. Several types of defects were traced to the grinding operations

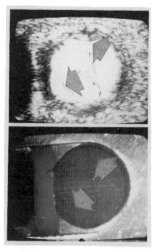

Figure 10. An Optical and Acoustic Micrograph of Damage Produced by
            Centerless Grinding

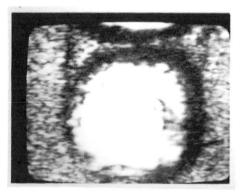

Figure 11. An Acoustic Micrograph of Damage Produced by the Final
            Surface Grind After the Welding Operation

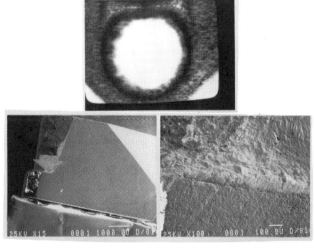

Figure 12. Acoustic Micrograph of Insert (Top). Electron Micrographs
            (Bottom) Show the Separation of the Metallization at
            Interface

143

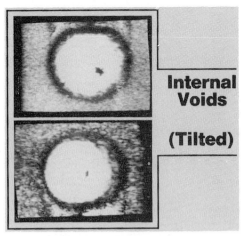

Figure 13.   Acoustic Micrograph of Internal Void in Ceramic;   Bottom
Image Shows Defect Shift When Test Fixture was Tilted

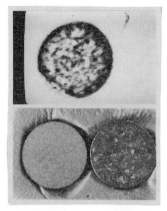

Figure 14.   Mottled Acoustic Pattern Produced by Abnormal Grain Size.
The Optical Micrograph (Below) Compares the Etched
Appearance of a Normal (Left) and Overfired (Right)
Ferrite

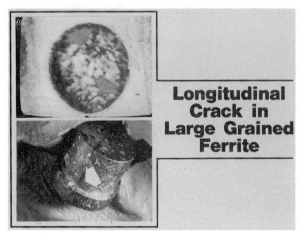

Figure 15.   Acoustic Micrograph of a Large-grained Ferrite Containing
a Thermal Shock Crack.   The Actual Appearance of the Crack
is Shown in the Lower Photograph

144

identified in Figure 2. The first O.D. grind forms a rod from a fired
blank, and Figure 10 presents both an optical and an acoustic image of
damage produced by centerless grinding. After the ferrite headers
have been electron-beam welded into the housing subassemblies, the
mating surfaces are given a final grind to establish flatness. Damage
done to the ferrite insert by this operation appears as multiple short
radial cracks around the periphery of the insert, as shown by the
acoustic image, Figure 11. In both cases, grinding parameters were
modified to reduce or eliminate these defects.

SLAM also detected another form of defect in which
vapor-deposited copper metallization had separated from under the lip
of the ferrite insert. The acoustic image in Figure 12 shows a dark
shadow band radiating a finite distance in from the outer edge of the
ferrite insert. After the ferrite insert was removed and
cross-sectioned, the scanning electron microscope was used to examine
the interface region. The electron micrographs, also in Figure 12,
clearly show the separation of the metallization at the step. This
type of defect is not regarded a technical problem.

Intrinsic defects in the ferrite material were also detected with
the SLAM during the course of this investigation. The presence of an
internal void in the ceramic is shown in Figure 13. Tilting the test
fixture slightly to the left, shifts the shadow of the defect to the
left (see lower acoustic image in Figure 13). Measuring the tilt
applied to the test specimen and the distance traversed by the defect
image permits the axial location of the void to be calculated.[5]

A number of inserts from specific ferrite lots were found to
produce acoustic images which could be characterized as "mottled",
Figure 14. Destructive cross-sectioning yielded no defects which
could account for this kind of acoustic signature. Only after the
ferrite material was etched with an acid did the origin of the mottled
images become apparent. This distinctive acoustic signature was
produced by abnormally large grains (>200 microns) which result when
the ferrite material is incorrectly fired. The lower part of Figure 14
compares the etched appearance of normal (left) and overfired (right)
ferrite.

The presence of large grains in the ferrite material poses
significant technical problems in fabricating the material into its
final shape. The strength of the ceramic is inversely proportional to
grain size[6] and Figure 15 shows an example of a crack in a
large-grained ferrite ceramic insert which had been thermally shocked.
The SLAM provided evidence that ferrite firing conditions were not
properly controlled. A grain size specification was subsequently
imposed and the material supplier took corrective action.

CONCLUSION

Scanning laser acoustic microscopy was used as a non-destructive
evaluation technique for the detection and characterization of
internal defects. This capability allowed a systematic study of
manufacturing processes, at both the piece part and next assembly
levels, and established direct correlation of acoustic images with
specific physical defects.

ACKNOWLEDGEMENTS

The authors wish to acknowledge the technical assistance provided by E.K. Beauchamp, Sandia National Laboratory-Albuquerque and B.L. Milbourn, Allied Corporation, Bendix Kansas City Division.

REFERENCES

1. A. Goldman, "Understanding Ferrites," Bulletin of Amer. Ceramic Soc. 63 (4), 582-590 (1984).

2. F. J. Schnettler, Microstructure and Processing of Ferrites, in: "Physics of Electronic Ceramic," L. L. Hench and D. B. Dove, ed., Marcel Dekker, New York (1972).

3. Sonomicroscope Users Manual, "Theory of Operation," Vol. 2.

4. D. E. Yuhas and L. W. Kessler, "Defect Characterization by Means of the Scanning Laser Acoustic Microscope (SLAM)," SEM Inc. 385-391 (1980).

5. Ibid.

6. W. D. Kingery, H. K. Bowen, and D. R. Uhlmann, "Introduction to Ceramics," John Wiley, New York (1976).

INTERACTION BETWEEN VOID PARAMETERS AND THE OUTPUTS OF THE SCANNING LASER

ACOUSTIC MICROSCOPE

G.Z. Al-Sibakhi*, M.G.M. Hussain**, and M.M. Sadek*

*Mechanical Engineering Department
**Electrical and Computer Engineering Department
Faculty of Engineering and Petroleum
Kuwait University, P.O. Box 5969 Kuwait

ABSTRACT

The Scanning Laser Acoustic Microscope (SLAM) has been used to detect and identify the geometrical configuration of void irregularity in epoxy resin bonded joints to ensure the integrity of such joints and to relate these defects to the joint strength. Practical problems were encountered in accurately defining the size and depth of voids resulting from imperfections during the bonding process. Aluminum and plastic samples with drilled holes of various geometries and shapes were tested using the SLAM. The data obtained from the SLAM are in the form of colored photos showing discrepancies in the samples, velocity pattern variation, and readings of the sonometer. Data obtained from various samples were analyzed to extract information regarding size, depth, shape, and position of the simulated defects in the samples.

In this paper, the experimental results obtained using the SLAM will be presented. A comparison between theoretical and experimental results is given. Practical problems encountered in the use of the SLAM are discussed.

INTRODUCTION

The strength of bonded joints is considerably dependent on the homogeneity of the bond line. Taking all possible precautions during the application of the bond, the presence of voids and air bubbles cannot be avoided, thus resulting in non-homogeneities in the bond line. This will cause considerable reduction in the joint strength.

It is therefore necessary to establish a relationship between the amount of non-homogeneities in the joint as measured by non-destructive testing and that measured by material testing. The stages of testing will be as follows:

1. Non-destructive testing of the bonded joint after the joint is fully cured.

2. Material testing using the SLAM.

3. Measuring the surface of the fracture line in the bond after the joint has broken using the electron microscope.

The aim of this paper is to deal with the development of the first stage, i.e. the non-destructive testing of bonded joints using the Scanning Laser Acoustic Microscope system model 130. In this technique, a magnified image of the internal structure of the tested sample is generated using very high frequency ultrasound. The objective of this investigation is to determine the geometry and location of any defects in the bonded joints. The characteristics of defects in the bond are porosity, delamination (lack of bonding), cracks, and voids. Nearly all non-destructive testing (NDT) methods are used for the inspection of adhesive bonds in industry. The majority of these techniques are applicable for the determination of the location and sizing of the gross flaws in the bond layer. These techniques will be discussed separately, along with their usefulness to detect defects in adhesive bonds:

 i. Layer Capacitance Measurement [1]. Layer capacitance measurements have been carried out on small bonded areas with some success. It does not discriminate between void and solid bond lines. It has been found to be ineffective for large surface areas, though recent investigations show that promising development can be expected.

 ii. Thermal inspection methods [2]. These are based either on thermal conductivity, emissivity, or thermal capacity. All of these are related to film thickness and density.

 iii. Acoustic Emission [3]. Elastic movements of atoms create sound at sonic and ultrasonic frequencies which can be detected with highly sensitive modern transducers. In the present application, a pre-loading stress is generated using a rig which is developed to exert a calibrated force on a well-defined bond area and which records the acoustic emission during the loading cycle. The level of loading must be chosen so that the emitted energy does not exceed a certain level. Microphones are placed at strategic positions on the component, thus identifying the acoustic emission contours.

 iv. Ultrasonic Inspection. This is a similar method to that used for inspecting solid metal components. It can take the form of transmission, pulse-echo, or ultrasonic resonance. At present, such inspection could not be used for large components. [4,5,6]

 v. Detection of Structural Failure by Vibration Monitoring. This technique is mainly developed for off-shore platforms. It is based on detecting the damage by monitoring the changes of vibration characteristics, modal shapes, and frequencies associated with severance of individual submerged structural members. [7,8,9]

The Scanning Laser Acoustic Microscope is a type of ultrasonic inspection technique. It produces a magnified image of the internal structure by impinging very high frequency ultrasound upon the object tested. Acoustic energy is transmitted through the sample by using a high frequency piezoelectric element called a transducer. After passing through the sample the sound hits the bottom of the coverslip causing ripples in the gold layer. These ripples are detected by a rapidly scanning focused laser beam. The sound is then reflected onto a photodiode which measures the intensity of the laser beam. This is turned into an electrical signal producing the acoustic image as that illustrated in

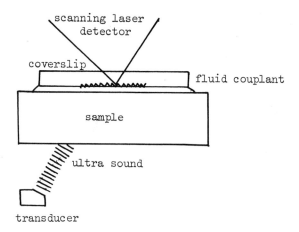

Fig. 1 SLAM Operation

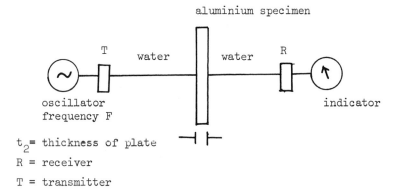

Fig. 2   Transmission through thin plate, t_2
         sample thickness, T transmitter,
         R receiver.

Fig. 1. There are two types of acoustic micrograph images: the first is the normal amplitude image in which the bright regions of the micrograph denote good acoustic transmission and dark ones to poor transmission. The second type of acoustic micrograph image is the interferogram image. This is characterized by vertical straight fringes, the spacing of which is uniform if the velocity of sound is uniform in a flat specimen. In this paper, samples of aluminum strips with defects of prescribed shapes and sizes deliberately inserted have been experimented upon to establish the capabilities of the SLAM. Correlation between experimental and theoretical data has been carried out for the relative transmission through aluminum strips, with drilled holes simulating the defects. Moreover, analysis of the interferogram images was carried out to identify the variation of the velocity pattern in relation to the sample thickness and depth of holes. This system works on either 30 MHz or 100 MHz frequencies but the former has been selected, since high penetration through the sample is obtained with lower frequency, though higher frequency will result in higher resolution. It is the objective of this paper to establish parameters which can decide the geometry and location of the voids or non-homogeneity.

GENERAL BACKGROUND OF TRANSMISSION THROUGH THIN PLATES [10,11,12,13]

A sound wave passing through a layer with a thickness which is less than the wavelength of the impinging sound will create interference. This interference results from multiple reflections within the layer causing the superposition of waves having different phases. The magnitude and direction of the interference effects depend on the phase of the waves. The effects are greatest under the following conditions: when the attenuation losses are low, when there is impedance mismatching of the surfaces i.e. poor coupling, the duration of the incident signal is less than $2t_2/v_2$. The transmission through a thin aluminum plate is demonstrated in the sketch shown in Fig. 2, which is self-explanatory. The relative transmission Tp through the sample is given in terms of the sample thickness t_2 and the wavelength λ_2, neglecting losses such as reflection and absorption:

$$\frac{W_t}{W_i} = Tp = \frac{1}{1 + \frac{1}{4}(r - \frac{1}{r})^2 \sin^2 \frac{2\pi t_2}{\lambda_2}} \tag{1}$$

It can be seen in the equation that the transmission maxima are repeated at a plate thickness of integral half wavelengths and the minima at odd quarter wavelengths.

Manipulating equation (1) will result in:

$$\frac{t_2}{\lambda_2} = \frac{1}{2\pi} \sin^{-1}\left(\sqrt{\frac{Tp}{\frac{1}{4}(r - \frac{1}{r})^2 \, Tp}} \right) \tag{2}$$

where Tp is the relative transmission.

PATTERN RECOGNITION [14,15]

In this investigation, the SLAM has been used to investigate the feasibility of identifying the geometrical configuration of drilled holes simulating prescribed non-homogeneities in bonded joints. For this purpose, the acoustic micrograph image (the normal amplitude image) has been obtained for strips of aluminum of 4 mm thickness, 25.4 mm width, and 77.01 mm length. In these experiments the ultrasonic frequency of 30 MHz

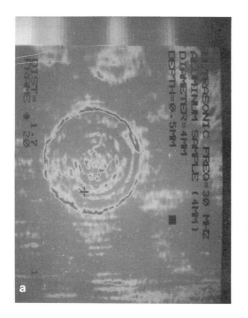
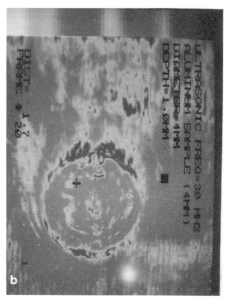
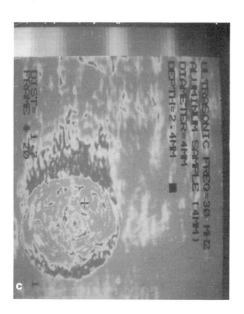
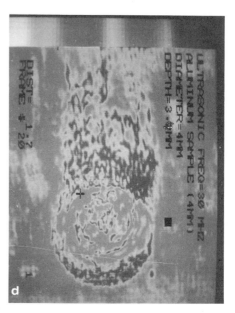

Fig. 3 Acoustic micrograph (30 MHz) of aluminum samples having
holes of diameter 4 mm. The depths of holes are:

a) 0.5 mm   b) 1.6 mm   c) 2.4 mm   d) 3.4 mm

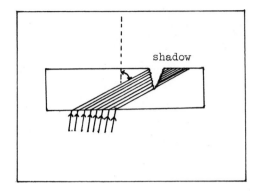

Fig. 4 Schematic diagram illustrating isonofication of
a sample having a deep surface crack. A characteristic shadow is produced in the acoustic image
whose dimensions are determined by the crack depth
and the angle of isonofication.

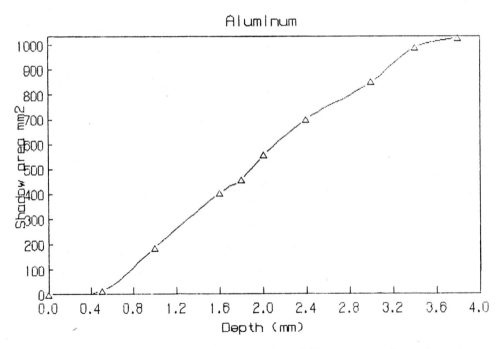

Fig. 5 The depth of the holes for differenct samples of
aluminum vs. the area of the shadow of the image.

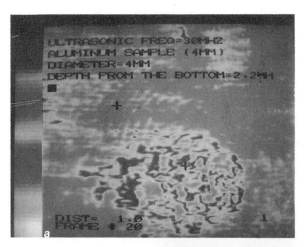

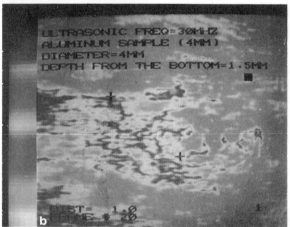

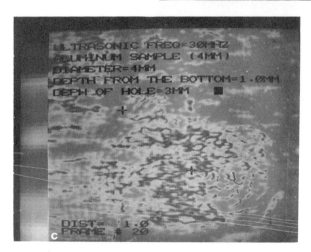

Fig. 6 Acoustic micrograph (30 MHz) of aluminum sample having holes of
        diameter 4 mm. The depth of holes at bottom is:

        a) 1.8 mm   b) 2.5 mm   c) 3.0 mm

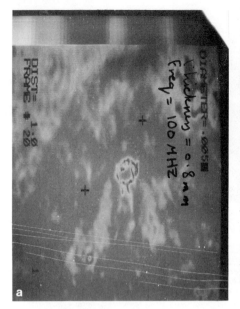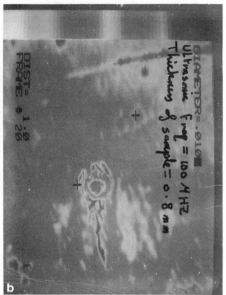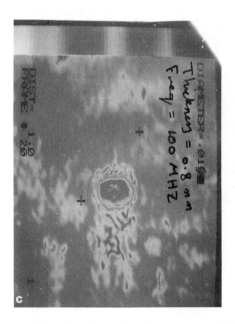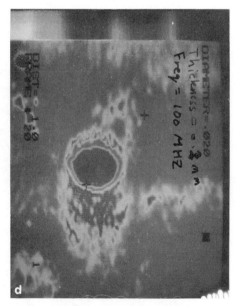

Fig. 7 Acoustic micrograph (100MHz) having through holes of
ceramic samples of thickness = .2 mm. The diameters are:

a) 0.005 mm   b) 0.01 mm   c) 0.015 mm   d) 0.02 mm

is used and the horizontal and vertical fields of view are 12 and 9 mm respectively. The normal amplitude acoustic micrograph images are obtained for a circular hole of 4 mm diameter, with various depths ranging from 0.5 mm to 3.8 mm in 8 steps. Samples of these micrograph images are presented in Fig. 3.

DEPTH MEASUREMENTS [16,17,18,19]

The characteristics of a crack, revealed by acoustic microscopy, depend on its orientation relative to the direction of sonic propagation, degree of tightness, the roughness of the crack surface, and its extension beneath the surface. The acoustic transmission loss across the fracture interface may vary from almost total attenuation revealed by a shadow (as shown in Fig. 4) to only slight variations in acoustic contrast. A relation between the depth of the hole and the shadow of the image has been determined by measuring the shadow area which is plotted versus the depth as shown in Fig. 5. This graph shows that the shadow area increases with the increased depth of the hole. The images are illustrated in Fig. 3. There is some scatter which is caused mainly by the inaccuracy in measuring the shadow area manually. This will be improved by the software of the digital image analyzer and by image enhancement. In this series of tests, the holes were facing the laser beam. In another series of tests, the sample was put upside down in such a way that the holes were not directly facing the laser beam. The images obtained in this case are shown in Fig. 6. Measuring the shadow area in the same manner as previously explained results in the same relationship between the depth of the hole and shadow area. In a similar manner, ceramic samples with different holes were tested but with 100 MHz frequency. The images of these are shown in Fig. 7 giving a similar relationship between the shadow area and depth. This feasibility study shows that the depth of the hole which is simulating the void in the bonded joints can be identified by this shadow of the image.

AN ANALYSIS OF THE SCANNING LASER ACOUSTIC MICROSCOPE INTERFEROGRAM IMAGES [15,20,21,22,23]

An acoustic interferogram of equal phase wavefronts is produced with the Scanning Laser Acoustic Microscope. The interferogram contains information about the thickness of the specimen and the depth of the hole. The vertical interference lines represent equal phase wavefronts after the sound wave has passed through the specimen. This spatial map of the phase information of the acoustic wave yields the relative sound velocity variations in the specimen. In Fig. 8, a sample of aluminum having a thickness of 4 mm and velocity greater than that of the surrounding distilled water is the reference medium. This has been positioned horizontally across the center of the interferogram. Note that the interference lines are shifted to the left as they pass from the reference medium. The horizontal and vertical fields of view with 30 MHz transducer are 12 mm and 9 mm vertically. There are approximately 30 interference lines, which are spaced at intervals of about 85 MHz. In the spatial domain technique (sdt), the specimen's velocity is determined from the following expression:

$$C_x = (c_o/\sin \theta_o)\sin \left[\tan^{-1}(1/\tan \theta_o - n \lambda_o/t \sin \theta_o) \right] \qquad (3)$$

The normalized fringe shift N is determined by dividing the horizontal through which the fringe has moved from the reference medium by the horizontal distance between the interference lines. Samples of aluminum with different thickness and without holes have been tested by the SLAM using the interferogram images. The normal fringe shift N is determined from the images by:

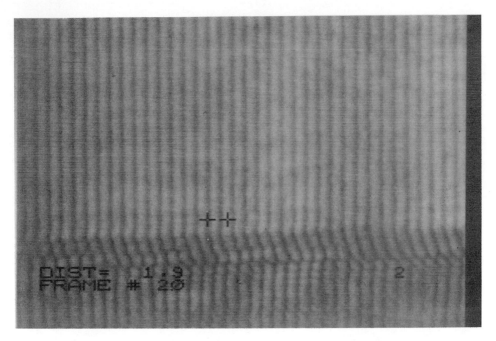

Fig. 8 Interferogram photograph from the SLAM monitor specimen is aluminum with thickness of 4 mm and it is positioned horizontally in the middle of the screen. Field of view is 12 mm horizontally by 9 mm vertically.

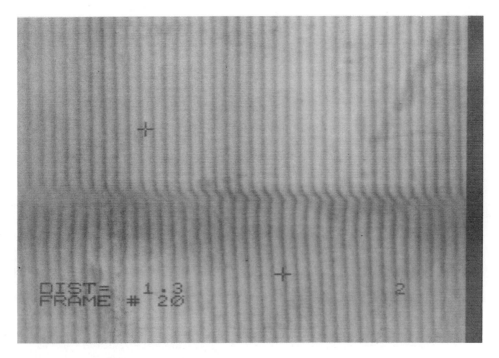

Fig. 9 Interferogram photograph from the SLAM monitor specimen of aluminum with thickness of 1 mm and it is positioned horizontally in the middle of the screen. Field of view is 12 mm horizontally by 9 mm vertically.

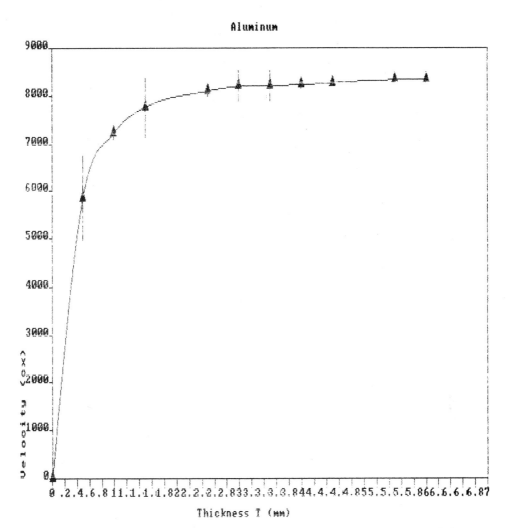

Fig. 10 The velocity cs of different samples of aluminum with minimum
and maximum standard deviation by taking the readings 10
times at every thickness vs. the thickness of all samples.

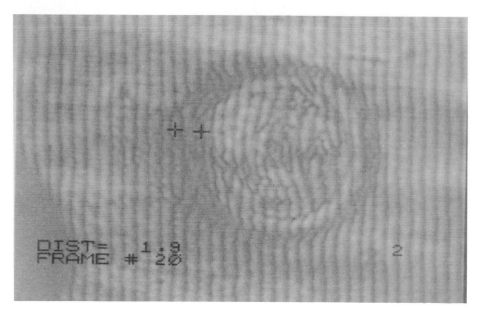

Fig. 11   Interferogram photograph from the SLAM monitor for aluminum
specimen of thickness 4 mm and hole having diameter 4 mm
and depth 1.6 mm.

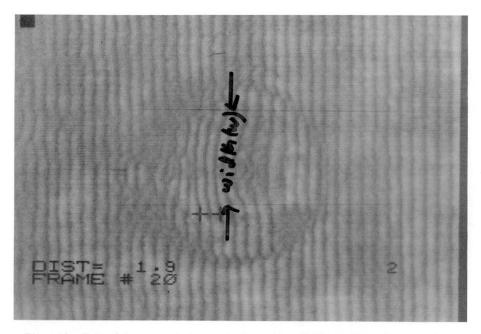

Fig. 12   Interferogram photograph from the SLAM monitor for aluminum
specimen of thickness 4 mm and hole of diameter 4 mm and
depth 1 mm.

$$N = \frac{\text{normalized distance that the fringe has shifted}}{\text{horizontal distance between the interference lines}}$$

while the velocity C_x is determined using equation 3. A sample of the interferogram image for 1 mm thickness aluminum specimen is shown in Fig. 9.

This shows that the velocity is constant until it reaches the aluminum sample where the velocity increases. The lower level represents the velocity in medium 1 (water), the higher level represents the sound wave in the aluminum sample. For each specimen, each test is repeated ten times, the velocity of the sound wave in the aluminum specimen is determined and the average, together with the minimum and maximum standard deviation, were plotted. This velocity plot is shown in Fig. 10. This figure superimposes at each point the standard deviation as represented by the vertical lines. This shows that the velocity increases linearly with the thickness of 0.5 mm, after which the rate of increase is reduced until it reaches an asymptotic value of approximately 3500 m/s. Further tests were carried out on aluminum samples of 4 mm thickness, having drilled holes of different depths, and the interferogram images were obtained. Samples of these are shown in Fig. 11. It is noticed from this image that the interferogram lines shift to the left when passing from reference media and shift even further when passing through the hole. The difference in this shift can be determined by measuring the width of the interferogram line through the hole, as shown in Fig. 12-13. The variation of the interferogram width has been plotted versus the depth is shown in Fig. 14. This shows that as the width increases, the depth increases. This study shows that the sudden changes in the interferogram lines can be considered as a measure for the depth of the hole when the laser beam is facing the hole. The size of the diameter of the hole can also be determined by a magnifying factor which can be determined.

COMPARISON OF EXPERIMENTAL AND THEORETICAL DATA FOR THE RELATIVE TRANSMISSION VERSUS DEPTH USING SLAM [24,25]

The relative transmission through an aluminum thin sheet (medium 2) of thickness t_2 surrounded with water (medium 1) as shown in Fig. 2, can be predicted theoretically from Equation 1. It can be detected experimentally using the sonometer of the SLAM, as will be explained later. Using Equation 1, it can be shown that the relative transmission "Tp" can be predicted if the thickness of aluminum sample "t_2" is less than either 0.84 mm or 0.25 mm for ultrasonic frequencies of 30 MHz or 100 MHz respectively. The variation of the relative transmission "Tp" is plotted in Fig. 15 versus the thickness t_2 in the non-dimensional form $t_2/0.5\lambda_2$ for 30 MHz. This curve contains loops, the maximum of which is at -30 dB and the minimum zero. It should be noted that this curve is for an angle of incident of 90° being the case which has been experimented upon. Experimentally, the relative transmission factor is measured by the sonometer by correlating the strength of the reflected signal, which is represented by the strength of the ripples in the gold layer of the coverslip sensed by the laser beam to that of the incident signal. The measurements are graphically displayed within the data window by being superimposed upon the acoustic image. Both the acoustic image and the sonometer output are simultaneously visible. In this experiment aluminum samples of thickness ranging from 0.1 to 0.9 mm were measured using the sonometer of the SLAM at a frequency of 100 MHz. The measured value of the average transmission "Tp" in dB is plotted in Fig. 15 versus the thickness of the aluminum strip in the non-dimensional form "$t_2/0.5\lambda_2$". The experimental points are superimposed on the figure by triangles. This shows reasonable correlation between theoretical and experimental points. It can thus be concluded that measurement of the transmission can be used to determine the thickness of the strips.

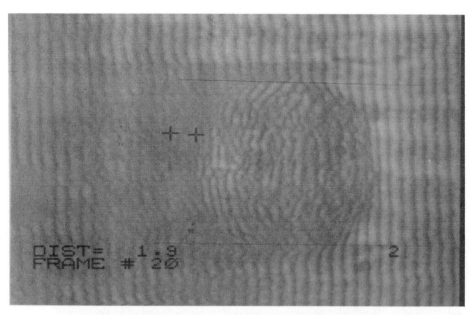

Fig. 13 Interferogram photograph from the SLAM monitor for aluminum specimen of thickness 4 mm and the hole of diameter 4 mm and depth of 3.5 mm.

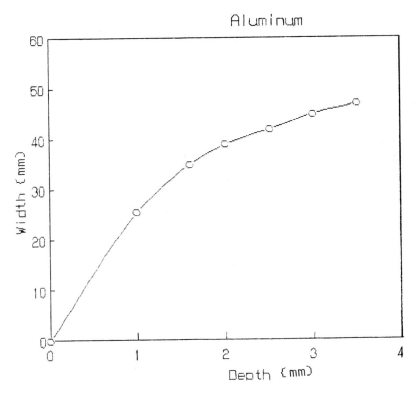

Fig. 14 The depth of holes for samples of aluminum vs. the width
of the interferogram lines through the hole.

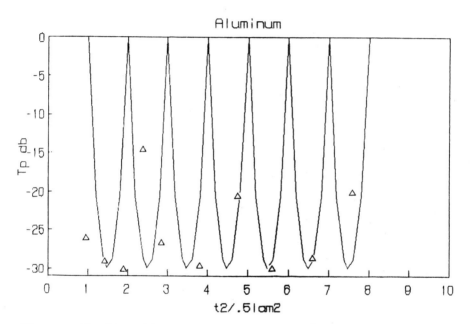

Fig. 15 The theoretical values of the transmission Tp vs. the thickness of different samples t_2 of aluminium over the wavelength of AL ($t_2/0.5\ \lambda_2$). The experimental values for the same plot are superimposed on the theoretical values by triangular points.

CONCLUSION

In this paper, a feasibility study has been carried out to explore the scope of identification of voids or non-homogeneities in samples. Four techniques were used. These are: pattern recognition, shadow area, the interferogram images, and finally the measurement of the relative transmission using the sonometer. The parameters required to be identified for any voids are the location of the void, the void area, and the depth of the void in the material. These are simulated in the experiments conducted in this investigation by the depth and area of the drilled hole and the thickness of the sample. The pattern recognition and the shadow area were used successfully in determining the depth of the hole, while the analysis of the interferogram images was successfully used by measuring the velocity changes and width of the interferogram line through the hole respectively. The thickness was measured by the measurement of transmission. The inaccuracy in these measurements is due to visual identification. The sensitivity of measurements can be greatly improved by using the combination of image analysis and enhancement. The next stage in this work will be to develop a technique for identifying the voids in bonded samples.

NOTATIONS

ρ = density kg/m^3
v = velocity of the wave m/sec
λ_2 = v_2/f = wavelength of medium 2
w_i = incident beam intensity
w_t = transmitted beam intensity
z_1 = impedance of medium 1 = $\rho_1 v_1$
z_2 = impedance of medium 2 = $\rho_2 v_2$
r = z_2/z_1
c_o = velocity in the medium surrounding the specimen
θ_o = angle from the normal of the acoustic mean in the reference medium
N = normalized fringe shift
T = specimen thickness
λ_o = wavelength of the sound in the surrounding medium

ACKNOWLEDGMENT

This work has been carried out within the large grant EM 023 awarded to the third author by Kuwait University Research Council.

REFERENCES

1. D. Hudson, "The Use of Automatic Dielectroetry of Auto-Clave Molding of Low Void Composites", Composites, Vol. 4, May 1974, p.247.
2. P.R. Vettito, "A Thermal I.R. Inspection Technique for Bond Flaw Inspection", Appl. Polymer Symp., No.3, p. 435, Interscience Publication, New York, 1966.
3. W. Brockmann, "Acoustic Emission - A Test Method for Metal Adhesive Bonded Joints", Material Prufung, 19, October 1977, p. 430.
4. R.S. Sharpe, "Non-Destructive Ultrasonic Testing", C.M.F., I.Mech.E., London, June 1979, pp. 65-68.
5. B.H. Lidington, M.G. Silk, "Ultrasonic Testing Measurements of the Depth of Fatigue Cracks", A.E.R.E., Report, No R.8190, 1976.
6. J. Szilard, "Ultrasonic Testing. Non-Conventional Testing Techniques", John Wiley & Sons, 1982.

7. P. Lepert, J.Y. Heas, P. Narzul, "Vibro-Detection Applied to Off Shore Platforms, presented at the 12th Annual O.T.C., Houston, Texas, May 5-8, 1980, pp. 627-634.

8. R.M. Kenley, C.J. Dodds, "Detection of Damage Response Measurements", presented at 12th Annual O.R.C., Houston, Texas, May 1980, pp. 111-118.

9. R.A. Allen, "Structural Integrity Monitoring Off Shore Steel Platforms", presented at Integrity Off Shore Structures Conference, Glasgow, July 1981.

10. R.D. Ford, "Introduction to Acoustics", Elsevier Publishing Co., Ltd.,Amsterdam, London, New York, 1970.

11. Kinsler, Frey, Coppers, and Sanders, "Fundamentals of Acoustics", Wiley, 1982.

12. V.P. Severdenko, V..V. Klubovich, and A.V. Stepanenko, "Ultrasonic Rolling and Drawing of Metals", special report translated from Russian.

13. Robert C. McMaster, "Non-Destructive Testing Handbook", Edited for the Society of Non-Destructive Testing, Vol.II, The Tonald Press Co., NY, 1959.

14. Abdullah Atalar, "Penetration Depth of the Scanning Acoustic Microscope", IEEE Transactions on Sonics and Ultrasonics, Vol. SU-37, No.2, March 1985.

15. K.M.U. Tervola and William D. O'Brien, Jr., "Spatial Frequency Domain Technique: An Approach for Analyzing the Scanning Laser Acoustic Micro-scope Interferogram Images", IEEE Transaction on Sonics and Ultrasonics, Vol. SU-32, No.4, July 1985.

16. Glen Wade, "Acoustical Holography", Vol 4, California, April 1972.

17. M. Karen, R.K. Muller, and J.F. Greenleaf, "Acoustical Imaging", Vol. 13, Plenum Publishing Corp., 1984.

18. F.J. Fry, "Methods and Phenomena 3. Ultrasound: Its Applications in Medicine and Biology", Part I, Elsevier Scientific Publishing Co., 1978.

19. F.J. Fry, "Methods and Phenomena 3. Ultrasound: Its Applications in Medicine and Biology", Part II, Elsevier Scientific Publishing Co., 1978.

20. Paul M. Embree, K.M.U. Tervola, Steven G. Foster, and William D. O'Brien, Jr., "Spatial Distribution of the Speed of Sound in Biological Materials with Scanning Laser Acoustic Microscope", IEEE Transactions on Sonics and Ultrasonics, Vol. SU-32, No. 2, March 1985.

21. K.M.U. Tervola, M.A. Gummer, J.W. Erdman, Jr., and W.D. O'Brien, Jr., "Ultrasonic Attenuation and Velocity Properties in Rat Liner as a Function of Fat Concentration. A Study at 100 MHz Using Scanning Laser Acoustic Microscope", J. Acoust. Soc. Am., 77 (1), January 1985.

22. Charles A. Edwards and William D. O'Brien, Jr., "Speed of Sound in Mammalian Tendon Treads Using Various Reference Media", IEEE Transactions on Sonics and Ultrasonics, Vol. SU-32, No. 2, March 1985.

23. S.A. Goss and W.D. O'Brien, Jr., "Direct Ultrasonic Velocity Measurements of Mamalian Collagen Threads", Bioacoustics Research Laboratory, University of Illinois, Urbana, August, 1978.

24. M.G. Oravecz, "Quantitative Scanning Laser Acoustic Microscopy: Attenuation", Sonoscan, Inc., Bensenville, Dec. 1985.

25. Thomas, Graham Havens, "An Ultrasonic Design Tool and Experimental Test Bed for the Prediction of Adhesive Bond Strength", 1979.

TIME-OF-FLIGHT APPROXIMATION FOR MEDICAL ULTRASONIC IMAGING

Steven R. Broadstone and R. Martin Arthur

Biomedical Computer Laboratory and
Department of Electrical Engineering
Washington University, St. Louis, Missouri 63130

ABSTRACT

Solutions of the acoustic wave equation linearized using the Born assumptions require determination of the time of flight from the transmitting transducer to the field point and back to the receiver. The time of flight within a plane in a medium that has a constant phase velocity is found by taking the square root of the sum of squares of scaled range and azimuth distances. The paraxial approximation significantly reduces calculation time and provides good estimates near the propagation axis of the transducer and at long range. We matched the time-of-flight profile with a second-degree, two-dimensional polynomial containing 9 coefficients that was more accurate than the paraxial approximation. Coefficients were found using the method of moments. In its forward difference form the polynomial can produce a new time-of-flight approximation in a single-addition time. If the division in the paraxial approximation is replaced by a scale factor, the result can also be expressed as a forward difference and evaluated in a single-addition time. Assuming a 1500 m/s background velocity, the maximum errors over a 3 x 3 cm region on the transducer axis fell below 1/3 the wavelength of a 3.5 MHz signal for ranges greater than 3 and 22 cm, for the first-difference form of the moments and paraxial approximations, respectively. At angles up to 30 degrees off axis the moments error fell below 1/3 wavelength beyond a range of 6 cm. At an angle of 15 degrees, the usual limit for the paraxial approximation, its maximum error was more than 10 wavelengths at distances up to 30 cm.

INTRODUCTION

The objective of the inverse scattering problem is extraction of physical properties of the medium of interest using scattered energy. Three approaches to solving problems of this type exist. Solutions can be found iteratively (generalized inversion), by exact direct inversion, or by approximate direct inversion[1]. In this paper we analyze two methods for finding approximate direct solutions to two-dimensional inverse scattering problems as they apply to medical imaging using ultrasound. Specifically we compare the well-known paraxial approximation[2,3] to a new approximation based on the method of moments for accuracy and speed.

Speed of digital image generation based on solution of the linearized wave equation is proportional to the number and type of arithmetic operations required to backproject received ultrasonic energy. Ideally, a pixel value should be modified by backscattered signals in the time of a single arithmetic operation. We compared both paraxial and moment approximations in forward-difference formulations which can be computed in a single-addition time.

APPROXIMATE DIRECT SOLUTION OF THE ACOUSTIC WAVE EQUATION

In a constant-density elastic medium the homogeneous wave equation can be written as[2,4]

$$\nabla^2 p = \frac{n^2}{c_0^2} \frac{\partial^2 p}{\partial t^2} \quad , \tag{1}$$

where p represents the field-pressure and $n = c_0/c$, the ratio of the background velocity to the phase velocity of the medium.

By conservation of energy in a passive system, the total pressure is the sum of the incident and scattered pressures:

$$p = p_i + p_s \approx p_i \quad . \tag{2}$$

Using the identity for n^2 which defines q the scattering potential within the region of support

$$n^2 = 1 + \frac{c_0^2 - c^2}{c^2} = 1 + q \quad , \tag{3}$$

and the Born assumption[5-7] that the variance of c is small, so that the scattering potential is small and therefore $p_s \ll p_i$, allows us to write the wave equation as

$$\nabla^2 p_s - \frac{\partial^2 p_s}{c_0^2 \partial t^2} \approx \frac{c_0^2 - c^2}{c_0^2 c^2} \frac{\partial^2 p_i}{\partial t^2} \quad . \tag{4}$$

The solution of the linearized wave equation is given by[5,6,8]

$$p_s \approx \frac{1}{c_0^2} \int_V \frac{c_0^2 - c^2}{c^2} g_0 \, p_i \, dV \quad . \tag{5}$$

The term g_0 is the Green's function associated with a point source in a homogeneous medium. Inversion of the approximate direct solution in the equation above yields q the scattering potential given in equation (3) from measurements of the scattered pressure p_s.

APPROXIMATE DIRECT SOLUTION METHODS

Inversion of the approximate direct solution of equation (5) in the time domain results in the mapping of backscattered signals to spatial locations which are consistent with measured times of flight[5,6]. Application of this geometrical or ray-tracing interpretation of equation (5) requires computing the time of flight T from the ultrasonic transmitter to a field point of interest and back to the receiver.

$$T = \frac{1}{c_0} (r_1 + r_2) \quad , \tag{6}$$

where c_0 is the background velocity[9],

$$r_1 = \sqrt{(x - x_t)^2 + (z - z_t)^2} \quad , \tag{7}$$

$$r_2 = \sqrt{(x - x_r)^2 + (z - z_r)^2} \quad , \tag{8}$$

166

(x_t, z_t) is the location of the transmitting transducer and (x_r, z_r) the location of the receiving transducer.

If ultrasonic energy is transmitted from and detected by the same transducer, arcs of constant time of flight in two dimensions are circles centered at the transducer[5]. When the transmitting and receiving transducers are not the same, arcs of constant time of flight are ellipses whose foci are the locations of the transmitting and receiving transducers.

Unfortunately, because of the number of arithmetic operations which includes computation of squares and square roots, calculation of the time of flight is a time-consuming task. To obtain faster reconstructions the time-of-flight computation has often been approximated to reduce the number and complexity of arithmetic operations. The cost of using simpler expressions for the time of flight is, of course, inaccuracy in the time-of-flight calculation.

Paraxial Approximation

The exact form of the time of flight can be rewritten as a binomial expansion. Truncation of the expansion after the constant and linear terms yields the paraxial approximation[2,3] to the time of flight,

$$\hat{T}_p = \frac{1}{c_0} \left(z + \frac{x^2}{2z} \right) \ . \tag{9}$$

Although possessing fewer arithmetic operations than the exact expression for the time of flight, the computation of T using equation (9) still involves division, multiplication, and addition. Furthermore, as its name implies, it is accurate only near the propagation axis of the transducer. Accuracy of the approximation can be improved by keeping higher-degree terms in the expansion at the expense of more arithmetic operations.

Method-of-Moments Approximation

An alternate time-of-flight approximation possessing arbitrary accuracy can be found by expanding the time-of-flight profile in terms of its spatial moments. Using this approach, time-of-flight profiles can be approximated by a two-dimensional polynomial in azimuth and range

$$\hat{T}_m = \frac{1}{c_0} \sum_{k=0}^{K} \sum_{l=0}^{L} a_{k,l} \, x^k \, z^l \ , \tag{10}$$

where x is the azimuth and z the range. Coefficients are found by minimizing the mean-square error between the polynomial and the exact time-of-flight profile.

$$a_{k,l} = \int_{Z_1}^{Z_2} \int_{X_1}^{X_2} x^k \, z^l \sqrt{x^2 + z^2} \ dx \ dz \ . \tag{11}$$

For most regions of interest the time-of-flight profiles are relatively smooth. Consequently, a second-degree polynomial containing 9 coefficients was chosen for the method-of-moments approximation in this study.

$$\hat{T}_{m_{deg2}} = \frac{1}{c_0} \left(a_{2,2} \, x^2 z^2 + a_{2,1} \, x^2 z + a_{2,0} \, x^2 + a_{1,2} \, xz^2 + a_{1,1} \, xz + a_{1,0} \, x + a_{0,2} \, z^2 + a_{0,1} \, z + a_{0,0} \right) \ . \tag{12}$$

Forward-Difference Expansion

Because the method-of-moments polynomial possesses positive integer powers in x and z, it can be formulated as a stable forward-difference equation in the discrete case with a finite number of terms. Given an nth degree polynomial, $(n+1)^2$ forward differences describe the time-of-flight profile. We previously described an application-specific integrated circuit which could implement a 9-coefficient, forward-difference equation in a single-addition time[10]. The circuit must be seeded with forward differences (9 for a second-degree polynomial). It then calculates approximate values for round-trip times as sample addresses within memory containing backscattered signals. Because only 9 coefficients are required to map a 512 x 512 field of view, this structure requires loading 262,135 fewer terms than a look-up table to map the same region.

This circuit can also implement the paraxial approximation in a single-addition time if the paraxial approximation is modified to contain a finite number of integer powers of x and z. A simple way to modify the paraxial approximation for single-addition operation is to eliminate the division by the range variable.

$$\hat{T}_p \approx \frac{1}{c_0} (z + \frac{x^2}{2 z_0}) \; . \tag{13}$$

With the denominator of the second term in the paraxial approximation fixed, the paraxial approximation can be formulated as stable forward-difference equation containing 4 terms. These terms are a constant, first differences in x and in z, and a second difference in x.

Including the next higher-degree term from the binomial expansion in the time-of-flight approximation requires expanding the parallel structure required to implement the forward-difference equation in a single-addition time[10]. The structure must be increased from 9 to 25 coefficients because the resultant approximation is fourth degree in x.

APPROXIMATION ERRORS

For comparison, each approximation was implemented for regions of interest centered about the range axis and for regions 15 degrees off the range axis (see Figure 1). Fifteen degrees is often assumed to be the off-axis limit of the paraxial approximation.

A typical center frequency for medical ultrasonic transducers is 3.5 MHz. A 512 x 512 image reconstructed from backscattered signals sampled at 12.5 MHz, which is a suitable aperiodic sampling rate for a 3.5 MHz transducer, is about 3 x 3 cm. The maximum error in the spatial location of a backprojected sample, was determined over 3 x 3 cm regions of interest for distances from 0 to 30 cms.

Figure 2 shows the maximum approximation error over 3 x 3 cm regions of interest on axis and at 15 degrees for the paraxial approximation and for the method of moments using a second-degree, two-dimensional polynomial. Although the method of moments is a consistently better approximation, we note that the error for both approximations is largest for regions near the transducer. An aperiodic sample rate of 1/3 the center frequency wavelength provides a measure of the useful range of the approximations. Note that the 1/3 wavelength line in all figures is its backprojected value which is half that in the medium. This criterion leads

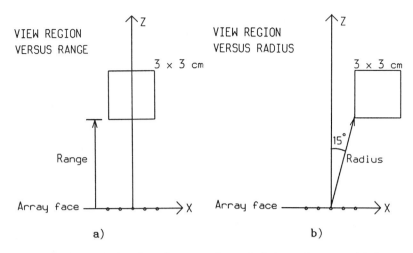

Fig. 1. Location of the 3 x 3 cm region of interest over which
approximations were tested. On axis (a) approximation errors
were found for ranges from 0 to 30 cms. At 15 degrees (b) the
errors were found from 0 to 30 cm where the distance was measured
from the origin to the nearest point in the region.

to a fully useful range where the error is below the 1/3 wavelength value.
The range is fully useful in that the criterion is met for every pixel in
the 3 x 3 cm region of interest.

In Figure 3 we compare the forward-difference forms of the paraxial
and method-of-moments approximations, i.e., forms that can be implemented
in single-addition times. The forward-difference form of the moments
approximation with a minimum fully useful range of 3 cm on axis and 6 cm at
15 degrees is the same as that shown in Figure 2. In the forward-
difference form the paraxial useful range increased to 22 cm on axis and
was > 30 cm at 15 degrees. These results for the forward-difference forms
of the approximations are summarized and extended to 30 degrees in Table 1.

Enlarging the extent of the time-of-flight profile increases the error
for both approximations as shown in Figure 4. The error in the moment
approximation with a second-degree polynomial is, however, more than 12 dB
better than the paraxial approximation for a 12 x 12 cm region of interest
on axis. Note that the wavelength in the minimum-range criterion has been
increased to a value consistent with demodulating a 3.5 MHz transducer
signal. Otherwise the number of pixels in the image must be increased from
512 x 512 to 2048 x 2048, i.e., the sample rate must be maintained to avoid
aliasing.

The mean value of the moment-approximation errors for both 3 x 3 and
12 x 12 cm regions of interest was zero. RMS values of the error are
plotted in Figure 5. If the 1/3 wavelength criterion is applied to the RMS
error the minimum useful range for a 3 x 3 cm region is about 1 cm. It is
about 3 cm for a 12 x 12 cm region of interest.

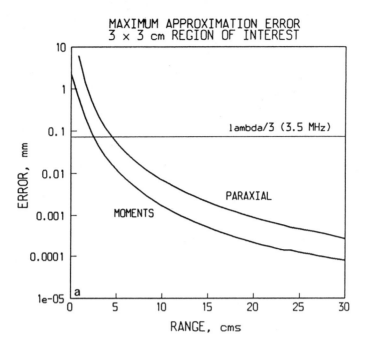

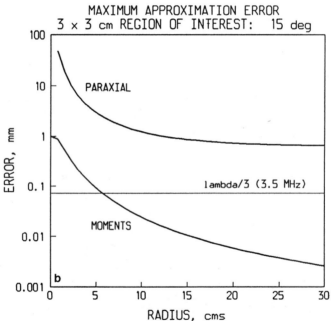

Fig. 2. Maximum approximation errors (a) on axis and (b) at 15 degrees
for the paraxial and moments approximations over a 3 x 3 cm
region of interest. The paraxial approximation error was
calculated using the constant and linear terms in the binomial
expansion of the time-of-flight profile. A second-degree
polynomial in azimuth and range was used to fit the time-of-
flight profile on a least-squares-error basis for the method-of-
moments approximation.

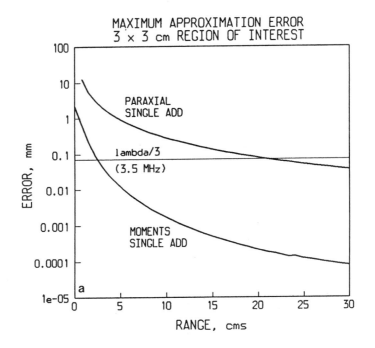

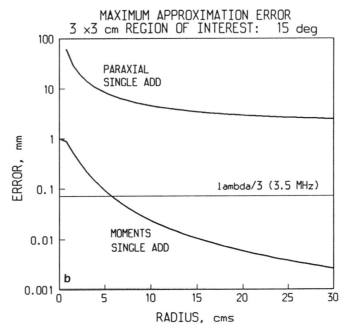

Fig. 3. Maximum approximation errors (a) on axis and (b) at 15 degrees
for the forward-difference forms of the paraxial and moments
approximations over a 3 x 3 cm region of interest. The paraxial
approximation error was calculated with the division by range in
the linear term of the binomial expansion of the time of flight
replaced by a scale factor. A second-degree polynomial in
azimuth and range was used to fit the time-of-flight profile on a
least-squares-error basis for the moments approximation.

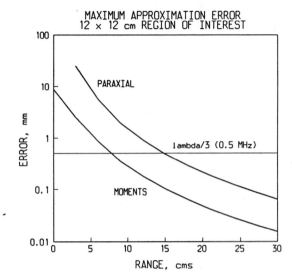

Fig. 4. Maximum approximation errors on axis for the paraxial and moments approximations over a 12 x 12 cm region of interest. The paraxial approximation error was calculated using the constant and linear terms in the binomial expansion of the time-of-flight profile. A second-degree polynomial in azimuth and range was used to fit the time-of-flight profile on a least-squares-error basis for the method-of-moments approximation.

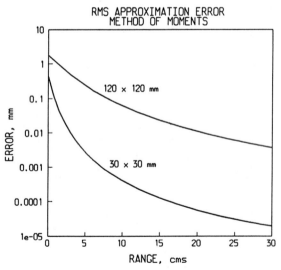

Fig. 5. RMS approximation errors on axis for the moments approximation over 3 x 3 and 12 x 12 cm regions of interest. Mean value of the moments error was zero at all ranges for both size regions. A second-degree polynomial in azimuth and range was used to fit the time-of-flight profile on a least-squares-error basis for the method-of-moments approximation.

Table 1. Minimum fully-useful[1] range: transducer to 3 x 3 cm view region

	On axis	15 degrees	30 degrees
Method of moments: Second-degree polynomial[2]	3.0	6.0	6.0 cm
Paraxial approximation[2]	22.0	> 30	-----

[1] Maximum error less than 1/3 wavelength at 3.5 MHz
[2] Forward-difference form (single-addition time)

CONCLUSIONS

The paraxial and moments approximations are comparable for small regions of interest on axis. For off-axis regions of interest, however, the method-of-moments error is not much worse than the on-axis case, whereas the paraxial approximation degrades rapidly with off-axis angle. The method-of-moments approximation with a second-degree polynomial therefore in general provides a much more accurate representation of the exact time of flight than the paraxial approximation.

New time-of-flight values can be found in a single-addition time using the forward-difference forms of both the paraxial and method-of-moments approximations. The cost of reducing the paraxial computation to a single-addition time is a substantial increase in error. Method-of-moments approximations of a given degree can be found in single-addition times with no increase in error compared to a least-squares-error fit of the time-of-flight profile.

ACKNOWLEDGEMENTS

This work was supported in part by the NIH under Grants RR01362 and RR01380 from the Division of Research Resources and by Washington University.

REFERENCES

1. F. Calogero and A. Degasperis, "Spectral Transform and Solitons: Tools to Solve and Investigate Nonlinear Evolution Equations," Vol. I, North-Holland Publishing Company, New York (1982).

2. J. F. Claerbout, "Imaging the Earth's Interior," Blackwell Scientific Publishing Company, London (1985).

3. P. D. Corl, G. S. Kino, C. S. DeSilets, and P. M. Grant, A digital synthetic focus acoustic imaging system, in: "Acoustic Imaging," 8:39-53, A. F. Metherell, ed., Plenum Press, New York (1980).

4. P. M. Morse and K. U. Ingard, "Theoretical Acoustics," McGraw-Hill, St. Louis (1968).

5. J. A. Fawcett, Inversion of N-Dimensional Spherical Averages, SIAM J. of Appl. Math., 45:336-341 (1985).

6. A. Ozbek and B. C. Levy, Inversion of Parabolic and Paraboloidal Projections, IEEE Trans. on Acoust., Speech and Sig. Proc., in press (1987).

7. M. Kaveh, M . Soumekh, and R. K. Mueller, A Comparison of Born and Rytov Approximations in Acoustic Tomography, in: "Acoustic Imaging," 11:325–335, J. P. Powers ed., Plenum Press, New York (1982).

8. P. Laiily, Migration methods: Partial but efficient solutions to the seismic inverse problem, in: "Inverse Problems of Acoustic and Elastic Waves," pp. 182–214, SIAM, Philadelphia (1984).

9. D. de Vries, and A. J. Berkhout, A note on the effect of velocity errors in computerized acoustic focusing techniques, J. Acoust. Soc. Am., 74:353–356 (1983).

10. S. R. Broadstone and R. M. Arthur, An approach to real-time reflection tomography using the complete dataset, in: "IEEE 1986 Ultrasonics Symposium Proceedings," 2:829–831, ed. B. R. McAvoy, IEEE No. 86CH2375-4, New York (1986).

MULTIPLANE DECONVOLUTION IN ORTHOGRAPHIC

ULTRASONIC TRANSMISSION IMAGING

Zse-Cherng Lin, Philip S. Green, Joel F. Jensen, and Marcel Arditi[*]

SRI International
333 Ravenswood Avenue
Menlo Park, California 94025

ABSTRACT

Orthographic ultrasonic transmission imaging has received renewed interest and demonstrated significant potential clinical applicability during the past few years. The ultrasonic camera developed at SRI for the Orthopaedic Clinic of the University of Müenster in West Germany is an example of a real-time ultrasonic transmission imaging system. Using a spatially and temporally incoherent method of insonification, which eliminates diffraction artifacts and smoothly blurs out-of-focus structures, the SRI camera produces sharply focused images within the focal plane. Although well-focused images can be obtained in orthographic transmission imaging, the overlapping of out-of-focus structures frequently detracts from image quality. To overcome this difficulty and to extend the applicability of ultrasonic transmission imaging, we proposed to apply image-processing techniques to remove out-of-focus structures through multiplane deconvolution. We used a linear model for the imaging process and investigated different algorithms for the multiplane deconvolution. Our simulation results show that an algorithm based on the minimum mean square error appears to be the most effective. We have also applied the algorithm to improve the images collected with an experimental ultrasonic transmission imaging system. In this paper, we briefly describe the principles of multiplane deconvulution and present simulation and experimental results that illustrate its effectiveness in improving the image quality.

INTRODUCTION

Orthographic ultrasonic transmission imaging can provide unique diagnostic information in a number of clinical applications, especially in orthopedics, pediatrics, breast cancer screening, and abdominal imaging.[1-5] It has received renewed interest during the past few years. The ultrasonic camera developed at SRI International for the Orthopaedic Clinic of the University of Müenster in West Germany[4-5] is an example of a real-time transmission imaging system. The SRI camera uses a spatially and temporally incoherent method of insonification, which eliminates diffraction artifacts and blurs out-of-focus structures, to produce sharply focused images within the focal plane.

---

[*]M. Arditi is now with Battelle Memorial Institute, Geneva, Switzerland.

In general, the propagation of ultrasound through an imaged object is very complicated. Energy absorption, refraction, scattering, and reflection can each contribute to the attenuation of ultrasound. In an orthographic transmission imaging system, it is usually assumed that the displayed images represent—in some way—the relative attenuation of ultrasound waves traversing the object focal plane. For an object with longitudinal extent, the out-of-focus object planes also contribute to the ultrasound attenuation. Although well-focused images can be obtained in orthographic transmission imaging, the overlapping of out-of-focus structures frequently makes the images hard to interpret.

To overcome this difficulty and to extend the applicability of ultrasonic transmission imaging, we proposed to apply image-processing techniques to remove the out-of-focus structures through multiplane deconvolution. Our general approach is as follows: We experimentally determined the system-transfer functions of our laboratory transmission imaging system, using a linear, time- and space-invariant model of the system. By means of the transfer functions and a linearized model for the propagation of ultrasound through objects of finite thickness, we were then able to write a system of linear equations to describe the formation of transmission images. In this way, we have reduced the problem of multiplane deconvolution to the simpler problem of solving a system of linear equations.

It has been shown that in imaging systems using incoherent radiation, input and output intensities are linearly related.[6] Therefore, we assumed that, in terms of its property of blurring out-of-focus planes, a transmission imaging system can be modeled as a linear, space-invariant system that is characterized by two-dimensional transfer functions. We also modeled the object field as a concatenation of many thin planes, each plane contributing only small attenuation variations. It is then reasonable to assume that a transmission image is a linear superposition of the focal-plane structure and defocused structures.

Finding the inverse of the matrix that describes the linear equations is the most direct approach to solving a system of linear equations. However, the existence of noise often makes the matrix-inversion method unrealistic. By including an additional noise term in the problem formulation, another approach based on the minimum mean square error can be used to find an optimal solution. Both computer simulation and experimental study have shown that the minimum-mean-square error algorithm appears to be the most effective in deconvolving out-of-focus planes. To apply this method to a real object, we collect a series of images of an object, with the focal plane of each image parallel to but displaced from the previous plane by several millimeters. Once this set of images has been collected, we are able to improve the clarity of each image by deconvolving the effects of surrounding image planes.

In the next section, image formation in orthographic transmission imaging is formulated under the linear-system assumption, and the problem of multiplane deconvolution is discussed. The deconvolution algorithm, based on the minimum mean square error, is then described. We then present some simulation and experimental results to show the improvement in images accomplished by multiplane deconvolution.

IMAGE FORMATION AND MULTIPLANE DECONVOLUTION

As described in the introduction, it is very complicated to formulate exactly the process of image formation in ultrasonic transmission imaging. However, under the assumption that the imaging process can be modeled as a two-dimensional linear system, the focal-plane image $q_i(x,y)$ of plane i can be written as

$$q_i(x,y) = \sum_{j=1}^{N} o_j(x,y) * h_{ij}(x,y) \quad , \tag{1}$$

where $o_j(x,y)$ denotes the ultrasound transmission distribution of the object at plane j, $h_{ij}(x,y)$ is the point spread function of a point at plane j when plane i is in focus, the notation * denotes a two-dimensional convolution, and N is the number of planes used to model the object. The point spread function includes all linear space-invariant effects of the imaging system. In Eq. (1), we also assume that the system is noise-free.

The Fourier transform of Eq. (1) is

$$Q_i(u,v) = \sum_{j=1}^{N} O_j(u,v) H_{ij}(u,v) \quad , \tag{2}$$

where all upper-case letters represent the Fourier transforms of the corresponding functions with lower-case letters. A system of linear equations can be written by varying the subscript i to describe all N different planes. Using a matrix representation, it can be expressed as

$$Q(u,v) = H(u,v)O(u,v) \quad , \tag{3}$$

where $Q(u,v)$ and $O(u,v)$ are column vectors consisting of the Fourier transforms of the detected images and object distributions respectively, and $H(u,v)$ is an N×N matrix denoting the system-transfer matrix with $H_{ij}(u,v)$ as its elements. From Eq. (3), we see that the inverse of the system-transfer matrix must be found to solve for the object distribution. The Fourier transform of the object distribution can be described as

$$O(u,v) = H^{-1}(u,v)Q(u,v) \quad , \tag{4}$$

where $H^{-1}(u,v)$ is used to denote the inverse of $H(u,v)$. All the Fourier transforms of the object distribution at different spatial frequencies (u,v) can be obtained according to Eq. (4). The ultrasound transmission distribution of the object can then be determined by an inverse Fourier transform.

Although it seems straightforward that the multiplane deconvolution can be accomplished by a matrix inversion, the reconstruction is usually very poor in practice. The inaccuracy in determining the system-transfer matrix and the noise in the detected images often produce undesirable artifacts. In the next section, we describe an algorithm that takes the noise and inaccuracies into account to provide better deconvolution.

MINIMUM-MEAN-SQUARE-ERROR ALGORITHM

Assuming that a noise term is included, we can rewrite Eq. (1) as

$$q_i(x,y) = \sum_{j=1}^{N} o_j(x,y) * h_{ij}(x,y) + n_i(x,y) \quad , \tag{5}$$

where $n_i(x,y)$ denotes the additive noise. Because we have modeled the imaging formation as a linear process, it is reasonable to assume that an optimal estimate of plane i can be obtained by linearly filtering the detected images. We further assume that the best estimate of plane i will be the one that minimizes the mean square error, i.e. the estimated object distribution at plane i can be described as

$$\hat{o}_i (x,y) = \sum_{j=1}^{N} q_j (x,y) * f_{ij} (x,y) \quad , \tag{6}$$

with the condition:

$$< | \hat{o}_i (x,y) - o_i (x,y) |^2 > \text{ is minimum} \quad , \tag{7}$$

where $f_{ij} (x,y)$ is the impulse response of the optimal filter for reconstructing the plane i from image j. The notation $< >$ denotes the ensemble average or expectation value.

To simplify the problem, the noise in Eq. (5) is assumed to be uncorrelated with the object planes. The cross-correlations of different planes are assumed to be constant. Under these assumptions and based on the orthogonality principle,[7] it can be derived that the impulse responses of the optimal filter must satisfy the following equation:[7–8]

$$\sum_{m=1}^{N} \sum_{n=1}^{N} S_n H_{mn} F_{im} H_{jn}^* + R F_{ij} - S_i H_{ji}^* = 0 \quad , \tag{8}$$

where S_i and R are power spectral densities of $o_i (x,y)$ and $n (x,y)$, and F_{ij} is the Fourier transform of the impulse response $f_{ij} (x,y)$. Here we have assumed that the constant cross-correlations between different object planes, which only affect the zero spatial frequency, are zero. The noise power spectral density was assumed to be the same for different planes. Note that we have dropped the spatial variables (u,v) to simplify the equation. In terms of matrix representation, the system-transfer matrix of the optimal filter must satisfy

$$\mathbf{FM} = \mathbf{W} \quad , \tag{9}$$

in which the elements of matrices $\mathbf{F}$, $\mathbf{M}$, and $\mathbf{W}$ are F_{ij}, M_{ij}, and W_{ij} as shown in the following:

$$M_{ij} = \sum_{n=1}^{N} S_n H_{in} H_{jn}^* + \delta_{ij} R \quad , \tag{10}$$

$$W_{ij} = S_i H_{ji}^* \quad . \tag{11}$$

The first step of multiplane deconvolution is thus to find the transfer matrix of the optimal filter from Eq. (9) or

$$\mathbf{F} = \mathbf{W} \mathbf{M}^{-1} \quad . \tag{12}$$

After the optimal filter has been determined, the Fourier transform of the object distribution can be estimated by using the Fourier transform of Eq. (6), i.e.

$$\hat{O}_i(u,v) = \sum_{j=1}^{N} F_{ij}(u,v) \, Q_j(u,v) \quad . \tag{13}$$

As shown in Eqs. (10) and (11), the determination of the optimal filter requires knowledge about the object and the noise power spectral densities S_i and R. Although they are usually unknown, a reasonable signal-to-noise ratio can be assumed to find the transfer matrix of the optimal filter. It is interesting to note that, when there is no noise, the transfer matrix F (u,v) converges to the inverse matrix of H (u,v) in Eq. (4).[8] In other words, the algorithm is the same as the matrix-inversion method if we assume the imaging system is noise-free.

SIMULATION AND EXPERIMENTAL RESULTS

To investigate the effectiveness of multiplane deconvolution described in the previous sections, we performed both simulations and experiments. We first derived a series of system-transfer functions for the transmission imaging system in our laboratory. For the simulation, an object of three planes with simple structures was generated by the computer. The system-transfer functions were then applied to this object to simulate transmission images. Three focal-plane images were generated in this way. Multiplane deconvolution using the matrix-inversion method and the minimum-mean-square-error algorithm were performed to improve the images. Following the simulation, we fabricated a three-plane test sample similar to what had been simulated. Transmission images of the sample at three different focal planes were collected. The images were deconvolved to examine how effectively the out-of-focus structures could be removed.

In this research, we used an existing computer-controlled mechanical raster scanner and a special pair of transducers to acquire orthographic transmission images. The transmitting transducer array consists of 19 hexagonally packed disk-shaped elements, which are driven by a computer-controlled array of 19 pulsers. All of the transmitting elements are located on a spherical surface, which is confocal with the receiving transducer, to produce spatially and temporally incoherent insonification that is localized to the receiver focus. The receiving transducer is a single spherical element. At each pixel location, the 19 transmitting transducers are fired sequentially with carefully chosen time separation between each firing so that the individual received pulses do not overlap each other. The 19 received pulses are each squared and then integrated to represent the intensity of the image pixel. The firing and scanning of the transducer is controlled by a host computer. By positioning the object along the central axis of the transducer pair, focal-plane images of different planes can be acquired.

For the purpose of easy comparison, our simulated object contained three non-overlapping thin planes, with each plane consisting of a circular disk embedded in a uniform background. The ultrasound attenuation within the disk was 20 percent greater than within the background. To simplify the computation, we assumed that the imaging system was circularly symmetric. Because the system functions were well-defined and the computational error due to the finite-word-length effect of the computer was very small, the matrix-inversion algorithm could remove the out-of-focus structure quite successfully. Figure 1 shows the simulation results. The top row shows three computer-generated transmission images; it can be seen that the images contain both focused and defocused structures. The bottom row shows the results of deconvolution by means of the matrix-inversion method.

The matrix-inversion method worked quite well for the simulated images, as shown in Fig. 1. However, for an algorithm to be practically useful, it must provide satisfactory results

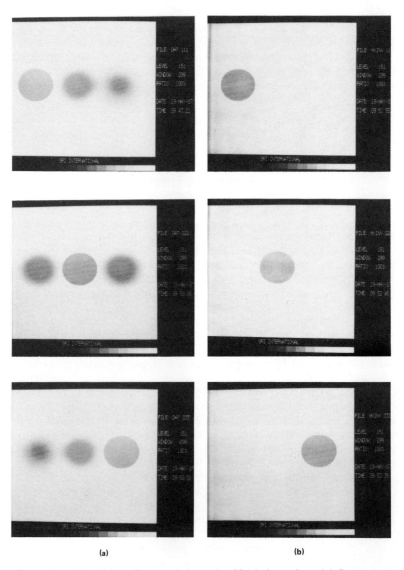

(a)   (b)

Fig. 1. Simulation of Multiplane Deconvolution using Matrix-Inversion. (a) Computer-
generated transmission images of a three-plane object. (b) Images after deconvolution
using the matrix-inversion method.

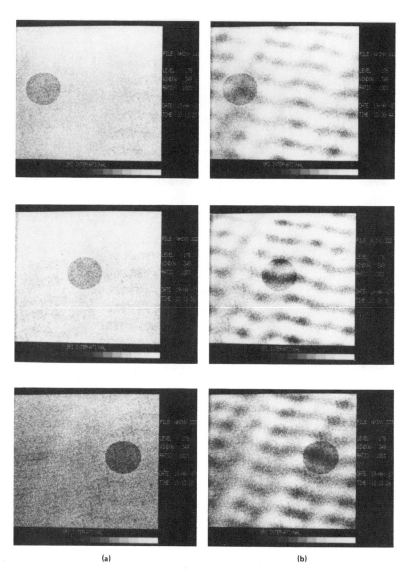

(a)   (b)

Fig. 2.   Simulation of Multiplane Deconvolution with Noisy Data. Images reconstructed from
noisy data which were simulated by adding random noise to images of Fig. 1(a).
(a) Images after deconvolution using the matrix-inversion method. (b) Images after
deconvolution using the minimum-mean-square-error algorithm.

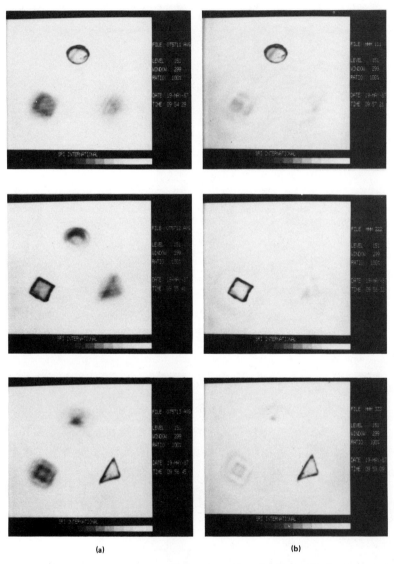

(a)   (b)

Fig. 3.   Experimental Results of Multiplane Deconvolution. (a) Transmission images collected
from an experimental imaging system for a three-plane object. (b) Images after
deconvolution using the minimum-mean-square-error algorithm.

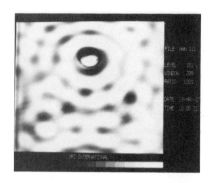

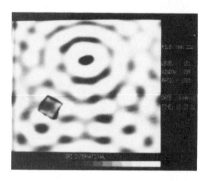

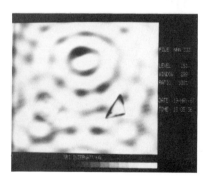

Fig. 4.   Experimental Results of Multiplane Deconvolution. Images after deconvolution using the matrix-inversion method.

when the images contain some noise, as all real images will. To test for noise sensitivity, we added a small amount of random noise to the simulated images, then attempted to reconstruct them with each of the two algorithms. The results are shown in Fig. 2. Figure 2(a) shows that the minimum-mean-square-error algorithm removed almost all of the out-of-focus structures, although the images were still noisy. The results of the matrix-inversion method, however, contained many artifacts as shown in Fig. 2(b). The magnitude of the random noise added to the images had a uniform distribution. The noise had zero mean and its standard deviation was 10 percent of the background level.

In the real experiment, we used a very thin Mylar film as the background for each of three planes. The distance between two adjacent planes was 5 mm. Different shapes cut from a polyethylene film were glued to the Mylar films. The object was positioned in such a way that focal-plane images of each plane were collected. The images were then deconvolved by both the minimum-mean-square-error algorithm and the matrix-inversion method. The images collected in the experiment are shown in the top row of Fig. 3. The results of deconvolution using the minimum-mean-square-error algorithm are presented in the bottom row. We can see that although the reconstruction was not perfect, the out-of-focus structures had been greatly removed. The results of the matrix-inversion method are shown in Fig. 4. Many artifacts appear in the reconstructed images. The images are not recognizable. It should be noted that because the actual impulse responses of our imaging system are only approximately circularly symmetric, our symmetry assumption is probably a source of significant error.

CONCLUSION

The problem of multiplane deconvolution in ultrasonic transmission imaging has been investigated in this research. To simplify the problem, we have assumed a linear model for the process of image formation. The principles of the matrix-inversion method and the algorithm based on minimum mean square error have been presented. We compared the results of deconvolution by both simulation and experiment. The results showed that the minimum-mean-square-error algorithm is an effective approach for removing the out-of-focus structures.

ACKNOWLEDGMENT

This work was supported by the National Institutes of Health under PHS Grant CA42681.

REFERENCES

1. L.M. Zatz, Initial clinical evaluation of a new ultrasonic camera, *Radiology,* 117:399 (1975).

2. L.M. Zatz, K.W. Marich, P.S. Green, M.J. Lipton, J.R. Suarez, and A. Macovski, Real-time imaging with a new ultrasonic camera: Part II, Preliminary studies in normal adults, *J. Clin. Ultrasound,* 3:17 (1975).

3. V.R. Hentz, K.W. Marich, and P. Dev, Preliminary study of the upper limb with the use of ultrasound transmission imaging, *J. Hand Surgery,* 9A:198 (1984).

4. H. Woltering, H.H. Matthias, P.S. Green, and D. Klein, Ultraschalltransmission, *Munch. Med. Wschr.,* 126:1431 (1984).

5. H. Woltering, H.H. Matthias, P.S. Green, P. Edmonds, and J.C. Taenzer, Ultrasonic transmission imaging camera, a new method of function analysis in orthopedics, *EURO-SON 84, 5th Congress of the European Federation of Societies for Ultrasound in Medicine and Biology, Strasbourg,* (1984).

6. J.W. Goodman, pp. 113-115 *in*: "Introduction to Fourier Optics," McGraw-Hill, New York, (1968).

7. A. Papoulis, pp. 168-170 *in* "Probability, random variables, and stochastic processes," McGraw-Hill, New York, (1965).

8. G.S. Laub, G. Lenz, and E.R. Reinhardt, Three-dimensional object representation in imaging systems, *Optical Engineering,* 24:901-905 (1985).

REAL-TIME ENHANCEMENT OF MEDICAL ULTRASOUND IMAGES

Tim Blankenship

Department of Electrical Engineering
University of Missouri-Rolla
Rolla, MO 65401

INTRODUCTION

Ultrasound provides a useful means of obtaining real-time images of internal body tissues with non-ionizing radiation. The real-time capability of ultrasound imaging allows the operator to select desired images interactively. The images obtained represent the actual tissue structure being imaged quite well; and, since ultrasound is not ionizing radiation, it is safe. There are, however, some problems inherent to ultrasound imaging that require special processing. The large dynamic range of the return signal presents a problem that can be alleviated by using some type of logarithmic amplifier. Another common problem is echo attenuation as signals are returned from deeper portions of the imaging area. This problem is solved by increasing amplification as the depth increases (time/gain compensation). Additionally, an ultrasound image usually has low contrast, therefore details in the darker regions are often not perceptible. The contrast of the image can be stretched using any of a number of contrast-enhancement methods such as histogram modification. Finally, the ultrasonic beam, having a depth-variant width, causes the image to be somewhat blurred. The image may be high-pass filtered in order to sharpen the blurred features. Techniques for deblurring include inversion and local feature enhancement. The latter two problems will be addressed in this paper.

The images used here were obtained with a surgical ultrasound unit designed and manufactured by Linscan Systems, Inc., of Rolla, Missouri. A linear probe was used with a 10MHz transducer. With this probe, the transducer is moved linearly along the tissue surface a total distance of 27mm. The 6-bit image data was downloaded from a frame buffer and processed on an IBM PC-compatible image processing system. The image improving methods discussed, while only a small portion of the techniques available, are considered because of the greater possibility of their inclusion in the real-time medical imaging system used to obtain the images shown in this paper. Most real-time implementations of processing methods will require insertion in the video pipeline. The video pipeline processes the pixel stream as it is clocked out of the frame buffer at the pixel clock rate toward the video D/A converter. Several different processing techniques could be implemented in the pipeline. The final stage, however, should be the "gamma" curve mapping RAM that improves contrast and expands the data to 8 bits.

HISTOGRAM HYPERBOLIZATION

As mentioned above, contrast enhancement can be used to improve the contrast of most ultrasound images. The techniques that modify the gray scale of a given image in order to improve contrast can be divided into two types (1). The first is histogram transformation in which the goal is to change the gray scale to increase contrast. An example of this technique is the direct transformation of each pixel to a new value according to a fixed mapping function. This type of operation stretches the contrast quite well as long as the transforming function is matched to the specific type of image being processed. This method is also the easiest and least expensive to implement. The disadvantage of this technique is that, if the statistics of the image change, direct mapping may not increase the contrast significantly. The second type of contrast enhancement, histogram modification, is actually just a special case of the first. Here, the gray scale is transformed to give the image a specified histogram. Two examples are: histogram equalization, which attempts to produce an image with a uniform distribution; and histogram hyperbolization, which attempts to produce an image with a distribution that appears to be uniform to the human eye. These techniques overcome the disadvantage of direct histogram transformation in that they are "normalized;" that is, each produces approximately the same gray level distribution for all images. From numerous processing attempts it was found that histogram hyperbolization highlights more image information than does equalization. Therefore, only hyperbolization will be considered for producing contrast improvements.

With histogram hyperbolization, an image is transformed to have a gray level distribution of hyperbolic form. The equation that describes this transformation is:

$$J(I) = (G - 1) * c * (\exp((\log(1 + 1/c)/(M * N)) * \sum_{i=0}^{I} h(i)) - 1),$$

where G is the total number of gray levels to be output, (M * N) is the total number of pixels in the image, and the sum is the cumulative histogram for a pixel with gray level I. The parameter c can be changed to adjust the contrast for different images or situations. According to Frei (2), application of this transformation should cause the perceived images to have a uniform distribution.

The histogram hyperbolization operation was performed with many different parameter values. Only values of c between about 0.05 and 1.0 gave useful results. Figure 1 compares the linear mapping of an image (Figure 1a) with the histogram hyperbolization of an image with c equal to 0.1 (Figure 1c). This image, a carotid artery with a thyroid gland adjacent to it, shows the improvement in contrast that is obtained. The histogram for the linear-mapped image is given in Figure 1b and the hyperbolized histogram is given in Figure 1d. For these and all the images shown in this paper, G is 256. The 6-bit data is expanded to 8 bits. All images presented later in this paper will be histogram-hyperbolized with c equal to 0.1 after they are processed.

Perhaps the most important aspect of histogram hyperbolization is that it can be applied to many different types of images with the resulting contrast of each being approximately the same. Thus, regardless of the processing that is performed on the image, the contrast would not change appreciably if hyperbolization always followed the processing. The use of parameters in this operation provides another degree of flexibility, since to some extent the amount of modification can be specified.

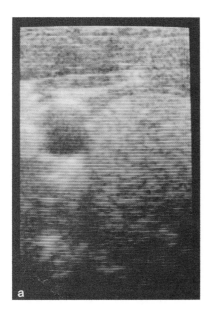

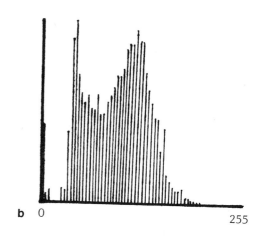

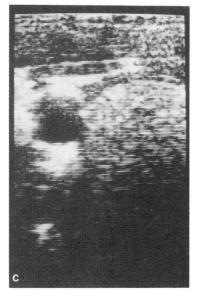

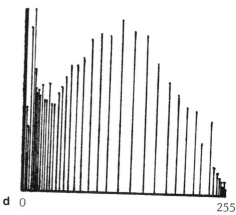

Figure 1. Carotid Artery and Thyroid (a.) Linear mapped; (b.) Linear
mapped histogram; (c.) Hyperbolized, c = 0.1; (d.) Hyperbolized histogram.

Since obtaining a histogram requires that the entire image be examined, contrast enhancement methods requiring histogram information can only be near real-time operations. In normal medical diagnostic use, gray level distributions change slowly enough from frame to frame that a new "gamma" curve computed from past frames will provide nearly correct contrast improvement for current frames. This mapping curve would be loaded into a look-up table that is in the video pipeline. Pixels coming down the video pipeline act as the address to the contrast mapping look-up table which ouputs a new pixel value. The histogram hyperbolization must be done as the last processing step so that it can adjust contrast for any previous processing performed. An implementation such as that shown in Figure 2 would be required. The histogram must be computed with hardware because of the high pixel-clock frequency. When the histogram has been completed, the CPU gains access to the RAM containing the histogram (during vertical blanking) and computes the mapping function. After being computed, the mapping function is loaded into the look-up table during a later vertical-blanking period. A further degree of flexibility is added to this design by including an input port which indicates to the CPU the type of histogram modification to perform. With the inclusion of this port in the circuit, the user could select different contrast stretching schemes as desired.

STATISTICAL DIFFERENCING

While contrast enhancing methods, for the most part, work well on a global scale, techniques for sharpening must be done on a local scale, since the blurring is depth-variant. It should be noted that any attempt to sharpen edges is some form of high-pass filtering and thus will emphasize the noise as well as high-frequency image information. One method that could be used is some sort of inversion or deconvolution. This requires rather complete knowledge of the distortion that caused the blur-

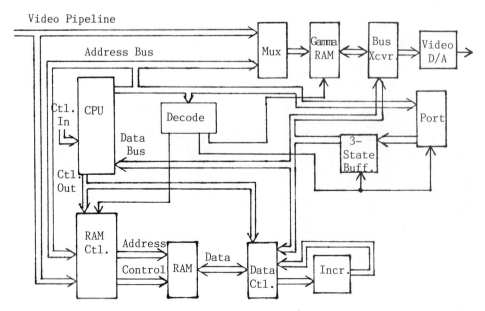

Figure 2. Possible Implementation to Perform Histogram Hyperbolization or Other Histogram Modification Techniques.

ring. It is quite difficult to find the distorting function in a clinical instrument. Even if the distortion was known, the number of computations necessary to perform deconvolution would limit it to freeze frame use.

Another technique that provides locally varying sharpening, but does not require knowledge of the distortion, is called statistical differencing by Pratt (3). This operation alters a pixel's gray level according to equations comparing local and global image statistics. Thus, this operation can be considered to be a type of local feature enhancement. Because of its flexibility and sharpening capability, statistical differencing will be examined here.

With the statistical differencing method, a pixel's gray level is adjusted by a formula that compares the local mean and standard deviation to some desired mean and standard deviation. In fact, the process can be viewed as a gain and bias applied to the center pixel of a neighborhood where the gain is a function of the local standard deviation and the bias is a funtion of the local mean (4,5). The equation to do this is:

$$g' = (g - \mu)/((\sigma/\sigma_d) + 1/A) + \alpha(\mu_d - \mu) + \mu,$$

where g' is the new gray level of the center pixel, g is the original gray level of the center pixel, μ and σ are the local mean and standard deviation, μ_d and σ_d are the desired mean and standard deviation, A is a gain factor, and α is a proportionality factor that controls the ratio of edge to background composition. Including the neighborhood dimensions, R and C (row and column size), there are five parameters. The global mean and standard deviation will be used as the desired values μ_d and σ_d, leaving three parameters that can be varied.

With the number of parameters that can be changed, many processed images could be generated. Here only the most useful cases will be presented. Because of the way it is used in the algorithm, 1/A is varied instead of A. This gain factor is allowed to range between zero and one with smaller values of 1/A increasing sharpness and boosting the noise. The proportionality factor, α, also ranges between zero and one, but here larger values increase sharpness and boost noise. Figure 3 shows an image processed with 1/A of 0.2, α of 0.3, and neighborhood of 3 x 3 (R x C). This particular image displays substantial sharpening along with a tolerable increase in noise. Parameter values can be chosen that produce much less noise, but also much less sharpening.

The last parameter to be investigated is the neighborhood size. The neighborhood size is important because it dictates how the statistical differencing will be implemented. For this comparison, α is set equal to 0.3, 1/A is set equal to 0.2, and the neighborhood is varied. From numerous processing attempts, it was found that neighborhoods with more rows than columns sharpen horizontal edges more, while neighborhoods with more columns than rows sharpen vertical edges more. Figure 4 shows the image processed with a 2 x 2 neighborhood. This square neighborhood results in sharpening image features both vertically and horizontally. There is very little difference between this image and the image of Figure 1. Thus, the 2 x 2 neighborhood would be preferred, since it would allow a simpler, faster implementation.

A further reduction in bits to be processed could be accomplished by dropping a bit from each pixel in the neighborhood. When the least significant bit is ignored during processing with a 2 x 2 neighborhood, the image appears as shown in Figure 5. This image compares favorably with the regular 2 x 2 processed image of Figure 4, therefore, this reduction could be used to simplify implementations.

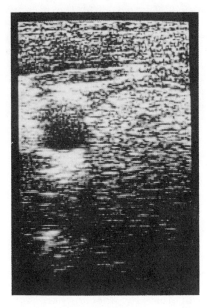

Figure 3. Image Processed with Statistical Differencing, 1/A = 0.2,
α = 0.3, R x C = 3 x 3.

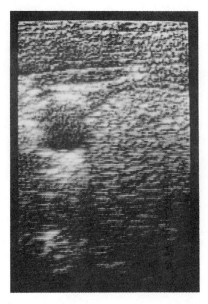

Figure 4. Image Processed with Statistical Differencing, 1/A = 0.2,
α = 0.3, R x C = 2 x 2.

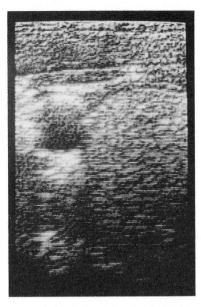

Figure 5. Image Processed with Statistical Differencing and Least
Significant Bit of Each Pixel Dropped, $1/A = 0.2$, $\alpha = 0.3$,
R x C = 2 x 2.

Although statistical differencing is a rather complex process, it can
be easily implemented using a look-up table. This is possible since, for
every combination of pixels in the neighborhood, the algorithm produces a
new center pixel value. All possible values can be computed and loaded
into EPROM. Then, when addressed by correctly ordered pixel values, the
EPROM will provide the new pixel value. This is most simply implemented
for neighborhoods with only one row such as, 1 x 3 or 1 x 5, since just a
few latches are needed to set up the pixels for addressing the memory.
For two dimensional neighborhoods, additional circuitry is needed to store
rows of the image.   A block diagram illustrating this for a 2 x 2 neighborhood is shown in Figure 6. Larger neighborhoods would require more
image-line memories.

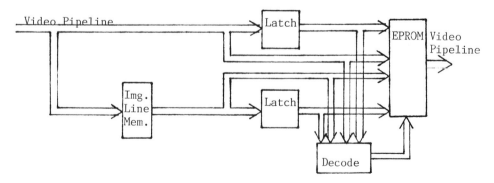

Figure 6. Look-Up Table Implementation of Statistical Differencing
for a 2 x 2 Neighborhood.

There are two problems with the look-up table implementation: large amounts of memory are needed and there is no control over the processing parameters. The size of the memory can be somewhat reduced by choosing smaller neighborhoods and dropping insignificant bits as mentioned above; but a 2 x 2 neighborhood (about the smallest useful one), with the least significant bit dropped from each 6-bit pixel, still requires 1M byte of EPROM (32 27128's or 16 27256's). The equation parameters could be varied by either bank switching or using RAM instead of EPROM and loading new values, both of which require even larger amounts of memory. Thus, the look-up table implementation will become more feasible as memory densities increase, but only for cases where processing parameters need not be changed much.

All of the look-up table problems can be overcome by directly implementing the statistical differencing equation with hardware in the video pipeline. A block diagram for this type of implementation is shown in Figure 7 for a 2 x 2 neighborhood. Again, an image-line memory is needed to store the previous row of the image. The mean and standard deviation are first computed. Then the bias and gain are computed using the mean, the standard deviation, the parameter values, and the current pixel value. The answer is then put back on the video pipeline. The entire process takes approximately 10 pixel clock periods. While this permits parameter value changes, the circuit is much more complicated and requires about 60 IC's.

CONCLUSION

This paper has demonstrated a few of the many image processing techniques available for improving medical ultrasound images. Histogram hyperbolization is seen to provide an "adaptive" contrast enhancement that can be used after any other type of processing scheme to produce images with a gray level distribution that appears uniform to the human eye.

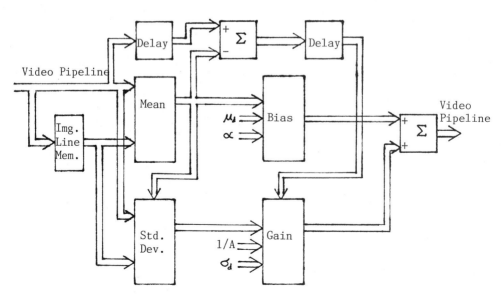

Figure 7. Direct Hardware Implementation of the Statistical Differencing Equation for a 2 x 2 Neighborhood.

This technique can be implemented using a dedicated CPU that could actually perform any contrast enhancement method involving histograms. The sharpening technique discussed, statistical differencing, is a space-variant process that relies on local image statistics to increase the high frequency content of the image. Two implementations were proprosed to perform this processing. Both had advantages and disadvantages. The look-up table method, while simpler, requires large amounts of memory and fixed parameters. The direct implementation of the formula provides parameter flexibility, but with an increase in circuit complexity. While either implementation provides the desired results; the one chosen for use depends on the needs of the system.

ACKNOWLEDGMENTS

This work was supported with grants from Linscan Systems Inc., Rolla, Missouri and the University of Missouri Research Assistance Act.

REFERENCES

1. A. Rosenfeld and A. Kak, "Digital Picture Processing," 2nd Ed., Vol. 1, Academic Press, Inc., Orlando, Florida (1982).
2. W. Frei, Image Enhancement by Histogram Hyperbolization, Computer Graphics and Image Processing, 6:286-294 (1977).
3. W. K. Pratt, "Digital Image Processing," John Wiley and Sons, Inc., New York, New York (1978).
4. D. J. Ketcham, Real-Time Image Enhancement Techniques, Proceedings of SPIE/OSA Conference on Image Processing, 74:120-125 (1976).
5. R. Wallis, An Approach to the Space Variant Restoration and Enhancement of Images, Symposium on Current Math. Problems in Image Science, pp. 107-111 (1976).

A SURGICAL ULTRASOUND SYSTEM

USING AN EMBEDDED PERSONAL COMPUTER

William David Richard

Chief Engineer
Linscan Systems, Inc.
Rolla, Missouri 65401

General Motor's Fellow
Department of Electrical Engineering
University of Missouri-Rolla
Rolla, Missouri 65401

INTRODUCTION

The field of diagnostic medical ultrasound has established itself
as an integral and essential part of diagnostic imaging. This is
indicated by the fact that the number of sonographers more than tripled
between 1975 and 1982 [1]. In February of 1984, the National Institute
of Health estimated that one-third to one-half of all pregnant women in
the United States have ultrasound evaluation at some point [2].

A less widely known use of ultrasonic imaging is its use operatively as an aid during surgery. Pioneers in this field include R. J. Lane,
F.R.A.C.S., of the Royal North Shore Hospital, Sydney, N.S.W. Lane and
his colleagues have used ultrasound operatively to successfully locate
insulinomas within the pancreas that were not demonstrated correctly by
repeated angiography, computerized tomography, endoscopic retrograde
cholangiopancreatography, or transcutaneous ultrasound [3]. Lane has
also used operative ultrasound to explore the common bile duct [4] and
to obtain images of the carotid vessels after endarterectomy [5]. In
the latter application, 60 operative scans revealed 9 abnormalities, 4
of which were of haemodynamic significance [5].

Here, we can only touch upon a few of the applications of operative
ultrasonic imaging. The field, while relatively new, is expanding
rapidly. One of the major advantages of this type of imaging is speed.
The time required for an operative choledochosonography, 3 to 5 minutes,
is significantly shorter than the 10 to 15 minutes required for an
operative cholangiograph [4]. Another, possibly more important,
advantage is the ability to image tissue not discernible with other
imaging techniques [3].

Several companies have adapted their ultrasonic imaging systems for
operative use. Most of these systems are large, bulky units originally
intended for use by radiologists. None of them feature menu-driven
operation: they all use knobs and switches for system adjustment.

An operative ultrasonic imaging system, the Linscan (TM) Imaging System, has been developed by Linscan Systems, Inc., of Rolla, Missouri, in cooperation with the University of Missouri-Rolla.* This system, which is based on an IBM PC/XT (TM) bus-compatible mother board, is shown in Figure 1.** Several images produced using this instrument are shown in a companion paper [6]. The goal used to guide the development of the Linscan Imaging System was that a surgeon should be able to operate the resulting instrument during surgery without assistance of an ultrasound technologist. Instruments similar in function to the Linscan Imaging System have been developed, but they have found little use in surgical applications because of their complexity and lack of "friendliness".

To achieve our goal of obtaining a "surgeon-friendly" instrument, a hierarchical system of menus, implemented in FORTH and controlling a computer-based instrument, was designed. Menu-driven software has been found to be an effective way of interacting with less technical users [7]. The resulting operative ultrasound imaging system is designed so that the instrument can be operated by users who have a minimal amount of knowledge about the system, while simultaneously providing a system that can grow with the user as his/her knowledge of the instrument's operation expands.

Fig. 1. The Linscan Imaging System and associated
operator keypad.

---

* Linscan is a trademark of Linscan Systems, Inc.
** IBM PC/XT is a trademark of International Business Machines.

ULTRASONIC IMAGING OVERVIEW

The acoustic imaging modality used in the Linscan Imaging System is reflection or echo imaging. Reflection imaging is made possible by the relatively slow propagation velocity of ultrasound through tissue: about 1500 m/sec. Ultrasonic energy is propagated through the body causing reflected waves to occur at various discontinuities along the path of the propogated beam. Since the round-trip time through 25 cm of tissue is approximately 333 usec, it is relatively simple for modern electronic circuitry to distinguish reflections at different depths with good resolution.

The reflection imaging arrangement used in the Linscan Imaging System is shown in Figure 2. With the switch thrown in the transmit (T) position, the pulse waveform p(t) excites the ultrasonic transducer. This causes ultrasonic energy to propogate into the tissue as shown. Immediately following the pulse transmission, the switch is thrown to the receive (R) position. When the transmitted wavefront hits a discontinuity (change in acoustic impedance), a reflected (scattered) wave is produced. This reflected wave is detected by the transducer. The resultant signal is then processed and displayed. Processing consists of amplification, envelope detection, temporal filtering, gain control, lateral filtering, and gamma correction.

The display format used in the Linscan Imaging System is the B scan, or B mode. The signal E(t) represents the reflectivity, as a function of time, along the axis of the transducer. The return echo corresponding to a given position of the transducer forms a single display line, one line being formed each time the transducer is pulsed. This process is repeated at regular intervals as the transducer is moved linearly, as is done with the linear probe, or rotated, as is done with the sector probes, to construct a complete image of a 2-D slice of anatomy. In this mode of operation, the brightness of a pixel is used to represent the reflectivity at a given point in the body.

More information about ultrasonic imaging techniques, including a mathematical development of the subject, can be found in Macovski [8].

SYSTEM ARCHITECTURE

The Linscan Imaging System is an operative B mode ultrasonic imaging system that uses high frequency sound pulses to create real-time gray scale images of internal body tissue. A block diagram of the Linscan Imaging System is shown in Figure 3. The main system components are: (1) the mechanical scanning device, which houses the transducer; (2) the analog front end, which consists of a linear preamplifier, a logarithmic amplifier, and a digitizer; (3) the position sense module; (4) the frame grabber module and frame buffer; (5) the image and character display subsystem; (6) the motor control module; (7) the user keypad controller and associated keypad; (8) the EEPROM nonvolatile storage module; and (9) the single board computer system. Each of these units and its function is described further below.

The scanning device is either a linear probe or one of three different types of sector probes. The linear probe moves a single 10 MHz transducer linealy along the skin or organ surface at a near constant velocity (the total linear travel is 27 mm), while the sector probe pivots either a 5, 7, or 10 MHz transducer about a fixed point (the total angle swept is 50 degrees). For a given probe, the ultra-

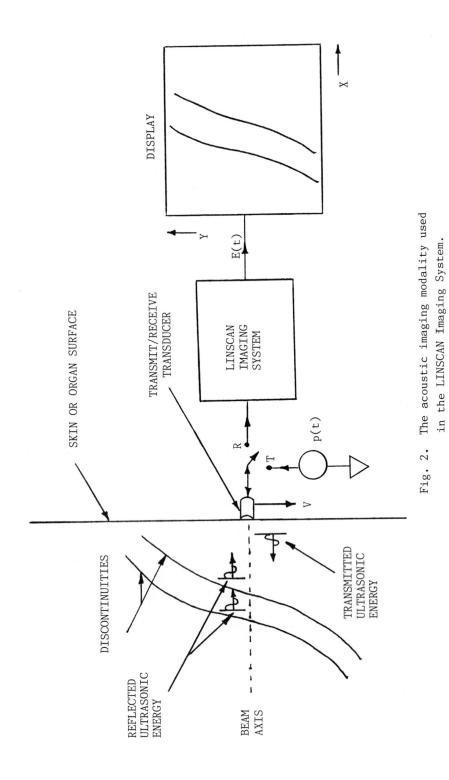

Fig. 2. The acoustic imaging modality used in the LINSCAN Imaging System.

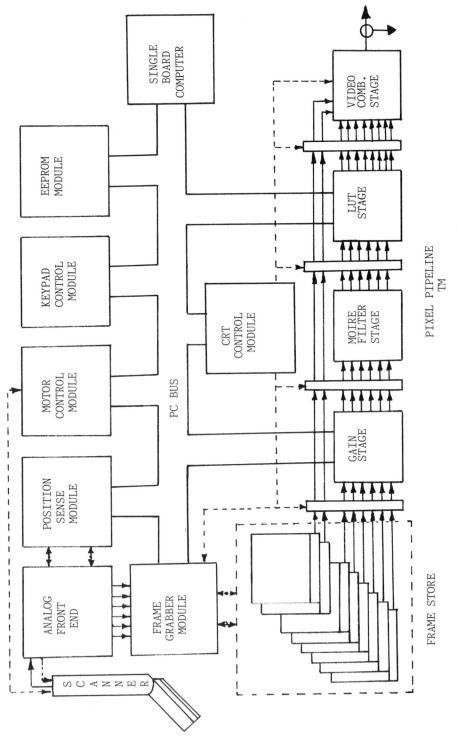

Fig. 3.   Block Diagram of the LINSCAN   Imaging System.

201

sonic transducer is excited by control signals from the position sense module. A return echo is detected by the same transducer that generated the initial burst. The detected signal is fed to a logarithmic amplifier to compress its dynamic range, envelope detected, digitized via a flash converter, and then fed to the frame grabber module.

The position sense module controls the digitization of the image data and acts as the interface between the front end circuitry and the frame grabber. The position sense module uses proprietary devices to monitor the positions of the transducers in the various ultrasound probes. When a new valid position is detected, the position sense module pulses the ultrasonic transducer, initiates signal digitization, and informs the frame grabber module of both the availability of data and the position of this data (x-position for the linear probe, 0-position for the sector probes).

The frame buffer is a 65536 byte dynamic RAM memory module (organized as a 256 x 256 byte array) composed of TI 4161 video RAMs [9]. These video RAMS have a built-in 256-bit shift register which may be loaded in parallel from any of the 256 rows of the memory array. Each chip implements one of the eight bit planes that form the frame buffer. Six of these bit planes are dedicated to image data, while the remaining two are used for graphic overlays.

The frame buffer is essentially a three port memory accessed by the CPU, the video display logic, and the digitizer. The video display is driven by the on-chip shift registers. This significantly reduces the number of memory cycles required. The frame buffer is controlled by a fast state machine which arbitrates among buffer access requests and performs refresh cycles as required. This state machine is also responsible for storing digitized image data from the digitizer into the frame buffer when the system is in the scan mode of operation. Frame-to-frame averaging is performed with the aid of a second state machine and frame buffer.

The image and character video display subsystem is responsible for creating the composite video signal that is fed to the display monitor. An integrated CRT controller is used in conjunction with the system DMA controller (all under software control) to generate the character video signal. The CRT controller also creates all the timing signals necessary for the 60 Hz, non-interlaced display format used by the system, e.g., the dot clock, horizontal and vertical sync signals, horizontal and vertical blanking signals, etc. A three stage arithmetic pipeline clocked at the dot clock rate is used to process the image data before it is displayed. The character, image, and sync signals are combined by a video D/A converter, which drives the display monitor. A proprietary circuit is used to perform this combination so that graphics and characters overlay the ultrasonic image.

The three stage video pipeline performs gain, filtering, and mapping operations on each detected pixel value before presenting it to the output D/A converter. This pipeline has a clock frequency of 6.000 MHz, which corresponds to a clock period of 166.7 ns. The entire pipeline is implemented with low-power Schottky TTL logic.

The first stage of the display pipeline performs a gain function on the digitized image. This proprietary circuit uses gain values specified by the control software. These values are changed when the user requests a new gain setting from the keyboard. The gain function is implemented so that the gain of the system can be changed when the system is in either the "scan" or the "freeze" mode of operation.

The second stage of the pipeline performs a lateral spatial filtering operation on the pixel stream. This proprietary circuit is used to eliminate Moire "holes" in sector images caused by the $(r,0)$ to (x,y) conversion process required when an image in $(r,0)$ format is placed in an (x,y) format frame store.

The last of the three pipeline stages performs a mapping operation on each pixel in the pixel stream. When this type of operation is used to correct for photographic film response, it is termed "gamma correction". Even though this mapping operation is used to correct for the response of the human eye, the term "gamma correction" is often used [10]. The mapping function is performed using a proprietary circuit that allows the mapping function to be changed when a new mapping function is requested by the user. Currently, four different gamma curves are available for selection by the user.

An electrically-erasable programmable read-only memory (EEPROM) module is included in the system for nonvolatile storage. System parameters (gain, range, starting depth, etc.) are stored here so that at power up the system can return to its previous state. This feature is especially useful to specialized surgeons who normally only perform one type of surgery. Sixteen sets of factory specified parameters are also stored here for use with the Diagnostic Specific Menu, which is discussed below.

Other functions performed by a combination of hardware and software include speed control of the transducer motor and operator input via a 10 key keypad.

The system is controlled by a single board computer based on the Intel iAPX 88 CPU that is bus compatible with the IBM PC/XT. In addition to the features of the IBM PC/XT, this board contains two serial ports, a parallel port, and a floppy disk controller. The IBM PC bus was chosen in order to take advantage of the availability of the large number of low cost CPU and peripheral boards currently being offered.

No attempt was made to make the system software compatible with the IBM PC/XT disk operating system (PC DOS). The single board computer runs under the control of a locally developed FORTH operating system based on the FORTH 79 standard. All target system code is written in FORTH or in assembly language.

THE FORTH SOFTWARE DEVELOPMENT ENVIRONMENT

The FORTH system used in the development of the system software was developed by Mr. Kurt Hambacker of the Electrical Engineering Department of the University of Missouri-Rolla. The version of FORTH used was adapted for the single board computer by the author, having run originally on the CRL88 boards developed by Mr. Hambacker. This version of FORTH is based on the FORTH 79 standard and has extensions for the 8088 processor on which it runs. It has editor, assembler, and debugger vocabularies in addition to the FORTH vocabulary. Target system code was developed using these vocabularies and executed interactively (any FORTH procedure, called a "word", could be called interactively from the developement terminal). More information on FORTH can be found in Brodie [11].

Initially, the Linscan Imaging System consisted of the single board computer along with its associated power supply and terminal. The first system prototype served as the development system. Software development began as soon as the FORTH system became operational. As additional cards containing various subsystems were added, the drivers for these subsystems were written and tested. Proceeding in this manner, it was possible to have the program segments necessary for system operation completed when the hardware became fully functional. Additional soft- ware development was required to implement the interactive menu-driven user interface described below.

Once a version of the target software was frozen, it was compiled and transferred to EPROM. A small change was made to the power up procedure so that a target operating system was loaded into RAM and executed upon power up, rather than the FORTH text interpreter. During development, the underlying FORTH machine could be reached through windows in the target system software. Development could then continue with the previous target software precompiled and resident in the FORTH dictionary.

The FORTH development system described above was used extensively in the development of the target software for the Linscan Imaging System. A low-cost cassette tape unit and a standard floppy disk drive compatible with the FORTH system were used for program storage during development. A FORTH-driven EPROM programmer was developed for use with the host to allow easy transfer of compiled target operating systems to EPROM.

THE MENU-DRIVEN OPERATING SYSTEM

The menu-driven operating system, as seen from an operator's point of view, consists of a hierarchical system of menus (a menu tree) that may be reached from the initial state. The various levels of the menu tree (excluding the initial state) are: (1) the Owner's Manual, (2) the Diagnostic Specific Menu, (3) the Neurological Menu, (4) the Vascular Menu, (5) the Abdominal Menu, (6) the Other Organs Menu, (7) the Param- eters Menu and the associated help screens, and (8) the Engineering Menu. The relationship between each of these levels is illustrated in Figure 4. Major system functions, such as "gain up", "gain down", "scan/freeze", and "zoom", are available on a 10-key keypad. These functions are all that are needed in many cases. Complete control over the system is provided through the hierarchical menu system without requiring the myriads of knobs and switches that can dumbfound all but the most experienced user.

The mode of operation assumed by the Linscan Imaging System upon power up, for a given probe, depends on the system state when power was last removed. System parameters (range, gain, starting depth, etc.) are stored in EEPROM so that at power up the system can return to its previous state. In the initial state, as well as at any level of the menu tree, functions on the keypad are available to the user.

The [HELP ON/OFF] key is used to branch to the online Owner's Manual from the initial state. This online manual, the Linscan OnLine Assistant, or L.O.L.A., was not intended to be a complete description of the system. It was designed to serve both as a quick reference for the surgeon and as an introduction for the first time user. It may, how- ever, be all that is required by users already familiar with ultrasonic imaging techniques. This manual, which currently contains 409 lines of 80 characters, includes a dictionary of key function definitions and a description of the hierarchical menu system. Other topics range from probe draping techniques to information on troubleshooting.

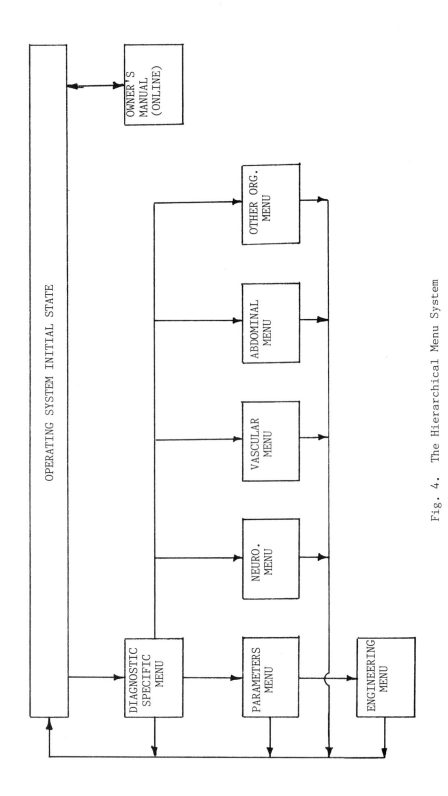

Fig. 4. The Hierarchical Menu System

205

The Diagnostic Specific Menu can be reached from the initial state by pressing the [MENU ON/OFF] key. This branch of the menu tree lists four lower menus, the Neurological, Vascular, Abdominal, and Other Organs Menus, that can be used to initialize the system for one of sixteen specific applications. Other selections allow the current depth mode to be changed or the image to be inverted (for linear scanners). A selection leading to the Parameters Menu also exits.

Each of the four more specific menus reached from the Diagnostic Specific Menu allow a user to select one of four specific applications. For example, the Neurological Menu has the selections: (1) Small Incision (Burr Hole) - Superficial, (2) Small Incision (Burr Hole) - Deep, (3) Large Incision (Flap) - Superficial, and (4) Large Incision (Flap) - Deep. Once a selection is made, the system, via control from the single board computer, initializes the imaging hardware for the chosen application using factory specified parameter sets.

The Parameters Menu, which can be reached from the Diagnostic Specific Menu, is an even lower branch of the menu tree. This menu allows the user to modify any or all of the system imaging parameters. Available parameters include the Time Gain Control (TGC) Delay, the TGC Slope, the Near Field Attenuation, the Far Field Attenuation, and the Gamma Curve. Once a user has customized a parameter set, a "hook" in the Diagnostic Specific Menu allows the user to replace one of the factory parameter sets with the customized parameter set.

The Engineering Menu can be reached from the Parameters Menu via a special key sequence. This lowest branch of the menu tree lists the values of the parameters stored in EEPROM that define the system config-uration. The Engineering Menu was initially developed as a developement tool, but it is also valuable on production units. Including this menu on production units allows things like motor speed to be adjusted in the field, without the need for specialized test equipment.

The Linscan Imaging System is designed to accept several different linear and sector probes. This versatility is achieved via a general-ized hardware interface and an adaptive FORTH-driven operating system. If a probe is not present at power up, a special dummy operating system is loaded by the FORTH kernel. This operating system displays the system logo and instructs the user to insert a probe.

Each probe is keyed with a special code that can be read by the FORTH operating system. This code is used by the FORTH system kernel to load the correct operating system for a given probe. This is done automatically at power up if a probe is inserted prior to applying power to the system. In this case, the dummy operating system is bypassed. Each individual operating system monitors the probe code and transfers control to another operating system if probes are changed.

The availability of multiple operating systems, stored in EPROM, greatly enhances software maintainability. It also makes expansion very easy when new probes are offered by the manufacturer. The user inter-face is maintained by each operating system. The underlying control routines, however, differ considerably for different types of probes. Rather than try to write and maintain one very large, overly complicated operating system, we have chosen to write several small, dedicated systems. These systems are relatively easy to write and maintain. The cost of this approach is the unavoidable duplication of some software.

CONCLUSIONS

The Linscan Imaging System, which is currently being marketed by Cavitron Surgical Systems of Stamford, Conneticut, as the Intrascan (TM) Imaging System, has received a favorable response from the medical community.*** It is the first ultrasonic imaging system designed specifically for operative use. It is also the first system that uses a menu-driven operating system. Its size, relatively low price, and versatility define a distinct place for it in the market. The unique scanner designs and the menu-driven operating system add significantly to the system's versatility. The system can also accept several different types of probes. The menu-driven operating system is designed so that the instrument can be used by a surgeon during surgery without technical assistance. This feature, which is not found on any of the other surgical systems currently available, represents a major contribution to the field of operative ultrasonic imaging.

The use of FORTH as a high-level design language has proven very effective (both in terms of time and effort) in the development of the target software for the interactive medical ultrasound diagnostic tool described above. Significant savings were realized because the development system and the target system were actually the same unit. Since FORTH code is very modular in nature, it is easily maintained. The target operating system, with its hierarchical system of menus, provides the user with complete control over the instrument without requiring a myriad of knobs and switches. The ease with which this complex operating system was developed is due largely to the use of FORTH as the high-level design language for the instrument. The interactive FORTH operating system was very valuable as a debugging tool during development, and it is also used to test and debug all manufactured systems.

The single board computer used in the Intrascan Imaging System functions mainly as an intelligent controller. There are certainly more powerful (and more expensive) single board computers available. We felt, however, that this structure represented a good compromise between cost and performance for our application. The choice of the IBM PC/XT bus for the Intrascan Imaging System has insured a supply of relatively low cost single board computers for use with the Linscan Imaging System. Two different boards have actually been used in the past, and a port of the FORTH operating system kernel to a third board is currently under way. This port is required because the FORTH kernel, like the Basic Input/Output System (BIOS) used with all computers, is hardware dependent. An operating system based on PC DOS would not require such a port when a new board is used. By the same token, a generalized and readily available operating system makes copying the system software much easier. The FORTH operating system, unlike PC DOS, is not disk based, even though a disk is supported. This eliminates the need for a disk drive, or some type of disk emulator, in every unit. Overall, it is felt that the choice of the IBM PC/XT bus and FORTH was a good one.

ACKNOWLEGEMENTS

The work reported in this paper was supported by grants from the Missouri Research Assistance Act and Linscan Systems, Inc. I would like to thank Mr. Kurt Hambacker, Dr. Hardy Pottinger, Proffesor Tom Herrick, and Mr. William Lindgren, president of Linscan Systems, for their direction and encouragement.

*** Intrascan is a trademark of Cavitron Surgical Systems.

REFERENCES

[1] B. Anderhub, "Manual of Abdominal Sonography", University Park
     Press, Baltimore (1983).
[2] U.S. Department of Health and Human Services, Public Health
     Service National Institues of Health, "Diagnostic Ultrasound
     Imaging in Pregnancy", NIH Pub. No. 84-667.
[3] R. J. Lane and G. A. E. Coupland, The Interoperative Ultrasonic
     Features of Insulinomas, "American Journal of Surgery", (1981).
[4] R. J. Lane and G. A. E. Coupland, The Ultrasonic Indications to
     Explore the Common Bile Duct, "Surgery", (1981).
[5] R. J. Lane and M. Appleberg, Real Time Intraoperative Angiosono-
     graphy Post-Carotid Endarterectomy, "Surgery", 92:5 (1982).
[6] T. Blankenship, Real-Time Enhancement of Medical Ultrasound Images,
     in: "Acoustical Imaging Volume 16", Plenum Publishing Company,
     New York (1987).
[7] W. M. Newman and R. F. Sproull, "Principles of Interactive Computer
     Graphics", McGraw-Hill, New York (1979).
[8] A. Macovski, "Medical Imaging Systems", Prentice-Hall, Englewood
     Cliffs, New Jersey (1983).
[9] M. C. Whitton, Memory Design for Raster Graphics Displays, "IEEE
     Computer Graphics and Applications", 4:48 (1984).
[10] W. K. Pratt, "Digital Image Processing", John Wiley and Sons,
     New York (1978).
[11] L. Brodie, "Starting FORTH", Prentice-Hall, Englewood Cliffs, New
     Jersey (1981).

IN–VIVO AND IN–VITRO MEASUREMENTS

OF TURBULENT BLOOD FLOW CHARACTERISTICS

USING A MULTI–DIMENSIONAL ULTRASONIC PROBE

A.R. Ashrafzadeh, J.Y. Cheung*, K. J. Dormer**, and
N. Botros***

University of Detroit, Detroit, MI
*University of Oklahoma, Norman, OK
**University of Oklahoma, HSC, OKC, OK
***S. Illinois University, Carbondale, IL

I. INTRODUCTION

Many studies have been reported on the Doppler Ultrasound Blood
Flowmeters, which is now in daily use in clinical settings. However, less
attention has been paid to quantitative analysis of Doppler waveforms. The
quantitative capability of conventional Doppler ultrasound systems are limited
by the effects of acoustics, time, and system parameters, as well as its
inability to obtain precise measurement of the Doppler angle, which in most
cases must be estimated [1–3].

In this study, we have utilized a multi–dimensional Doppler
ultrasound system for direct calculation of the Doppler angle and therefore
accurate determination of the components of the blood flow velocities.
Determining blood flow velocities in different direction enabled us to
calculate precisely not only the blood flow volume rate but turbulent blood
flow characteristics such as level of turbulency and shear stress. These
characteristics can be related to physiological changes inside the vessel.
Something which is not possible to obtain with the conventional single
transducer systems. Another important advantage of the developed model
compare with the conventional system is that by displaying the velocity
profiles in at least two channels the more accurate blood flow pattern can
be obtained. This is very crucial in studying the behavior of the blood flow
in major arteries such as coronary and carotid.

In this paper, first, the theoretical investigations of multi–
dimensional ultrasound sensor model is presented. Then the fluid dynamic
aspects of the work is briefly discussed. Finally the results of in–vitro and
in–vivo experiments are presented.

II. THEORETICAL INVESTIGATION

In a conventional single transducer pulsed–Doppler ultrasound, the relationship between the received Doppler signal and the directional velocity of the blood flow can be obtained by the following expression [4]:

$$\hat{V}_R = k \cdot F_1/\cos\theta \qquad (1)$$

where k is the constant depending upon different acoustical, system and physiological parameters, F_1 is the Doppler frequency shift and Θ is the angle between the direction of the flow and ultrasound beam (Doppler angle). The conventional system is only capable of presenting data for F_1. Therefore   should be calculated based on assumption for Doppler angle Θ. To overcome this "assumption" problem, we have developed the multi-dimensional sensor.

The new probe can be configured as "Dual" for simpler tasks such as obtaining the Doppler angle and velocity components or as "Triple" for more complicated tasks as producing an image of blood flow and the artery under investigation. A typical configuration for a dual transducer system is shown in Figure 1.

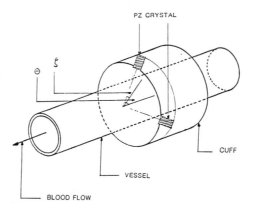

Fig. 1   A typical configuration for a dual transducer model.

To implement the dual model, a pulsed–Doppler instrument with at least two channels is required. Using both channels and thismodel Equation 1 can be written in the following forms:

$$\hat{V}_R = k \cdot F_1/\cos\theta \qquad (2)$$

$$\hat{V}_R = k \cdot F_2/\cos(\alpha-\theta) \qquad (3)$$

where F_1 and F_2 are Doppler output signals from channel 1 and channel 2, respectively and α is the angle between the probes.

Note that there are three unknowns: $\hat{V}_R$, α and Θ and only two measurable quantities F_1 and F_2 in Equations (2) and (3). Fortunately, the angle between the probes, α, is fixed at the time when the probe is built.

If α is known by the process of manufacturing or by calibration, then there are two equations and only two unknown. Therefore the Doppler angle can be calculated directly from the following expression:

$$\theta = tg^{-1} \left(\frac{F_2 - F_1 \cos\alpha}{F_1 \sin\alpha} \right) \tag{4}$$

This accurate measurement of Doppler angle has a number of advantages including precise measurement of arterial diameter which is of great importance in obtaining blood flow volumerate.

The interesting point is that velocity at the center point of the artery can be obtained from F_1 and F_2 without explicit measurement or calculation of Doppler angle Θ as following

$$\hat{V}_R = \sqrt{\frac{k^2 F_2^2 (1 - 2\cos\alpha) - k^2 F_1^2 \cos 2\alpha}{\sin^2\alpha}} \tag{5}$$

This is the primary advantage of dual transducer model over a single transducer model.

For the special case ($\alpha = 90°$), velocity estimation can be greatly simplified as:

$$\hat{V}_R \bigg|_{\alpha=90°} = k \sqrt{F_1^2 + F_2^2} \tag{6}$$

In order to demonstrate the effect of uncertainty in Doppler angle estimation, error analysis have been conducted comparing two models. For the case of single tranducer consider the velocity $\hat{V}_R$ in Equation (1) as estimated value, the actual velocity V_R can be described by the following equation:

$$V_R = k \cdot F_1 / \cos(\theta + \zeta) \tag{7}$$

where ζ is the discrepency between the actual and assumed Doppler angle. The ratio of measured ($\hat{V}_R$) to calculated (V_R) velocity is then

$$\frac{\hat{V}_R}{V_R} = \frac{\cos(\theta + \xi)}{\cos(\theta)} \tag{8}$$

and the normalized velocity error (E_1) can be expressed as:

$$E_1 = 1 - \frac{\hat{V}_R}{V_R} = \frac{\cos\theta - \cos(\theta + \xi)}{\cos\theta} \tag{9}$$

The percentage of normalized velocity error is calculated from Equation (9) and a plot is generated for this percentage when the uncertainty in the Doppler angle (ζ) is varied from $-45°$ to $+45°$, for Doppler angle (Θ) varying between $0°$ to $75°$ and is shown in Figure 2.

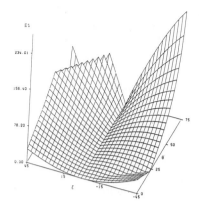

Fig. 2   The percentage of normalized velocity errors for single
transducer Pulsed-Doppler ultrasound level.

For the case of dual transducer is considered as error in calibration caused
by manufacturing process in obtaining angle between two transducers (α) not
Doppler angle (Θ).

Consider V_R in Equation (5) is measured velocity, then the actual
velocity can be obtained from the following expressions:

$$F_1 = \frac{V_R}{k}\cos\theta \qquad (10)$$

$$F_2 = \frac{V_R}{k}\cos(\alpha - \theta - \xi) \qquad (11)$$

where V_R is the actual value of the velocity

The ratio of the measured to actual velocity can be then
expressed as:

$$\frac{\hat{V}_R}{V_R} = \frac{\sqrt{[\cos^2(\alpha-\theta-\xi)](1 - 2\cos\alpha) - \cos^2\theta\cos2\alpha}}{\sin\alpha} \qquad (12)$$

and therefore the normalized velocity error is

$$E_2 = 1 - \sqrt{\frac{[\cos^2(\alpha-\theta-\xi)](1 - 2\cos\alpha) - \cos^2\theta\cos2\alpha}{\sin^2\alpha}} \qquad (13)$$

One obvious configuration is to make the two transducers orthogonal to one
another. In this case, the normalized velocity error can be neatly simplified
by substituting $\alpha = 90°$ into Equation (13) as following:

$$E_2\Big|_{\alpha=90°} = 1 - \sqrt{\cos^2\theta + \sin^2(\theta+\xi)} \qquad (14)$$

212

A plot is generated for percentage of normalized error for different value of Θ, in this case from 0° to 360°, while z varies between −45° and +45° and is shown in Figure 3.

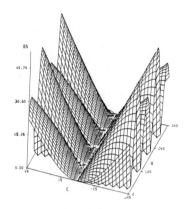

Fig. 3   The percentage of normalized velocity error for dual transducer Pulsed-Doppler ultrasound model.

It should be mentioned that the error in the dual model is controllable by careful calibration of the angle between the two transducer, while the error in the single model which is caused by assumption, is not controllable. Despite the above fact a simple coparison between two models indicates dramatic improvements in the calculation of blood flow velocity by the dual transducer model over the single transducer system. For example a false estimation of 10° in Θ causes 18.08 percent error in velocity in single transducer model while a false estimation of 10° in α causes 8.213 percent error in velocity in dual transducer model.

III.   FLUID DYNAMICS ASPECT OF THE STUDY

One obvious advantage of the developed model is its ability to obtain transverse and longitudinal components of the velocity, which can be used for both laminar and turbulent blood flow.

As it was mentioned earlier because of the simplicity in calculation and problems in the design of the conventional systems, in most of the cases investigators have assumed the blood flow has laminar pattern. The laminar approximation can not be validated for all the cases especially in and around the heart [5]. The dual and multi-dimensional transducer models are the proper tools for verifying the thru blood flow pattern.

In the case of turbulent blood flow, however, multi-dimensional ultrasound system is the only system capable of providing proper data for blood flow analysis.

For the simple case (two-dimensional), flow in turbulent region is moving in a direction parallel and also in a direction perpendicular to the axis of the pipe. Therefore, there are at least two components of the velocity v an u that they need to be detected as they are expressed in the following Equation [6]:

$$\hat{v} = \bar{v} + v'$$

(15)

$$\hat{u} = \bar{u} + u'$$

(16)

where $\hat{v}$ and $\hat{u}$ are instantaneous velocities, $\bar{v}$ and $\bar{u}$ are average velocities, and v' and u' are fluctionation components of the velocity. In order to obtain true value of the fluctuations, phase average of the velocities were calculated and then they are subtracted from instantaneous velocities.

By calculating the fluction components of the velocities one can obtain different characteristics of the blood flow. The most important characteristic of the blood flow in turbulent regions of the blood is shear stress. Shear stress is a function of different flow parameters such as structural parameters of the vessel which can be used to obtain the percentage of narrowing in stenoid vessel. The shear stress can be obtained by the following expression:

$$\tau_0 = -\rho <u'v'> \tag{17}$$

where $<u\,v>$ is the phase average of the product of the fluctuation velocities and   is the density of the flow.

One important application of shear stress is determining stenosis using geometrical analysis of the artery [8] as it is shown in Figure 4.

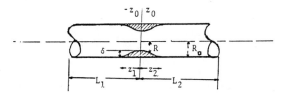

Fig. 4   A geometric presentation of stenosis in the artery.

Ratio of the shear stress at the wall (τ_n) to shear stress at the stenosis point (τ_0) can be expressed by the following equation:

$$\frac{\tau_0}{\tau_n} = \frac{1}{1-3\frac{\delta}{R_0} + 3(\frac{\delta}{R_0})^2 - (\frac{\delta}{R_0})^3} \tag{18}$$

where τ_n and τ_0 can be calculated by scanning through the diameter of the artery using Equation (17).

III.   EXPERIMENTAL RESULTS

Experimental study of blood flow fields have been augmented by theoretical model using computational fluid dynamics. In order to justify the theoretical models Doppler angle, blood flow velocity, volume rate, velocity profile, shear stress, etc., were calculated for both in-vitro and in-vivo experiments.

In this paper the results of in-vivo experiments in obtaining Doppler angle, velocity and volume rate for different condition of the hearts are presented. And for in-vitro experiments turbulency was generated to calculate turbulent blood flow characteristics.

For in-vivo experiments the constructed dual transducer model was implanted on the carotid artery of donor dogs and rabbits. Acute surgical technique was used and under barbiturate anesthesis, the common carotid arter was exposed. The experiments were conducted at three different ranges and three different heart rates. The results for normal condition of the heart (180 bpm) are presented in Table 1.

Table 1   The Doppler angle, flow velocity and flow
volume rate for normal condition of the heart

Range	Doppler Angle (θ) (degrees)	Velocity (mm/sec)	Diameter (mm)	Volume Rate (ml/min)
a	50.32	154.20	3	65.49
a	52.83	151.18	2	28.54
a	52.20	132.95	1	6.27
a	49.33	101.76	0.5	1.20
b	54.33	154.36	3	65.56
b	57.78	151.01	2	28.50
b	51.03	147.99	1	6.98
c	50.99	148.98	3	63.27
c	54.91	125.70	2	23.73
c	48.63	125.02	1	5.90

The important role of the dual transducer in calculation of Doppler angle Θ is more evidence from these results, since the mean value of the Doppler angle (for different arterial diameters ranging from 0.5 mm to 3 mm) is 51.04° and the normalized error varies between 1.4 percent to 3.3 percent. The results indicate that a larger reduction in the diameter of the artery causes a noticeable change in the Doppler angle.

For in-vitro experiments a hardware setup consisting of different elements to generate pulsatile flow was used. First Doppler angle and blood flow volume rate were calculated for various situations such as different speed of the pulsatile pump and different sizes of tubing, and then different statistical parameters such as mean, standard deviation and variance were determined.

To determine the statistical parameters, the following calculations are performed:

1. The cross sectional area of the artery under investigation using the measured diameter.

2. The Doppler angle using the output Doppler signals.

3. Average velocity and velocity components for each channel using the calculated Doppler angle.

4. Volume rate for each trigger using the velocity and cross sectional area.

5. Average volume rate, standard deviation and variance for each setup and compare the results with the measured volume rate.

The artery used in this in-vitro experiment was isolated carotid artery from a donor rabbit with diameter of 2.7 mm in normal condition. The effect of different volume rate caused by changes in the size of tubing or speed of the pulsatile pump were reflected on the amplitude of the

Doppler signals and were used for qualitative assessment of the test's condition. A plot presenting the mean measured volume rate vs. mean calculated volume rate was generated from experimental results and is depicted in Figure 5.

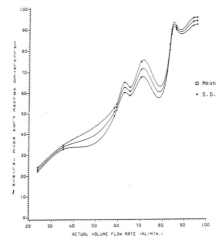

Fig. 5 The mean and standard deviation of calculated volume rate versus measured volume rate

The average percentage of error is 4.38 percent, ranging from 0.03 percent to 12.42 percent. The higher rate of error could be associated with sudden change in the speed of the pump and it seems to be the result of unsteady flow.

In the last part of the in-vitro experiments the turbulent flow was generated. The purpose of this experiment was to explore the effect of the stenosis upon the characteristics of turbulent flow. The turbulent shear stresses were obtained using the dual-transducer model. The experiment was performed using a turbulent-producing orifice at 1 mm diameter and approximately 5 mm axial length, which is inserted at the distal end of the artery as illustrated in Figure 6.

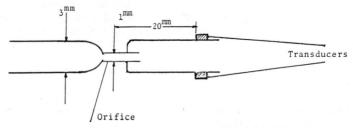

Fig. 6 Turbulent Producing Orifice

At appropriate Reynolds number, there is laminar flow upstream from the orifice, and turbulent flow downstream from the orifice. The random fluctuating velocities associated with the turbulent flow are qualitatively detectable on the display of chart recorder. The analog signal was digitized to obtain phase average of the variation of the velocity and distribution of the turbulent shear stress. Time average of shear stress was also calculated for normal and three different degrees of stenosis (25) percent, 50 percent and 75 percent. The artery under investigation is a specimen of the carotid of a donor dog. Time average values of shear stress for normal and above three conditions were 93, 75, 48, 15 which are in great agreement with generated degree of stenosis. Although the time average shear stress might show reasonable results in a particular test, because of random nature of turbulency, it is not a reliable approach for analysis of turbulent flow and clarification of stenosis. The phase average approach and power spectrum presentation of turbulent characteristics can be more reliable.

216

SUMMARY

In the theoretical portion of the present study a dual transducer model was developed as the first stage of multi-dimensional probe, to calculate the most important parameter (Doppler angle) in Doppler formula. The calculation of Doppler angle (instead of assumption) allowed the accurate measurement of the blood flow velocities. The superiority of the developed model in obtaining the flow velocity was verified both theoretically (through error analysis) and expreimentally comparing it with the conventional single transducer model.

The application of the developed model in both laminar and turbulent blood flow regions were outlined and it was shown that in fact the new model is an effective tool for quantitative assessments of the Doppler signals. Also advantages of the new model in obtaining turbulent blood flow characteristics such as turbulent intensity and shear stress were discussed since at least two or more components of the velocities are required to obtain such characteristics.

The in-vivo and in-vitro experiments were conducted using the developed model. The results show that in the case of Doppler angle and volume rate the model performs accurately and for the case of turbulent characteristics, although, the results are promising more experiments should be conducted, especially with the aid of microprocessor control systems, appropriate hardware should be designed to accomodate the developed sensor model for turbulent and image processing applications.

REFERENCES

1)   Baker, D.W., "Pulsed Ultrasonic Doppler Blood-Flow Sensing," IEEE Transactions on Sonics and Ultrasonics, Vol. SU-17, No. 3, July 1970.

2)   Ashrafzadeh, A.R., "A Dual-pulsed Doppler Ultrasound and Its Applications in Determining Turbulent Blood Flow Characteristics," in a Ph.D. Dissertation, University of Oklahoma, October 1985.

3)   Cheung, J.Y., Ashrafzadeh, A.R., Dormer, K.J., "Determining Stenosis in Turbulent Blood Flow Using a Dual-Transducer Pulsed Doppler Ultrasound System," IEEE Ultrasonic Symposium, November 1986.

4)   Kahlilollahi, A. and Sharma, M.C., "Numerical Analysis of Blood Flow Through Stenosed Arteries," Proc. of IASTED International Conference, ACTA Press, pp. 167-169, 1984.

5)   Letelier, M.F. and Leutheusser, H.J., "Analytical Deduction of the Instantaneous Velocity Distribution, Wall Shear Stress and Pressure Gradient from Transcutaneous Measurements of the Time-Varying Rate of Blood Flow," Medical and Biological Engineering and Computing, pp. 443-436, July 1981.

6)   Welty, J.R., Wicks, C.E., and Wilson, R.E., in "Fundamentals of Momentum-Heat and Mass Transfer," John Wiley & Sons, Inc., 1976.

7)   Hussian, A.K.M.F. and Reynolds, W.C., "The Mechanics of an Organized Wave in Turbulent Shear Flow," Journal of Fluid Mechanics, Vol. 41, Part 2, pp. 241-158, 1970.

8)   Young, D.F., "Effect of a Time-Dependent Stenosis on Flow Through a Tube," Journal of Engineering for Industry, Vol. 90, pp. 248-154, 1968.

ULTRASOUND TWO-DIMENSIONAL FLOW MAPPING

USING FAST SPECTRAL ANALYSIS

Piero Tortoli, Fabio Andreuccetti and Carlo Atzeni

Electronic Engineering Department, University of Florence

v.S.Marta, 3   50139   Firenze, Italy

INTRODUCTION

Ultrasound systems have been recently introduced capable of providing real-time two-dimensional color-coded maps of blood flow |1|. Cross sectional images are constructed according to linear or sector formats as in conventional tomographic systems. Blood velocity is detected by exploiting the Doppler effect. A difference in color is used to distinguish the opposite flow directions, while intensity is related to the absolute value of the velocity.

Due to pulsatility of blood flow, the interpretation of doppler images requires frame rates in the range of 10-30 per second. Each image is usually composed of a number of raster lines just like echo images. The main differences are that a single line is obtained through several subsequent pulse repetition intervals (PRIs) and that the processing of the received signals is needed in order to extract the Doppler information.

Real time operation of 2-D blood flow imaging systems was realized through the development of multigate pulse doppler instruments capable of instantaneously measuring blood velocity profiles across the beam axis. Several systems have been experimented in the last few years, either based on parallel or serial processing of doppler signals. Different techniques for doppler frequency extraction have been reported, including zero-crossing counters |2| and moving target indicators (MTI) |3,4|. First real time clinical images, presented in 1982 |5|, were obtained through a multigate system employing an autocorrelation technique. The success of this method was mainly due to the low sensitivity to the signal to noise ratio, along with the capability of operating on sampled points obtained by a limited number of PRIs.

In this paper, the application to flow imaging of Fourier Transform (FT) based on Surface Acoustic Wave (SAW) dispersive devices is described. The processing speed is demonstrated adequate to perform real time doppler spectral analysis of all the lines constituting the two-dimensional map. The peculiarity of SAW-FT processors to manage a

limited number of points is shown capable of meeting the typical
requirements of a 2-D flow imaging system. The technique is first
demonstrated in a multigate instrument displaying the image of flow along
a single line of sight. Preliminary two-dimensional doppler images
reconstructed off-line are then presented.

PRINCIPLES OF OPERATION

In 2-D flow imaging systems, the frame rate (FR), the number of
raster lines (NL) and the number of consecutive pulses (NP) fired for each
line of sight, are related by

$$PRF = FR \times NL \times NP$$

Since PRF = 1/PRI is fixed by the maximum exploration depth, the
other three terms have to be accordingly selected. Due to the pulsatility
of the observed events, frame rates in the range of 10-30 per second are
needed. On the other hand, a high number of lines (from 32 to 128)
greatly improves image quality. Consequently, the only parameter which
can be reduced is the number of pulses NP. For example, typical values
such as PRF=10 KHz, FR=15 f/s, NL=32, yield to about 20 consecutive pulses
for each line. Low values of NP, on turn, involve a corresponding poor
frequency resolution, given by Df = PRF/NP.
Another aspect to be considered in the processing system design is
that each line of the image is composed of a number of pixels which are
at intervals of a quantity equal to that of resolution. As the imaged
structures can result extremely small, resolution lower than 1mm can be
necessary. For an analysed depth of several centimeters, as required for
heart imaging, a number of pixels M higher than 100 can be requested.
Correspondingly, when serial processing has to be performed, real time
operation requires a throughput rate not lower than M x PRF, which
represents the acquisition rate.
As an example, assume that a depth D=10 cm has to be imaged. The
maximum allowable PRF is therefore PRF = v/2D = 7.7 KHz (where v is the
velocity of ultrasound in the tissues). A minimum depth resolution of 1mm
implies that at least 100 pixels are acquired for each imaged line. Thus,
there is an acquisition rate of the order of several hundreds KHz,
imposing severe constraints to the subsequent processing chain.
The mentioned features of 2-D doppler systems, high processing rate
along with poor frequency resolution, well matches the characteristics of
SAW-based spectrum analysers. The typical configuration of a SAW analyser
is shown in fig.1. The input signal is multiplied by the linear FM chirp
waveform generated by a first (expansion) SAW filter. The product is sent

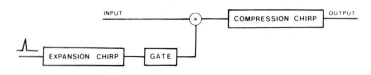

Fig.1 - Basic configuration of a SAW spectrum analyser

to a further (compression) SAW filter having an opposite FM slope. The RF output is then amplitude modulated by the input signal spectral amplitude. It can be demonstrated that if SAW filters with time-bandwidth product TxB = N are used, the analyser is equivalent to an N/4 point FFT unit. As an example, the SAW filters employed in the prototype system presented here, have 3 MHz bandwidth and T = 30 μs impulse response, thus providing, at a rate equivalent to 750 KHz, about 20 transform points. This represents a very low number when compared to the 64- or 128- point digital FFT units used in commercial CW or single-gate pulsed-doppler instruments. Despite of this fact, significant results have been obtained both "in vitro" |6| and "in vivo" |7|.

Dispersive delays T of SAW filters are limited within a few tens of μs, while bandwidths of several MHz are usually provided. Typical time-bandwidth products are thus of the order of some tens, and a frequency analysis of 20-30 points is exploited at MHz's rate.

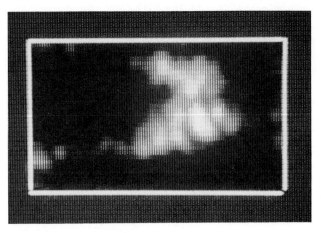

Fig.2 – Velocity profile detected in real time from a c.c. artery
(X-axis: velocity, Y-axis: depth, Z-axis: spectral amplitude)

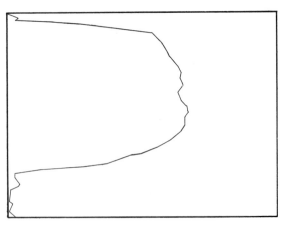

Fig.3 – Mean velocity profile detected from the spectral pattern
shown in fig.2

EXPERIMENTAL RESULTS

A prototype multigate system was developed, employing a SAW spectrum analyser for doppler frequency detection. It essentially consists of 4 blocks: an analogical transmitter-receiver unit, an 8-bit digital memory for data buffering, the SAW-FT unit, and a digital post-processing unit.

Doppler data from 32 independent sample volumes located along the ultrasonic beam, are collected over subsequent PRIs and stored in digital RAMs. Samples relative to each sample volume are then sequentially read out, feeding the SAW analyser according to a serial processing approach. Spectral analysis of each data block is completed in about 30 μs, so that 32 blocks are processed in less than 1 ms. Analog to digital conversion is then accomplished to allow the extraction of mean frequency to be performed in digital form for each spectrum. Further details of the system implementation are given in |6,7|.

Doppler data available at the SAW analyser output can be directly presented on an analogical display, according to the format shown in fig.2. It reports depth on vertical axis and doppler frequencies between 0 and PRF/2 on horizontal axis. Each line of the image displays spectral amplitudes coded in colors of different brightness. The velocity profile shown in fig.2 is relative to a human carotid artery and has been obtained by "freezing" the display during the systolic peak of the cardiac cycle.

Real time detection of mean frequency from each imaged line provides mean velocity profiles as the one shown in fig.3. This forms the basis of two-dimensional doppler images which come out when the ultrasonic beam is scanned along a given cross section. "In vitro" experiments have been conducted by impressing a mechanical motion to the probe and acquiring subsequent profiles on a computer. The scanned area was a rectangular cross section of a plastic tube with variable diameter, merged in water. A fluid with the same viscosity of the blood was forced to flow through

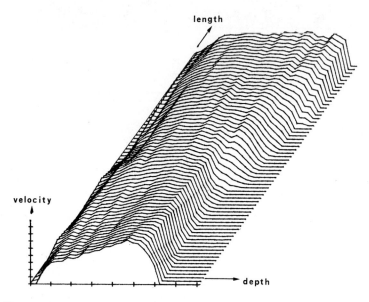

Fig.4 - Mean velocity profiles detected by scanning through parallel lines a tube with variable diameter

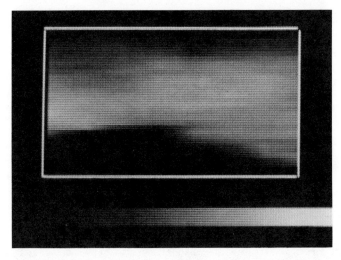

Fig.5 - Two-dimensional image obtained by the profiles shown in fig.4

the tube at a steady rate. Parallel scanning lines spaced about 1mm apart were subsequently taken, giving the profiles displayed in fig.4. This 3-D plot highlights the diameter reduction in the tube with a corresponding increasing of mean frequency in the middle of the lumen. In the 2-D color image shown in fig.5, this increase involves a corresponding major color brightness.

The system can also work in a bidirectional mode in order to show flow inversions. An illustrative experiment is reported in fig.6, relative to a stenosis simulated by fixing an iron obstacle in the center of the tube. The ultrasonic beam was scanned here in a two-dimensional linear format, allowing the color image in fig.6b to be reconstructed. Blue color corresponds to flow **towards** the probe, while red color highlights flow inversions arising distal from the simulated stenosis. Maximum velocity can be observed in correspondence of the maximum reduction of the vessel lumen.

CONCLUSION

Preliminary 2-D doppler images reconstructed off-line after impressing a mechanical motion to an ultrasonic probe have been presented. They were obtained by a multigate system based on a SAW-FT unit capable of performing doppler spectrum analysis at all points of a two-dimensional map in real time. Although presently limited in dynamic range by the 8-bit format of the buffer memory, the system was shown adequate for "in vitro" analysis and "in vivo" analysis of major vessels. Application of this kind of analysis to conditions where a lower S/N ratio and larger dynamic range must be managed requires a higher number of bits to be considered. Results obtained by the actual prototype system, however, appear encouraging in view of its possible clinical application.

LINES OF SIGHT

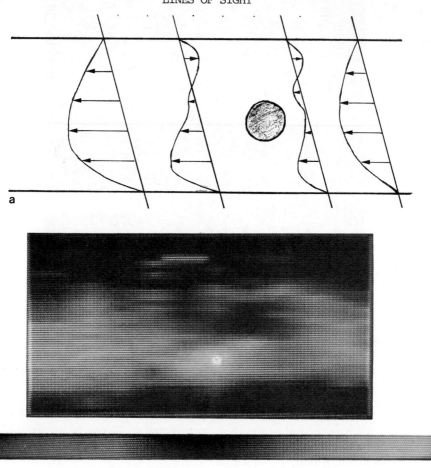

Fig.6 - 2-D Color image of a plastic tube with a simulated stenosis
(a - estimated profiles,   b - experimental profiles)

REFERENCES

|1| Roelandt J., ed.1986, "Color Doppler Flow Imaging", Martinus Nijhoff
    Pub., Dordrecht.
|2| Hoecks A.P., Reneman R.S., Peronneau P., 1981, "A multigate pulsed
    doppler system with serial data processing", IEEE Trans. on Sonics
    and Ultrason., SU-28, 242-247.
|3| Brandestini M., 1978, "Topoflow - A digital full range Doppler
    Velocity meter", IEEE Trans. on Sonics and Ultrason., SU-25, 287-
    293.
|4| Nowicki A., Reid J.M., 1981, "An infinite gate pulse Doppler",
    Ultrasound in Med. & Biol., vol.7, 41-50.
|5| Namekawa K., Kasai C., Tsukamoto M., Koijano A., 1983, "Realtime
    bloodflow imaging utilizing autocorrelation techniques", in:

Lerski R.A., Morley P. (ed.s), Ultrasound '82, 203-208.

|6| Tortoli P., Manes G. and Atzeni C., 1985, "Velocity profile reconstruction using ultrafast spectral analysis of doppler ultrasound", IEEE Trans. on Sonics and Ultrason., SU-32, 555-561.

|7| Tortoli P., Andreuccetti F., Manes G. and Atzeni C., 1985, "A range-doppler flowmeter for imaging of blood velocity profiles in arteries", IEEE Ultrasonics Symposium, 959-962.

PERFORMANCE EVALUATION OF PHASE-ONLY TECHNIQUE

FOR HIGH-RESOLUTION HOLOGRAPHIC IMAGING

Hua Lee and Jen-Hui Chuang

Department of Electrical and Computer Engineering
University of Illinois at Urbana-Champaign
Urbana, Illinois 61801 USA

Abstract- Phase-only image reconstruction is a widely used technique in holography. It utilizes only the phase information of the complex amplitude of the resultant wave-field for image formation. In many cases, the phase-only approach is advantageous over conventional methods in terms of data acquisition, storage, and computation complexity. In addition, it is capable of high-resolution imaging at extremely low bit-rates.

However, the resolving capability of the phase-only technique has been demonstrated for only the cases that the sources are consisting of a few point scatters or are limited to small localized regions. Suggestions are made that the performance and limitation of the phase-only method should be evaluated for generalized cases.

The purpose of this paper is to provide a quantitative performance evaluation of the phase-only imaging technique. We will first introduce the formulation to measure the estimation error of the phase-only method in both source and wave-field domains. Then we perform the error analysis and compare the resolving capability of image reconstructions by amplitude-only method as well as the standard backward propagation of the complete wave-field information. This approach for performance evaluation cannot only be used to predict image quality and limitations of phase-only or amplitude-only methods, but also be applied to optimize wave-field signal representation and quantization bit-rate distribution.

Introduction

Phase-only technique has been widely applied to image reconstruction and spectrum estimation[1,2]. This is to perform image reconstruction or signal restoration without the knowledge of the amplitude distribution. The signal available for processing only contains the phase information. This is also known as one of the topics in the area of signal reconstruction with partial information. The key objectives of the phase-only methods include data acquisition bit-rate reduction and computation simplification[3-7].

Previous research results pointed out that the phase information of the resultant wave-field is important to the resolution of the reconstruction, and is relatively more important than the amplitude information[3]. In addition to high-resolution reconstruction, the phase distribution is also capable of image enhancement especially for point scatters[4]. However, these findings are mainly based on demonstrations with very specific target distributions. Recently it has been suggested by researchers that the phase-only method needs generalized and quantitative study regarding the fundamental performance and limitation of this approach.

In this paper, we introduce the formulation of the normalized error as an indication of the performance of the phase-only technique. For comparison purpose, the normalized error for the amplitude-only reconstruction is also formulated. The formulation of the normalized error is achieved by introducing a complex scaling factor to the process. This wave-field domain scaling factor does not change the quality of the reconstruction due to the linearity of the image formation algorithm. This performance evaluation approach is based on the assumption that the quality of the reconstructed images has direct relationship to the wave-field detection error.

Detection Error and Image Quality

Consider a typical linear imaging system that the resultant wave-field distribution due to a radiating source can be represented by

$$y(x) = \int s(\alpha)\, h(x-\alpha)\, dx \tag{1}$$

where x is a three-dimensional position vector, $s(x)$ is the active source distribution, $y(x)$ is the resultant wave-field, and $h(x)$ is the wave-propagation kernel commonly known as the Green's function. Image reconstruction is to estimate the source distribution based on the detected wave-field signal. When the complete wave-field is available over the aperture, the quality of the reconstruction is mainly governed by the parameters such as the aperture size, wavelength, range distance, and noise level, which is often referred to as the Rayleigh criterion. When only partial wave-field information is available for image reconstruction due to the limited capacity of the data acquisition devices, additional image degradation is expected.

Image degradation can be regarded as the uncertainty or ambiguity of the estimate of the source distribution which is caused by the loss of information content during data acquisition. These degradation factors include the loss of spatial-frequency components due to the attenuation of the evanescent waves, the loss of the space-domain wave-field distribution due to the finite size receiving aperture, noise level, sampling effects, and quantization error. In this paper, we consider the degradation due to additional information loss of the amplitude or phase variations. We represent the resultant wave-field over the aperture as

$$y(x) = A(x)e^{j\theta(x)} \tag{2}$$

where $A(x)$ is the amplitude distribution and $\theta(x)$ denotes the phase variation. The information loss caused by finite frequency bandwidth, finite aperture size, noise, quantization, and sampling can be all characterized by linear truncation operations. However, the loss of amplitude or phase information is a nonlinear multiplicative process, and modifications will be needed for the representation of the detection error.

Phase-Only Data Acquisition

Phase-only approach is a holographic image reconstruction technique that only the phase variation of the resultant wave-field is used for the image formation. The amplitude information of the wave-field distribution is not available often because the coding bit-rate has been devoted to only the phase information. This unusual distribution of the coding bit-rate is largely due to the emphasis of the importance of the phase information in the resolution of the image reconstruction. In addition, phase-only methods are often applied for computation simplification purpose. For phase-only data acquisition systems, the available wave-field for image reconstruction is in the form of

$$\hat{y}(x) = e^{j\theta(x)}. \tag{3}$$

Due to the loss of the amplitude information, the detection error cannot be defined as the square error without modifications. It has been recognized that a reconstructed image represents a relative variation of a function and a scaling factor does not significantly change the information content of the image. Because of the linearity of the reconstruction algorithm, a scaling factor can be applied to either the wave-field or the resultant image. Therefore, we introduce a scaling factor k to the phase-only wave-field for the formulation of the detection error for the phase-only approach. Then the detection error can be evaluated as

$$E = \int_R |y(x) - k\hat{y}(x)|^2 dx = \int_R |A(x)e^{j\theta(x)} - ke^{j\theta(x)}|^2 dx$$

$$= \int_R |A(x) - k|^2 dx, \tag{4}$$

where R denotes the receiving aperture. We should also select a proper scaling factor corresponding to the minimum detection error. This can be achieved by

$$\frac{\partial}{\partial k}(\text{Detection Error}) = 0. \tag{5}$$

Then we have the condition

$$\frac{\partial}{\partial k} \int_R |A(x) - k|^2 dx = \int_R 2(k - A(x)) \, dx = 0. \tag{6}$$

And the optimal k value can be written as

$$\hat{k} = \frac{1}{D} \int_R A(x) \, dx, \tag{7}$$

where $D = \int_R dx$ is the total aperture size. For this particular case, the scaling factor k is a real and positive value which is mathematically the average of the amplitude variation. Accordingly, the minimum detection error is

$$E_{min} = \int_R |A(x) - \hat{k}|^2 dx = \int_R A^2(x) dx + \int_R \hat{k}^2 dx - 2\hat{k}\int_R A(x) dx$$

$$= \int_R A^2(x) dx + \hat{k}^2 D - 2\hat{k}^2 D = \int_R A^2(x) dx - \hat{k}^2 D$$

$$= \int_R A^2(x) dx - \frac{1}{D}[\int_R A(x) dx]^2. \tag{8}$$

It can be seen that range of the detection error depends mainly on the total wave-field energy. To produce a generalized parameter for comparison purpose, we normalize the error by the total energy of the wave-field.

$$\hat{E}_{min} = \frac{\text{Detection Error}}{\text{Total Wave-Field Energy}} = \frac{\int_R A^2(x) dx - \frac{1}{D}[\int_R A(x) dx]^2}{\int_R A^2(x) dx}$$

$$= 1 - \frac{[\int_R A(x) dx]^2}{D \int_R A^2(x) dx}, \tag{9}$$

The normalized detection error is bounded between 0 and 1.

Amplitude-Only Data Acquisition

In the previous section, we provided the formulation of the normalized detection error for phase-only techniques by introducing a scaling factor to the wave-field which can be used for quantitative measure of the image degradation. This measure will serve as the fundamental indicator of the performance of the phase-only image reconstruction. And this concept can be also applied to various cases of image formation with partial wave-field information.

One of the most interesting topics in signal processing and image reconstruction is the relative importance of the amplitude and phase information. Previous research results have been limited to demonstrations with very specific cases. In this section, we take the opportunity to quantitatively formulate the normalized detection error based on the same concept. For amplitude data acquisition systems, the detected wave-field can be simply represented as

$$\hat{y}(x) = A(x). \tag{10}$$

This is very common for data acquisition systems with power detectors. Similarly, we insert a scaling factor for the evaluation of the detection error

$$E = \int_R |y(x) - k\hat{y}(x)|^2 \, dx = \int_R |A(x)e^{j\theta(x)} - A(x)k|^2 \, dx$$

$$= \int_R A^2(x) |e^{j\theta(x)} - k|^2 \, dx. \tag{11}$$

The optimal value for the scaling complex scaling factor can be computed by differentiating the detection error with respect to k

$$\frac{\partial}{\partial k}(\text{Detection Error}) = \int_R A^2(x)[k^* - e^{-j\theta(x)}] \, dx$$

$$= 0. \tag{12}$$

Then we find the k value in closed form

$$\hat{k} = \frac{\int_R A^2(x)e^{j\theta(x)} \, dx}{\int_R A^2(x) \, dx}. \tag{13}$$

We should also point out that the optimal k value for the phase-only cases is a real and positive number, and the optimal k value for the amplitude-only case is a complex number with the magnitude smaller than one. Mathematically, we can describe this complex k value as the weighted average of the phase distribution of the wave-field over the receiving aperture. With the k value, we can subsequently compute for the minimum detection error for the amplitude-only cases as

$$E_{min} = \int_R A^2(x) |e^{j\theta(x)} - \hat{k}|^2 \, dx$$

$$= \int_R A^2(x)[1 + |\hat{k}|^2 - \hat{k}e^{-j\theta(x)} - \hat{k}^*e^{j\theta(x)}] \, dx$$

$$= (1 + |\hat{k}|^2)\int_R A^2(x) \, dx - \hat{k}\int_R A^2(x)e^{-j\theta(x)} \, dx - \hat{k}^*\int_R A^2(x)e^{j\theta(x)} \, dx. \tag{14}$$

From the relations

$$\hat{k}\int_R A^2(x)e^{-j\theta(x)} \, dx = \hat{k}\hat{k}^*\int_R A^2(x) \, dx$$

$$= |\hat{k}|^2\int_R A^2(x) \, dx \tag{15}$$

231

and

$$\hat{k}^* \int_R A^2(x)e^{j\theta(x)}\,dx = \hat{k}^*\hat{k} \int_R A^2(x)\,dx$$

$$= |\hat{k}|^2 \int_R A^2(x)\,dx, \tag{16}$$

we formulate the detection error in closed form as

$$E_{min} = (1 - |\hat{k}|^2) \int_R A^2(x)\,dx. \tag{17}$$

Normalizing with the total energy of the wave-field over the aperture, we now have the normalized detection error for the amplitude-only cases

$$\hat{E}_{min} = (1 - |\hat{k}|^2)$$

$$= 1 - \left[\frac{| \int_R A^2(x)e^{j\theta(x)}\,dx |}{\int_R A^2(x)\,dx} \right]^2. \tag{18}$$

The simulations are generated based on 100 uniformly-spaced wave-field samples with the half-wavelength sample spacing. For each set of simulation, we have three cases corresponding to the range distance of 25λ, 50λ, and 75λ. The source distributions are generated randomly with various source sizes. Fig.(1) shows the normalized detection errors in the wave-field domain of the phase-only and amplitude-only data acquisition formats. Fig.(2) shows the normalized estimation error of the phase-only and amplitude-only reconstructions with respect to the backward propagated images with complete wave-field information. And Fig.(3) shows the normalized estimation error distributions of the backward propagated reconstructions of phase-only, amplitude-only, and complete wave-field with respect to the exact source distributions.

Conclusion

The main purpose of this paper is to provide a quantitative performance evaluation of the phase-only holographic technique. The analysis is based on the assumption that the image quality is directly related to the wave-field detection error. This is because the estimation error of the source distribution is linear related to the wave-field error by the backward propagation reconstruction filter. Therefore, the source-domain performance evaluation of phase-only method can be alternatively performed in the wave-field domain by evaluating the wave-field detection error.

Because the phase-only or amplitude-only data acquisition formats are nonlinear operations, we need to modify the definition of detection error with a complex scaling factor. This scaling factor does not change the image quality due to the linearity of the reconstruction process. As a result, we formulated the normalized detection error

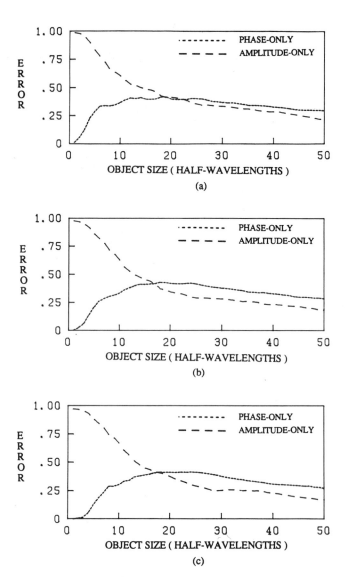

Fig. 1. The normalized detection errors in the wave-field domain of the phase-only and amplitude-only data acquisition formats for range distances equal to (a) 25λ, (b) 50λ, and (c) 75λ.

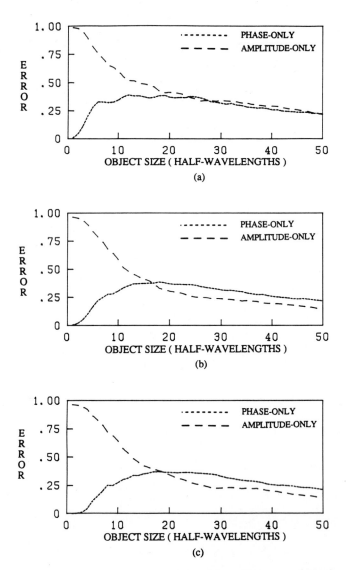

Fig. 2. The normalized estimation error of the phase-only and amplitude-only reconstructions with respect to the backward propagated images with complete wave-field information for range distances equal to (a) 25λ, (b) 50λ, and (c) 75λ.

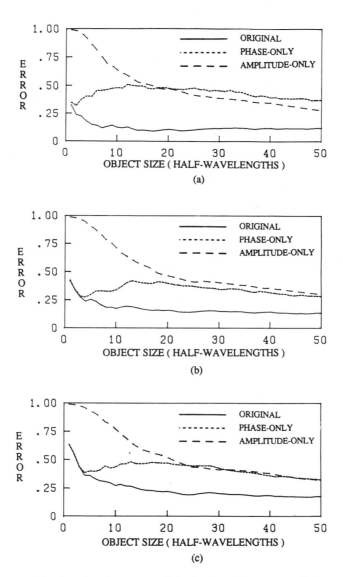

Fig. 3. The normalized estimation error of the backward propagated reconstructions of phase-only, amplitude-only, and complete wave-field with respect to the exact source distributions for range distances equal to (a) 25λ, (b) 50λ, and (c) 75λ.

of the phase-only data acquisition format in closed form which can be used as an indication for performance evaluation. In addition, the normalized detection error for the amplitude-only approach is also formulated for comparison purpose. It can be seen that the detection error is signal-dependent. The simulations indicate that the performance of the phase-only method is consistent and better than the amplitude-only method especially for limited-size source distributions.

This approach for performance evaluation can be also applied to the study of optimal wave-field coding or quantization schemes. In addition, the limitation of phase-only or amplitude-only spectral estimation can be also analyzed based on this concept.

Acknowledgment

This research is supported by the National Science Foundation under Grants IST-8409633 and ENG-8451484, Motorola Inc., and Hughes Aircraft Company. The authors also wish to thank Professor John Powers of the Naval Postgraduate School for his valuable suggestions related to this research topic.

References

1. A. F. Metherell, "The Relative Importance of Phase and Amplitude in Acoustical Holography," *Acoustical Holography,* vol. 2, Metherell Ed., Plenum Press, Chapter 14, 1969.

2. John Power, John Landry, and Glen Wade, "Computed Reconstructions from Phase-Only and Amplitude-Only Holograms," *Acoustical Holography,* vol 2, Metherell Ed., Plenum Press, Chapter 13, 1969.

3. A. V. Oppenheim and J. S. Lim, "The Importance of Phase in Signals," *IEEE Proceedings,* vol. 69, no. 5, pp.529-541, May 1981.

4. Hua Lee and Glen Wade, "Resolution Enhancement on Phase-Only Reconstructions," *IEEE Transactions on Sonics and Ultrasonics,* vol. SU-29, no. 5, pp. 248-250, September 1982.

5. Hua Lee and Glen Wade, "Evaluating Quantization Error in Phase-Only Holograms," *IEEE Transactions on Sonics and Ultrasonics,* vol. SU-29, no. 5, pp. 251-254, September 1982.

6. Hua Lee and Glen Wade, "Analysis and Processing of Phase-Only Reconstruction in Acoustical Imaging," *Proceedings on Physics and Engineering in Medical Imaging,* IEEE Computer Society Press, pp. 240-246, 1982.

7. Hua Lee and Jen-Hui Chuang, "Computation Simplification for High-Speed Acoustical Image Reconstruction," *Acoustical Imaging,* vol. 15, H. Jones Ed., Plenum Press, New York, 1987.

DIFFRACTION TOMOGRAPHY ALGORITHMS USING

THE TOTAL SCATTERED FIELD

Marc L. Feron and Jafar Saniie

Department of Electrical and Computer Engineering
Illinois Institute of Technology
Chicago, IL 60616

INTRODUCTION

Diffraction tomography based on the back–scattered field and the concurrent use of the transmitted and back–scattered field have received little attention. Using the total scattered field may improve image resolution, which is due to an increased coverage of the wavenumber domain. Contour information (i.e. a bandpass filtered version of the image) is located in a ring of the spatial Fourier domain and can be extracted from the backscattered field only, while grey level information (i.e. a lowpass filtered version of the image) is located in a disk centered at the origin and is accessible from the forward scattered field. Therefore, using only the forward scattered field amounts to discarding some of the image data and results in images of inferior quality.

A possible reason why the backscattered field has often been ignored is that it usually contains less energy than the forward scattered field due to the assumption of small index variation. The backscattered information has the potential to add to the intelligibility of the reconstructed image through the availability of higher frequencies in the spatial Fourier data. Moreover, one can argue that such information could not be obtained by ad-hoc signal processing (e.g. extrapolation techniques, window design...)

In this report, we present an interpretation of the Fourier diffraction theorem using the Born approximation for an arbitrary insonifying angle. This interpretation is supported by simulations using *exact* solutions for the total scattered field of homogeneous cylinders. The performance of the Fourier diffraction theorem and its ability to reconstruct different sizes of cylinders and indices of refraction is discussed.

THEORETICAL BACKGROUND

High resolution images in diffraction tomography are obtained when the Fourier domain frequency components are reproduced accurately, especially for the higher frequency components that contain the detail information. Diffraction tomography algorithms are derived (e.g. [1,2]) by expressing the wave equation in an integral form suitable for application of a small perturbation method. The wave function, $U(r)$, is measured on a detector placed in the far field of the object and the relative refraction index $n(r)$ is calculated in every point r within the object. Images can then be obtained by displaying the refraction index. If k_0 is the wavenumber of the incident wave, the Helmholtz wave equation can be written:

$$(\nabla^2 + k_0^2 n^2(r))U(r) = 0 \tag{1}$$

where ∇^2 is the Laplacian operator. Let U_0 represent the field that would exist if no object were present and $U_s = U - U_0$ represent the scattered field component of the actual wave function U. Then,

$$(\nabla^2 + k_0^2)U_s(r) = -k_0^2(n^2(r) - 1)U(r) \tag{2}$$

The integral form of this wave equation, using the Green function g_{k_0} associated with equation (2), becomes

$$U_s(r) = \int_{object} I(r')U(r')g_{k_0}(r - r')dr' \tag{3}$$

where the quantity $I(r) = -k_0^2(n^2(r) - 1)$ is our unknown "image function". Note that $I(r) = 0$ outside the object.

Direct inversion of this equation is not possible, however, using the small perturbations approximation (Born or Rytov) results in approximate solutions for $I(r)$ [1,2]. This type of solution yields the Fourier Diffraction Theorem. The exact expression of the Fourier diffraction theorem is dependent on the detector geometry and we present the derivation of the two dimensional Fourier diffraction theorem utilizing the Born approximation for an arbitrary angle of insonification Ψ, as shown in Fig. 1. Consequently, Eq. 3 becomes:

$$U_s(r) = \int_{object} I(r')U_0(r')g_{k_0}(r - r')dr' \tag{4}$$

where [3],

$$g_{k_0}(r - r') = \frac{j}{4\pi} \int_{-\infty}^{\infty} \frac{e^{j[\alpha(x-x') + \beta(\alpha)|y-y'|]}}{\beta(\alpha)} d\alpha \tag{5}$$

$$\beta(\alpha) = \pm\sqrt{k_o^2 - \alpha^2}$$

$$r = (x, y); \qquad r' = (x', y')$$

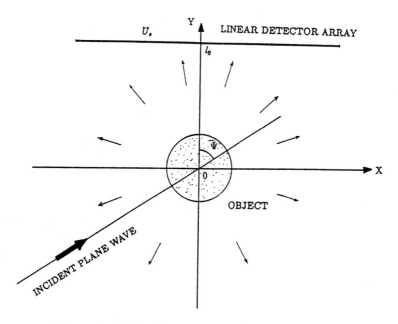

Figure 1: Diffraction tomography with an
arbitrary detector angle Ψ.
The object is rotated after each measurement.

Eq. 5 expresses a cylindrical wave made up of an infinite sum of plane waves. Each
plane wave is of wavenumber k_0 and propagates at angle $\phi = \tan^{-1} \frac{\beta(\alpha)}{\alpha}$. It should
be noted that for $|\alpha| > k_0$ an evanescent wave is obtained because β is imaginary.
By placing the detector in the far field of the object, these evanescent waves can be
ignored. The equation of the incident plane wave is:

$$U_0(r') = e^{jk_0(x' \sin \Psi + y' \cos \Psi)} \tag{6}$$

An infinite linear detector array is placed at $y = l_0$ in the far field of the object and
parallel to the x axis. The derivation of the Fourier diffraction theorem requires the
calculation of the scattered field detected at $r = (x, l_o)$ of this array. After some
manipulation, the Fourier diffraction theorem for an arbitrary angle of insonification
Ψ is obtained:

$$\tilde{I}(k - k_0 \sin \Psi, \sqrt{k_0^2 - k^2} - k_0 \cos \Psi) = -2j \sqrt{k_0^2 - k^2} \exp(-j \sqrt{k_0^2 - k^2} l_0) \tilde{U}_s(k, l_0) \tag{7}$$

In these equations $\tilde{U}_s$ represents the one–dimensional spatial Fourier transform of the
measured wave function U_s (measured on an infinite detector line at distance l_0 from
the object) and $\tilde{I}$ represents the two–dimensional spatial Fourier transform of the
image function.

239

INTERPRETATION OF THE FOURIER DIFFRACTION THEOREM

The Fourier diffraction theorem presented in Eq. 7 was derived by using an arbitrary insonifying angle Ψ and a linear detector. The two–dimensional Fourier transform of the image function is measured on a semicircular contour of center $C: (-k_0 \sin\psi, -k_0 \cos\psi)$, radius k_0, base parallel to the detector and convexity turned towards the detector, as shown in Figure 2. It must be noted that the proportionality factor $-2j\sqrt{k_0^2 - k^2}\exp(-j\sqrt{k_0^2 - k^2}l_0)$ goes to zero towards the extremities of the semicircle ($k = \pm k_0$). The reconstructed components of the image function in this area are likely to be very small and possibly unreliable since the scattered field U_s has a finite energy. Because of this irregular weighting pattern, certain angles Ψ will provide a more reliable reconstruction of certain wavenumbers than others, and the data that can be considered reliable is projected in the central region of the semicircle.

Low spatial frequency information is obtained by placing the detector array behind the object ($\Psi = 0$) to measure transmitted waves. The middle of the semicircle is positioned on the origin and thus corresponds to information with low spatial frequency contents (for clarity, see Fig. 2b). This will yield a low pass filtered version of the cross–section being imaged and therefore will not give much detail, such as boundary or high contrast information. Furthermore, ripple artifacts are likely to appear at sharp object boundaries due to the truncation of the object's two–dimensional Fourier transform at a wavenumber $\sqrt{2}k_0$.

Higher wavenumber information is extracted by positioning the array on the same side as the source ($\Psi = \pi$) to measure backscattered waves. The backscattered data supplies information in a ring of the wavenumber domain of inner radius $\sqrt{2}k_0$ and outer radius $2k_0$ and corresponds to the detail, boundary and high contrast information (Fig. 2c). The bandwidth of the backscattered information is limited, so it is unlikely that meaningful images can be obtained from backscattered data alone. Nevertheless, the concurrent use of the forward and backscattered field permits the bandwidth of the reconstructed image to be improved by the thickness of the ring (an improvement of 40% over the bandwidth attainable with the transmitted wavefield only).

The maximum spatial frequency information can be recovered by placing a single detector array parallel to the direction of the incident wave ($\Psi = \frac{\pi}{2}$ or $-\frac{\pi}{2}$); in theory the complete $2k_0$ wavenumber disk can be obtained with a single detector parallel to the insonifying wave. This may seem very attractive from the analytical point of view: if the detector is of infinite length, then all of the object wavenumber information can be recovered with a single detector (Fig. 2d). In practice, however, small weights towards the extremities of the semicircle and a finite detector length may cause the data to be unreliable at the extremities of the semicircle. The consequence would be inaccurate low frequency contents which can result in large image artifacts. Consequently, the use of a near-normal angle may turn out to be a better alternative.

Complete coverage of the wavenumber domain can be obtained by rotating the object relative to the source and detector geometry between measurements. The image function $I(r)$ can then be retrieved via a Fourier domain interpolation followed by a two–dimensional inverse Fourier transform. It should be noted that by using proper

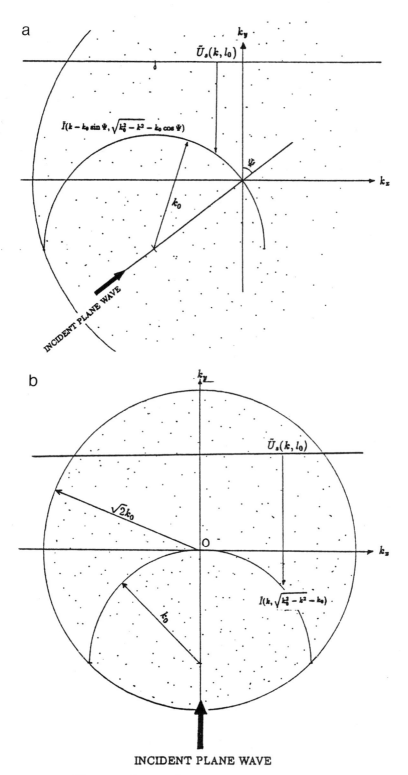

Figure 2: Fourier domain interpretation of the
Fourier diffraction theorem:
a) arbitrary insonification angle Ψ;
b) $\Psi = 0$ (classical or forward scattering case).

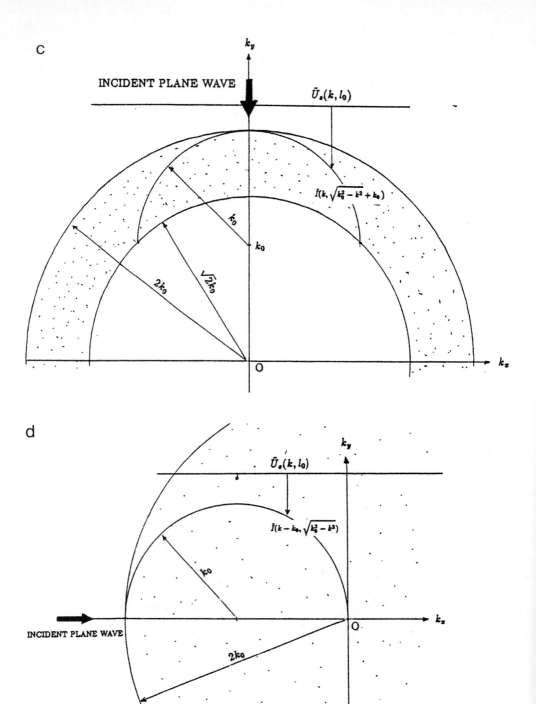

Figure 2: Fourier domain interpretation of the
Fourier diffraction theorem:
c) $\Psi = \pi$ (back scattering case);
d) $\Psi = \frac{\pi}{2}$ (Total scattered field).
The regions of availability of the Fourier data
correspond to the darker areas.

scanning procedures [4], the need for interpolation in the spatial Fourier domain can be virtually eliminated. Also, the use of a Hankel transform instead of a two–dimensional Fourier transform for circularly symmetric objects yields an interpolation free image cross–section.

Single frequency diffraction tomography has an inherent resolution limit because information about the Fourier transform of the image function for wavenumbers greater than $2k_0$ cannot be obtained. Therefore, the best possible resolution is on the order of $\frac{\pi}{2k_0}$. As k_0 increases, the coverage of the wavenumber domain increases and a better resolution is achieved. On the other hand, as frequency increases, the algorithm's tolerance to index variation decreases significantly and, from the practical point of view the effects of attenuation may become intolerable.

EFFECTS OF FINITE DETECTOR LENGTH

It is clear from Fig. 1 that reducing the detector length will result in the non- detection of certain scattered waves issued from the object, which causes a loss of information in the reconstruction and further degradation in resolution. One method to obtain a full coverage of the Ewald circle is to intercept all the scattered waves by surrounding the object section with a piecewise linear detector and applying the expression of the Fourier diffraction theorem given in Eq. 7 repeatedly for different values of Ψ. This technique also has the advantage that unreliable portions of the semicircle can be discarded since different values of Ψ will give better estimates of the discarded sections.

Another interpretation of the effects of finite detector length is obtained if the Fourier transform $\tilde{U}_s(k, l_0)$ of the scattered field $U_s(x, l_0)$ is calculated using only the scattered field data available on the detector between points $(-a, l_0)$ and (a, l_0) and assuming that the field is zero elsewhere (since it cannot be measured). Then, the Fourier transform $\tilde{U}_s^T$ of the truncated field is

$$\tilde{U}_s^T(k, l_0) = \int_{-a}^{a} U_s(x, l_0) e^{-jkx} \, dx \tag{8}$$

the relation between the measured scattered field and the actual image function can be found from Eq. 4 and Eq. 8:

$$\tilde{U}_s^T(k, l_0) = \frac{ja}{2\pi} \int_{-\infty}^{\infty} \frac{e^{j\beta(\alpha)l_0}}{\beta(\alpha)} \tilde{I}(\alpha - k_0 \sin \Psi, \beta(\alpha) - k_0 \cos \Psi) \text{sinc}((\alpha - k)a) d\alpha \tag{9}$$

The above equation is essentially the convolution of Eq. 7 with a sinc function related to the finite detector length. This convolution has a smoothing effect on the scattered field Fourier transform and will introduce image errors, principally where the transform image data is small (i.e. around $k = \pm k_0$). In the limit, as the detector length approaches infinity, the sinc function will tend to a delta function and the smearing errors will decrease. Direct implementation of Eq. 9 seems impractical for the determination of the image function since the computational advantage of the Fourier diffraction theorem is lost. Nevertheless, it is helpful in assessing the consequences of the use of the Fourier diffraction theorem in the finite length case since it permits the calculation of $\tilde{U}_s$ for a finite length detector from the knowledge of the spatial Fourier transform of the object. Replacing the actual $\tilde{U}_s$ by $\tilde{U}_s^T$ in Eq. 7, we obtain:

$$\tilde{I}_{\text{reconstructed}}(k - k_0 \sin \Psi, \sqrt{k_0^2 - k^2} - k_0 \cos \Psi) = \frac{a}{\pi}\sqrt{k_0^2 - k^2}e^{-j\sqrt{k_0^2-k^2}l_0}$$

$$\int_{-\infty}^{\infty} \frac{e^{j\beta(\alpha)l_0}}{\beta(\alpha)}\tilde{I}(\alpha - k_0 \sin \Psi, \beta(\alpha) - k_0 \cos \Psi)\text{sinc}((\alpha - k)a)d\alpha \qquad (10)$$

The above equation provides a basis for evaluating the distortion of the reconstructed image due to the finite length of the detector.

METHOD OF EVALUATION

Diffraction tomography algorithms using the Fourier diffraction theorem and various detector geometries often cause reconstruction errors which are function of the object size and index as well as the detector length, spacing, angle, etc... Proper evaluation of reconstruction errors can only be carried out if an exact solution to the scattered field of an object exists. However, such a solution is generally not available, except in the case of homogeneous cylinders. The diffraction pattern of a homogeneous cylinder insonified by a plane wave can be found by solving the Helmholtz wave equation using a wave decomposition method and applying appropriate boundary conditions at the surface of the cylinder [5–7].

Let a homogeneous cylinder of infinite length, index $n = n_1$ and radius a be placed at the origin, and an insonifying plane wave $U_0(x, y)e^{-j\omega t} = Ae^{j(k_0 x - \omega t)}$ travel in the OX direction in the propagating medium of index $n = 1$. Due to the geometry of the problem, it is convenient at this point to use cylindrical or polar coordinates. The Helmholtz wave equation becomes:

$$\frac{1}{r}\frac{\partial}{\partial r}\left(r\frac{\partial U(r, \theta)}{\partial r}\right) + \frac{1}{r^2}\frac{\partial^2(U(r, \theta)}{\partial \theta^2} + n^2 k_0^2 U(r, \theta) = 0 \qquad (11)$$

It is convenient to separate the field U into the incident field $U_0 = Ae^{jk_0 r \cos \theta}$ and the unknown scattered field U_s such that $U = U_0 + U_s$. The general solution for the scattered field is obtained by the method of separation of variables:

$$U_s(r, \theta) = \sum_{m=0}^{\infty} \beta_m \cos(m\theta)H_m^{(1)}(k_0 r) \qquad (12)$$

where $H_m^{(1)} = J_m + jN_m$ is the Hankel function of type 1 and order m, noted H_m from this point on. Using the continuity of the pressure and normal particle velocity at the cylinder boundary, the coefficients β_m can be shown to be:

$$\beta_m = \xi_m A j^m \frac{\frac{n_1 \rho_0}{\rho_1} J_m'(n_1 k_0 a) J_m(k_0 a) - J_m'(k_0 a) J_m(n_1 k_0 a)}{J_m(n_1 k_0 a) H_m'(k_0 a) - \frac{n_1 \rho_0}{\rho_1} J_m'(n_1 k_0 a) H_m(k_0 a)} \tag{13}$$

Through computer simulations, it has been observed that the coefficients β_m seem to decay very rapidly for orders m greater than the circumference of the cylinder expressed in wavelengths. This characteristic implies that: i) the scattered field is spatially band limited so that aliasing will not occur if the detector spacing is adequately small; and ii) the scattered field can be reconstructed *exactly* with a finite sum (this is critical for computer simulations).

The above expression depends on three cylinder parameters: $k_0 a = 2\pi \frac{a}{\lambda_0}$ which is equal to the cylinder circumference expressed in wavelengths, the relative index n_1 of the homogeneous cylinder, and the relative density of the cylinder with respect to the medium $\frac{\rho_1}{\rho_0}$. The solution to Eq. 12 and Eq. 13, although computationally heavy, is much preferable to the methods making use of the small perturbation approximations commonly used for wavefield simulation.

In summary, our simulation method is based on Eqs. 7 , 12 and 13 and consists of the following steps: i) calculation of the exact scattered field on the detectors; ii) discrete Fourier transform of the scattered field; iii) calculation of the projections on the semicircle; iv) inverse two–dimensional Fourier transform, preferably without any interpolation or Hankel transform if the object has circular symmetry; and v) error characterization.

SIMULATION RESULTS

Computer simulations were carried out without any approximation or interpolation except for the two approximations inherent to the Fourier diffraction theorem: the Born approximation in the integral equation (Eq. 4) and the far field approximation (Eq. 5). It was observed that the reconstructions obtained with the standard Fourier diffraction theorem yield good results for index variations of less than 3% but break down for higher values of the index variation. For example, a cylinder of diameter 16 wavelengths and index 1.01 is reconstructed with negligible error (see Fig. 3a) but a reconstruction of the same object with an index of 1.03 results in unacceptable distortion (see Fig. 3b). Since the index variation found in medical applications is usually larger than this threshold, a worthwhile goal for future research seems to be a more precise method of approximations for the Fourier diffraction theorem. It was also observed that in certain cases, the far field requirement of the Fourier diffraction theorem is not very stringent. For example, acceptable reconstructions have been obtained with the detector directly in contact with a cylinder of diameter 6 wavelengths.

The exact cylinder radiation patterns were examined in all our simulations and results show that the size and number of the sidelobes increased with the index, the size of the object and its relative density. The radiation patterns reveal that there is very little backscattered field for cylinders with small index and density variations (see Fig. 4a). However,the backscattered field increases significantly when either the index or the relative density increases (see Figs. 4b and 4c), and an improved diffraction tomography algorithm should take those fields into account.

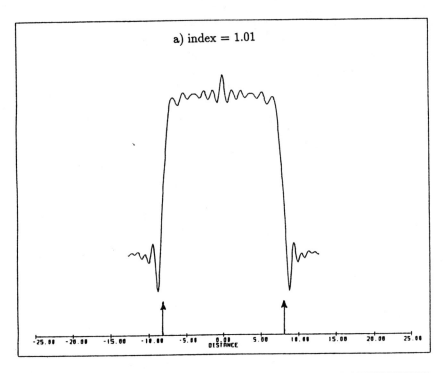

a) index = 1.01

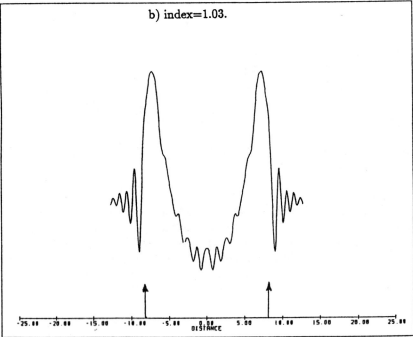

b) index=1.03.

Figure 3: Reconstruction of two cylinders of diameter equal to 16 wavelengths, density ratio = 1

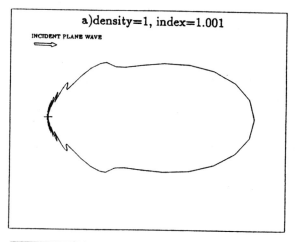

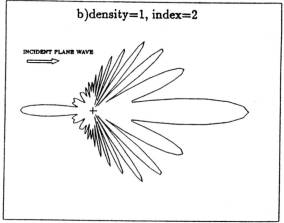

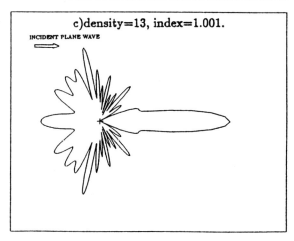

Figure 4: Cylinder radiation patterns
(diameter = 6 wavelengths)

Present simulations taking both the forward and backscattered field into account were not successful, largely because the data resulting from the forward scattered field of a cylinder of large index causes the Born approximation to break down. It was also noticed that small density variations have relatively little influence on the reconstruction quality. Further, and despite the fact that the Fourier diffraction theorem does not take density variations into account, the boundary of a cylinder of diameter equal to 6 wavelengths, index 1.01 and relative density 13 (mercury) was identifiable, although the low spatial frequency information was in error (i.e. the cylinder was not reconstructed properly).

REFERENCES

[1] R.K. Mueller, M. Kaveh and G. Wade, "Reconstructive Tomography and Application to Ultrasonics", Proceedings of the IEEE, Vol. 67 No. 4, pp. 567-587, 1979.

[2] M. Slaney and A.C. Kak, "Imaging with Diffraction Tomography", Research report TR-EE 85-5, School of Electrical Engineering, Purdue University, February 1985.

[3] P.M. Morse and H. Fehsbach, "Methods of Theoretical Physics", McGraw Hill, New York, 1953.

[4] D. Nahamoo, S.X. Pan and A.C. Kak, "Synthetic Aperture Diffraction Tomography and its Interpolation-Free Computer Implementation", IEEE Transactions on Sonics and Ultrasonics, Vol. SU-31, pp. 218-229, 1984.

[5] M. Azimi, "Tomographic Imaging with Diffracting Sources", PhD. Thesis, Purdue University, 1985.

[6] P.M. Morse and K.U. Ingard, "Theoretical Acoustics", McGraw Hill, New York, 1953.

[7] H.C. van de Hulst, "Light Scattering by Small Particles", Wiley, New York, 1957.

EFFICIENT SAMPLING FOR NEARFIELD ACOUSTIC HOLOGRAPHY

S. M. Gleason and Anna L. Pate

Department of Engineering Science and Mechanics
Iowa State University
Ames, Iowa 50011

ABSTRACT

The purpose of this investigation was to determine an efficient means of sampling and windowing data for the digital signal processing that is required for nearfield acoustic holography (NAH). In particular, a method for choosing the best grid spacing for data in the hologram plane was considered.

The pressure distribution over the surface of a baffled piston calculated by using the back-projected nearfield holography was of particular interest in this research. The effect of two sampling parameters, namely the aperture size and the sampling rate, were studied numerically. The recommended values for the efficient sampling parameters were developed.

A second aspect of this research concerned the location of the grid of the source. A power-of-two FFT algorithm is often used for the transformations, thus necessitating an even number of points in the sample. Centering this grid directly above the circular piston source left no data point on the axis of the piston. The nearfield of a piston vibrating at high frequencies had sharp peaks or depressions on the axis, so a centered grid with a relatively small spacing still missed a substantial portion of the on-axis maximum or minimum value. A routine was used to calculate the pressure on a grid that contained the on-axis value.

INTRODUCTION

Several researchers [1-4] have demonstrated that NAH is a powerful technique that can be used for the imaging of acoustic sources. NAH has been developed as an extension of generalized acoustic holography (GAH). However, NAH is distinct in that it attempts to preserve the fast-decaying evanescent waves generated by acoustical sources. In addition, NAH provides means of efficient back-propagation of these evanescent waves, along with the waves propagated into the far field.

The evanescent waves have been ignored in GAH because there are several difficulties associated with them. One of the most important difficulties is that the evanescent waves decay very fast when propagating away from a source; for example, in planar holography the decay is exponential. Therefore, reliable measurements of the evanescent waves need to be performed in the nearfield of a source, and a large dynamic range of the data acquisition system is required. In addition, great care is required in signal processing of the evanescent waves. Significant errors can arise from the back-propagation of the evanescent waves, since the back-propagator in this case is described by a real, exponentially growing function. Several ways of dealing with this problem by filtering in a spatial frequency domain have been presented in the literature [2].

NAH is a relatively new technique; therefore, the efficient sampling required in this method is not completely understood yet. Several excellent experimental studies [1,3,4] have provided recommendations for the appropriate choices of some sampling parameters, such as the aperture size and the sampling rate. However, these were usually developed for a specific source and not generally valid for other sources. In addition, it is not known how sensitive the holographic results would be to changes in the sampling parameters.

This paper reports on a numerical investigation of the back-projected NAH used for a baffled piston. The baffled piston was chosen for convenience in calculations and also for its directivity properties, which could be changed easily with the frequency. The effects of several sampling parameters were investigated in order to establish recommended values for the aperture size and the sampling rate. Several examples were shown to demonstrate how the piston image depended on the number of samples, the size of the aperture, and the spatial frequency and the real space windowing.

Theoretical Formulations of the Back-projected Reconstruction

The theory of NAH was applied to a baffled, rigid piston radiating into a free half-space, where the homogeneous Helmholtz equation was satisfied as

$$\nabla^2 p(\bar{r}) + k^2 p(\bar{r}) = 0 \tag{1}$$

where $p(\bar{r})$ is the acoustic pressure at location $\bar{r}(x,y,z)$ in the half-space. By using the basic theory of acoustic holography [1], one can relate the pressure in the plane of the piston (x,y,z_o) and the pressure in any parallel plane (x,y,z_H) as

$$p(x,y,z_o) = F^{-1}\{\hat{P}(k_x,k_y,z_H) \cdot \hat{G}(k_x,k_y,z_o - z_H)\} \tag{2}$$

where

F^{-1} denotes the inverse Fourier transformation

$\hat{P}(k_x,k_y,z_H)$ is the Fourier transform of the pressure in the hologram plane

$$\hat{G}(k_x,k_y,z_o - z_H) = \exp(jk_z(z_o - z_H)) \tag{3}$$

is the Fourier transform of the half-space Green function, and

$$k_z = \sqrt{k^2 - k_x^2 - k_y^2} \ .$$

The evanescent components of the spectrum occur when k_z is imaginary since $k_x^2 + k_y^2 > k_z^2$. The Fourier transform of the Green's function then becomes a function that increases exponentially for the negative distance of transformation (back-projection).

In this paper the back-projection with NAH was used for a baffled piston of 0.05-m radius, as shown in Fig. 1. The piston vibrated with sinusoidal velocity magnitude of $U_o = 1$ m/s and frequency of 1085.44 Hz (ka = 1) or 10,854.4 Hz (ka = 10). The pressure data in the hologram plane (x,y,z_H) can be calculated using the Rayleigh integral [5]

$$p(x,y,z_H) = j \frac{\rho_o ck}{2\pi} U_o \int_S \frac{e^{-jkR}}{R} dS \tag{4}$$

where ρ_o is the air density, c is the sound speed in air, k is the wave number, S is the surface of the piston, and R is the distance from points on the piston to a given field point.

Equation (4) was converted into a single integral with fixed limits using the method presented by Archer-Hall et al. [6]. In addition, the same method was used to calculate the pressure at the surface of the piston $p(x,y,z_o)$. These pressure values were calculated for comparison purposes and were called exact herein.

The objective of NAH was to calculate the discrete pressure values in the piston plane $p(x_i,y_i,z_o)$ from the discrete pressure data in the hologram plane $p(x_i,y_i,z_H)$ using the back-projection described in Eq. (2). The distance of transformation was chosen as $d = z_o - z_H = \lambda/10$. This value of d provided that the hologram plane was always close enough to the source to allow using the NAH method.

Numerical Investigations of the Sampling Parameters

Two different sampling parameters were investigated: the aperture size (2D) and the sampling rate in the real space (Δ). Only square arrays of N × N data points were considered with the square spacing of $\Delta(\Delta_x,\Delta_y)$. The square arrays were chosen because of the symmetry of the investigated source (the circular piston shown in Fig. 1).

Maynard et al. [1] described the problem of the wraparound error due to the spatial replication. In the FFT calculations performed according to Eq. (2), the input data $\hat{P}(k_x,k_y,z_H)$ are assumed to be periodic. Therefore, the repeated images of the original source are multiplied by the Green's function in the reconstruction routine and consequently contribute incorrectly to the final result. The typical method used to avoid the wraparound errors is to use the large aperture and to add zeros in the input data.

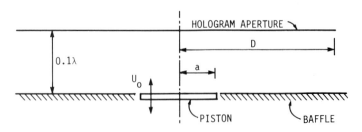

Fig. 1. Geometry of the source.

In this numerical investigation we started with the large aperture size and very fine sampling rate to produce high-quality back-propagated images. We then reduced the aperture size until the errors became unacceptable. In each case, zero padding in the real space was used. The quality of the image was evaluated by the averaged error between the image pressure and the exact pressure in the projection plane.

Similarly, the sampling rate was investigated by observing the image degradation when the sampling rate was reduced from the very fine values to coarsely sampled points in the same aperture.

RESULTS

The back-projected images for two sources were calculated by using the NAH technique. Two different baffled pistons were considered; these were of the same radius (a = 0.05 m) but vibrated at frequency 1085.44 Hz (ka = 1) or 10854.4 Hz (ka = 10). The low frequency case (ka = 1) simulated a nondirectional source that required a large aperture. In contrast, the high frequency case (ka = 10) simulated a quite directional source; the size of the cross-section of the radiation beam was comparable with the size of the source.

The significance of the NAH method is the inclusion of the evanescent waves in the image calculations. Thus, the two sources chosen for this study represented two different cases in terms of the content of the evanescent waves. The pressure magnitudes generated at both sources are shown in Fig. 2. The magnitudes of the pressure spectra are shown in Fig. 3. The low frequency source generated a very large amount of the

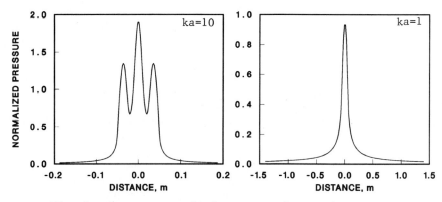

Fig. 2. Pressure magnitude generated at each source.

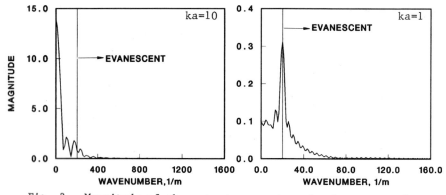

Fig. 3. Magnitude of the pressure spectrum in the source plane.

evanescent waves. The integrated spectrum within the evanescent range of frequencies amounted to 75% of the integrated spectrum within the propagating frequencies. On the contrary, the high frequency source generated a relatively small amount of the evanescent waves. In this case the integrated evanescent spectrum was only 5% of the propagating spectrum. The pressures calculated for these two sources without including the evanescent waves are shown in Fig. 4. It is evident that the quality of the image of the high frequency source is quite good, while the image of the low frequency source is very poor. Consequently, the pressure near the high frequency source could be calculated using GAH. However, the pressure calculations over the low frequency source would require the nearfield holography technique.

Pressure images obtained through the back-projected NAH method (projections from the distance $\Delta = \lambda/10$ to the source itself) are presented in Fig. 5. In this figure several sampling rates are included while the size of the aperture is kept constant; the aperture size was chosen such that the pressure magnitude over the aperture decayed up to 99% of the maximum value, which occurred at the center of the aperture. It is evident that this aperture size provided extremely accurate results when

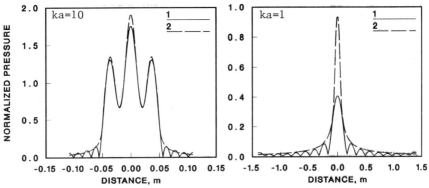

Fig. 4.   Pressure reconstructed at the source:
          (1)   Without evanescent waves;
          (2)   Including evanescent waves.

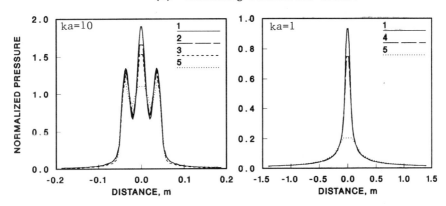

Fig. 5.   Pressure images obtained with the aperture including 99%
          of the peak magnitude:
               (1) $\Delta = \lambda/16$ (exact);
               (2) $\Delta = \lambda/4$, no spectral filtering;
               (3) $\Delta = \lambda/4$, including spectral filtering;
               (4) $\Delta = \lambda/8$, including spectral filtering;
               (5) $\Delta = \lambda/2$, no spectral filtering

253

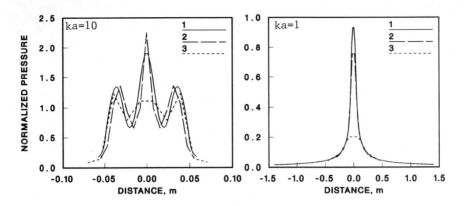

Fig. 6. Pressure images obtained by keeping only 90% of the maximum
         pressure magnitude within the aperture:
                  (1) Δ = λ/16;   (1) Δ = λ/16;
                  (2) Δ = λ/4;   (2) Δ = λ/8;
                  (3) Δ = λ/2;   (3) Δ = λ/2.

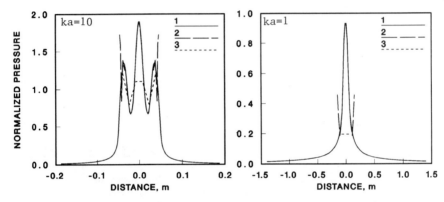

Fig. 7. Result of using a too-small aperture:
                  (1) Exact;
                  (2) Δ = λ/16;
                  (3) Δ = λ/2.

sampling was fine; however, when sampling equaled λ/16 the array size was
very large (N = 192). On the other hand, the more crude sampling rate of
$\Delta_{10} = \lambda_{10}/4$ and $\Delta_1 = \lambda_1/8$ (where subscript 1 refers to ka = 1 and subscript 10 to ka = 10) provided quite acceptable images. In addition,
results were better when no filtering in the spatial frequency was used
with this coarse sampling. The Nyquist sampling rate produced poor
results. In Fig. 6 the images are shown for the aperture covering
pressure magnitude up to 90% of its maximum value. In this case, results
were excellent when the sampling interval was Δ = λ/16. However, errors
occurred when the sampling interval became larger. Acceptable results
were obtained in this case when $\Delta_{10} = \lambda_{10}/4$ and $\Delta_1 = \lambda_1/8$. The Nyquist
sampling rate resulted in even more distortions than in the case with the
larger aperture size. Finally, the aperture size that included pressure
magnitude up to 73% of the maximum was inadequate, as shown in Fig. 7.
In this case the results, even with the fine sampling rate, were quite
distorted; when the sampling rate became smaller the images were totally
inaccurate.

An additional procedure, called asymmetric sampling, was studied for the on-axis point located just above the center of the piston. The asymmetric sampling grid was achieved by shifting the symmetric sampling grid by a distance of a half-sampling interval ($\Delta/2$). The comparison between the symmetric and the asymmetric sampling was made on the basis of errors in back-propagated images calculated with the same sampling rate. It was found that the error using the completely symmetric data was only slightly smaller than from the symmetric case. Therefore, retaining the on-axis amplitude for reconstruction was not critical, even when spacings were coarse and a significant portion of the peak was missing. At lower frequencies the true center value was of little consequence since the pressure field for such cases was more dome-like. Similar conclusions were reached by Veronesi and Maynard [2].

CONCLUSIONS

Accurate reconstructions were shown to result from fairly coarse grid spacings that allowed a reduction of the aperture size to components equal to 10% of the maximum value. Windowing in the real space was necessary in this case in order to minimize the wraparound error.

We demonstrated that as the sampling became more coarse, it became necessary to retain all evanescent components within calculated spectra. Consequently, better results were obtained when no filtering in the spatial frequency was used. However, in the case of fine sampling, the frequency filtering improved results.

The amount of evanescent waves generated by a source determined the proper sampling rate required in NAH. Fields that contain more evanescent waves required a smaller sampling interval with respect to the wavelength (i.e., $\lambda/\Delta = 8$ when $ka = 1$ but $\lambda/\Delta = 4$ when $ka = 10$).

Other researchers [7] have shown previously that a tradeoff existed between the sampling interval and the aperture size. When the aperture size became smaller, higher sampling rates were required to accomplish the same quality of the reconstruction. Moreover, the Nyquist sampling rate was inadequate in all cases considered. However, errors in results obtained with the Nyquist sampling were considerably smaller for the high frequency source.

All cases in this study were calculated with a propagation distance of 0.1λ. This condition required that the hologram plane was located at a very small distance from the source. This distance was only 3 mm when $ka = 10$. Future research should be directed toward developing methods to reconstruct sources from greater distances while still retaining acceptable resolution.

ACKNOWLEDGMENT

This work was supported by a National Science Foundation grant (# MSM-8508694) and by the Engineering Research Institute of Iowa State University.

REFERENCES

1. J. D. Maynard, E. G. Williams, and Y. Lee, "Nearfield acoustic holography, I. Theory of generalized holography and the development of NAH," J. Acoust. Soc. Am. 78(4):1395-1413 (Oct. 1985).

2. W. A. Veronesi and J. D. Maynard, "Nearfield acoustic holography (NAH), II. Holographic reconstruction algorithms and computer implementation," J. Acoust. Soc. Am. 81(5):1307-1322 (May 1987).

3. William Y. Strong, Jr., Experimental Method to Measure Low Frequency Sound Radiation-Nearfield Acoustic Holography, Master's Thesis, Penn State (1982).

4. E. G. Williams, H. Dardy, and K. Washburn, "Generalized nearfield acoustic holography for cylindrical geometry: theory and experiment," J. Acoust. Soc. Am. 81(2):389-407 (Feb. 1987).

5. Allan D. Pierce, Acoustics: An Introduction to its Physical Principles and Applications, McGraw-Hill, New York, p. 215 (1981).

6. J. A. Archer-Hall, A. I. Bashter, and A. J. Hazelwood, "A means for computing the Kirchhoff surface integral for a disk radiator as a single integral with fixed limits," J. Acoust. Soc. Am. 65(6):1568-1570 (June 1979).

7. H. Lee and G. Wade, "Sampling in digital holographic reconstruction," J. Acoust. Soc. Am. 75(4):1291-1293 (April 1984).

VELOCITY-BASED NEARFIELD ACOUSTIC HOLOGRAPHY

M. Robin Bai and Anna L. Pate

Department of Engineering Science and Mechanics
Iowa State University
Ames, Iowa

ABSTRACT

Numerical investigations of nearfield acoustic holography in a contaminated acoustic field were performed. In this study, a baffled piston vibrating at low frequencies was chosen as the primary source. In addition, two types of disturbing sources were also considered. These sources were monopoles and plane waves propagating either perpendicularly or parallel to the hologram plane. Because pressure-based nearfield holography (PBNAH) was quite sensitive to the disturbing sources, a new method called velocity-based nearfield holography (VBNAH) was developed. The backpropagation images of the baffled piston obtained by using the VBNAH technique were compared to the images provided by the PBNAH technique. In the presence of the disturbing sources, VBNAH provided, in general, significantly better images than PBNAH. Errors caused by the disturbing sources depended on the type of those sources and their location with respect to the hologram plane.

INTRODUCTION

Acoustic holography has become a powerful source imaging technique[1] that is capable of reconstructing both the acoustic field variables (such as pressure, velocity, intensity, energy density, etc.) and source mechanical variables (velocity, structural intensity, power, etc.).

The most common applications of acoustical holography are based on the sound pressure data measured in the hologram plane[2]. In contrast to the conventional approach, this paper proposes a new nearfield acoustic holography technique that applies a different transformation basis. Instead of measuring the sound pressure data, the partical velocity was recorded in the hologram plane. Based on the information of the partical velocity, either forward propagation or backward reconstruction can be performed.

The motivation for velocity-based nearfield holography (VBNAH) comes from investigations of the performance of conventional acoustic holography when disturbing sources were present. Significant errors occurred with pressure-based nearfield holography (PBNAH) when data were measured in a noisy environment with acoustical disturbances coming from many directions. In this case, several assumptions required in the derivations of the holography method were not satisfied. Specifically, not all acoustical sources were confined under the hologram plane, and in addition both the outgoing waves and the incoming waves were present. However, even with these violations in place, the VBNAH method was capable of discriminating against some type of the disturbing sources.

The objective of this paper is to present the theoretical formulation for VBNAH. In addition, numerical simulations of images obtained in the presence of the disturbing sources were of interest. For every investigated case, the PBNAH simulations were performed for comparison purposes.

VELOCITY-BASED NEARFIELD ACOUSTICAL HOLOGRAPHY (VBNAH)

A two-dimensional acoustical source generates a pressure field $p(\bar{r},t)$ within a three-dimensional space that satisfies the homogeneous wave equation

$$\nabla^2 p(\bar{r},t) - \frac{1}{c^2}\frac{\partial^2}{\partial t^2}\,p(\bar{r},t) = 0 \qquad (1)$$

where ∇^2 is the Laplacian operator and c is the sound propagation speed. Performing Fourier transform of Eq. (1) yields the Helmholtz equation

$$\nabla^2 \tilde{p}(\bar{r},\omega) + k^2 \tilde{p}(\bar{r},\omega) = 0 \qquad (2)$$

with the wave number $k = \omega/c$; $\bar{r} = (x,y,z)$. Taking the Fourier transform with respect to variables x and y of Eq. (2) results in[3]

$$\frac{d^2}{dz^2}\,\hat{p}(k_x,k_y,z) + k_z^2\,\hat{p}(k_x,k_y,z) = 0 \qquad (3)$$

where

$$k_z = \sqrt{k^2 - k_x^2 - k_y^2}$$

The ordinary second-order differential equation, Eq. (3), in the variable z may be solved for $\hat{p}(k_x,k_y,z)$ as

$$\hat{p}(k_x,k_y,z) = Ae^{jk_z z} + Be^{-jk_z z} \qquad (4)$$

Because the propagation of only outgoing waves is assumed, $B = 0$.

The values of A can be determined by substituting the pressure data $\hat{p}_H(k_x,k_y,z_H)$ measured on the hologram plane into Eq. (4), as

$$A = \hat{p}(k_x,k_y,z_H)e^{-jk_z z_H} \qquad (5)$$

Combining Eqs. (4) and (5) gives

$$\hat{p}(k_x,k_y,z) = \hat{p}_H(k_x,k_y,z_H)e^{jk_z(z-z_H)} \qquad (6)$$

258

Based on Eq. (6), the particle velocity $\hat{\underline{v}}(k_x, k_y, z)$ can be readily found from Euler's equation, as

$$\hat{\underline{v}}(k_x, k_y, z) = \frac{1}{\rho\omega}(k_x \underline{e}_x + k_y \underline{e}_y + k_z \underline{e}_z)\hat{p}(k_x, k_y, z) \qquad (7)$$

where ρ is the density of medium. In particular, the normal component of the particle velocity equals

$$\hat{v}_z(k_x, k_y, z) = \frac{k_z}{\rho\omega}\hat{p}(k_x, k_y, z) \qquad (8)$$

Equations (6) and (8) formed the basis of the PBNAH and VBNAH, as shown schematically in Fig. 1. For the PBNAH calculations, the following equations were used (paths 1 and 2 in Fig. 1):

$$\hat{p}(k_x, k_y, z) = \hat{p}(k_x, k_y, z_H)e^{jk_z \cdot d} \qquad (9)$$

$$\hat{v}_z(k_x, k_y, z) = \frac{k_z}{\rho\omega}\hat{p}(k_x, k_y, z) \qquad (10)$$

On the other hand, for the VBNAH calculations, the following equations were used (paths 3 and 4 in Fig. 1)

$$\hat{v}_z(k_x, k_y, z) = \hat{v}_z(k_x, k_y, z_H)e^{jk_z \cdot d} \qquad (11)$$

$$\hat{p}(k_x, k_y, z) = \frac{\rho\omega}{k_z}\hat{v}_z(k_x, k_y, z) \qquad (12)$$

In addition to sound pressure and particle velocity, complex sound intensity and energy density can be calculated if desired.

NUMERICAL SIMULATIONS OF VBNAH

From the previous derivations, it may appear that PBNAH and VBNAH are only different formulations of an identical technique. However, in practical implementation, a different spatial transformation basis could provide different results when disturbing sources are present.

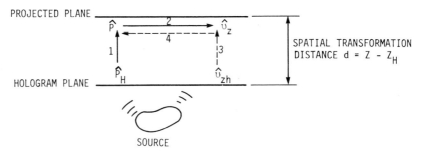

Fig. 1. Two types of nearfield acoustic holography:
a) pressure-based method (paths 1 and 2), and
b) velocity-based method (paths 3 and 4).

In order to investigate the effects of disturbing sources on near-field acoustic holography, a numerical simulation was conducted. The plane waves and a monopole were chosen as the disturbing sources for convenient analysis. The first two cases included the plane wave as a disturbing source. The plane waves were propagating either parallel or perpendicularly to the hologram plane, as shown in Figs. 2a and 2b. The remaining three cases included a monopole as the disturbing source. The monopoles were positioned above, along, and under the hologram plane, as shown in Fig. 3.

A baffled piston was used as the primary source in all simulated cases, with only backward reconstruction of concern. In all simulations, the pressure from the disturbing sources was chosen to be comparable to the maximum pressure from the piston in the hologram plane. The piston surface velocity was 1 m/s, its diameter d = 0.161 m, and its frequency . was 2200 Hz. In the first two cases, the hologram plane was chosen at z_H = 0.0051 m from the piston surface and the back propagation distance was d = 0.005 m. However, in the last three cases, z_H = 0.02037 m and d = 0.02 m. Because of limited space, only the reconstructed velocity magnitude (in dB) is shown.

RESULTS

As can be seen in Fig. 4, when the disturbing plane wave propagated parallel to the hologram plane, significant errors occurred in both

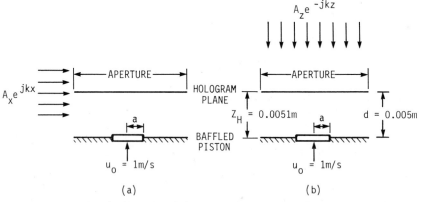

(a)   (b)

Fig. 2. Investigated cases of the NAH imaging of a baffled piston with plane disturbing waves. (a) parallel disturbing waves; (b) perpendicular disturbing waves.

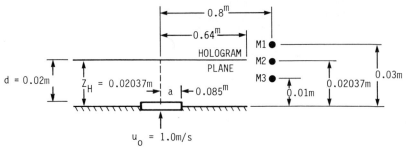

Fig. 3. The configuration of the nearfield acoustic holography when a monopole is used as the disturbing source. (M1, M2, M3 - monopole locations.)

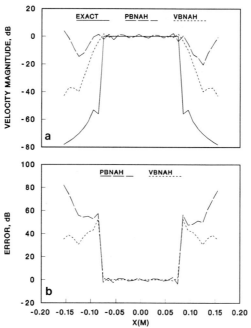

Fig. 4. (a) Velocity magnitude calculated by the PBNAH and
VBNAH techniques (parallel disturbing plane waves);
(b) error of velocity magnitude in (a).

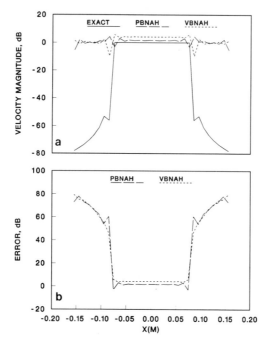

Fig. 5. (a) Velocity magnitude calculated by the PBNAH and VBNAH
techniques (perpendicular disturbing plane waves);
(b) error of velocity magnitude in (a).

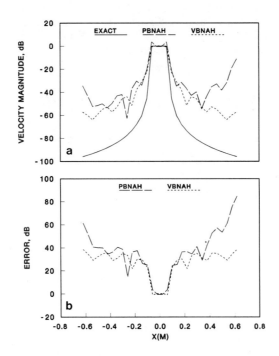

Fig. 6. (a) Velocity magnitude calculated by the PBNAH and VBNAH
         techniques with the disturbing monopole M2 (Fig. 3);
         (b) error of velocity magnitude in (a).

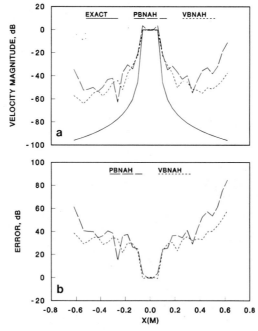

Fig. 7. (a) Velocity magnitude calculated by the PBNAH and VBNAH
         techniques with the disturbing monopole M1 (Fig. 3);
         (b) error of velocity magnitude in (a).

holography methods. However, the results obtained from VBNAH appear less susceptible to the disturbing source, especially outside the piston area. On the other hand, in view of Fig. 5, neither nearfield acoustic holography technique gains particular advantage if the disturbance propagated perpendicularly to the hologram plane. The true image was overwhelmed by the disturbing noise.

For more insight into the effects of disturbing sources, three more simulation cases were investigated using a monopole as the disturbing source. The results are shown in Figs. 6, 7, and 8 for a monopole positioned above, in, or underneath the hologram plane, respectively. It was observed from the simulation results that VBNAH and PBNAH produced different images. The case with the monopole positioned in the hologram plane demonstrated the superiority of VBNAH over PBNAH to the greatest extent. The influence of disturbing sources was manifested by the asymmetrical velocity magnitude. Interestingly enough, Fig. 8 shows that significant error still occurs even though the assumption that all the sources must be confined under the hologram plane is satisfied. This phenomenon can be accounted for by the fact that there is another assumption of nearfield acoustic holography that was violated. That assumption is that no sources (in this case, the disturbing monopole) should be located in the space between the hologram plane and the projected plane.

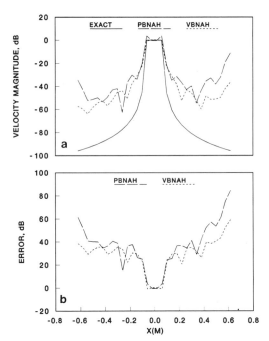

Fig. 8. (a) Velocity magnitude calculated by the PBNAH and VBNAH techniques with the disturbing monopole M3 (Fig. 3); (b) error of velocity magnitude in (a).

CONCLUSIONS

In view of previous results and discussions, a number of comments can be made. First of all, the disturbing sources degraded the performance of the nearfield acoustic holography because the assumption that all sources have to be confined under the hologram plane was violated. Secondly, VBNAH proved to be superior over PBNAH if the hologram plane was properly positioned, subject to the directivity of the disturbing source. More precisely, the smaller the velocity contribution from the disturbing source on the hologram plane became, the better the VBNAH performance. This can be analytically justified from the cases previously simulated. When the plane waves propagated parallel to the hologram plane, the velocity contribution from the disturbing source was

$$v_z(z_H) = \frac{1}{j\rho\omega} \frac{\partial}{\partial z} \left(A_x e^{jk_x x} \right) = 0 \tag{13}$$

However, for the plane waves propagated perpendicularly to the hologram plane, the velocity contribution is no longer negligible because

$$v_z(z_H) = \frac{-k_z}{\rho\omega} A_z e^{-jkz_H} \neq 0 \tag{14}$$

Applying similar reasoning to the monopole disturbing source, the velocity contribution can be evaluated as

$$v_z(z_H) = \frac{1}{j\rho\omega} \frac{z_H - z_m}{r} \frac{\partial}{\partial r} \left(S_m \frac{e^{jkr}}{r} \right) \tag{15}$$

where

$$r = \sqrt{(x - x_m)^2 + (y - y_m)^2 + (z_H - z_m)^2}$$

Note that when the monopole disturbing source was positioned in the hologram plane, $(z_H = z_m)$, $v_z(z_H) = 0$.

These derivations support the results obtained from the numerical simulations--that plane waves from the side and the monopole positioned in the hologram plane achieved the best performance among all cases.

In summary, when dealing with a contaminated acoustical environment, the velocity-based nearfield holography technique proved to be less susceptible to disturbing sources than its counterpart, the pressure-based nearfield holography technique. However, the position of the hologram plane with respect to the disturbing source was very important. The best location was such that the contribution from the disturbing source into the normal component of the particle velocity was minimized. The difference of the performance between these two approaches mainly resulted from the fact that sound pressure is a scalar, whereas particle velocity is a vector. Due to the field directivity formed by the interactions between the primary source and the disturbing sources, PBNAH and VBNAH techniques produced different projected images.

Despite the advantage of VBNAH demonstrated in the numerical simulations, one needs to recall the technical difficulties associated with the measurement of the particle velocity. The most common approximate methods, such as the finite difference two-microphone method, cause bias error due to finite spacing. In addition, the phase and magnitude sampling mismatch result in significant errors when estimating the pressure gradient[4,5,6]. The experimental implementations of the presented VBNAH method are currently under investigation.

ACKNOWLEDGMENTS

This work was supported by a National Science Foundation grant

264

(#MSM-8508694) and by the Engineering Research Institute, Iowa State University.

REFERENCES

1. H. M. Smith, "Principles of Holography," Wiley-Interscience, New York (1969).
2. J. D. Maynard, E. G. Williams, and Y. Lee, Nearfield acoustic holography: I. Theory of generalized holography and the development of NAH, J. Acoust. Soc. Am., Vol. 78, No. 4 (1985).
3. D. Dudgeon and R. M. Mersereau, "Multidimensional Digital Signal Processing," Prentice-Hall, Inc., pp. 359-363 (1984).
4. A. L. Pate and R. M. Bai, One-microphone sound intensity method, J. Acoust. Soc. Am., Suppl. 1, Vol. 80 (1986).
5. M. R. Bai and A. L. Pate, Experimental investigations of acoustic field based on one-microphone, sequential sampling technique, submitted for publication in Transactions of ASME (1987).
6. W. A. Veronesi and J. D. Maynard, Nearfield acoustic holography (NAH) II. Holographic reconstruction algorithms and computer implementation, J. Acoust. Soc. Am., Vol. 81, No. 5 (1987).

AN EXAMINATION OF APERTURE EFFECTS IN

CYLINDRICAL NEARFIELD HOLOGRAPHY

Timothy W. Luce (The Graduate Program in Acoustics)
Sabih I. Hayek (The Department of Engineering Science and Mechanics
and The Applied Research Laboratory)

The Pennsylvania State University
University Park, PA 16802

DEVELOPMENT OF PROPAGATORS

The development and implementation of cylindrical holography is derived from the time independent homogeneous Helmholtz equation for acoustic pressure, p, in cylindrical coordinates

$$(\nabla^2 + k^2)p(r, \phi, z) = 0, \tag{1}$$

where $k = \omega/c$. In solving this equation an $e^{-i\omega t}$ time dependence is assumed and the method of separation of variables is used. Furthermore, it is assumed that the source is confined to a finite cylindrical band of radius, a, and length, l, in an infinite cylindrical baffle. Also, free-space propagation exists external to a so that at any $r \geq a$

$$p(r, \phi, z) = \sum_{n=-\infty}^{\infty} e^{in\phi} \int_{-\infty}^{\infty} A_{nk_z} H_n^{(1)}(k_r r) e^{ik_z z} dk_z, \tag{2}$$

where $k_r = \sqrt{k^2 - k_z^2}$ and the Hankel function of the first kind represents outgoing waves. By applying a two-dimensional Fourier transform to both sides of (2) and evaluating at r_0, the hologram radius, one obtains

$$P_n(r_0, k_z) = \frac{1}{2\pi} A_{nk_z} H_n^{(1)}(k_r r_0). \tag{3}$$

The modal coefficents A_{nk_z} are easily evaluated from (3). Re-evaluating (3) at another radius, r, and then substituting the previously evaluated coeeficients in (3) yields the modal pressure at any radius, $r \geq a$, resulting from the modal pressure at r_0:

$$P_n(r, k_z) = P_n(r_0, k_z) \frac{H_n^{(1)}(k_r r)}{H_n^{(1)}(k_r r_0)}. \tag{4}$$

It is the ratio of Hankel functions that represents the pressure to pressure propagator in cylindrical coordinates. Applying the inverse Fourier transform gives the pressure at r from the measured pressure distribution at the hologram surface, r_0,

$$p(r, \phi, z) = \frac{1}{2\pi} \sum_{n=-\infty}^{\infty} e^{in\phi} \int_{-\infty}^{\infty} P_n(r_0, k_z) \frac{H_n^{(1)}(k_r r)}{H_n^{(1)}(k_r r_0)} e^{ik_z z} dk_z. \tag{5}$$

When $k_z^2 > k^2$ the argument of the Hankel function becomes imaginary and the modified Bessel function $K_n(k_r' r)$ with $k_r' = \sqrt{k_z^2 - k^2}$ must be substituted. These functions decay exponentially away from the source surface and hence represent evanescent waves. The symbols Z_n and k_r will be used to represent both propagating and evanescent waves.

The propagation and transformation of a pressure field to a velocity field is accomplished by recalling,

$$ikpc\vec{v}(r,\phi,z) = \vec{\nabla}p(r,\phi,z). \tag{6}$$

Using this relationship and the methods used in the pressure to pressure derivation while separating the radial, circumferential and axial components, results in expression for the pressure to radial velocity transform-propagator,

$$V_{n_{rad}}(r,k_z) = \frac{1}{2\pi} \frac{k_r}{ikpc} P_n(r_0,k_z) \frac{Z'_n(k_r r)}{Z_n(k_r r_0)}. \tag{7}$$

The circumferential and axial velocity transform-propagators do not have the derivative with respect to the argument and are essentially scaled versions of the pressure to pressure propagator.

IMPLEMENTATION ERRORS

In the implementation of these formulas, the fast Fourier transform (FFT) is utilized on finite, discretely sampled holograms. Errors are introduced by this discretization and the finiteness of the aperture. Inherent to the FFT is the axial spatial frequency increment $\Delta k_z = 2\pi/L$ where L is the axial aperture. This indicates that the largest wavelength calculated by the FFT is L. Due to periodicity, the circumferential aperture is essentially infinite in cylindrical scans which cover the complete 2π radians. The angular spatial frequency increment $\Delta k_\phi = 2\pi/2\pi = 1$ is inherent in the summation on n.

When using the FFT, the limits on circumferential and axial wavenumbers are far from the ideal $-\infty$ to $+\infty$. For $N = 128$ points in both scan directions, this limits the ϕ sumation to $n = \pm 64$ while the limits on k_z are $-\pi/\Delta z$ to $\pi/\Delta z$. Since large k_z values correspond to highly evanescent waves which quickly decay below the dynamic range of the recording system, very little information is lost in the truncation of these sumations during the reconstruction of experimental data. Similarly, the very high mode orders represented by large absolute values of n are seldom realized in real systems. Synthetic holograms, with their dynamic range limited only by the computer's word length, can be forced to exhibit much more evanescent behavior.

In planar holography the error arising from the use of a piecewise constant field and FFTs has been shown by Veronesi and Maynard (1987) to be

$$\frac{\tilde{F}(k_x,k_y)}{F(k_x,k_y)} = \frac{\Delta k_x \Delta k_y}{4\sin(\Delta k_x/2)\sin(\Delta k_y/2)}, \tag{8}$$

where $F(k_x,k_y)$ is the two dimensional Fourier transform of the continuous field and $\tilde{F}(k_x,k_y)$ is the Fourier transform of the piecewise field. Analogously, in cylindrical holography for the k_z factor

$$\frac{\tilde{F}(r,k_z)}{F(r,k_z)} = \frac{\Delta k_z}{2\sin(\Delta k_z/2)}. \tag{9a}$$

It should be noted that this ratio is practically unity for small Δk_z when $\sin(\alpha) \approx \alpha$. In order to keep Δk_z small, the largest practical aperture L should be used. A scan of less than 2π in the circumferential dimension will introduce a comparable error term

$$\frac{\tilde{F}(r,k_\phi)}{F(r,k_\phi)} = \frac{\Delta k_\phi}{2\sin(\Delta k_\phi/2)}. \tag{9b}$$

For a full scan of 2π this leaves $\Delta k_\phi = 1$ and this term is a constant factor equal to 1.0429.

Undersampling in wave-number-space is a cause of many aliasing problems. The simplest method of reducing this error is to augment the hologram with a guard band of N zeros before performing the FFT. Then all operations are carried out on the extended $2N$ data set. This is only needed in the axial direction for cylindrical holography. The natural periodicity in the circumferential direction removes the need for this in cylindrical scans. If, however, a partial angular scan is used the angular data should be buffered with zeros to achieve a full 2π scan.

A WINDOW FUNCTION

Compounding these errors is the exponential growth associated with back propagation. Since evanescent waves fall off exponentially in the outward propagation, noise and errors will at some point overwhelm the information in those waves. In back propagation, a window function is applied to the k-space data-set to filter those components that fall below the noise threshold. The current axial window function is derived from the two-dimensional planar holography window suggested by Veronesi and Maynard.

$$
\begin{aligned}
W(k_z) &= 1 - \frac{1}{2} e^{-\left|\frac{|k_z|}{k_{zmax}} - 1\right| s} \qquad ; k_z \le k_{zmax} \\
&= \frac{1}{2} e^{-\left|\frac{|k_z|}{k_{zmax}} - 1\right| s} \qquad ; k_z > k_{zmax},
\end{aligned}
\tag{10a}
$$

where at r, the current reconstruction radius,

$$
k_{zmax} = \sqrt{k^2 + \left[\frac{\frac{log(2.)+.05\alpha}{2\pi log(e)}}{r_0 - r} \right]^2}.
$$

The angular components, while not so susceptable to errors due to a finite scan, are similarly windowed. Williams et al. (1987) showed that circumferential evanescent waves are dependent on radius and mode order, n. As we propagate outward from a source, for some frequency and specified angular wavelength, there is a radius where the spatial separation of that angular wavelength reaches the propagation wavelength of the medium and a formerly evanescent wave propagates. This evanescent behavior is exhibited as power law decay to the mode order or n^{th} power. In back projection this must be taken into account — at some radius what was a propagating wave at the hologram plane could become evanescent. The smallest radius of reconstruction is the source radius, a, so it is used in creating the non-dimensional limiting factor $k_{\phi max}$ necessary for filtering with respect to the mode order, n. This allows the angular window function to use the same form as the axial window such that

$$
\begin{aligned}
W(n) &= 1 - \frac{1}{2} e^{-\left|\frac{|n|}{k_{\phi max}} - 1\right| s} \qquad ; n \le k_{\phi max} \\
&= \frac{1}{2} e^{-\left|\frac{|n|}{k_{\phi max}} - 1\right| s} \qquad ; n > k_{\phi max},
\end{aligned}
\tag{10b}
$$

where for the current reconstruction radius, r,

$$
k_{\phi max} = a k_{zmax}.
$$

In both the above, α and s are adjustable parameters. The desired threshold level α in decibels is determined by the dynamic range of the scanning system and the digital word length used. The sharpness factor s is entered as a real number and effects the slope of the window and thus the smoothing of sudden transitions and discontinuities. Each of these window functions can be accessed separately or simultaneously to study their individual effects. The threshold level and sharpness are identical for both the axial and angular windows when the two are used simultaneously.

EXPERIMENTAL INVESTIGATION

Sources: In this investigation two analytically defined distributions were used to generate pressure data holograms. A one meter radius, a, by three meter long, l, free ended cylindrical pressure distribution in the (3,1) (m,n) mode was used to verify the pressure to pressure reconstructions where

$$
p = \cos(m\pi z/L)\cos(n\phi).
$$

269

Likewise, to verify the velocity reconstructions, a one meter radius, a, by one meter long, l, free ended cylindrical radial velocity distribution in the (2,1) mode where

$$v = \sin(m\pi z/L)\cos(n\phi),$$

was transform-propagated. Both of these distributions were propagated at 1.0 and 12000.0 Hz. to create ka values of .004 and 50.9 with wavelengths of 1481 m. and 12.34 cm. respectively. The pressure (amplitude and phase) for each was recorded at a radius of $r_0 = 1.05$ m. with an axial aperture of 6 m. for the pressure derived holograms and 3 m. for the velocity derived ones. Both the axial and angular directions were sampled at 128 points. This created an axial sample spacing of 4.724 cm. for the pressure holograms and 2.362 cm. for the velocity derived holograms. The circumferential spacing for each was 2.8125 deg. or 5.154 cm. at the hologram plane.

Standard Reconstructions: The full holograms were reconstructed at the surface radius with various window thresholds and sharpnesses. It was found that varying the threshold level altered the reconstructions only slightly as would be expected from synthetic data. Therefore, the threshold level, α, was set at 12 dB for all subsequent tests. This level was chosen from prior experience with a 12-bit A/D converter system where the worst case dynamic range between propagating and evanescent waves is 12 dB (Eshenberg & Hayek, 1986). The most suitable sharpness, s, was found to be .7 for velocity generated holograms (Figs. 1,2) and .1 for pressure generated holograms (Figs. 3,4). At these values the reconstructed free edges of the disributions had the least rounding and oscillation. Although there is a slight frequency dependence on the sharpness factor, s, these window factors were used for both axial and angular windowing on all subsequent reconstructions unless otherwise noted. In the velocity generated reconstructions only the radial componant is shown, since only the radial velocity was specified to generate the holograms. Associated angular and axial components exist, but the vector sum representation of these in-plane components by arrows is too detailed to be reproduced at the scale of these illustrations. These figures will be the standard for comparison in this study.

Window Effects: A surface reconstruction was also performed without any windowing on the 1 Hz. velocity generated hologram to verify the need to window the data. The plot (Fig. 5) reveals the obvious need for axial windowing but surprisingly the circumferential mode is still clearly discernable. The discontinuities at the free edges are clearly defined but, overall, there is no resemblance to the originating function. This led to tests with only the axial window on all the holograms. These reconstructions were done with the standard window parameters α and s at 12dB and .1 or .7, respectively in only the axial direction. At both frequencies there is excellent agreement with the standard reconstructions (Figs. 1-4). The 1.0 Hz. velocity reconstruction is shown in Fig. 6. These results should be expected since the windowing is an effort to control the finite aperture affects of the FFT and the periodicity in the circumferential direction creates an essentially infinite aperture. Also, the well defined mode order in the synthetic holograms eliminates high order terms which might need windowing. When performing tests on actual objects, high order harmonics due to angular discontinuities would be amplified during reconstruction without angular windowing.

Aperture Effects: Scanner vibration and frequency drift during hologram recording are major considerations in the implementation of holography. The expense and structural complexity of scanning arrays to accurately position the recording sensors increases dramatically with scanner size and the number of recording sensors. While a small scanner can be used in overlapping scans to create a larger hologram, this requires a very stable source and the accurate recording of the amplitude and phase of a hologram would be a difficult and time consuming task. Any method that lessens the number of points to be recorded or the size of the scanner to be constructed and still allows accurate reconstructions of the sound field should be investigated. Two obvious options exist: make measurements over a smaller aperture or sample at fewer points within the aperture. Each of these methods has resolution advantages. The smaller aperture with higher sampling density is more capable of resolving small wavelengths and details. Low sample density holograms with a large aperture are able to reconstruct and propagate the long wavelengths (up to the aperture size) necessary at low frequencies. However, at high frequencies, a low sample point density could undersample not only the acoustic wave, but also the structural bending waves.

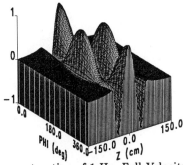

Fig. 1. Standard reconstruction of 1 Hz. Full Velocity Hologram
Z and ϕ Windows at 12 dB and .7 s.

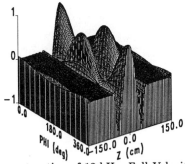

Fig. 2. Standard reconstruction of 12 kHz. Full Velocity Hologram
Z and ϕ Windows at 12 dB and .7 s.

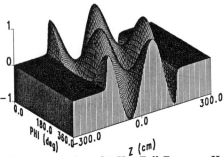

Fig. 3. Standard reconstruction of 1 Hz. Full Pressure Hologram
Z and ϕ Windows at 12 dB and .1 s.

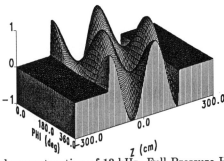

Fig. 4. Standard reconstruction of 12 kHz. Full Pressure Hologram
Z and ϕ Windows at 12 dB and .1 s.

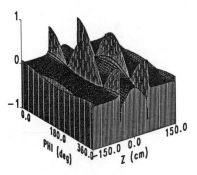

Fig. 5. No window on reconstruction of 1 Hz. Full Velocity Hologram

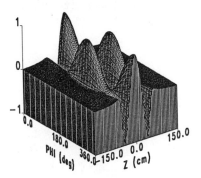

Fig. 6. Z-window reconstruction of 1 Hz. Full Velocity Hologram
Z Window at 12 dB and .7 s.

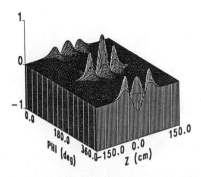

Fig. 7. Z-window reconstruction of 1 Hz. (0-180 deg.) Velocity Hologram
Z Window at 12 dB and .7 s.

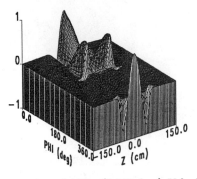

Fig. 8. Reconstruction of 1 Hz. (0-180 deg.) Velocity Hologram
Z and ϕ Windows at 12 dB and .7 s.

The effects of angular aperture were first investigated. A program was used that read each full sized hologram and removed either peripheral points or interspersed points to create holograms with smaller apertures or lower sample point densities. The original 128 x 128 point holograms were processed to remove half the data points in the angular scan. The removed data points were from 180 to 360 degrees causing the aperture edge to fall on antinodes of the distributions. The apertured hologram was then padded with zeros to complete the 2π scan expected by the processing algorithm. The hologram thus created corresponds to one taken over a portion of a vibrating cylinder (due to some physical constraints prohibiting a full 2π scan) after zero padding to a full scan. The first reconstruction was made with the 1 Hz. velocity generated hologram. The aperture of this hologram was from -1.5 m. to 1.5 m. axially with recorded data from 0 to 180 degrees and zeros to complete the scan.

While the neccesity of axial windowing has been shown, previous angular windowing has had little effect. Therefore, this first reconstruction (Fig. 7) was done with the standard window parameters of $\alpha = 12$ and $s = .7$ in the axial direction only. The edges of the measured data are defined and the axial mode is discernable but little relevent information is derived form this reconstruction. Reconstruction with both axial and angular windows at the standard settings is shown in Fig. 8. The peaks within the aperture are slightly distorted but the radial mode shape for the sampled portion of the cylinder is nearly identical to that of the full reconstruction (Fig. 1). While not shown, the vector velocity on the surface also corresponds well with the exception of those points at the aperture edge which flow into and out of the zeroed area. The field in the zeroed area is invalid since the use of zeros is based on the inaccurate assumption that the pressure field in that area of the hologram is zero. Similar results are seen at 12 kHz. although there are more severe angular oscillations due to the sudden discontinuity at the aperture edge. The 12 kHz. hologram was processed again with a higher sharpness value, $s = 1.5$ (Fig. 9). While this reduced the oscillations in the angular direction it also rounded the transition in the axial direction. Another partial angular scan was done at 12 kHz. using the data between 90 and 270 degrees while zeroing the rest. When this was reconstructed with the standard window (Fig. 10) there was no sign of the angular oscillations observed in the prior examples. The reason for this much smoother reconstruction at a lower sharpness value is the lack of discontinuities when the data started and ended on node lines.

Reconstructions were also done on similarly apertured (0 to 180 deg.) pressure generated holograms. The first reconstructions were made with the standard pressure hologram window parameters of $\alpha = 12dB$ and $s = .1$. The distribution at 12 kHz, although recognizable within the aperture, shows severe oscillations from the angular discontinuities. Raising the value of s to 1.5 alleviated some of this problem (Fig. 11) at the cost of rounding the free axial edge. The low frequency case reconstructed well with the window sharpness set to .7, since lower frequencies exhibit less oscillation at the discontinuities.

The axial component of the aperture also bears consideration. Each of the original holograms was processed to remove the points on either side of the active source area. For the pressure defined holograms this left a 3 m. hologram with 64 points axially and 128 points circumferentially. These half original sized holograms were processed with the standard pressure window parameters. As can be seen (Fig. 12), the mode shape is quite clear although there is a squeezing effect that pushes the peaks in from the free edges and rounds them slightly. The velocity holograms were likewise apertured to their source area. Since the sample points did not fall at exactly the source edges, the new holograms were 1.03 m. long with 44 points axially and 128 points circumferentially. Each frequency was processed with the standard velocity window parameters. Again these reconstructions show good agreement with the full sized holograms.

Sample Density Effects: The previous examples tried to lessen hologram recording time by restricting the aperture size and thus lessening the number of recorded points, however, a lower sample point density allows a faster scan time over a large aperture. To investigate the effects of reduced sample point density, we removed every other point from the original holograms. This created a 64 x 64 point hologram with one quarter the original sample density with the full aperture. These quarter density holograms have an axial sample spacing of 9.44 cm. for the pressure generated ones and 4.72 cm. for the velocity generated holograms. Circumferentially both types have a spacing of 5.625 degrees or 10.308 cm. at the hologram

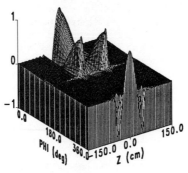

Fig. 9. Reconstruction of 12 kHz. (0-180 deg.) Velocity Hologram
Z and φ Windows at 12 dB and 1.5 s.

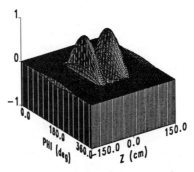

Fig. 10. Reconstruction of 12 kHz. (90-270 deg.) Velocity Hologram
Z and φ Windows at 12 dB and .7 s.

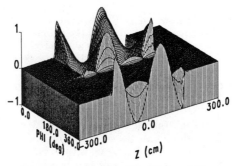

Fig. 11. Reconstruction of 12 kHz. (0-180 deg.) Pressure Hologram
Z and φ Windows at 12 dB and 1.5 s.

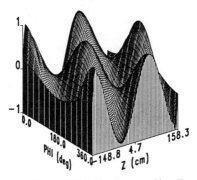

Fig. 12. Reconstruction of 12 kHz. Source Size Pressure Hologram
Z and φ Windows at 12 dB and .1 s.

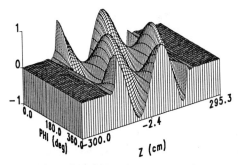

Fig. 13. Reconstruction of 12 kHz. Quarter Density Pressure Hologram
Z and ϕ Windows at 12 dB and .1 s.

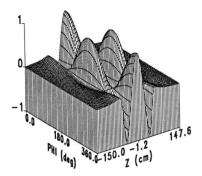

Fig. 14. Reconstruction of 1 Hz. Quarter Density Velocity Hologram
Z and ϕ Windows at 12 dB and .7 s.

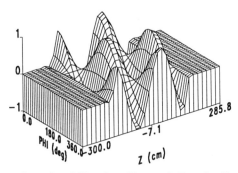

Fig. 15. Reconstruction of 12 kHz. One-Sixteenth Density Pressure Hologram
Z and ϕ Windows at 12 dB and .1 s.

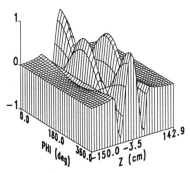

Fig. 16. Reconstruction of 1 Hz. One-Sixteenth Density Velocity Hologram
Z and ϕ Windows at 12 dB and .7 s.

recording radius. Once again the axial and angular window settings for the reconstructions were $\alpha = 12$ with $s = .1$ for pressure and $s = .7$ for velocity.

Both of the pressure derived holograms reconstructed almost perfectly. At 12 kHz. (Fig. 13) there is a slight, but noticeable, oscillation evident past the source area which can be removed by fine-tuning the window parameters. There is a difference noted between the slopes of the free edges on the negative and positive sides of center. This is due to the non-symmetry introduced by removing the even numbered points without recentering the aperture. The velocity reconstructions also suffer from the non-symmetry effect but to little degradation of the image (Fig. 14).

By removing all but every fourth point from the original holograms, we create a 32 x 32 point hologram with one sixteenth the original sample density. These pressure holograms have an axial sample spacing of 18.88 cm. which is approximately 1.5 times the acoustic wavelength at 12 kHz. The velocity derived holograms have a 9.44 cm. axial spacing. Both holograms' circumferential spacing is 11.25 degrees corresponding to 20.616 cm. on the hologram recording plane. This is considerably greater than the acoustic wavelength at 12 kHz.

The pressure reconstruction using the standard windows at 12 kHz. (Fig. 15) shows remarkable accuracy considering the sample spacing in both scan directions is considerably larger than the acoustic wavelength. Note the loss of clear edge definition due to sample points too far to either side of the actual edge. The test distribution had a structural wavelength much greater than the sample spacing and so it was still imaged with reasonable fidelity. However, any structural vibrations with a half wavelength less than the sample spacing would be falsely interpreted as one having a much longer wavelength. The 1 Hz. reconstruction exhibits similar fidelity. The radial mode shape in the velocity reconstructions is clearly discernable even at 1 Hz. (Fig. 16). The reconstructions have been shifted due to the non-symmetries introduced by the point removal program which are exaggerated by removing more points.

CONCLUSIONS

In this study the effects of both aperture size and sample point density on the reconstructive abilities of cylindrical nearfield holography were examined by specifying pressure and velocity distributions, creating associated holograms at a wide range of frequencies and then attempting to reconstruct the original distributions. While axial windowing is required for all reconstructions, it was shown that for smoothly varying full angular scans the effect of the angular window is negligible. Angular windowing becomes necessary with the introduction of any discontinuities or during a partial angular scan. It was also shown that partial angular scans and reconstructions are possible with only slight distortion at the aperture edges. This has great promise for many situations where physical constraints restrict a full angular scan of a source. Axially, the aperture may be shrunk to the source size with negligible loss in resolution if the longest wavelength to reconstruct is less than the aperture length in the direction of that wave. Reducing the sample density was also shown to be a viable method of reducing scan time. Although low densities were incapable of finely resolving the free edges, they did accurately describe the modal shapes. Scans with a sample spacing greater than the acoustic wavelength, while not recommended, were shown to reproduce the structural field as long as the structural Nyquist rate is observed when sampling.

REFERENCES

Eshenberg, K. E. and S. I. Hayek, "Measurement of Submerged-Plate Intensity Using Nearfield Acoustic Holography," Proceedings Inter-Noise '86, pp. 1229–1234.

Veronesi, W. A. and J. D. Maynard, "Nearfield Acoustic Holography (NAH) II. Holographic Reconstruction algorithms and Computer Implementations," JASA, 81 (5), May 1987, pp. 1307–22.

Williams, E. G., Henry D. Dardy and Karl B. Washburn, "Generalized Nearfield Acoustical Holography for Cylindrical Geometry: Theory and Experiment," JASA, 81 (2), February 1987, pp. 389–407.

SCANNING ACOUSTIC MICROSCOPE FOR QUANTITATIVE

CHARACTERIZATION OF BIOLOGICAL TISSUES

N. Chubachi, J. Kushibiki, T. Sannomiya, and N. Akashi
Department of Electrical Engineering, Tohoku University
Sendai 980, Japan
M. Tanaka and H. Okawai, Research Institute for Chest
Diseases and Cancer, Tohoku University, Sendai 980, Japan
F. Dunn, University of Illinois, 1406 W. Green St., IL
61801

INTRODUCTION

In the field of medical and biological sciences, an acoustic microscope is being developed as a practical research tool [1-3]. This paper
describes a system of reflection-type scanning acoustic microscopy which
can measure quantitatively visco-elastic properties of biological tissues
on a microscopic scale. This system operates in both amplitude and phase
modes with the following two functions; an acoustical imaging function
for displaying both amplitude and phase images with high spatial resolution, and a quantitative measurement function for determining both
velocity and attenuation at arbitrary points in an imaging area. The
operating frequency is variable so that it is possible to make spectro-
scopic analyses of tissues through acoustic properties. The spatial
change of acoustic properties in tissues due to pathological degeneration
can be extracted easily by an additional function of false color imaging.
Experiments have been performed on dog cardiac infarcted tissue in the
frequency range 100 to 200 MHz.

MEASUREMENT SYSTEM

A block diagram for a scanning acoustic microscope system, in
reflection mode, for biological tissue characterization is shown in Fig.
1. The system operates simultaneously in both amplitude and phase modes.
In the phase mode, using an electronically variable phase shifter [4] the
interferogram is used to determine the velocity.

In this system, an RF burst pulse (100-500 nsec duration) with an
on-off ratio higher than 140 dB is generated by the RF oscillator
through RF pulse modulator. The RF pulse signal is supplied to the
focusing ultrasonic transducer through a directional coupler, and the
resulting focusing ultrasonic beam radiates into the specimen.

The focusing transducer and the specimen are scanned mechanically in
a raster pattern to compose acoustic images. The transducer is vibrated
at a frequency of 60 Hz for the X-scan. A specimen mounted on a glass
plate is moved by a stepping-motor stage for the Y-scan. The time

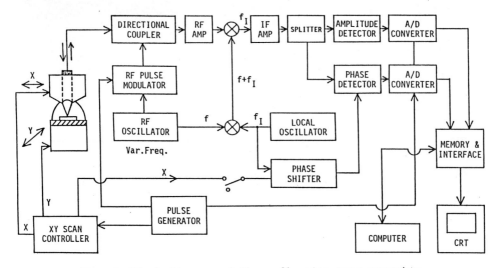

Fig. 1. Block diagram of the reflection-type scanning
acoustic microscope system.

required for a frame scan is 8 sec. The frame size can be selected to
be 0.5mm x 0.5mm, 1mm x 1mm, or 2mm x 2mm.

The operating frequency in the system is variable between 100 and
200 MHz in 1MHz steps. The focusing transducer is a ZnO film transducer
fabricated on one end of sapphire rod with an acoustic lens at the other
end. The operating center frequency of the transducer is 150 MHz. The
lens used for the present study has a radius of curvature of 1.25mm and
an opening half angle of 30deg. An acoustic matching layer of cal-
chogenide glass is formed on the lens surface. Lateral resolution in
water is approximately 7.5 μm at 150 MHz.

The received signal reflected from the specimen is amplified and
time-gated so that only the reflection from the object is detected.
After the reflected signal is heterodyned down to an IF frequency, it is
supplied to the amplitude and phase detectors. The output signals from
those detectors are fed to sample-and-hold circuit and A/D converter. As
the ultrasonic beam is scanned in raster pattern, the signals are stored
into the digital frame-memories, so that the amplitude image and the
phase image can be obtained on the CRT display, respectively. Those
image data are also transferred to a computer to obtain the velocities
and attenuations. The velocity and attenuation distributions are displayed on the CRT monitor as quantitative false color images.

In the present study, the thin tissue specimen was mounted on a
glass substrate. To obtain quantitative images, the numerical processing
was carried out assuming the simple model of a transmission line where
viscosity and multiple reflections are neglected in the specimen [3,5].
The acoustic properties of the specimen were determined using liquid of
water as the reference. With this simple model, the reflected wave from
the specimen is approximately the sum of only the wave S_1, reflected from
the water-specimen boundary, and the round trip wave S_2, as shown in Fig.
2.

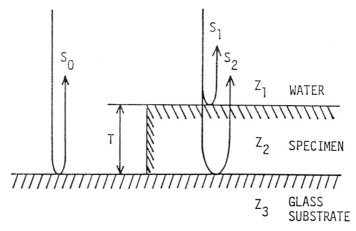

Fig. 2. Simple acoustic transmission line model with
three components of reflected waves, to determine
acoustic properties of tissue. Liquid of water
is the reference medium.

The sound velocity V_s is given as

$$V_s = 1/(1/V_w - \Delta\phi/4\pi fT),\tag{1}$$

where V_w is the sound velocity in the reference water, f is the
ultrasound frequency, $\Delta\phi$ is the difference in phase of the wave travelling in the water with and without the specimen, and T is the thickness of
the specimen.

The attenuation constant α_s of the specimen is given by

$$\alpha_s = -(1/T)10 \cdot \log[\,|S_1+S_2|/|S_0|\,] + \alpha_w,\tag{2}$$

where S_0 is the reference wave reflected from the glass substrate without
the tissue, and α_w is the attenuation constant of water.

The phase difference $\Delta\phi$ and the normalized reflection coefficient R_s
$(=(S_1+S_2)/S_0)$ of the specimen depend on the acoustic frequency, the sound
velocity, the attenuation, and the thickness of the specimen. The
specimen thickness, sound velocity and attenuation of the specimen can be
determined by computer-fitting technique employing the frequency charac-
teristics of $|R_s|$ and $\Delta\phi$ [5,6].

EXPERIMENTS

The scanning acoustic microscope system was applied to investigate
infarcted cardiac tissues taken from the left ventricle of a dog's heart.
Samples were prepared with a thickness of about 10 μm by a microtome from
a section of formalin-fixed and paraffin-embedded tissues, in which
artificial infarction had been developed. Several tissue samples were
mounted on glass slides and the paraffin was removed using xylene and
ethyl alcohol. One was processed by the Elastica Masson stain for optical observation. The other was used for the acoustical observation with
this system.

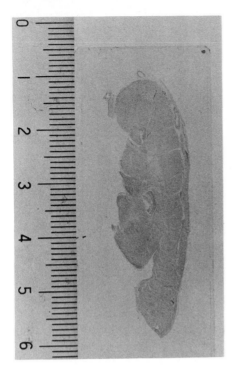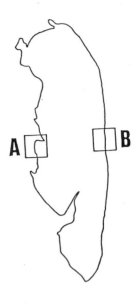

Fig. 3. Photograph of dog's cardiac infarcted tissue processed
by the Elastica Masson stain and a sketch indicating
the two regions, A and B, studied in the experiments.

An optical image of a stained tissue is shown in Fig. 3.
Experimental results are presented here for two typical regions studied,
namely, A and B, depicted in Fig. 3. The region A contains an infarcted
area, while the region B corresponds to a normal area. Both regions were
examined by the scanning acoustic microscope as well as with an optical
microscope. The results are shown in Figs. 4 and 5. By this staining
the normal heart muscle was stained reddish brown and the collagenous
fiber tissue, which replaced the degenerated tissue produced by the
infarction, was stained green. Acoustical images in the amplitude and
phase modes were taken every ten megahertz in the frequency range from
100 MHz to 200 MHz over a viewing area of 2mm x 2mm. According to the
experimental procedure, the measured data were analyzed as a function of
frequency. The thickness of the specimen in experiments was determined
to be 12 μm. With this value, the attenuation and the velocity in the
specimen were calculated. Two-dimensional quantitative images obtained
for the two regions at 130 MHz are shown in Figs. 4 and 5, together with
the optical images and typical A-mode profiles along each scanning line
indicated with arrows. The attenuation and velocity images in the
figures are quantitatively displayed with the scales listed in Table 1.

In region A, the infarcted area, the structure is so complex as seen
in Fig. 4(a) that the remarkable changes in the distribution of the
attenuations and velocities are also observed in Figs. 4(b) and 4(c).
For the collagenous fiber tissue, the measured velocities vary in the
range from 1650 m/s to 1800 m/s, and the measured attenuations vary in
the range from 200 dB/mm to 450 dB/mm.

(a) Optical

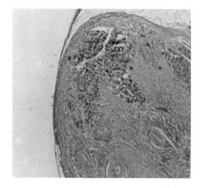

REGION A

(b) Acoustical (attenuation)

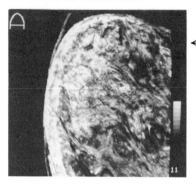

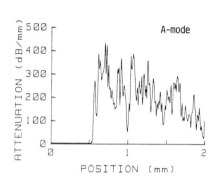

(c) Acoustical (velocity)

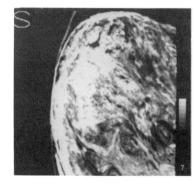

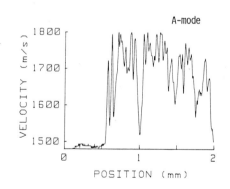

Fig. 4. Acoustical and optical images of region A.
Acoustic frequency is 130 MHz.

(a) Optical

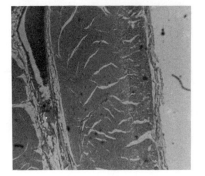

REGION B

(b) Acoustical (attenuation)

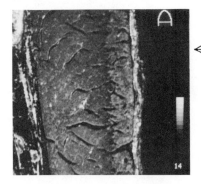

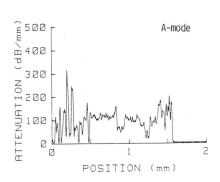

(c) Acoustical (velocity)

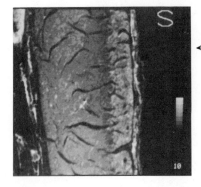

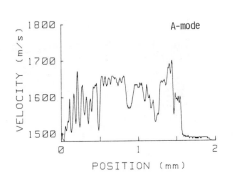

Fig. 5. Acoustical and optical images of region B.
Acoustic frequency is 130 MHz.

Table 1. Color scales in the acoustical images.

Color	Attenuation(dB/mm/MHz)	Velocity(m/s)
red	≥2.375	≥1765
magenda	2.25 ±0.125	1750 ±15
orange	2.00 ±0.125	1720 ±15
brown	1.75 ±0.125	1690 ±15
yellow	1.50 ±0.125	1660 ±15
green	1.25 ±0.125	1630 ±15
olive green	1.00 ±0.125	1600 ±15
cyan	0.75 ±0.125	1570 ±15
royal blue	0.50 ±0.125	1540 ±15
blue	0.25 ±0.125	1510 ±15
black	<0.125	<1495

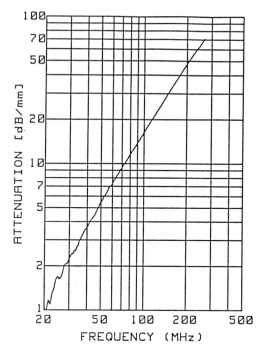

Fig. 6. Typical frequency dependence of attenuation
in fresh dog cardiac muscle tissue measured
at 22 °C by the bulk measurement method.

On the other hand, in region B, the normal muscle tissue has the relatively homogeneous distributions of the acoustic properties compared to the results obtained in the region A. The velocities are 1630–1660 m/s, and the attenuations are 100–130 dB/mm. The acoustic properties for water, in the viewing area external to the tissues, was measured to be 3.7 dB/mm at 130 MHz in attenuation and 1490 m/s at 23 °C in velocity, and were both displayed in the images with black in the color scale.

DISCUSSIONS

In the above experiments, the acoustical images were taken for so-called formalin fixed tissues. The acoustic properties are considered to be quite different from those of fresh cardiac muscle tissues. In order to discuss this argument, the bulk attenuation and velocity of fresh dog cardiac muscle tissues were investigated by the bulk method using an RF pulse [7]. The frequency dependence of the attenuation of the fresh tissue is shown in Fig. 6. The velocity is 1561 m/s, at the same position. The slope of the dependence is about 1.55, in the experimental frequency range. At 130 MHz, the attenuation is 24.5 dB/mm. Because the acoustic properties depend on the wave propagation direction associated with the muscle structure and orientation and the position, several experiments were performed with many samples. It was obtained for the normal tissues at 130 MHz that the attenuations are around 20–25 dB/mm and the velocities are around 1525–1565 m/s. In the fixed normal tissues of the region B, the velocities obtained by the scanning acoustic microscope are greater by about 100 m/s over those of the fresh tissues, while the attenuations around 120 dB/mm are about five times in decibels greater than those of the fresh tissues. It is not completely clear why such great differences were measured in the values of acoustic properties between the fixed tissues examined by the acoustic microscope and the fresh tissues examined by the bulk RF pulse method, especially in attenuation. The following reasons might be considered for the differences; changes in acoustic properties due to formalin fixation of tissues as represented by protein cross-linking [8], high absorption loss due to paraffin remaining in tissues, and/or increase of scattering loss due to fixation. This result presents a very important problem on the establishment of specimen preparation for acoustic microscopical examination.

CONCLUSION

A reflection-type scanning acoustic microscope system for biological tissue characterization has been developed. This system operates in both amplitude and phase modes, and can measure acoustic properties, viz., velocity and attenuation, of tissues in two-dimensional color imaging, as well as at arbitrary points in an imaging area. A demonstration for dog cardiac infarcted tissues prepared from a formalin-fixed, paraffin-embedded sample has been made. Spatial changes of acoustic properties in tissues due to pathological degeneration were displayed clearly in color imaging, and quantitative differences in acoustic properties between the normal cardiac muscle tissues and the collagenous fiber tissues were determined.

It has been further pointed out that there are great differences in acoustic properties between a formalin fixed normal muscle tissue and fresh normal muscle tissue. This may suggest that preparation of specimens is one of the most important problems in the quantitative characterization of biological tissues by means of acoustic microscopy.

ACKNOWLEDGMENTS

This work was supported in part by the Research Grant-in-Aids from the Japan Ministry of Education, Science & Culture and by a grant from the Japan-United States Cooperative Science Program from the JSPS (Japan) and the NSF (US). The authors are grateful to Honda Electronics Co., Japan, for invaluable technical assistance.

REFERENCES

1. C. F. Quate, A. Atalar, and H. K. Wickramasinghe, Acoustic microscope with mechanical scanning —— a review, Proc. IEEE, 67:1092 (1979).
2. S. D. Bennett and E. A. Ash, Differential imaging with the acoustic microscope, IEEE Trans. Sonics & Ultrason., SU-28:59 (1981).
3. N. Chubachi, J. Kushibiki, T. Sannomiya, M. Tanaka, H. Hikichi, and H. Okawai, Observation of biological tissues by means of a scanning acoustic microscope, Technical Reports of IECE, MBE81-37:9 (1981) (in Japanese).
4. N. Chubachi and T. Sannnomiya, Acoustic interference microscopes with electrical mixing method for reflection mode, Proc. IEEE Ultrasonics Symp., pp.604-609 (1984).
5. H. Okawai, M. Tanaka, N. Chubachi, and J. Kushibiki, Non-contact simultaneous measurement of both thickness and acoustic properties of a biological tissue by using focused wave in the scanning acoustic microscope, Jpn. J. Appl. Phys., (in press) (1987).
6. N. Akashi, J. Kushibiki, and N. Chubachi, Fundamental study on quantitative measurements of acoustic properties of biological tissues in acoustic microscopy —— Theoretical and experimental considerations using plane wave model ——, Technical Reports of IECE, US86-33:1 (1986) (in Japanese).
7. J. Kushibiki, N. Akashi, N. Chubachi, M. Tanaka, and F. Dunn, A method of ultrasonic tissue characterization at high frequency —— Measurement method and system ——, Reports of Spring Meeting Acoust. Soc. Japan, pp.689-690 (1986) (in Japanese).
8. J. C. Bamber, C. R. Hill, J. A. King, and F. Dunn, Ultrasonic propagation through fixed and unfixed tissues, Ultrasound in Med. & Biol., 5, pp.159-165(1979).

ESTIMATION OF ACOUSTIC ATTENUATION COEFFICIENT

BY USING MAXIMUM ENTROPY METHOD

Sun I. Kim and John M. Reid

Biomedical Engineering and Science Institute
Drexel University, Philadelphia, PA. 19104

ABSTRACT

The acoustic attenuation coefficient of soft tissue has been observed to increase as an approximately linear function of frequency. The amount of center frequency shift is related to the severity of some diseases. In diagnostic ultrasound experiments, the conventional periodogram approach cannot be applied directly to estimate small spectral changes due to its inherent performance limitations. We applied the Maximum Entropy Method (MEM) to estimate the spectrum of computer simulated echo signals with multiplicative random noise. It is observed that simple 2nd order MEM gives a weighted mean frequency value of spectral distribution. Using this, a more accurate center frequency estimation than the periodogram approach can be obtained. The MEM also can reduce data length and number of independent data segments, thereby reducing the size of the resolution cell as well as the number of experiments. In the case of a small object, this small resolution cell can avoid spectral smearing by the spectrum of adjacent tissue.

INTRODUCTION

Because of the frequency dependent attenuation of propagating acoustic waves in soft biological tissue, the returning ultrasound signal shows a downward spectral shift of its center frequency. The amount of shift is a function of travel distance as well as tissue characteristics. The slope of the acoustic attenuation coefficient with frequency, denoted by β, shows particular clinical utility. The value of β has been observed varying with the severity of some diseases [1].

There are several methods for estimating the acoustic attenuation from reflected ultrasound echo signals. Because of the convenience and noise sensitivity, frequency domain method is more widely used than the time domain method. Among them, the spectral shift approach, which measures the β value from the downward shift in the power spectrum of the propagating pulse as it penetrates the soft tissue, is commonly used. Only the center frequency location needs to be estimated to determine a β value, rather than the entire spectrum. Kuc reported that the spectral shift approach gives more accurate estimation than the spectral difference method in most practical cases [2].

Because of the computational advantage, the periodogram approach using the fast Fourier transform (FFT) has been applied in most spectral analysis of frequency dependent attenuation estimation. In diagnostic ultrasound experiments, however, the returning echo signals from either soft tissues or phantoms are not well behaved and fluctuate randomly. Thus the conventional periodogram approach using FFT gives biased and spuriously fluctuated spectral estimation in a small number of experiments. To determine small spectral shifts, the periodogram approach needs a number of independent and large data sets to perform ensemble averaging to get smooth and consistent spectra. However, it is an unrealistic assumption to have such large and independent data sets in practical situations and to have the same attenuation characteristic throughout the whole target. Also, if the data segment is short, which is true in most ultrasound experiments, the spectrum is determined mainly by the data window shape, rather than the actual spectrum.

The Maximum Entropy Method (MEM) to estimate the spectrum is an alternative method. Generally the MEM has several advantages over the conventional periodogram approach. These are; 1) improved smoothness of spectral estimation with fewer independent data sets 2) freedom from the side-lobe leakage of data windows and 3) increased frequency resolution even with short data records. Thus, this approach provides not only accurate estimation of the attenuation coefficient but reduction of the number of independent experiments, as well as of the size of each data record.

It is known that the lower order MEM yields smooth spectral estimation [3]. But with the lower order of MEM, it still can not be used to estimate the central frequency of the noise corrupted signal. In the extreme case, as low as 2nd order, we found that the MEM realization gives a single peak at the weighted average value of the spectral distribution. This is a very useful fact to estimate the center frequency from the noise-corrupted ultrasound echo signal. We sucessfully applied this 2nd order MEM to estimate the central frequency shift between the near and far region of computer simulated noisy ultrasound signals.

THEORY

The returning echo signal from soft tissue can be divided into small data segments. The signal can be considered as wide-sense stationary within each short segment [2]. If the propagating medium is linear and homogeneous, the frequency dependent attenuation process can be thought as a linear space-invariant system which has linear-with-frequency attenuation characteristic. The ultrasound echo signal returning from the near region is an input to the system, and the signal from the far region is an output. Applying the result from linear system theory, the spectrum of the reflected far region pulse is equal to;

$$Po(f) = | H(f) |^2 Pi(f) \qquad\qquad (1)$$

where $H(f)$ is the power transfer function of the soft tissue between the two regions and $Pi(f)$ and $Po(f)$ are the power spectral density, or spectrum, of the propagating ultrasonic pulse $Pi(t)$ and $Po(t)$ in near and far region, respectively. It is assumed that the linear-with-frequency attenuation transfer function can be expressed as

$$| H(f) |^2 = Exp(-4 \ \beta \ f \ D), \qquad\qquad (2)$$

where f is the frequency in Hz, β is the slope of the acoustic amplitude attenuation coefficient with unit of Np/cm-MHz, and D is a distance between the near and far region. The factor 4 results from the total pulse propagating distance between the two regions which is equal to 2D and the attenuation coefficient τ. Let the reference input ultrasonic pulse be a sinusoid modulated by a Gaussian envelope with time constant . Then the Fourier amplitude spectrum also has a Gaussian envelope of the form;

$$Pi(f) = Ai \; Exp[-(f-fi)^2 /B^2],   \hspace{2cm} (3)$$

where Ai is a constant, B is a measure of pulse bandwidth which is equal to $1/2 \; \pi\tau$ and fi is the carrier frequency, the centroid of the spectrum. After the pulse propagates through the soft tissue, the output pulse will experience a downward shift in its center frequency, and will yield the spectrum Po(f) after substuting Eqs. (2) and (3) into Eq. (1);

$$Po(f) = Ao \; Exp[-(f-fo)^2 /B^2],   \hspace{2cm} (4)$$

where Ao is a constant and the frequency shift fo is;

$$fo = fi - 2 \; \beta \; DB^2 .   \hspace{4cm} (5)$$

In other words, the output spectrum maintains the same Gaussian form as that of the input, but has been shifted to a lower center frequency fo. The value of β can then be determined from the downward shift in the center frequency;

$$\beta = 4.343 \; (fi - fo)/(DB^2). \quad [dB/cm\text{-}MHz]   \hspace{1cm} (6)$$

Having established a mathematical model of the attenuation process in soft tissue, it is possible to get a better understanding of the process with various kinds of practical measurement limitations with the aid of computer simulation. Results can be extended to the real situation of experiments. An appropriate digital filter of linear-with-frequency attenuation characteristic is necessary to perform simulation on a digital computer. The minimum phase filter proposed by Kuc well represents the physical mechanisms which produce a linear loss function with frequency, and fits well with experimental data [4]. A minimum phase system is one that has all its poles and zeros inside unit circle in the z plane. Even though there are several other models available, this minimum phase filter model gives quite good results. We used this model to simulate the frequency dependent attenuation process.

In practical ultrasound, however, the signals reflected from the body organs cannot have the ideal pulse shape as above. The reflecting scatterers located within the soft tissue are randomly shaped, oriented and disrtibuted in range. Thus, we need another model to simulate the ultrasound signals reflected from randomly shaped interfaces for the completeness of the computer simulation. Let's assume that the observed data segment from a particular region is produced by a convolution of the propagating pulse waveform in that region with a sequence of white Gaussian random noise, which representing the random scatterers in that region. This model is appropriate here because it generates distorted spectra, which appear similar to those observed experimentally.

To estimate the spectrum of the propagating pulse in the near region from the noise corrupted observed data Ri(k), we will calculate

the Fourier transform of Ri(k), denoted by Ri(f).   Using   the
convolution theorem, we have;

$$Ri(f) = G(f) \ Pi(f) \tag{7}$$

where G(f)is a spectrum of random noise generated by the random
reflectors and Pi(f) is the spectrum of the pulse in the near region,
as observed at the receiver.   This process generates multiplicative
random noise in the frequency domain. Because the magnitude of the
noise power spectrum can be thought as constant throughout the
frequency band, the estimated signal power spectrum will have same
shape as the power spectrum of pulse after a number of ensemble
averages.   Thus, even though the spectral estimation of ultrasonic
signal at certain segment is randomly distorted, we can still use it to
provide estimates of the tissue attenuation. In the same manner, the
power spectrum at the far region can be estimated from the measured
ultrasound signal;

$$Ro(f) = G'(f) \ Po(f). \tag{8}$$

2ND ORDER MAXIMUM ENTROPY METHOD AND COMPUTER SIMULATION

Estimation of the power spectrum of discretely sampled data is
usually based on procedures utilizing the FFT, namely the periodogram.
This approach to spectrum analysis is computationally efficient and
produces reasonable results. However, when this periodogram is applied
to finite data sets of stochastic processes, it may result in
statistically inconsistent estimations if no ensemble averaging is
performed. Welch has introduced a modification of the averaging
periodogram procedure that is well suited to direct computation of a
power spectrum estimate using FFT.   By windowing the data segments
before computing the periodogram, he achieved the variance reduction of
the original periodogram and at the same time achieved smoothing of the
spectrum with the concomitant reduction of resolution [5].   To
accomplish this modified average periodogram, one needs large numbers
of independent data sets. For several reasons we cannot either prolong
the data segments or increase the number of independent data sets in
diagnostic ultrasound experiments;   1) for a long data segment, the
signal within the segment becomes non-stationary because of the
frequency dependent attenuation, 2) the long segment will include the
spectra of unwanted adjacent organs and 3) the target organ should have
large a area of constant characteristics to provide a number of
equivalent and independent data sets.

It is important to note that in the case of a zero mean, white
Gaussian process, the variance of the periodogram approaches a non-zero
constant and that the spacing between spectral samples with zero
covariance decreases as the data length increases. Thus, as the record
length becomes longer, the rapidity of the fluctuation in the
periodogram increases [6]. It is also true that for the non-white but
Gaussian process, for example the returning echo from the randomly
distributed scatterers with Gaussian shape excitation pulse, the
variance of the periodogram is proportional to the square of the
spectrum as the data length increases.   Thus in general, the
periodogram is not a consistent estimator and it can be expected to
fluctuate rather wildly about the true spectrum value [7]. As a
result, estimating the attenuation coefficient by periodogram method is
practically very difficult and sometimes gives false values.

The MEM is a good tool for spectrum analysis from short data
records with relatively good noise rejection and with no windowing

effect. One needs not assume that the measured process is zero or cyclic outside the measurement interval. Thus the need for a window can be eliminated along with its distorting impact, without sacrificing the quality of estimation. MEM also can reduce data length and number of independent data segments, thereby reducing the size of the resolution cell as well as the number of experiments. With fixed speed of sound and limited sampling rate, this means reducing the resolution cell. In the case of small objects, this small resolution cell can avoid spectrum smearing by the spectrum of adjacent tissue.

It was observed that, for the signal containing two closely spaced frequencies, the lower order MEM realization will not yield a sufficient resolution of the corresponding power to reveal both components, but instead gives just one peak at the mean position with a power estimate equal to the sum of the power of the two components [8]. This is true for the case in which many frequency components are distributed over the entire spectral domain. The lower the order is, the less the peaks appear at intermediate positions of the local frequencies. We found that the extremely low order of the 2nd order MEM realization gives a single peak at a position corresponding to the weighted average value of the whole spectral distribution. This is very useful fact to estimate the central frequency from the noise-corrupted ultrasound echo signal. Each realization yields one peak at a mean frequency that is unbiased, and can be used to estimate the central frequency through ensemble average.

We have to give a credit to Kuc and Li who first applied the 2nd order autoregressive approach to estimate the center frequency [9]. However, it is known that the forward-only linear prediction method they used gives more false peaks and greater perturbations of spectral peaks from their correct frequency locations than our forward-backward MEM estimation approach. The forward-only linear prediction leads to autoregressive (AR) parameter estimates with greater sensitivity to noise [10].

With an input driving sequence g(n) and the output sequence x(n), the second order AR process can be written as;

$$x(n) = - a(1) x(n-1) - a(2) x(n-2) + g(n).$$ (9)

where a(1) and a(2) are AR coefficients. The corresponding transfer function in the z plane is given as;

$$H(z) = \frac{G(z)}{1 + a(1) z^{-1} + a(2) z^{-2}}$$ (10)

By noting the pole zero pattern in the z plane, we can relate the AR coefficient to the position of poles;

$$H(z) = \frac{G(z)}{1 - 2 r(\cos w_o) z^{-1} + r^2 z^{-2}}.$$ (11)

The resonance frequency, wo, is equal to;

$$w_o = \cos^{-1} \left(\frac{a(1)}{2\sqrt{a(2)}} \right),$$ (12)

and the radius from the origin to pole position is;

$$r = \sqrt{a(2)} .$$ (13)

Thus, from the two AR coefficients we can get the resonance frequency and the corresponding power level.

For the computer simulation, the input pulse sequence pi(n) in the near region before it is affected by attenuation, is generated. The pi(n) is modelled as a sinusoid modified by a Gaussian envelope;

$$pi(n) = \sin(2 \pi fi \ n \ Ts) \ Exp[-(n-11)^2 \ Ts^2 \ /(2 \ \tau^2 \ )]$$ (14)

for n = 1,2,.....,21

where the time constant τ = 0.2 micro sec., sampling interval Ts = 0.1

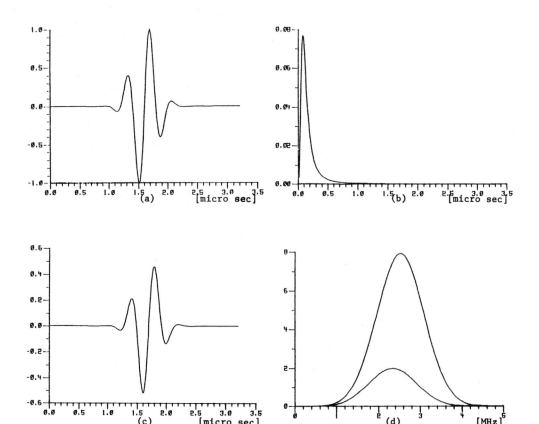

Fig. 1. Simulated pulse shape and spectral wave forms. a) Near region pulse sequence. Center frequency is 2.5 MHz and time constant is 0.2 micro sec. b) Impulse response of minimum phase filter, having linear-with-frequency loss characteristic with slope equal to -0.5dB/cm-MHz. c) Far region pulse sequence after 5cm of propagation. This is obtained by convolution of a) and b). d) Ideal spectrum of near (upper) and far (lower) region pulses.

micro sec (10 MHz) and carrier frequency fi (center frequency) is 2.5 MHz. The far region pulse is generated by convolving the near region pulse sequence with impulse response of attenuation transfer function given in Eq. (2). The finite-duration impulse response of the transfer function is obtained by inverse Fourier transforming the complex transfer function, which is characterized as a minimum phase filter. Each near and far region pulse sequence is then convolved with an independent set of Gaussian noise to generate simulated ultrasound echo signals.

In this simulation, we assumed the attenuation slope β as −0.5 dB/cm-MHz, which is a value of normal liver, and the total travel distance 2D as 5 cm (distance between the two region is 2.5 cm). Fig. (1) shows simulated near and far region pulses, and their Fourier transforms. The central frequency of the far region spectrum is downward-shifted about 0.182 MHz from the center frequency (2.5 MHz) of near region. Fig. (2) shows one example of a computer generated near region echo signal and its periodogram estimation using FFT, and 25th order MEM spectral estimation by least-square algorithm. Data length is 128 points which corresponds to 1 cm of range gate size for a sound velocity equal to 0.1500 cm/micro sec at 10 MHz sampling rate. As we

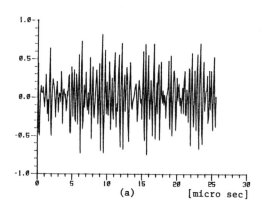

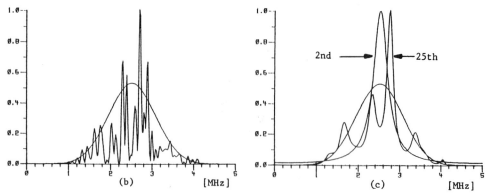

Fig. 2. a) Computer generated ultrasound echo signal. b) Spectrum of one 128 point segment of a) using FFT. Ideal spectrum is superimposed for the convenience of comparison. c) 25th order MEM (fluctuating waveform) and 2nd order (one high peak) with ideal spectrum.

noticed, it is difficult to determine the center frequency from either periodogram or ordinary MEM spectral estimation because of the fluctuations.

The 2nd order MEM spectral estimation is shown in Fig. (3) for the case of 12 realizations of each 64 point (0.5 cm of range gate at 10 MHz sampling rate) near and far regions. As we see in this figure, every estimation is highly concentrated near the true value of center frequency. The deviation of estimation, we believe, is purely due to the uneven distributions of noise spectrum, and not due to the bias of estimation. Thus simple averaging of each estimated value produces a nice result, a center frequency estimation which is very close to the true value. After the average of 64 realizations, we got 2.498 MHz for the near region (true value is 2.5 MHz), and 2.319 MHz for the far

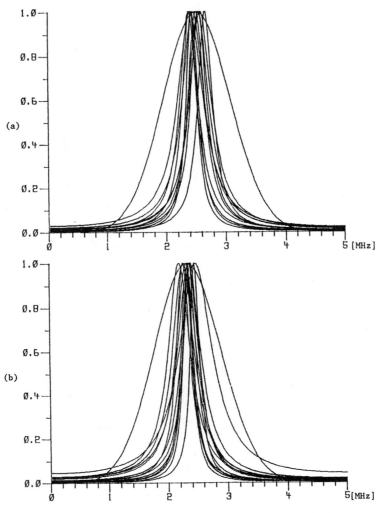

Fig. 3. 12 independent realizations of 2nd order MEM of a) near and b) far region echo signal, superimposed with ideal spectrum.

294

region (calculated value = 2.318 MHz). We found that the 2nd order MEM method generally gives reasonable results at about 50 realizations. This is a remarkable reduction of the number of independent realizations compared to Kuc's more than 1,000 times [12].

CONCLUSION

We examined the well known Burg algorithm as well as the least-square (LS) algorithm. It has been observed that the peak locations in Burg's MEM spectral estimate depends on the phase of the signal. The phase dependence of the estimate increases as the data length decreases. The LS algorithm is reported to have the least dependence on signal phase [11]. Even though these two algorithms yield similar results, the LS algorithm yields slightly better estimation than Burg's method.

As pointed out earlier, the quality of Welch's modified periodogram estimation is depend on the length of each data segment and number of independent data sets. To determine the center frequency from the computer simulated data, Kuc had to use more than 1,000 sets of independent realizations of 128 point Hamming windowed data [12]. For clinical application, this requires that the target under examination should be large and have the same attenuation characteristic throughout the whole target. If we accept the assumptions, the usage of this method is quite limited and can be applied only to large and homogenuous organs. Moreover, the periodogram estimation not only suffers from sidelobe leakage of the data window, but also requires a large set of experiments which is tedious and time consuming.

One thing that should be noticed here is the power level of estimated spectrum. In computer simulation, the power level of input signal is known, thus the each spectrum can be normalized to the same scale and can be ensemble-averaged to get the smooth spectrum. However in practical cases, the power of returning echo signal is changing as the acoustic wave propagates through the soft tissue and is generally unknown. Thus the ensemble average of power spectral components of different magnitudes can produce meaningless results. This problem can be avoided, because 2nd order MEM estimation yields only one mean frequency value from each realization regardless of the power level of input signal. Simple algebric averaging of these mean frequencies will generate a result which is equivalent to the ensembling average of whole spectra, thus saving computational time.

The advantage of 2nd order MEM applied to estimate the attenuation of soft tissue can be summarized as; 1) a great reduction of the size of each data set and the number of independent data sets, 2) improvement of the quality of central frequency estimation in terms of unbiasedness and consistency, 3) no need of prior knowledge of input and output signal power level, 4) simplicity and swiftness of operational procedure, and 5) estimation independent of the acoustic signal bandwidth. There is a weak trend that the more the estimation of center frequency deviates from the true value, the closer is the pole position to the origin in z plane(i.e. small radius in Eq. (13)). We found the correlation is not significant. However, still the exclusion of estimations with very small radius will help to reduce the variance of the averaging process.

We found that the MEM approach overcomes some practical limitations of ultrasound experiments, and provides more accurate noise-immune estimation of attenuation than the periodogram using FFT.

This technique also can be extended to other areas of ultrasound tissue characterization, for example real time Doppler to get mean Doppler-frequency shift.

ACKNOWLEDGEMENT

We wish to thank Dr. F. Yarman-vural for helpful discussion and encouragement. This work was supported in part by Grant 30045 from the National Institutes of Health.

REFERENCES

[1] J.G. Miller,"Ultrasonic Tissue Characterization: Properties and Applications," IEEE Ultrason. Symposium Proc. 1983

[2] R. Kuc,"Estimating Acoustic Attenuation from Reflected Ultrasound Signals: Comparison of Spectral-Shift and Spectral-Difference Approaches," IEEE Trans. Acoust. Speech, and Signal Processing, vol.ASSP-32, No.1, pp.1-6, Feb. 1984.

[3] J.Makhoul,"Spectral Linear Prediction: Properties and Application," IEEE Trans. Acoust., Speech, Signal Processing, vol.ASSP-23, No.3, pp.283-296, Jun. 1975.

[4] R. Kuc,"Digital Filter Model for Media Having Linear with Frequency Loss Characteristics," J.Acoust.Soc.Am., vol.69(1), pp.35-40, Jan. 1981.

[5] P.D. Welch,"The Use of Fast Fourier Transform for the Estimation of Power Spectra: A Method Based on Time Averaging over Short, Modified Periodograms," IEEE Trans. Audio and Electroacoust., vol.AU-15, pp.70-73, Jun. 1967.

[6] A.V. Oppenheim and R. W. Schafer, Digital Signal Processing. Englewood Cliffs, NJ: Prentice-Hall, Ch. 11. 1975.

[7] G.M. Jenkins and D.G. Watts,"Spectral Analysis and Its Application," Sna Francisco, Holden-Day, Inc., 1968.

[8] S.J. Johnsen and N. Andersen,"On Power Estimation in Maximum Entropy Spectral Analysis," Geophysics, vol.43, No.4., pp.681-690, Jun. 1978.

[9] R. Kuc and H. Li,"Reduced-Order Autoregressive Modeling for Center-Frequency Estimation," Utrasonic Imaging vol.7, pp.244-251, 1985.

[10] D.N. Swingler,"A Comparison between Burg's Maximum Entropy Method and a Nonrecursive Technique for the Spectral Analysis of Deterministic Signals," J. Geophysical Res., vol.84, pp.679-685, Feb. 1979.

[11] S.L. Marple,Jr.,"A New Autoregressive Spectrum Analysis Algorithm," IEEE Trans. Acoust., Speech, Signal Process., vol.ASSP-28, pp.441-454, Aug. 1980.

[12] R. Kuc,"Estimating Reflected Ultrasound Spectra from Quantized Signals," IEEE Trans. Biomed. Eng., vol.BME-32, No.2., pp.105-112, Feb. 1985.

CHARACTERIZING ABNORMAL HUMAN LIVER BY

ITS IN-VIVO ACOUSTIC PROPERTIES

N. Botros[1], J.Y. Cheung[2], W.K. Chu[3], and A.R. Ashrafzadeh[4]
[1]S. Illinois U., Carbondale, IL [2]U. of Oklahoma, Norman, OK
[3]U. of Nebraska, Omaha, NE   [4]U. of Detroit, Detroit, MI

I. INTRODUCTION

The goal of this study is to investigate the possibility of implementing digital signal analysis technique to differentiate between normal and abnormal soft tissue in in-vivo environment. In this study, we focus on human liver. Tissue differentiation means the ability to differentiate between normal and abnormal tissues. To achieve the goal of this study, we have developed a model describing the average ultrasound backscattered energy from normal livers. To examine any human liver, we record the backscattered energy from that liver and, by comparing the model of this backscattered energy with the model of normal liver tissue, we can tell whether this liver is normal or abnormal. This comparison is based on the frequency dependency of the acoustic attenuation and backscattering coefficients, these two acoustic coefficients are the main parameters that control the shape of the backscattered energy.

Although ultrasonic imaging has been widely employed in modern ultrasonic diagnostic equipment, researchers started, with the recent advances in the digital technology, to employ quantitative techniques on the ultrasound backscattering signal in order to extract more information that can help in characterizing the human tissue. Nicholas[1], Kuc[4], Maklad[5].

Our guidelines for achieving reliable results are as follows:
1. A large number of samples of the backscattered data should be taken. The scattering process is a statistical one, so obtaining reliable results necessitates the collection of large number of samples.
2. The noise must be filtered out as much as possible.

This filtering process consists of a digital process to filter out the undesired frequency components from the power spectrum of the backscattered data and a formation of frequency histograms to exclude the odd values from calculations of the attenuation and backscattering coefficients. An example of the odd values of the attenuation or backscattering coefficient would be negative values.

3. Careful adjustment of the gain of the amplifiers that amplify the backscattered signal should be set to avoid saturation in these amplifiers. Saturation, if it happens, results in inaccurate measurement of the backscattered signal and consequently, inaccurate calculation of the attenuation and scattering coefficients results. On the other hand, this gain should not be very low, otherwise the signal to noise ratio of the backscattered signal will be reduced.

II. A THEORETICAL MODEL FOR THE BACKSCATTERING PROCESS

The input of the theoretical model is the power spectrum of the transmitted pulse from the ultrasound transducer W_t. The output of the model is the power spectrum of the gated backscattered ultrasound signal from selected region of the liver, W_r. $H(f)$ is the Fourier transform of the power transfer function, it can be expressed as follows:

$$H(f) = W_r(f)/W_t(f) \tag{1}$$

To develop the analytical formula for $W_r(f)$, some assumptions must be made. The following is a summary of the assumptions made.
1. There are no shear waves.
2. There is no flow or movement inside the scattering region under investigations.
3. All analyses are made in the Fraunhofer region, i.e., in the far field.
4. The waves can be approximated as plane waves in the far field.
5. The medium where the waves travel is a non-dispersive medium.
6. The square of the received echo amplitude is a linear measure of the scattered energy.
7. The number of scatterers per unit volume is small enough so that the interference between the scattered waves can be neglected.
8. Time Gain Control (T.G.C.) is disabled, i.e., the amplifier gain is time-independent.
9. The scattered pressure is weak relative to the incident pressure.

Following the same guidelines shown in Nicholas et al.[1,2,3], the received ultrasound power W_r from a depth X_1, in the frequency domain, can be written as:

$$W_r(X_1) = \frac{A*(X_1)^2}{8*R^4*k^2}*U_d*(f_0/f)^6*(1/U)*e^{-2*U*X_1} *$$

$$[1-e^{-U*C*T}]*[1-e^{-\left(\frac{k*f*r_s}{f_0*x}\right)^2}]*W_t \qquad (2)$$

Where:

A is the aperture of the face of the transducer.
k is the constant depends on the type of the transducer.
U is the attenuation coefficient of the tissue medium.
R is the diagonal distance from the transducer center to the scatterer.
f_0 is the fundamental frequency of the ultrasound beam.
C is the speed of sound in soft tissue.
T is the gate duration.
r_s is the radius of the transducer.

Let us first simplify the model described by (2). All of our analyses have been done in the far zone. In this zone, $(r_s/X_1) << 1$. Also, K, the transducer constant, is generally less than one Nicholas et al. [1,2,3], and within the frequency bandwidth of the pulse echo diagnostic machines, $.5 < f/f_0 < 1.5$. From the above, we can assume that $[k*f*r_s/(X_1*f_0)]^2 << 1$ and hence, we can write:

$$e^{-\left(\frac{k*f*r_s}{f_0*x}\right)^2} \cong 1-\frac{(k*f*r_s)^2}{(X_1*f_0)^2} \qquad (3)$$

inserting (3) into (2), we get:

$$W_r(X_1) = \frac{A*(X_1)^2}{8R^4*k^2}*U_d*(f_0/f)^6*(1/U)*e^{-2U*X_1} *$$

$$[1- e^{-U*C*T}]*\frac{(k*f*r_s)^2}{(X_1*f_0)^2}*W_t \qquad (4)$$

or:

$$W_r(X_1) = \frac{A*(r_s)^2}{8R^4}*U_d*(f_0/f)^6*(1/U)*e^{-2U*X_1} *$$

$$[1-e^{-U*C*T}]*W_t \qquad (5)$$

The above equation describes the backscattered power W_r from depth X_1 inside the human liver within the frequency range from $.5 < f/f_0 < 1.5$ Mhz. Now, if we keep the window width $(C*T)$ constant, and select another depth X_2, then (5) can be written as:

$$W_r(X_2) = \frac{A*(r_s)^2}{8R^4} *U_d*(f_0/f)^6*(1/U)*e^{-2U*X_2} *$$

$$[1-e^{-U*C*T}]*W_t \qquad (6)$$

Dividing (5) by (6), we get:

$$\frac{W_r(X_1)}{W_r(X_2)} = e^{-2U*X_1} / e^{-2U*X_2} \qquad (7)$$

Note that in the above equation, it is not necessary to know the absolute value of X_1 or X_2, but only the difference between them. Rearrange (7), we obtain:

$$U = \frac{1}{2*dx} \ln[\frac{W_r(X_1)}{W_r(X_2)}] \qquad (8)$$

with $dx = X_2-X_1$ is the difference in centimeters between the two depths.

Equation (8) is the same as the familiar form of the attenuation coefficient of ultrasound waves propagating in fluids except the factor "2" is added because the backscattered ultrasound waves must travel round trip.

To complete and verify the developed theoretical model of the backscattered signal, we need to measure the attenuation coefficient U. Investigation of (8) shows that to measure U, we have to measure the backscattered power W_r. To measure W_r, we have to capture the backscattered signal, digitize it, store it, and then analyze it. To perform these operations, we have built a data acquisition analysis system. This system is capable of capturing a selected portion of the backscattered signal, digitizing this portion of the signal with a sample rate up to 16 MHz, and then storing this digitized information on diskfiles for further analysis. A block diagram of the system is shown in Fig. 1. Details of this system can be found in Botros et al. [6],[7].

III. EXPERIMENTAL PROCEDURES

The data acquisition and analysis system developed by the authors was used to collect the data of the ultrasound backscattered signal. Data was collected from 24 volunteers,

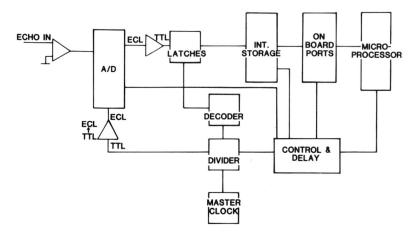

Fig. 1.   Block diagram of the system.

with no history of liver diseases, and from 15 patients with
any   kind   of   abnormalities   in   their   livers.   These
abnormalities were detected by the ATL Diagnostic Machine
using   the   conventional   imaging   detection   and   biopsy
examinations. From every subject, the data was taken 8 times.
We collect data for a period of 32 microseconds and then store
this   data   in the diskfile under the name of the volunteer.
Each file consisted of 512 data points. The data was taken at
different depths: 5.0, 6.5 and 8.0 cm. A software threshold
was inserted in the data acquisition segment of the software.
This   threshold   screens out any "false file".   False files
happen when the transducer is not firmly attached to the body
of the patient or when the transducer itself is broken.

     To   transfer   the   data from   time   domain   to frequency
domain, we apply the Fast Fourier Transform (FFT) algorithm to
these   data. A Fast Fourier Transform program was written   in
the software of our system.   We apply this program on the 512
data points of each file, the output of the FFT is 256 points
represent the power spectrum of the 512 points. The frequency
range   of this power spectrum extends from 0 to 8   MHz.   To
investigate the validity of this range, we have plotted the
power   spectrum of the incident wave, see Fig. 2.   After
investigating this power spectrum, we found that   frequencies
less   than   1.5   MHz or greater than 4.5 MHz have very small
amplitudes,   and hence,   these components can be excluded from
the power spectrum.   Accordingly, we applied a software
digital filtering process on the 256 frequency points of the
power spectrum of each file and all components outside   the
range extending from 1.5 to 4.5 MHz were excluded. After the
filtering   process,   each   file   consists   of   75   points
representing the power spectrum of the backscattered signal
in frequency range from 1.5 to 4.5 MHz.   Investigation of
Files taken from the same subject shows that these files are

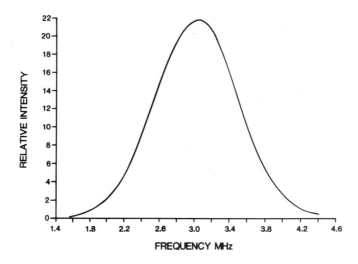

Fig. 2. Power spectrum of the transmitted wave.

not identical. This discrepancy between the 8 files can be explained as follows:

First, the backscattering process itself is a statistical process rather than a deterministic process. This means that the data taken from same person may not be identical. Secondly, the random noise, introduced by our system and the ultrasound machine, tends to distort randomly the stored data in our microcomputer. Therefore, data taken from the same person may not be identical.

To minimize the discrepancy between the files, we formed histograms from the files representing the power spectrum of the backscattered data. Fig. 3 shows a portion of the histogram representing the average power spectrum of the backscattered data from normal liver volunteer. Each point in the histogram represents the center of the corresponding interval.

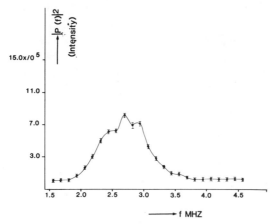

Fig. 3. Histogram of the power spectrum of normal tissue.

Since the primary goal of this study is tissue differentiation, we focus our attention on the attenuation coefficients rather than on the comprehensive model that describes the backscattering process. The attenuation coefficient is calculated by the use of (8). As shown in this Equation, to calculate the attenuation coefficient of human liver, we first calculate the backscattering power at two different depths and then apply (8). To be sure that all of our calculations are done in the far field, only 5.0, 6.5 and 8.0 cm. depths are to be considered here.

We follow the above steps to calculate the attenuation coefficient from backscattered data of depths 5.0 and 6.5 cm. using (8), then we perform the same calculation for depths 6.5 and 8.0 cm. We take the average value of the two above calculations to obtain the final average value of the attenuation coefficient as a function of frequency. To curve-fitted these average values, we tried different kinds of functions using the least square fitting technique. The function that gave the lowest error was found to be as follows:

$$U = a*f*e^{-b*(f-f_e)^2} \qquad (9)$$

Equation (9) can be seen as a product of two functions. The first function is a linear function of the form a*f, and the second function is a Gaussian function of the form of EXP[-b(f-f_e)^2]. The values of a, b and f that give lowest error were found by trial and error. These values are as follows:

a = .05, b = .6179 and f_e = 3.128 MHz.

Fig. 4 graphs (9). It is to be mentioned here that Equation

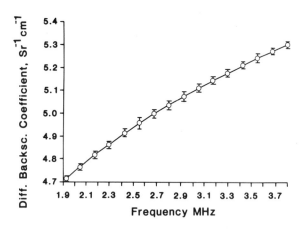

Fig. 4. Attenuation coefficient of normal tissue

(9) can be approximated to a linear relationship between the attenuation coefficient and the frequency for frequencies close to f_e.

We calculate the attenuation coefficients of abnormal liver cases by following the same steps as in the case of normal liver subjects. Fig. 5 shows the attenuation coefficient of normal liver subjects along with 4 abnormal liver subjects. From this Figure, we conclude that differentiation between normal and abnormal liver tissues can be done by investigating the relationship between the attenuation coefficient of the tissue and the frequency.

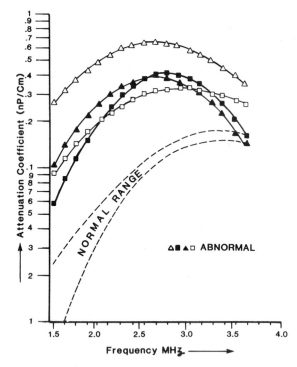

Fig. 5. Attenuation coefficients for normal & abnormal tissues.

V. SUMMARY

Our theoretical investigations of the backscattered signal showed that the attenuation coefficient is one of the main parameters that control the backscattering process.

To investigate the frequency-dependency of this parameter, we developed a theoretical model describing the backscattering process of ultrasound waves from human liver tissue. From this model, we derived simple formulas for the measurement of the attenuation coefficient (8) To calculate the frequency-dependency of this coefficient experimentally, we designed and constructed a comprehensive data-acquisition-analysis system. Through Fortran Programs, the system calculated the attenuation coefficient as functions of frequency. Our results showed that it is possible to differentiate between normal and abnormal liver cases through investigating the frequency-dependency of the attenuation Finally, we would like to emphasize the fact that in all of our study we incorporated the use of a microcomputer, which is in contrast to most previous reports which employed main-frame or mini-computer. Since most state-of-the-art ultrasound equipment contain microprocessors or their equivalents, it is out hope to have our data-storage and analysis system incorporated as in integral part of the ultrasound equipment. Then clinical ultrasound data can be obtained in true digital format and on a real-time basis.

REFERENCES

1. D. Nicholas, O.R. Hill and D.K. Nassiri, "Evaluation of Backscattering Coefficient for Excised Human Tissues: Principles and Techniques." Ultrasound in Med. & Biol., vol. 8, no.1, pp. 7-15, 1982.
2. D. Nicholas, "Evaluation of Backscattering Coefficients for Excised Human Tissue: Results, Interpretation and Associated Measurements." Ultrasound in Med. & Biol., vol. 8, no. 1, pp. 17-28, 1982.
3. D. Nicholas, "Orientation and Frequency Dependence of Backscattered Energy and its Clinical Application." Recent Advances in Ultrasound Biomedicine, vol. 1, pp. 29-54, 1977.
4. R. Kuc, "Processing of Diagnostic Ultrasound Signals." IEEE ASSP Magazine, pp. 19-23, January, 1984.
5. N.F. Maklad, J. Ophir and V. Balsara, "Attenuation of Ultrasound in Normal Liver and Diffuse Liver Diseases in Vivo." Ultrasound Imaging, vol. 6, pp. 117-125, 1984.
6. N. Botros, "Tissue Differentiation Through Digital Signal Processing." Ph.D. Dissertation, University of Oklahoma, 1985.
7. N. Botros, W.K. Chu, J.C. Anderson, T.J. Imray and J.Y. Cheung, "A Microprocessor-Based Tissue Differentiation Analysis". Proceedings of the IEEE Seventh Annual Conference of the Engineering in Medicine and Biology Society, vol. 1, pp. 217-221, 1985.

NON-LINEAR PARAMETER IMAGING WITH REFINEMENT TECHNIQUES

W. Hou, Y.Nakagawa, A. Cai, N. Arnold and G. Wade

Department of Electrical and Computer Engineering
University of California
Santa Barbara, Ca 93106

ABSTRACT

The parameter β is a function of the acoustic nonlinearity of a medium. We previously reported making tomograms of spatial variations in β by using the concept of the parametric acoustic array. We have reconstructed tomograms from two nonlinearly generated secondary waves: 1) a difference-frequency wave and 2) a second-harmonic wave.

The quality of the images previously obtained showed evidence of the detrimental effects of refraction from various medium boundaries not taken in to account by our theory. We have recently attempted to correct for this refraction by refining the scanning technique in such a way as to obtain more accurate projection data from both primary and secondary waves. Using this technique, the image quality has improved to the extent it now compares with that predicted by the non-refraction theory. Other techniques have also been applied to compensate for inaccuracies introduced by attenuation.

INTRODUCTION

When finite-amplitude acoustic waves (i.e waves of small enough amplitude to avoid shock formation) propagate through nonlinear reactive media, they generate a variety of secondary waves depending upon the nonlinearities they encounter. By measuring the secondary waves we obtain a measure of the nonlinearities. Several different methods are currently being employed to carry out this type of measurement. A system to produce tomograms of variations in the nonlinearity may be useful in diagnostic medicine because it could possibly produce a sensitive indication of the state of health of biological tissue. In this paper we describe the latest results from our research on a tomographic method based on the concept of a parametric acoustic array [1].

Our imaging system makes use of the nonlinearly generated difference-frequency (DF) wave and second-harmonic frequency (SHF) wave [1,2,3,4]. When two primary waves at different frequencies are transmitted through a nonlinear reactive medium, the above mentioned secondary waves are generated. The pressure amplitude of each of these secondary wave at the receiving end of the system is proportional to a line integral of the distribution of the nonlinear parameter β (defined later) along the propagation path. An analytical formulation involving the line integral indicates that the distribution of the nonlinear parameter in a bounded region can be reconstructed tomographically.

We have reconstructed nonlinear parameter tomograms from both DF-wave and SHF-wave data for various test phantoms. The initial results were unsatisfactory because of severe refraction effects at the medium boundaries. In some cases, the nonlinear parameter variations in the phantom were not apparent as a result of refraction.

Because the effect of refraction is to scatter the waves, it can sometimes be minimized by employing a receiving transducer large enough to compensate for the scattering. But using a large transducer increases the possibility of phase cancellation on the transducer surface and thus introduces a new type of error in the measurement [5]. To avoid phase cancellation we use a needle-type hydrophone whose reception is confined to a highly-localized region. As a consequence, wave energy dispersed by refraction will not be intercepted by the hydrophone. To minimize errors due to scattering, we have refined our scanning technique by adjusting the lateral position of the hydrophone so that it tracks the peak pressure amplitude at the receiving end. This is sometimes done also in acquiring data for attenuation tomograms [5]. The quality of the images thus obtained is much better than before. The new images display the nonlinear parameter distribution in the phantom with substantially better accuracy.

In this paper, we briefly present the theoretical basis for nonlinear parameter tomography. We show several sets of reconstructed images of different phantoms. We compare the quality of tomograms obtained with primary waves (attenuation tomography) and secondary waves (nonlinearity tomography), with and without correcting for errors due to refraction. We also compare the images of β obtained with DF waves and SHF waves. With the refined scanning technique, the image quality has improved to such as extent that it compares to that predicted by the non-refraction theory.

THEORY

When an acoustic wave is transmitted through a medium with a reactive non-linearity, it will be distorted along the propagation path. Burgers' equation can be used to calculate the quadratic secondary wave generated when a planar incident wave is assumed [6].

$$\frac{\partial P_s(x)}{\partial x} = \frac{\beta(x)}{\rho_0 C_0^3} P_p(x) \frac{\partial P_p(x)}{\partial \tau} + \frac{\alpha_s}{\omega_s^2} \frac{\partial P_s^2(x)}{\partial \tau^2} \tag{1}$$

where P_p and P_s are the sound pressure amplitudes of the primary and secondary waves, $\beta(x)$ is the nonlinear parameter distribution defined in reference [6], $\tau = t - \frac{x}{C_0}$ is the propagation time, α_s is the attenuation coefficient for the secondary wave, ρ_0 is the medium density and C_0 is the sound speed in water.

Imaging from Difference-Frequency Wave Data

Assuming the two primary waves are well collimated and taking primary wave attenuation into account, we can write an expression for the primary wave $P_p(x)$ [6]:

$$P_p(x,\tau) = P_1(0)e^{-\int_0^x \alpha_1(s)ds} \sin\omega_1\tau + P_2(0)e^{-\int_0^x \alpha_2(s)ds} \sin\omega_2\tau \tag{2}$$

where $P_1(0)$ and $P_2(0)$ are the two primary wave pressure amplitudes just beyond the surface of the transmitting transducer, and α_1 and α_2 are the attenuation coefficients for the two primary waves. Substituting Eq. (2) into Eq. (1) and keeping only the relevant terms (i.e terms with $\omega_s = \omega_1 - \omega_2$), we have [7],

$$\frac{\partial P_s(x)}{\partial x} = \frac{\beta(x)\omega_s}{2\rho_0(x)C_0^3(x)} P_1(x)P_2(x) - \alpha_s(x)P_s(x) \tag{3}$$

308

Eq. (3) is a linear differential equation of first order where $P_{1,2}(x) = P_{1,2}(0)e^{-\int_0^x \alpha_{1,2}(s)ds}$. Solving Eq. (3) by imposing the proper boundary condition (i.e $P_s(0) = 0$), we obtain the secondary wave sound pressure amplitude for x=L

$$P_s(L) = \frac{P_1(0)P_2(0)\omega_s}{2\rho_0 C_0^3}\int_0^L \beta(x)e^{-\int_x^L \alpha_s(s)ds}\, e^{-\int_0^x (\alpha_1(s)+\alpha_2(s))ds}\, dx \qquad (4)$$

Let the primary wave amplitudes at the receiver be $P_{1,2}(L) = P_{1,2}(0)e^{-\int_0^L \alpha_{1,2}(s)ds}$. We can further simplify Eq. (4) to obtain

$$\frac{P_s(L)}{P_1(L)P_2(L)} = \frac{\omega_s}{2\rho_0 C_0^3}\int_0^L \beta(x)\left[e^{\int_x^L (\alpha_1(s)+\alpha_2(s))ds}\right]dx \qquad (5)$$

In Eq. (5) we neglect α_s by assuming $\alpha_s \ll \alpha_1 + \alpha_2$. We assume also that ρ_0 and C_0 do not change much with changes in $\beta(x)$. Eq. (5) shows that $\dfrac{P_s(L)}{P_1(L)P_2(L)}$ can be obtained as a result of integrating the product of $\beta(x)$ and an attenuation term along the propagation path. The nonlinear parameter variation in an object can therefore be reconstructed tomographically with proper attenuation corrections.

Imaging from Second-Harmonic Wave Data

Imaging with SHF wave data is very similar to that involving DF wave data. The SHF wave is generated in the medium when a primary wave $P_p = P_0 \sin\omega_p \tau$ propagates through the medium. By following a procedure similar to that for the DF case, we obtain the following for $P_s(L)$

$$P_s(L) = \frac{\omega_p}{4}P_1^2(L)\int_0^L \beta(x)e^{\int_x^L (2\alpha_p(s)-\alpha_s(s))ds}\, dx \qquad (6)$$

where α_p and α_s are the attenuation coefficients for the primary and secondary waves respectively, and ω_p is the primary frequency. This formulation appears to have an advantage when it is applied to soft biological tissue. In such a medium $\alpha_s \approx 2\alpha_p$, and Eq. (6) becomes:

$$\frac{P_s(L)}{P_1^2(L)} = \frac{\omega_p}{4}\int_0^L \beta(x)dx \qquad (7)$$

Using this equation, nonlinear parameter variation can be reconstructed directly by the tomographic method. Note that Eq. (7) is not a function of the attenuation coefficients. On the other hand Eq. (5) for the DF method is a function of attenuation. Thus attenuation compensation is essential in the DF case but not in the SHF case.

Attenuation Compensation for the DF Approach

We can use two different correction methods to compensate for attenuation. One is the so-called modified matrix method employed in single-photon emission tomography [8]. Eq. (8) gives the appropriate matrix:

$$C(x,y) = \left[\frac{1}{2\pi}\int_0^{2\pi} e\int_\theta \alpha_1(x,y) + \alpha_2(x,y) dl d\theta\right]^{-1} \qquad (8)$$

where θ is the projection angle and l_θ is the propagation path between the reconstruction point (x,y) and the receiver for angle θ.

The other method uses a pre-projection correction. For each projection, we calculate a correction term $q(z,\theta)$ as follows:

$$q(z,\theta) = \int_0^L e^{\frac{L}{x}\int \alpha_1(s) + \alpha_2(s) ds} dx \qquad (9)$$

We then find

$$\frac{P_s(z,L)}{P_1(z,L)P_2(z,L)q(z,\theta)} \qquad (10)$$

where z is the location of the receiver in the scanning direction. To compensate for attenuation, we use expression (10) rather than the left-hand side of Eq. (5) in reconstructing the tomogram.

In the next section, we show results from employing these two correction methods.

EXPERIMENTAL METHOD AND RESULTS

Fig. 1 shows the experimental system setup. The flowchart for data acquisition and image reconstruction is depicted in Fig. 2. Our transmitter was a Parametric transducer of 0.5 in. diameter and 5 MHz center frequency. The receiver was a needle-type wide-band hydrophone (manufactured by Nelson Twomey Inc.) of 0.5 mm diameter. Two cw primary waves of about equal amplitude and with frequencies of 5 and 6 MHz were launched from the transducer. The DF and SHF waves generated in the phantom were detected at a receiving line about 21 cm (17 cm in one case) away from the transducer. The phantom was placed about 7 cm from the transducer. This distance is outside the very near field region. The characteristics of the phantoms are shown in Fig. 3. We recorded the data in two ways: 1) by using a spectrum analyzer, 2) by using a bank of band-pass filters to select each frequency band and then measuring the amplitude.

As previously stated, the refraction from various regions within the phantom as well as between the phantom and the surrounding water has to be taken into account. To compensate for the effects of refraction, we refined our scanning technique by moving the hydrophone through a small distance at each measuring position to find the location of the maximum pressure point on the receiving line. The value measured there was then used rather than the value at the line-of-sight point. We found that the deviation of the maximum-pressure position from the line-of-sight position was usually less then the scanning increment (3mm), except at the very edge of the phantom. This indicates that a linear array could probably be used in acquiring the data in this kind of system. We observed also that the deviations of the maximum-pressure positions from the line-of-sight positions for the secondary waves were almost the same as for the primary waves

By following the above procedure we were able to reconstruct considerably better images than before. For comparison, Fig. 4 shows a β tomogram from line-of-sight data. We see a bright rim at the edge of the phantom caused by refraction from the boundary between the water and the phantom. Reconstruction artifacts due to the presence of large values at the edge tend to lower the reconstructed values inside the phantom. Fig. 5 shows an attenuation tomogram made by reconstructing data obtained

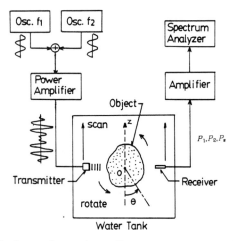

Fig. 1   System for nonlinear parameter tomography.

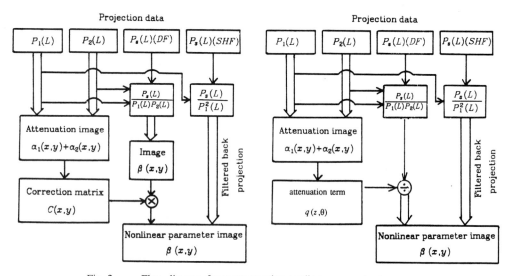

Fig. 2   Flow diagram for reconstructing nonlinear parameter tomogram.

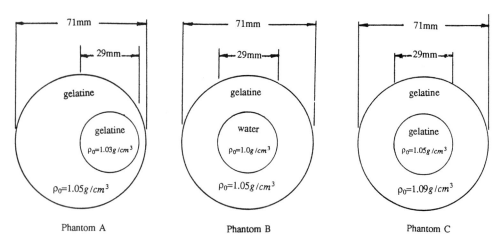

Fig. 3   Parameters for the phantoms.

311

with the refined scanning technique from measurements of one of the primary waves. Figs. 6, 7, 8 and 9 show nonlinear parameter images from secondary-wave data taken at the maximum-pressure points. Figs. 6 and 8 show the β tomograms of phantoms B and C using DF wave data. By comparing these two images, we can see that the gray-scale values for the inner-hole region of phantom C (Fig. 8) appear to be higher, on the average, than those for phantom B (Fig. 6). This is as it should be. Projection profiles for both images are shown in Fig. 10 and Fig. 11. From these two profiles, the fact that phantom C has higher β values than phantom B in the inner hole region is even more apparent.

Fig. 4   Nonlinear parameter tomogram of phantom A using DF data.

Fig. 5   Attenuation tomogram of phantom C.

Fig. 6   Nonlinear parameter tomogram of phantom B using DF data.

Fig. 7   Nonlinear parameter tomogram of phantom B using SHF data.

Fig. 8   Nonlinear parameter tomogram of phantom C using DF data.

Fig. 9   Nonlinear parameter tomogram of phantom C using SHF data.

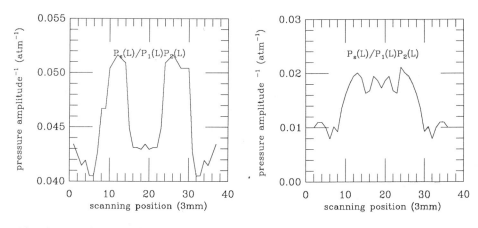

Fig. 10   Projection profile of phantom B.   Fig. 11   Projection profile of phantom C.

As shown in Fig. 12, the projection profile is relatively low and constant in regions A and E (which correspond to water). The profile increases gradually in regions B and D. These regions correspond to the β contribution from the relevant portions of the phantom indicated at the top of the plot. The shape of the projection profile is roughly what we should expect from such a phantom and demonstrates the validity of the theory.

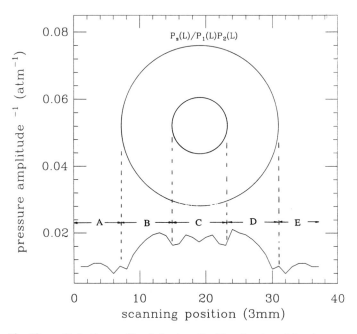

Fig. 12   Projection profile of phantom C with a drawing of the phantom.

We have not been successful in reconstructing high-quality nonlinear parameter tomograms with SHF data. These tomograms are substantially inferior to those reconstructed from DF data. Figs. 7 and 9 show the corresponding images. The projection profiles for these images are typically noisy and produce severely degraded tomograms. The problem arises perhaps from inaccurate data from the primary waves. In the DF case, the noise-like inaccuracies in $P_1(L)$ and $P_2(L)$ data are uncorrelated so they have relatively minor effect on the quotient $\dfrac{P_s(L)}{P_1(L)P_2(L)}$. In the SHF case, the inaccuracies in $P_1(L)$ data degrade the quotient $\dfrac{P_s(L)}{P_1^2(L)}$ to a greater extent, because of the second power of the $P_1(L)$ term.

Figs. 13 and 14 show respectively β tomograms from the DF data of phantom C using the two different attenuation-correction methods. Fig. 13 shows a β tomogram whose reconstruction employed the correction matrix approach [8]. Fig. 15 is an image of C(x,y) in Eq. (8). We can see a bright region in the center. Since C(x,y) is used to multiply the previously-reconstructed tomogram, the inner-hole value of the corrected tomogram will be brighter than that for the rest of the phantom after the multiplication. This method will not ordinarily change the image outline. By comparing Fig. 13 and Fig. 14 we can see that the major change is in the relative values of the shades of gray in each region.

Fig. 14 shows a tomogram resulting from the pre-projection method. This method involves dividing Eq. (5) by Eq. (9) as shown in expression (10). The attenuation term $q(z,\theta)$ is plotted as a function of z in Fig. 16. As far as the reconstructed values in the inner-hole region are concerned, the effect of this method is similar to that of the correction-matrix method. Because of the division process involved in this method, the corrected image looks noisier than that for the correction-matrix method. Further studies are needed to more accurately evaluate the effectiveness of these two attenuation correction methods.

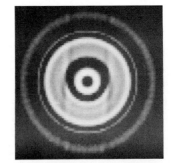

Fig. 13   Nonlinear parameter tomogram of phantom C using DF data after attenuation correstion.

Fig. 14   Nonlinear parameter tomogram of phantom C using DF data after pre-projection attenuation correction.

CONCLUSION

In this paper we have described a tomographic method of imaging the nonlinear parameter β with finite-amplitude waves. We have shown several tomograms of the β distribution reconstructed from DF and SHF data. By refining the scanning technique we have been able to reconstruct tomograms that compare in quality with those predicted by non-refraction theory. Nonlinear parameter images reconstructed from DF data are generally better than those reconstructed from SHF data.

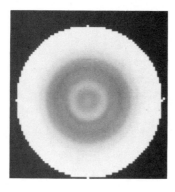

Fig. 15   Attenuation-correction matrix
for phantom C.

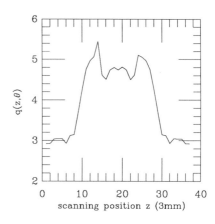

Fig. 16   Attenuation term q for phantom C.

We are planning additional experiments to further study the relationship between attenuation and nonlinear wave generation. We are also considering other ways of obtaining better nonlinear parameter images, such as by iteratively correcting for attenuation and by taking diffraction into account.

REFERENCES

1.   P. J. Westervelt, "Parametric Acoustic Array," J. Acoust. Soc. Am. 35, pp.535-537,1963

2.   Y.Nakagawa, M.Nakagawa, M.Yoneyama and M.Kikuchi, "Nonlinear parameter imaging computed tomography by parametric acoustic array," Proc. IEEE 1984 Ultrason. Symposium,pp.637-676,1984

3.   Y.Nakagawa, W.Hou, A.Cai N.Arnold and G.Wade, "Nonlinear parameter imaging with finite amplitude waves," proc. IEEE 1986 Ultrason. Symposium,pp.901-904,1986

4.   N.Arnold, A.Cai Y.Nacagawa W.Hou and G.wade, "Acoustic tomography for imaging the nonlinear parameter," Proc. SPIE. Vol.768, International Symposium on Pattern Recognition and Acoustic Imaging, 1987, (Accepted for Publication).

5.   K. M. Pan and C. N. Liu,"Tomographic Reconstruction of Ultrasonic Attenuation with Correction for Refraction errors," IBM J. RES. Develop. Vol.25 No.1, pp 71-82 (1981)

6.   O.V.Rudenko and S.I.Soluyan, "Theoretical Foundation of nonlinear acoustics," pp.39-61,pp.97-136,1977

7.   F. Dunn, W. K. Law and L. A. Frizzell, "Nonlinear Ultrasonic Wave Propagation in Biological Materials, Proc. IEEE Ultrasonic Symp., pp 527-532 (1981)

8.   Y.Morozumi, M.Nakajima, K.Ogawa and S.Yuta, "Attenuation correction method for single photon emission CT," Trans. IECE Japan J66-D, 10, 1130-1136,1983

SONO-ELASTICITY: MEDICAL ELASTICITY IMAGES DERIVED FROM ULTRASOUND SIGNALS IN MECHANICALLY VIBRATED TARGETS

Robert M. Lerner*, Kevin J. Parker**, Jarle Holen*, Raymond Gramiak*, and Robert C. Waag**

* Department of Radiology
**Department of Electrical Engineering
  University of Rochester Medical Center
  Rochester, New York 14642

INTRODUCTION

Cancers of the prostate have traditionally been detected by digital palpation which identifies increased stiffness (modulus of elasticity, hardness) of the abnormal tissue. Gray-scale ultrasound is insensitive to stiffness as an imaging parameter and often fails to reveal the extent or existence of a prostate cancer. Recent developments in technology and understanding have improved the diagnostic efficacy of transrectal ultrasound for detection of early prostate cancer, but even when coupled with digital rectal examination, a significant percentage of existing carcinomas may not be recognized. This statement is based on the lower detection rates for transrectal ultrasound and digital rectal examination in screening populations (0.1-4%) as compared to autopsy series prevalance rates of approximately 30%. Recent reports suggest that early cancers of the prostate may be characterized as hypoechoic areas in the peripheral zone of the prostate (1). Others suggest that the more advanced lesions have varied appearances ranging from hypo to hyperechoic with some even showing combinations of both (2,3). Our transrectal ultrasound experience from patients with palpable cancers (which include moderately to advanced disease), suggests a varied non-specific gray-scale appearance as well, despite the uniform impression that they are all firm on digital palpation.

Recent studies imply potential for curability of carcinoma of the prostate if it is detected before it reaches 1.0 cm^{3}in volume (presuming a spherical tumor, this corresponds to a 1.3 cm diameter)(4). It is suggested that biologically active carcinomas alter the supporting prostate stroma resulting in an increased modulus of elasticity which may be detected as a region of stiffness or hardness on digital rectal examination. Since not all lesions are within the range of the examining finger, nor are all lesions palpable for a variety of reasons, a sensitive and objective method for detecting abnormal regional elasticity in prostate should lead to improved detection of carcinoma of the prostate. The purpose of this research is to incorporate tissue stiffness features into ultrasound images (sono-elasticity). The concept is that stiff tissues (cancers) will respond differently to an applied mechanical vibration than normal tissues.

This approach combines external mechanical stimulation of target

tissues with Doppler ultrasound signal detection and processing to improve
lesion detection, provided the lesion creates an altered region of
elasticity. Stimulation of target tissues by a controlled mechanical
vibration causes regions of different elasticity to respond with different
displacements and velocities. The idea of characterizing tissue from the
motion or mechanical response is not new but has had only limited evaluation (5-11). Ultrasound detection schemes have utilized correlations between A-lines to detect motion of cardiovascular origin, and visual analysis
of M-mode waveforms have been performed to detect motion for a 1.5 Hz
external vibration source. None of these techniques has reached clinical
maturity because of a number of difficulties. The A-line or M-mode techniques rely primarily on echoes from speckle regions, with few discrete
specular reflections available to demonstrate motion unequivocally. The
changes in speckle pattern resulting from motion of the sample volumes are
complex and can be difficult to interpret. The correlations may suffer from
patient motion caused by respiration and other body movements. The use of
cardiovascular pulses to generate internal motion is problematic because the
"source function", the radial expansion of arteries, is generally of uncer-
tain strength, and has unknown coupling to the surrounding tissues. The
pulsatile movements near the arteries can be submillimeter in extent, which
is well below the resolution of conventional A-line or M-mode techniques.

    In comparison, our approach employs variable frequency, external,
periodic pulsations and incorporates range-gated Doppler ultrasound to
detect the periodic movements of tissue. The advantages can be summarized
as follows:

    i) External periodic pulsation applied to a specimen provides a known
stimulus which can be easily recognized as distinct from other velocity
(noise) sources. Sensitive coherent signal detection schemes may then be
applied to improve signal to noise ratios if indicated.

    ii) The Doppler detection technique is capable of resolving much
smaller displacements than A-line or M-mode techniques. This results from
the fact that Doppler ultrasound measures velocity (displacement times
frequency for sinusoidal motion) compared with techniques which measure displacement only. At "high" frequency (1 KHz) mechanical vibration, a conventional 2 MHz Doppler ultrasound time-frequency display is capable of resolving excursions (sinusoidal displacements) on the order of 0.06 mm (12).
Thus, very small vibrations in deep tissues can be measured.

    iii) The Doppler detection technique is sensitive in regions of low
speckle, as well as regions with specular reflectors. For example, color-
coded Doppler blood flow images are derived from regions where conventional
B-scan images show extremely low reflectivity. Doppler can also display
multidirectional information simultaneously, which may be of value in
identifying regions of relative motion.

    iv) Doppler ultrasound combined with mechanical stimulation is re-
latively insensitive to respiratory and other gross tissue movements. This
results from the examiners ability to identify the sinusoidal Doppler ultrasound output at a given known frequency of external vibration so that
motions which are not periodic at the drive frequency can be ignored.

    v) A variable frequency approach maximizes the likelihood of detecting
differences between normal tissue and tumor with concomitant altered
elasticity, because of the possibilities of exciting "mechanical
resonances". The mechanical properties of tissue include a high degree of
damping, and thus, strong resonance behavior (high Q) should not be encountered. Nonetheless, the ability to change stimulation frequency may add
additional information concerning the frequency dependent response of a

region, and this information can be compared with theoretical and experimental results to estimate the mechanical properties of the region.

As a result of these advantages, the combined external vibration/Doppler ultrasound detection approach appears to be a leading candidate for determining the elastic properties of discrete regions of tissue, which may permit early detection of prostate tumors and other focal abnormalities in soft tissue.

Doppler Detection of Vibration - Principles

In periodic, sinusoidal motion, an object's velocity, V, is proportional to displacement, e, times mechanical vibrational frequency, W, because velocity is the time derivative of displacement. Thus $V = e (W)$, and as mechanical vibration frequency increases, it is evident that equal velocities are generated from smaller displacements. This is the main reason why exceptionally small displacements (0.06mm) can be detected at 1 KHz (W) by conventional 2 MHz pulsed Doppler equipment.

The representation of velocities on the conventional time-frequency Doppler display is readily apparent only when low frequency mechanical vibration is employed. Specifically, when the period of mechanical vibration is longer than the FFT period of the Doppler spectrum analyzer time window, a positive and negative periodic sine wave is directly observed on the display, with the peak frequency shifts proportional to the peak velocity according to the Doppler formula. However, at high vibrational frequencies, when the period of mechanical oscillation is much shorter than the FFT window, analysis shows that a Fourier Series solution to the Doppler shifted signal is obtained, and the time-frequency display therefore shows continuous (in the time domain) equally spaced (in the frequency domain) bands of both positive and negative Doppler-shifted frequencies. In this regime, the peak vibrational velocity is indicated by the frequency shift of the highest band and peak displacement by the number of bands (12). The following quantitative considerations are germane to preliminary experiments which follow:

i) The minimum detectable velocity (of vibrations) is influenced by the high pass filter required on all Doppler outputs to eliminate strong DC signal and low frequency "noise". We have found 333 Hz to be a reasonable cut-off in experiments, and given the Doppler carrier frequency (2-5 MHz typical) the minimum velocity is proportional to the inverse of the vibration frequency. Some typical examples are: min velocity 12.5 cm/sec (2 MHz carrier) corresponding to 1 mm displacements at low frequency, 18 Hz vibration; or 5 cm/sec (5 MHz carrier) corresponding to 0.01 mm displacements at higher frequency 1 KHz vibration.

ii) The maximum detectable velocity in pulsed Doppler systems is limited by the well known requirement that the pulse repetition frequency (PRF) be twice the expected maximum Doppler frequency (13,14). The PRF is limited by the need for unambiguous depth resolution so a velocity depth product maximum can be established. For example, with a maximum depth of 8 cm expected, the maximum resolvable velocities are on the order of 70 cm/sec for 5 MHz carrier, to 175 cm/sec for 2 MHz carrier.

Preliminary Experiments - Doppler Detection of Vibration

A specially constructed, motorized, offset-cam plunger (1 cm diameter) was used to vibrate phantoms at 18 Hz, with a plunger excursion of $\pm$ 2.5 mm. A submerged, degassed, sealed, moderately stiff sponge was utilized in one set of experiments. The sponge had a 2 cm embedded region of RTV silicone (a relatively harder "tumor" region) which had been injected in a central position before hardening occurred. Fig. 1 shows a 2.5 MHz B-scan

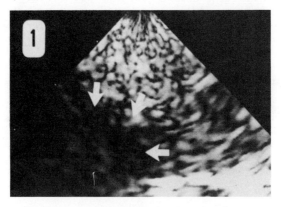

Figure 1. 2.5 MHz B-scan image of sponge
phantom with stiff, RTV "tumor" embedded
(arrow-heads). The higher attenuation
within the RTV results in some shadowing
(lower left). The overall configuration
corresponds to that of Figure 2.

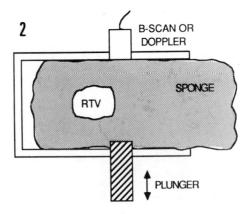

Figure 2. Schemetic illustration of inhomogeneous sample with mechanical vibration and
B-scan imaging or Doppler range gate measurement.

(Toshiba*) of the sponge RTV region. Shadowing occurs from the high attenuation within the RTV. However we verified that the 2 MHz Doppler transducer (Vingmed**) was able to detect motion within and distal to the
hypoechoic RTV region. The experimental configuration is shown in Fig. 2.
The axis of the Doppler probe-plunger was translated across the sponge in
1 cm steps and pulse Doppler data were recorded in 0.5 cm depth increments.
Typical output from the Doppler unit is shown in Fig. 3. The peak velocities were recorded as a function of range-gate position within the sponge-
RTV phantom. A gray scale rendition of these data is shown in Fig. 4 with
peak velocities within the sample varying linearly from white (2.5 KHz
Doppler shift) to black (< 300 Hz Doppler shift).

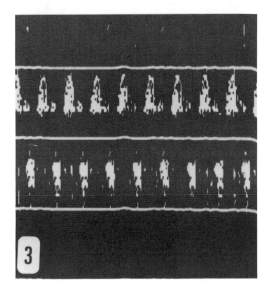

Figure 3. Output of range-gated Doppler display, with sample volume in
sponge phantom subjected to 18 Hz
vibration. Horizontal axis-time, 0.5
sec shown; vertical axis - Doppler
frequency shift with 1 KHz per division. For slow (less than 20 Hz)
sinusoidal mechanical vibrations, the
positive and negative (towards and
away from the Doppler probe) pulsations are clearly seen.

*Toshiba; Tustin, California
**Vingmed; Olso, Norway

Figure 4. Computer generated
image of sponge RTV phantom using
range-gated Doppler peak velocities measured as a function of
position. The geometry corresponds to Figures 1 and 2. Resolution is 7 blocks vertical
(5mm each) by 5 blocks horizontal (roughly 1cm transverse each),
thus the aspect ratio is different from Figures 1 and 2.
Gray scale brightness corresponds to measured peak velocities within the sample volume,
varying linearly from white (2.5
KHz Doppler shift) to black
(less than 300 Hz Doppler shift).
The bottom row is white due to
the measurement of plunger vibration. The "tumor" region appears dark because of low peak
velocities not because of its
lower backscatter. This is explained by the preferential
movement of the surrounding,
more compliant sponge.

The bottom row is white corresponding to plunger vibration. The orientation
of the computer image corresponds to Figs. 1 & 2, so the RTV "tumor" region
is in the lower left hand corner. The dark region here corresponds to de-
creased vibration of the RTV region (due to its higher stiffness; with
preferential compression of surrounding sponge), and not due to its lower
backscatter coefficient, as in traditional B-scan images. Furthermore, the
Doppler vibration image (sono-elasticity) (Fig 4) is quantitative, with the
bottom bright regions (in this configuration) attributable directly to the
plunger movement, with the gray scale in other regions representing local
velocities or displacements within the phantom.

Another series of experiments examined configurations of "hard" or

322

"stiff" gelatin (20% gelatin, 3% formalin) vs "soft" gelatin (10% gelatin, 2% formalin), with 0.5% suspended barium sulphate particles in both for backscatter. Composite blocks were made, with 1.5 cm thickness stiff layers. Figure 5 shows the configuration (from bottom to top) of plunger; stiff layer; soft layer; and B-scan origin; where the composite image shows a 3.5 MHz B-scan image and M-mode from a centered scan line. As can be seen, the M-mode representation is difficult to quantify in speckle regions, and requires vibrations of 1 mm or greater. Fig. 6 shows the opposite configuration: plunger; soft layer; hard (stiff) layer; B-scan transducer.

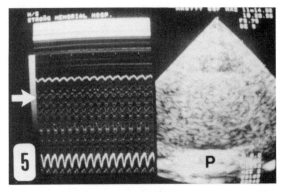

Figure 5. M-mode and B-scan images of a layered phantom. The image shown from bottom to top, the vibratory plunger (bright echo P), 2.5 cm of stiff gel and 1.5 cm of "soft" gel. The M-mode shows that the vibratory response at 2 cm distance (arrow) is of greater amplitude than in Figure 6 (below) where the vibratory stimulus first passes through a more "attenuating" 1.5 cm soft layer above the plunger.

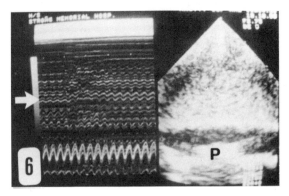

Figure 6. M-mode and B-scan images of layered phantom in reverse configuration of gel layers as compared to Figure 5. Plunger position (P) and 2 cm transmission distance (arrow) are indicated. Note that inadvertant movement of the M-mode transducer results in M-mode breakup and emphasizes difficulty, in resolving small vibrations within speckle regions by this method.

Comparing the two configurations, we see that the vibrations travel further, and with less diminuation in the hard (stiff) material than in soft, as expected from spring-mass model considerations. Fig. 6 also shows the deleterious effects of transducer movement on the M-mode - there is a pattern break-up in the center region caused by inadvertant movement of the ultrasound transducer.

In comparison, Fig. 7 shows range gated Doppler measurements of vibration within the samples, with 0.5 cm axial resolution and gray scale mapping, (corresponding to the configurations of Figs. 5 & 6). The Doppler "images" also show basic features such as the relatively uniform vibration within "hard", "stiff" layer, but rapid fall-off in the soft layer. However, the Doppler measurements are not restricted to large (1mm) displacements, and are not confused by regions of variable speckle as in the case of M-mode measurements. These preliminary results utilized prototype equipment ("Alfred"* Doppler range gate system) which did not ensure rigid Doppler probe positioning (probe vibrations are a source of noise), and used the Doppler display to visually estimate peak sinusoidal velocities.

Fig. 8 shows a halftone image of 2-D Doppler color flow image from a commercially available imager (Toshiba SSH65A). The same gelatin phantom as in Fig. 7 was scanned during low frequency vibration. The image demonstrates the feasibility of vibration imaging in real time. Different character of color coding in the two different layers was evident.

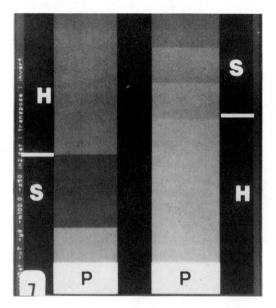

Figure 7. Gray scale image of velocity measured within vibrated, layered phantom . Doppler shift data from single scan lines through the phantom are shown (white = 2.5 KHz Doppler shift, black = below 300 Hz). Left configuration is (bottom to top) plunger (P bright sample), three range samples within "soft" gelatin layer (S), four range samples within "hard" layer (H). Each range sample corresponds to 0.5 cm distance. Right configuration is same phantom inverted.

*Alfred; Horten, Norway

324

In both configurations, the hard
layers transmit vibrations with little
loss, whereas the soft layers show rapid decay of vibrational amplitude
with distance from the plunger.

The first range band in the soft
layer on the left probably contains
plunger movement which gives it the
bright representation. The more distal samples in the soft layer show
low velocity representations. More
distally, in the hard layer (H) moderate velocities are represented. It
would appear that the low velocities
in the soft layer could result from
Doppler insensitivity to off axis velocities, since a considerable amount
of the plunger energy must result in
movements of the soft layer particles
laterally. We speculate that the re-
coil of the soft layer then imparts
energy to the hard layer via mode conversion. The sum total of this acti-
vity is that the hard layer is stimu-
lated to velocities which are higher
than those in the soft layer in the
direction of Doppler sensitivity.
This concept suggests that the movement of particles in a soft layer is
probably multi-directional under
these stimulus conditions and that detection of motion may be possible from
multiple angles.

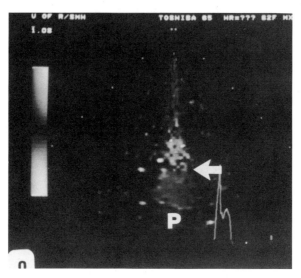

Figure 8. Halftone image representation of a
color coded Doppler image of vibrated layered
phantom obtained by a commercially available
color flow map imager (Toshiba SSH65A). This
imager is designed for cardiovascular imaging
and features real time display of blood flow

Figure 8. (Continued)
motion. The sector format is evident and the
relationship of the plunger (P), soft layer,
and hard layer is identical to that shown in
Figure 6 above. The arrow denotes the boundary between the soft layer and the hard layer. The color in the soft layer was a rather saturated red indicating that the data was
obtained during movement of the plunger towards the transducer located at the apex of
the scan. The homogeneity of color in the
soft layer indicates that the multiple
samples obtained in each beam position showed little variance in the detected flow velocity. In the hard sample, on the other-
hand, considerable yellow was intermixed
with the red indicating that the variance
was greater, a presentation which is also
seen when flow velocity is increased. Although this instrument is not optimized for
these investigations, it does demonstrate,
real time, features of vibration in tissues.

CONCLUSION

Preliminary work using an 18 Hz mechanical vibratory system and pulsed
Doppler detection has resulted in generation of prototype images of reflect-
or stiffness obtained from a sponge model containing a "tumor" nodule of
altered stiffness. Diminished vibratory motion in the nodule is clearly
evident. In another investigation, gelatin blocks of different stiffness
demonstrated recognizable differences in motion of contained reflectors as
well as differences in propagation of the low frequency mechanical energy
transmission. These preliminary results indicate that tissue stiffness can
be incorporated into gray scale ultrasound imaging.

These concepts are being applied to investigation of the prostate which
should result in improved detection of carcinoma since:

1) Areas of increased stiffness/hardness not favorably located for
digital rectal examination may be explored.

2) Detection of regions of altered elasticity resulting from carcinoma
of the prostate should complement conventional transrectal ultrasound imaging. The ultimate goal will be to provide objective quantifiable measures
of regional tissue elasticity analogous to the time honored data widely
accepted in clinical circles from digital rectal palpation but produced and
documented in a subjective format. Correlation of the data presented from
elasticity images to data presented from conventional transrectal ultrasound
and digital palpation should improve the diagnostic accuracy and confidence
for specific lesions. Extension of these conepts to other organ systems
follows naturally. For instance, estimates of liver fibrosis may be possible with this methodology as well as improved identification of metastatic
nodules. The method may also lend itself to pressure determination in
cardiac chambers and blood vessels.

REFERENCES

1.   Lee F, Gray JM, McLeary RD, Meadows TR, Kumasaka GH, Borlaza GS, Straub
     WH, Lee, Jr., F, Solomon MH, McHugh TA, and Wolf RM: "Transrectal
     ultrasound in the diagnosis of prostate cancer: Location, echo-
     genicity, histopathology, and staging", The Prostate 7:117-129,1985.
2.   Rifkin MD, Kurtz AB, Choi HY and Goldberg BB: "Endoscopic ultrasonic

evaluation of the prostate using a transrectal probe: Prospective evaluation and acoustic characterization", Radiology, 149:265-271, 1983.

3. Burks DD, Fleischer AC, Liddell H and Kulkarni MV: "Transrectal prostate sonography: Evaluation of sonographic features in benign and malignant disease", Radiology 149:265-271, 1983.

4. McNeal JE, Kindrachur RA, Freiha FS, Bostwick DG, Redwine EA, and Stamey TA: "Patterns of progression in prostate cancer", The Lancet pp 60-63, 1986.

5. Rockoff SD, Green, RC, Lawson, TL, Pett, Jr., SD, Devine,III, D, and Francisco, SG: "Noninvasive measurement of cardiac pressures by induced ventricular wall resonance: Preliminary results in the dog", Investigative Radiology Vol. 13:499-505, 1978.

6. Gore JC, Leeman S, Metreweli C, and Plessner NJ: "Dynamic autocorrelation analysis of A-scans in vivo", Ultrasonic Tissue Char. II, M. Linzer, ed. NBS special publ. 525:275-280, 1979.

7. Wilson LS and Robinson DE: "Ultrasonic measurement of small displacements and deformations of tissue", Ultrasonic Imaging 4:71-82,1982.

8. Dickenson RJ and Hill CR: "Measurement of soft tissue motion using correlation between A-scans", Ultras Med. Biol 8:263-271, 1982.

9. Eisensher A, Schweg-Toffler E, Pelletier G, and Jacquemard G: "La palpation echographique rythmee-Echosismographic", Journal de Radiologie 64:225-261, 1983.

]0. Tristam M, Barbosa DC, Cosgrove DO, Nassiri DK, Bamber JC and Hill CR: "Ultrasonic study of in vivo kinetic characteristics of human tissue", Ultrasound in Med. & Biol. Vol. 12:927-937, 1986.

11. Yamakoshi Y, Suzuki M and Sato T: "Imaging the elastic properties using low frequency vibration and probing ultrasonic wave", Japanese Meeting of Applied Physics, Spring 1987 Tokyo.

12. Holen J, Waag RC and Gramiak R: "Representations of rapidly oscillating structures on the Doppler display", Ultrasound in Med & Biol 11:267-272, 1985.

13. Baker DW: "Pulsed Ultrasonic Doppler Blood Flow Sensing", IEEE Trans Sonics Ultrs. SU-17(3) p 170-185, 1970.

14. Fish PJ: "Doppler Methods", in Physical Principles of Medical Ultrasonics, C.R. Hill, Ed, Ellish Horwood Ltd, Chichester, UK, 1986.

TISSUE CHARACTERIZATION BY SPECTRAL ANALYSIS OF

ULTRASOUND VIDEO IMAGES

G.D. Lapin, B.J. Sullivan and M.H. Paul

Division of Cardiology, Children's Memorial Hospital and
Department of Electrical Engineering and Computer Science
Northwestern University
2300 Children's Plaza, Chicago, IL 60614

I. INTRODUCTION

Tissue characterization utilizing the estimated spectra of reflected ultrasound beams has been performed almost exclusively with the radio frequency
(rf) ultrasound detected by the transducer. The advantages of this technique
include maximum frequency resolution, limited only by the digitizing hardware
utilized, and independence from most operator adjustable variables on the
standard ultrasound B-scan imager. The disadvantages include the need for
specially designed equipment and the difficulty with which it is operated.

In a clinical setting, ease of operation is a high priority. With this
in mind, a technique has been developed to perform tissue characterization
from reflected ultrasound spectra calculated from the video image produced by
a clinical ultrasound B-scanner. Since B-mode video images are readily a-
vailable in the clinical ultrasound laboratory and digital video image processing hardware is abundant, a technique for tissue characterization from
the B-mode video image would have greater utility.

Spectral analysis of backscattered ultrasound has been used for tissue
characterization in two basic ways. The frequency dependence of attenuation
of ultrasound by tissue has been shown to be related to tissue type[1,2].
Measurement of this property requires availability of the rf ultrasound
signal. Coherent interference of the backscattered ultrasound, called speckle, is another property that can be used for tissue characterization. Spectral analysis of amplitude demodulated rf ultrasound speckle pattern reveals
the characteristics of the microstructural features of tissue[3,4,5]. These
features are also evident in the video image produced by an ultrasound B-
scanner. We have developed an analytical method, named *Difference Cross-Covariance*, that is able to distinguish between video B-scan images of different tissue types[6].

II. ULTRASONIC SPECKLE

The speckle pattern generated by ultrasonic imaging consists of a combination of the components scattered from random and spatially regular microre-
flectors. Since the interaction of ultrasound with tissue is a linear process, the superposition principle holds and the analysis of backscattering can

be performed separately for the spatially regular and random components.

II.1 Video Image Spectral Tissue Signatures

A form of analysis has been developed that simulates the ultrasound beam paths on the digitized video image and generates power spectra along those lines. Although the ultrasound B-scan is inherently polar, a video image is typically displayed in rectangular format. Accordingly, the first step of this analysis reconstructs the directions of the ultrasound beam lines from the image.

The squared magnitude of the Fast Fourier Transform (FFT) is utilized as an estimate of the power spectrum of the image texture. The intervening tissue causes spectral noise and spectral power variations that are uncorrelated between beam lines and are lessened by averaging many power spectra from beam lines passing through different parts of a homogeneous section of tissue. The result constitutes the tissue signature, which is then compared to signatures from other tissue samples. Suppression of the DC component must be performed on the tissue signature before comparison in order to remove the effect of amplitude variations in the image so they do not influence the spectral comparison. This is accomplished by setting the zero frequency term equal to the second power spectral term.

The general shape of the tissue signatures is consistent throughout all samples of tissue examined. The same shape has been observed in the signatures from various phantom structures. The components of this shape have been described by Insana, Wagner et al.[4,5], whose work on spectra derived from rf ultrasound was progressing concurrently with this work. Components of the power spectrum were isolated and related to characteristics of the tissue microstructure. Their interpretation is that the power spectrum is composed of a set of frequency peaks related to the periodic spacing of the microstructure superimposed on an exponential shape which they have termed "Rician noise" (see Figure 1). (The Rician component is related to the

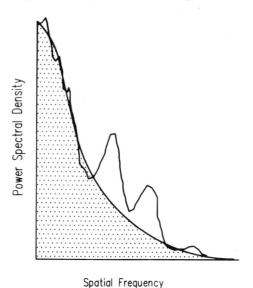

Figure 1: Simulated power spectrum of demodulated rf speckle showing the Rician component (shaded area) with superimposed spectral peaks from which the d parameter is obtained.

random scattering, which is a real tissue characterizing parameter, and is therefore considered here to be a Rician distributed random process rather than noise). From the frequencies of the significant peaks in the power spectrum that are determined after removal of the Rician component, the d parameter, or mean scatterer spacing, is derived. The autocorrelation, the inverse transform of the power spectrum, is used to determine two other parameters: r, the ratio of periodic to randomly spaced scattering components, and v, the fractional standard deviation of the periodic scattering component.

II.2 RF Logarithmic Amplification

The rf signal detected at the ultrasonic transducer is logarithmically amplified in the typical B-mode imager. The purpose of this stage is to compress the dynamic range of ultrasound reflections to allow both the major reflecting interfaces and the speckle pattern to fit into the available gray scale representation of reflectance. The dynamic range of ultrasonic reflections from biological tissue can be as high as 90 dB, which the logarithmic amplification stage compresses for visibility on the final image.

This nonlinear stage potentially precludes the use of a linear analysis technique. However, by applying the Born approximation, which states that scattering from tissue is very weak (in practice, at least -30 dB compared to a perfect reflector)[1,7], it can be concluded that microreflections are in the low end of the dynamic range of the logarithmic amplification function. Also, since the region of interest is constrained to fit into the transducer's focal zone, several centimeters of tissue attenuation act on both the incident and reflected energy. An example with cardiac tissue 5 cm below the skin and a 5 MHz ultrasound transducer gives approximately 47 dB of attenuation. Thus, a reflection from a microreflector within cardiac tissue will have a level at least 77 dB below a perfect reflector. Even with constructive interference increasing the detected reflection several times, the level remains in the lower portion of the logarithmic function. This can be approximated as a piecewise linear amplification factor.

Figure 2 illustrates a typical logarithmic response curve from a clinical ultrasound imager. If the energy reflected from a perfect reflector is taken to be 100 dB, the energy reflected from a microreflector will have a level of 23 dB. Allowing 7 dB for constructive interference, the speckle can

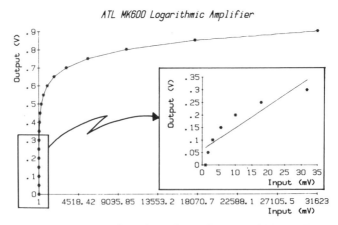

Figure 2: Transfer function of RF amplifier for ATL MK600 ultrasound imager. Inset is piecewise linear approximation for portion of the amplifier range corresponding to typical speckle amplitudes.

range from 0-30 dB, as illustrated in the inset of Figure 2. The logarithmic function for this case is fit with a least-squares linear regression from 0-30 dB. The piecewise linear approximation fits the actual logarithmic function in the new range with a goodness of fit of 0.91 out of 1.0. The standard error of the estimate is 0.046 volts and the maximum error between any point on the logarithmic curve and its linear regression estimate is 0.069 volts. These values imply that the piecewise linear approximation in the stated range is acceptable for performing linear analysis techniques on the signals.

III. THE DIFFERENCE CROSS-COVARIANCE

The comparison of tissue signatures is performed by using an algorithm that distinguishes between subtle shape differences in two similar waveforms. We developed the Difference Cross-Covariance as a signal comparison technique that addresses the problem of discrimination between the low resolution video power spectral tissue signatures[6]. Although visual comparison detects similarities and differences between the power spectra from various samples of tissue, it is grossly subjective and cannot recognize the subtle shape differences necessary for classification.

The Difference Cross-Covariance is defined as

$$d_{uv}(m) \triangleq \frac{c_{uv}(m)}{\sigma_u \sigma_v} - \frac{1}{2} \left[\frac{c_{uu}(m)}{\sigma_u^2} + \frac{c_{vv}(m)}{\sigma_v^2} \right] \qquad (Eq. 1)$$

where the covariances are calculated as

$$c_{uv}(m) = \frac{\sum\limits_{n=m+1}^{N} [u(n)-\bar{u}][v(n-m)-\bar{v}]}{N - |m|} \qquad (Eq. 2)$$

with the lag defined as integer m: $0 \leq m < N$ and similarly for negative values of lag. The mean values are calculated as

$$\bar{u} = \frac{1}{N} \sum_{n=1}^{N} u(n) \qquad (Eq. 3)$$

The standard deviations are calculated as

$$\sigma_u = \sqrt{\frac{\sum\limits_{n=1}^{N} u(n)^2 - N\bar{u}^2}{N-1}} \qquad (Eq. 4)$$

The Difference Cross-Covariance curve (herein referred to as *difference curve*) indicates tissue differences with a curve resembling one cycle of a sine wave. Zero amplitude indicates maximum similarity and increases in amplitude indicate differences. The difference curve passes through zero at or near zero lag. As the lag approaches its maximum positive or negative values, the value of the difference curve tends to increase greatly. These regions of the curve may be unreliable, however, since the number of terms in the sum of products is very small. Two examples of difference curves are shown in Figure 3, one indicating different tissue structures and the other indicating similar tissue structures.

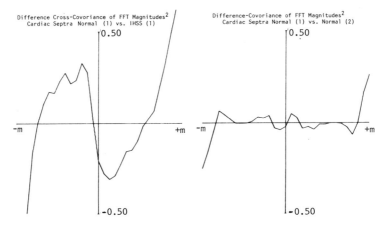

Figure 3: Typical Difference Cross-Covariance curves.
Left graph indicates large difference in tissue structure. Right graph indicates similar tissue structures.

The analysis of the d, r and v components of the rf speckle pattern described earlier helps to explain how the Difference Cross-Covariance measure
of tissue difference works. The reduced spectral resolution and increased
spectral noise encountered when analyzing the video tissue image rather than
the rf makes it difficult to delineate these parameters. For instance, the
sharp peaking of the rf power spectrum, from which the d parameter is obtained, appears as wide, rounded peaks which are often superimposed in the
video power spectrum. Though this representation precludes an exact specification of the d parameter, its value is reflected in changes of the curvature
of the power spectrum. The shape of the video power spectrum is similarly
modified by the r and v parameters. Differences in curvature of the video
power spectra are what the Difference Cross-Covariance detects, which is how
it responds to changes in the d, r and v parameters.

III.1 Characteristics of the Difference Cross-Covariance

The possible values of the Difference Cross-Covariance range from -2.0
to +2.0, with 0 indicating maximum similarity. The greater the deviation
from the zero axis, the greater the difference between spectra. Since a limited sampling is used, the reliability of the result falls off as the absolute value of the lag approaches the number of samples. In practice it was
found that comparison of power spectra from ultrasound examinations of tissue
yields Difference Cross-Covariance values in the range of ±0.5.

The following observations have been made about the shape of the difference curve: it has approximate odd symmetry around zero lag and has values
that grow very large as the absolute value of the lag approaches the data
array size, N, regardless of the amount of similarity.

Tissue comparisons with the Difference Cross-Covariance were presented
above qualitatively. A method of quantitation allows groups of tissue comparisons to be analyzed statistically. The root mean squared (RMS) is used
to quantitate the difference curve. The outlying two lags in both the positive and negative directions are disregarded. The value indicating the
amount of difference will be referred to as the *RMS difference measure*, or
simply, *difference measure*.

III.2 Response of the Difference Cross-Covariance to Video Modification

To test the response of the Difference Cross-Covariance to various modifications in video parameters, a digital image processor was utilized to simulate modification of the images. For instance, an image was brightened by adding 50 gray levels, out of a range of 256, to every pixel in the digitized image. Comparison of the brightened image to the original with the Difference Cross-Covariance method results correctly in a flat curve, indicating no difference regardless of modifications to the brightness of the image. In another type of image modification, the original image was changed by multiplying the level of all pixels by a constant, chosen as 1.5, which increased the pixel levels of the speckles in the region of interest without exceeding the maximum of 256 levels. This operation is equivalent to increasing the image contrast. When the increased contrast image is compared to the original, a flat difference curve is obtained. Thus, these two computer simulations indicate that modifications in either image brightness or contrast do not affect the accuracy of the DC suppressed Difference Cross-Covariance method of tissue characterization.

To further examine the response of the Difference Cross-Covariance method to image processing, a nonlinear mapping of pixel levels was performed. The range of pixels was mapped with a logarithmic function that compressed the levels in the region of interest to half of their original range. This highly modified image was compared to the original using the Difference Cross-Covariance and a small but definite difference was apparent. Nonlinear processing of an image appears to change the spectral characteristics of a speckle pattern, although only minimally. This test indicates that the Difference Cross-Covariance is minimally affected by logarithmic compression so signals that are not highly compressed should not change the Difference Cross-Covariance results.

IV. TISSUE EXPERIMENTS

IV.1 Animal Tissue *In Situ*

It has been demonstrated that the ultrasonic properties of *in vitro* tissue change as it becomes ischemic (reversible tissue effect of restriction of blood flow) and infarcted (irreversible tissue destruction)[7,8]. *In situ* imaging allows direct contact between the transducer and the living tissue. An experiment was designed to study the beating heart in the open chest of a sedated dog. This preparation has the advantages of removing the influence of any intervening tissue and facilitating the induction of myocardial (heart muscle) ischemia.

In order to properly image the myocardium from inside the chest cavity, it is necessary to space the transducer an appropriate distance from the tissue so that the focal point will fall within the region of interest. Spacing is accomplished with a plexiglas tube filled with degassed distilled water and with both ends of the tube closed off with rubber sheets that conform to the surfaces of the tissue and the transducer. There is minimal ultrasonic attenuation in the water tube, making it the closest approximation to placing the tissue in a water bath, which is the usual method of measuring objects *in vitro*.

A transthoracic ultrasound scan of the heart was obtained after the chest was shaved in an anesthetized dog. This image is used later to provide a limited control of *in vivo* imaging as compared to *in situ* imaging. The heart was exposed, the left anterior descending coronary artery (LAD), which supplies blood to the left ventricular (LV) myocardium, was isolated and a ligation loop loosely tied around it. The standoff unit was placed on the LV just above the apex with the transducer (5 MHz, medium focus) placed on the standoff unit and a longitudinal view of the LV was obtained. The depth of the standoff unit was chosen so that the myocardial tissue in the LV free

wall was centered in the focal zone of the transducer. Ultrasonic imaging
was repeated at one minute intervals after the LAD was tied and myocardial
ischemia progressed. Tissue signatures calculated from these images are with
the muscle in systole, or contracted.

IV.2 Clinical Images

An important test of the Difference Cross-Covariance is its ability to
distinguish between tissue types from clinical ultrasound images. Ultrasound
B-scan images of human pediatric age patient hearts are compared. Two types
of comparison are made from images of the intraventricular septum (IVS): a)
systolic, or contracted, vs. diastolic, or relaxed normal myocardium, and b)
normal vs. diseased systolic myocardium.

All ultrasound B-mode images are taken in the standard longitudinal
view with an ATL MK600 ultrasound scanner using a 5 MHz, medium focus transducer. The tissue regions of interest are horizontally centered in the ultrasound field and are imaged in the focal zone of the transducer (2-5 cm).

IV.3 Normal vs. Diseased Myocardium

Two types of diseased myocardium are studied: three cases of idiopathic
hypertrophic subaortic stenosis (IHSS) and one case which belongs to a class
of undefined cardiomyopathies. Diagnosis was additionally established from
cardiac function studies, drug testing, electrocardiography (ECG), cardiac
angiography and tissue pathology.

V. RESULTS

V.1 Dog Heart

Comparison of the tissue signature of normal myocardium to the signatures of the myocardium as it becomes increasingly ischemic is indicated by
the proportionality between the size of the difference curve and the duration
of ischemia. This function is plotted as the solid curve in Figure 4, which
is the RMS difference measure against time following loss of blood supply to
the myocardium.

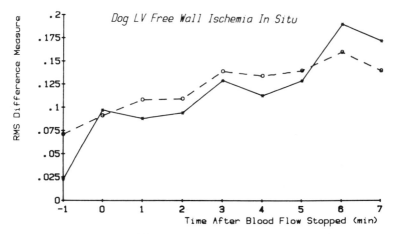

Figure 4: Difference measure vs. time for ischemic
dog myocardium. Solid line plots *in situ* comparisons. Dashed line plots *in situ* comparisons to an
initial *in vivo* image, as described in the text.

A similar series of difference curves examines the tissue signatures of increasingly ischemic myocardium imaged *in situ* compared to a tissue signature calculated from an *in vivo* ultrasonic image of the myocardium. The results are very similar to the *in situ* results, though it is interesting to note that the noise level of the difference curve increases when the *in vivo* image is used, indicating that spectral noise is added by the intervening tissue. The amount of difference is quantified and plotted as the dashed curve in Figure 4.

V.2 Clinical Images

The clinical ultrasound images described in the last section are compared to each other in several ways. All of the resulting difference curves are compared statistically after quantitation. The results of these comparisons, which are described below, are summarized in Figure 5.

As a control, Difference Cross-Covariance comparisons are made between tissue signatures of similar tissue types. This includes comparing eight normal septal B-scans to each other and four B-scans of diseased septa to each other. The mean difference measure of all similar comparisons is 0.063 ± 0.030.

Each of the tissue signatures from the four cases of diseased septa is compared to each of the tissue signatures from eight normal septa with the Difference Cross-Covariance. Differences are evident in all comparisons with a mean difference measure of 0.175 ± 0.029. Comparing this to the control values with the t-test results in a probability (p value) less than 0.0005, meaning that it is highly significant that they are distinguishable.

As muscles contract, the lengths and thicknesses of the muscle fibers change. Although this does not change the ratio of the random to spatially regular components since the reflectivity of the muscle fibers should not change, it does change the periodicity of the regular components. Clinical B-scans of seven normal beating hearts are examined. Tissue signatures of an area of the myocardium are taken at end diastole and end systole, as determined by both movement of the image and the ECG. Difference curves are generated for each of the seven pairs of tissue signatures. A difference between tissue structures is discernible in every pair, though the sizes of the

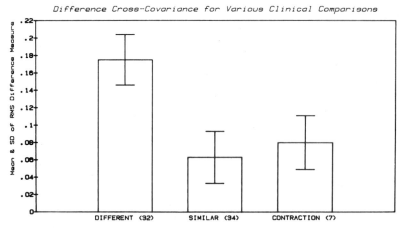

Figure 5: Mean and standard deviation of difference measures for clinical comparisons. Left bar is 32 comparisons of different tissue types. Center bar is 34 comparisons of similar tissue types. Right bar compares systole and diastole for 7 normal cardiac septa.

difference curves are relatively small. The mean difference measure is 0.080 ± 0.031, which is not significantly different from the interpatient difference of similar tissue.

VI. CONCLUSION

VI.1 Discussion

Utilizing B-mode video images to obtain ultrasound spectral tissue signatures necessitates a sensitive, yet robust, comparison technique to determine tissue differences. The Difference Cross-Covariance has been shown to be an effective comparator for this purpose. The reduced spectral resolution and increased noise caused by the conversion from rf ultrasound to the video B-mode image do not allow delineation of the tissue characterizing parameters that have been defined for rf ultrasound power spectra. Variations of these parameters are indicated in the video signatures by changes in the shape of the power spectra. The Difference Cross-Covariance detects the changes in shape yet is insensitive to changes caused by variations in the overall brightness and contrast of the video image.

The nonlinear processing of the echo signal does not have a great affect on the determination of tissue differences by this technique. One reason for this is that the signals of interest fall into a portion of the logarithmic transfer function which can reasonably be approximated with a linear transfer function. Additionally, the Difference Cross-Covariance, though affected by nonlinear amplification, is relatively insensitive to it. Therefore, the nonlinear amplification must severely modify the original signal of interest to affect the ability of the Difference Cross-Covariance method to determine tissue differences.

Tests with myocardial tissues, both *in situ* and *in vivo*, show that the Difference Cross-Covariance technique is capable of detecting differences in tissue structure due to ischemia and disease. It can detect these differences even in the presence of intervening tissue, as evidenced by comparisons between *in vivo* tissue signatures with those obtained from *in situ* tissue.

VI.2 Summary

Although additional experimental and clinical data are necessary before the Difference Cross-Covariance can be used as a technique for distinguishing tissue structural changes, the statistics presented in the clinical studies section are very encouraging. Future enhancements of the Difference Cross-Covariance technique will take into account the variations caused by use of different transducer frequencies and designs.

VII. REFERENCES

1.   F.L. Lizzi, M. Greenebaum, E.J. Feleppa, M. Elbaum and D.J. Coleman, "Theoretical framework for spectrum analysis in ultrasonic tissue characterization," J Acoust Soc Am, vol. 73, pp. 1366-1373, 1983.
2.   T.L. Rhyne, K.B. Sagar, S.L. Wann and G. Haasler, "The myocardial signature: absolute backscatter, cyclical variation, frequency variation, and statistics," Ultrasonic Imaging, vol. 8, pp. 107-120, 1986.
3.   L.L. Fellingham and F.G. Sommer, "Ultrasonic characterization of tissue structure in the *in vivo* human liver and spleen," IEEE Trans Sonics and Ultrasonics, vol. SU-31, pp. 418-428, 1984.
4.   M.F. Insana, R.F. Wagner, B.S. Garra, D.G. Brown and T.H. Shawker, "Analysis of ultrasound image texture via generalized Rician

statistics," in <u>Proc SPIE 556, International Conf on Speckle</u>, pp. 153-159, 1985.

5. R.F. Wagner, M.F. Insana and D.G. Brown, "Unified approach to the detection and classification of speckle texture in diagnostic ultrasound," in <u>Proc SPIE 556, International Conf on Speckle</u>, pp. 146-152, 1985.

6. G.D. Lapin, B.J. Sullivan and M.H. Paul, "Spectral analysis of tissue imaged with ultrasound sector scanning," in <u>Proc 1986 IEEE Ultrasonics Symposium</u>, pp. 941-944, 1986.

7. P.N.T. Wells, <u>Biomedical Ultrasonics</u>, London:Academic Press, 1977.

8. D. Nicholas, "Evaluation of backscattering coefficients for excised human tissues: results, interpretation and associated measurements," <u>Ultrasound Med Biol</u>, vol. 8, pp. 17-28, 1982.

NUMERIC EVALUATION OF BACKWARD WAVE PROPAGATION

IMAGE RECONSTRUCTION

J. F. McDonald, W. Coutts, G. Capsimalis*, P. K. Das,
N. A. Hijazi, D. J. Liguori, and K. B. N. Ratnayake

Center for Integrated Electronics
Department of Electrical, Computer and Systems Eng.
Rensselaer Polytechnic Institute
Troy, New York 12181

and

*The Binet Weapons Laboratory
Watervliet Arsenal
Watervliet, New York 12189

ABSTRACT

The Backward Wave Propagation (BWP) algorithm provides a means for reconstructing an image from spatial wave samples measured across an aperture at discrete transducer locations. The technique performs a spatial wave number decomposition on the aperture samples. This decomposition can be phase adjusted and retransformed back into the space domain taking into account the constraints of the wave equation. In this manner the image "scanned" by the aperture can be reconstructed using primarily Fast Fourier Transform (FFT) techniques. Use of parallel processing incorporating AMD 29500 series VLSI integrated circuits has been described previously to achieve image reconstruction at video rates with each frame derived from a single illuminating pulse. This paper explores the accuracy of the BWP image reconstruction by comparison with known interference patterns. The effects of finite word length arithmetic and noise corruption are examined. The conclusion suggests that the use of the new AMD 29325 series 32 bit floating point circuits will lead to a more robust image reconstruction. The previously reported system has been redesigned to incorporate these new circuits. These new building blocks illustrate once again the dramatic effect VLSI can have on the design of ultrasonic imaging systems.

INTRODUCTION

In a previous paper a digital architecture was presented
for the video rate reconstruction of ultrasonic images by
the Backward Wave Propagation (BWP) or Digital Holographic
method [1,2,3]. In this system, which is illustrated in
Figure 1, the target to be imaged is illuminated by signals
broadcast from an array of transducers. The broadcast
signals for each transducer are first placed in a set of CCD
analog shift register delay lines prior to transmission.
This permits a flexible choice of transmitter signals which
may be adapted for specific applications. Loading proceeds
by demultiplexing a set of digital to analog converters
drawing data from independent memory banks to groups of CCD
channels with 8 CCD shift registers in each group. This
parallel conversion permits rapid loading of the CCD's
because each group of 8 CCD channels can have one of its
channels loaded simultaneously with similar loading
operations occurring in all of the other groups. After
loading of the CCD's their clock rate is brought up to 10
MHz for transmission through high voltage amplifiers.

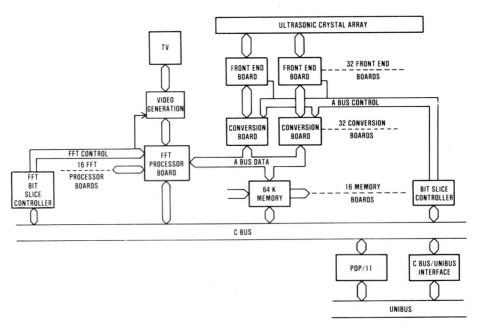

Figure 1. Overall system architecture for the BWP image
reconstruction system. Note that there are 16 A-buses
permitting parallel data transfer from the 16 DRAM's
to the 16 FFT processor boards. These same A-buses
service the 32 conversion boards.

During reception the process is reversed. Ramp amplified low voltage outputs from the same transducer elements are rewritten back into the CCD shift registers at the 10 MHz rate. Then the clock is slowed down by a factor of 80 to permit multiplexing the CCD shift registers (in groups of eight) to high precision analog to digital converters. These waveforms are then transferred from each converter to the independent 64K DRAM memory banks. The data thus captured from a single pulse from all of the transducers is then available for computational image reconstruction.

As mentioned in the earlier article [1] one advantage of this particular architecture is the ability to "freeze" the image by illuminating it only once with a single pulse (or "flash") for each image reconstruction. In this way, distortions resulting from target motion are minimized, and blurring resulting from attempted reconstruction using multiple pulse responses from the target as it moves from location to location is eliminated. All signals from the transducers are captured simultaneously from the response of the target to the single illumination pulse.

The parallel architecture of the data acquisition phase of the imaging operation permits extremely high effective data conversion rates. Analog to digital conversion by several converters simultaneously is a cost effective method to achieve high throughput and high conversion precision. The use of separate banks of storage memory for each converter also lends itself to digital access to this data during image reconstruction.

THE ALGORITHM

The Backward Wave Propagation (BWP) or Digital Holographic reconstruction algorithm relies on the Fourier angular spectral decomposition for the monochromatic spatial distribution of amplitude and phase for the backscattered wave which for the two dimensional problem is written as $U(x,z)$. If z is the distance from the point $P(x,z)$ to the center of the transducer array, then $U(x,0)$ is the value of this backscattered wave at the plane of the array.

Using only the properties of the wave equation we find the following pair of transforms:

$$A_0\left(\frac{\alpha}{\lambda}\right) = \int_{-\infty}^{\infty} U(x,0) \exp\left\{-2\pi j\left(\frac{\alpha}{\lambda}x\right)\right\} dx \qquad (1)$$

$$U(x,z) = \int_{-\infty}^{\infty} A_o \left(\frac{\alpha}{\lambda}\right) \exp\left\{ \frac{2\pi j}{\lambda} \sqrt{1-\alpha^2}\, z \right\}$$

$$\exp\left\{ 2\pi j \left(\frac{\alpha}{\lambda} x\right) \right\} d\frac{\alpha}{\lambda} \qquad (2)$$

This permits generation of the image for $U(x,z)$ along a "slice" or "lace" along a line for various x at a given location z using the measured value of $U(x,0)$ along the transducer array at z=0. Except for the multiplication by a phase factor (which is constant for a given value of z) in equation (2), the only operations required are Fourier transformations and their inverses. As a result of recent commercial developments in Very Large Scale Integration (VLSI) the discretized Fast Fourier Transform (FFT) operation can readily be generated with high computational throughput and low cost. Simultaneous computation of equation (2) by several independent parallel processors for different values of z can achieve a video rate of construction of one image every 1/30 second from each transmitted pulse. Use of even more parallel processors might permit rates fast enough to illuminate the target with several different sets of illumination waveforms followed by some form of interpulse post processing.

In fact the speed of reconstruction is limited only by the number of z slices in the image, the total number of pixel elements located along the x direction, and the number and speed of the parallel FFT elements.

Another important advantage of the architecture discussed here is the fact that the computation of (2) for various values of z amounts to bringing the transducer array to focus at each of these z values simultaneously to the maximum degree feasible for the given size of the array aperture. Consequently we can have the image everywhere in the best focus possible given the distance of that point from the array.

ROUNDOFF AND TRUNCATION ERRORS

The first generation of VLSI components suitable for implementation of the hardware FFT processor included the AMD 29501 register array and 16 bit arithmetic unit, the AMD 29540 butterfly address generator, and the AMD 29516 16 bit x 16 bit integer multiplier. However, direct implementation of the FFT processor using only these components can easily lead to overflow for the size of the transforms of interest. This problem was discovered during the development of the system discussed in the earlier paper [1].

The basic decimation in time (DIT) butterfly step used by the FFT repeatedly involves making the following computations:

$$Ar' = Ar + (Br * Wr(j) - Bi * Wi(j)) \qquad (3)$$

$$Ai' = Ai + (Bi * Wr(j) + Br * Wi(j)) \qquad (4)$$

$$Br' = Ar - (Br * Wr(j) - Bi * Wi(j)) \qquad (5)$$

$$Bi' = Ai - (Bi * Wr(j) + Br * Wi(j)) \qquad (6)$$

where (Ar, Ai) and (Br, Bi) are the complex FFT butterfly inputs, (Ar', Ai') and (Br',Bi') are the complex FFT butterfly outputs, and W(j) is the complex "twiddle" factor $exp(i2\pi j/N)$ where N is the number of data points to be transformed. Assuming the input quantities to all have the same value and that this value is equal to the largest number representable without overflow, then with j=N/8 we find that Ar' can be as large as 2.414 times as large as the largest number which can be represented without overflow. In other words, without carefully limiting the inputs overflows are likely to occur.

To cope with this problem one can scale the output results down by a factor of 4 for every use of the butterfly step. This amounts to shifting the integer results right by two bits and compensating this by an implicit scale factor at the end of the computation. A problem with this approach is that after only 8 butterfly passes over the data the data will have been shifted 16 times to the right producing in many cases fields of all zero quantities as a result of roundoff.

To alleviate this problem one might perform the right shift only when needed. However, this would require some form of exponent for each data item to indicate the degree of scaling which has taken place for each one. This would effectively be a floating point representation. The question is whether an intermediate representation which is less complicated than floating point gives satisfactory results.

An alternative approach is the so called "block" floating point method which would perform the shift only when required sometime during one butterfly pass over the data, and to then shift all results for that pass. Thus only one exponent would be needed for all the data, rather than requiring an exponent for each data item. In some cases, however, this can be just as bad as performing the shift under all circumstances since anytime a single overflow occurred the conditional shift would give the same effective precision as the mandatory shift, namely all the data would get shifted for that pass. The result would be severe leading bit truncation or low order bit roundoff depending on the treatment of the data. This effectively makes the computation operate somewhere between the precision of a 16 bit integer and no precision at all. The simulations presented in the next section will indicate that the only satisfactory solution is to use full floating point processing.

SIMULATION OF THE ALGORITHM

To evaluate the degree to which the aforementioned problem could cause a catastrophic failure in image reconstruction the algorithm was used to simulate the focal spot for the phased array. The number of effective bits of precision was varied to gain some insight into the robustness of the BWP to roundoff error. Block floating point was used in order to avoid consuming all of the accuracy with zero's. A 15 cm long array of 256 transducer elements was assumed. The frequency of operation was assumed to be 2.5 MHz and the image point was set at a distance of 15 cm from the array and steered to about 30 degrees off the beam axis.

The results are summarized in Figures 2-4 for 10, 9, and 8 bit effective precision. It can be seen that the focal spot disintegrates when the net effective precision is 8 bits. In conclusion, we see that as long as the data requires excessive shifting amounting to the loss of only 8 bits, then the image will be severely degraded in some cases. The only clean solution for this problem is the use of full floating point representation of the data.

A VLSI SOLUTION

Fortunately, the availability of the newest AMD part, the AMD 29325, has provided a very nice solution for the vexing problem of the truncation-roundoff encountered in the FFT butterfly operation. The 29325 is a full 32 bit floating point part capable of addition, subtraction, and multiplication along with several other useful operations. In addition to the 29325, AMD has also provided a 18 bit by 64 word register file numbered the 29334 and a fast sequencer, the 29331. These are shown in Figure 5. The block diagram for the 29325 part is shown in Figure 6. Each operation can be completed in one 100 nanosecond clock cycle in certain modes. Although the cost of these parts is still high, there is a reasonable expectation that they will decrease rapidly with time. This reduction in the cost of floating point addition, subtraction and multiplication, with large speed up for the

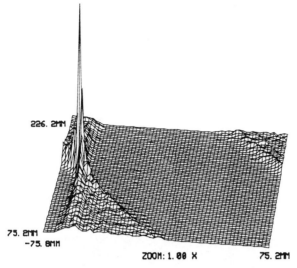

Figure 2a. Three dimensional plot for the magnitude of the focal spot of the diffraction pattern of a 15 cm 256 element array using 10 bits of effective precision.

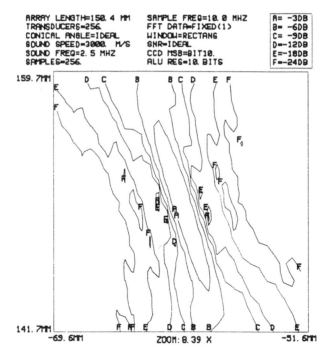

Figure 2b. Contour plot corresponding to Figure 2a.

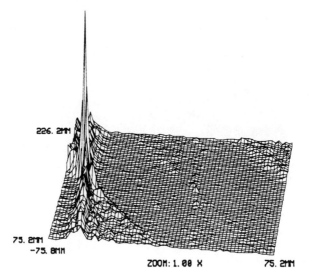

Figure 3a. Three dimensional plot for the magnitude of the focal spot of the diffraction pattern of a 15 cm 256 element array using 9 bits of effective precision.

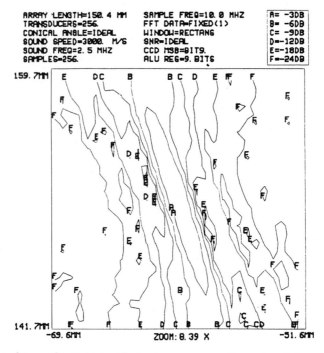

Figure 3b. Contour plot corresponding to Figure 3a.

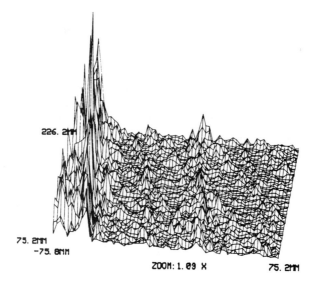

Figure 4a. Three dimensional plot for the magnitude of the focal spot of the diffraction pattern of a 15 cm 256 element array using 8 bits of effective precision.

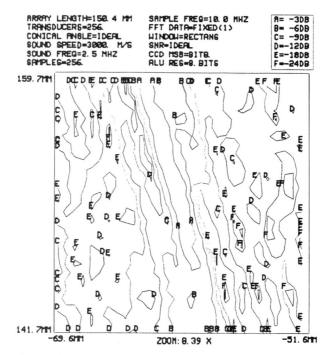

Figure 4b. Contour plot corresponding to Figure 4a.

multiplication illustrates once again the dramatic extent to which such developments in VLSI can affect the way designers of ultrasonic viewing systems can generate these images.

ORGANIZATION OF THE FFT PROCESSORS

The sixteen FFT processor boards shown in Figure 1 are organized as shown in Figure 7. The processor consists of two parallel AMD 29325 floating point units performing the real and imaginary computations simultaneously. The AMD 29540 address unit performs "twiddle" factor look-up operations and manipulates the data fetch operations from the AMD 99C328 32,768 x 8 bit static RAM working memory. Data transfers from the A-bus are not shown in this figure for simplicity. The AMD 29334 register file provides fast access storage for the floating point units.

Figure 5.   Pin grid array packages for the AMD 29325 floating point unit, the AMD 29334 18 bit by 64 word register file, andthe AMD 29331 microsequencer.

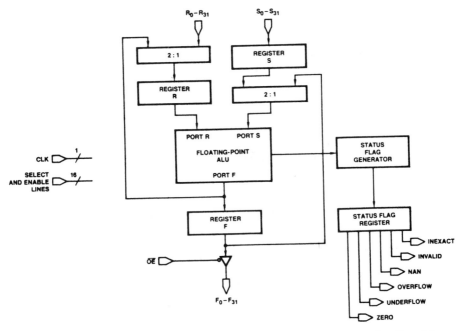

Figure 6.   The block diagram for the AMD 29325 floating point unit.

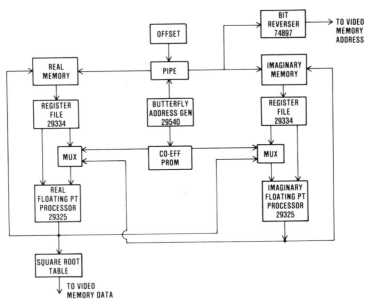

Figure 7.   FFT processor organization showing parallel real and imaginary floating point units.

349

The first transforms computed by the FFT boards convert the digitized received signals into the frequency domain. Following this conversion to spectral representation the frequency component selected for image reconstruction is transmitted to all of the other FFT processors simultaneously through the broadcast bus shown in Figure 8.

FFT PROCESSING SYSTEM ARCHITECTURE

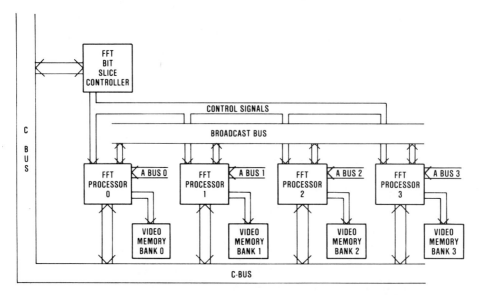

Figure 8.   Organization of a 4 FFT processor system showing the use of the broadcast bus and indicating the connections to the video display processor.

Once each FFT processor has its own copy of the spatial array of amplitude and phase information $U(x,0)$, each unit computes equation (1) in discrete form. Following this each unit then computes its share of the complex scalings and inverse FFT transforms implied by equation (2) for a collection of image "laces" corresponding to different z values. As shown in Figure 8 these laces are then made available to the video display in round-robin fashion. The timing of the system must be such that each FFT completes a group of laces prior to when the video display actually needs the results. For speed, the square root operation required for display of magnitude results is computed by table look-up as shown in Figure 7. Figure 9 shows the double buffering arrangement permitting the FFT processors to proceed to their next task after completing a set of laces by passing the information to the display without

requiring an additional move of the data. The FFT
processors can return image data to the supervising PDP-11
computer through the C-bus. In this way other image
processing steps can be performed such as feature selection
and pattern recognition.

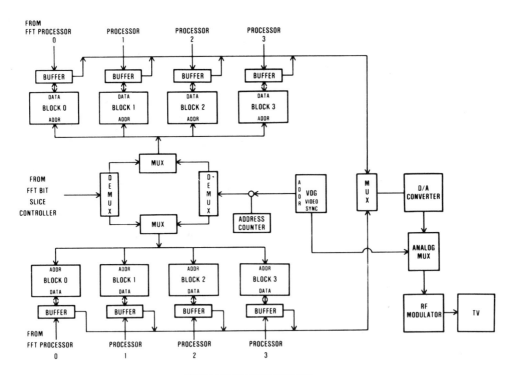

VIDEO GENERATION SYSTEM

Figure 9.   Architecture of the video display circuit for
             viewing the image as it is generated in real time.

SUMMARY AND CONSLUSIONS

     This paper has discussed some refinements of a digital
architecture previously published for generation of
ultrasonic images reconstructed using the backward wave
propagation or digital holographic algorithm. The
architecture permits reconstruction to proceed at a video
rate permitting real time viewing. Image generation is
possible from a single ultrasonic illuminating probe pulse.
The refinements have addressed a problem which arose in the
earlier design from roundoff and truncation effects in block
integer signal processing. Fortunately, improvements in the
VLSI building blocks available for constructing this refined
system have made the design changes extremely simple. This
illustrates once again the importance of VLSI in formulating
design decisions and in selecting the method for image
reconstruction.

ACKNOWLEDGEMENTS

The authors wish to acknowledge support by the U. S. Army under contract #DAAA22-81-C-0185. The authors also wish to thank Tom Needham, Steve Manship, Hitesh Merchant, and Mark Duckworth for assisting with this research at various stages of its development.

REFERENCES

[1] McDonald, J. F., Das, P. K., Laprade, K. C., Hidalgo, C. J., Goekjian, K. S., Jones, L., Shyu, H., and Capsimalis, G., "A High Spatial Resolution Digital System for Ultrasonic Imaging", _Acoustical Imaging_, Vol. 14 (Berkhout, A. J., Ridder, J., and van der Wal, L. F., Eds.) Plenum Publishing Corp., pp. 559-571, 1985.

[2] Yaroslavskii, L. P., and Merzlyakov, N. S., "Methods of Digital Holography", Consultant's Bureau Translation by David Parsons, Plenum Press, 1980.

[3] Goodman, J. W., "Introduction to Fourier Optics", McGraw Hill, pp. 49-53, 1968.

[4] Hildebrand, B. P., "Holograms, Ultrasound, and Computers to Detect FLAWS", Industrial Research and Development, Nov. 1982, (See also literature from Spectron Dev. Labs, Inc., Seattle, WA 98188).

[5] Farhat, N. H., and Chan, C. K., "Three Dimensional Imaging by Wave Vector Diversity", _Acoustical Imaging_, Vol. 8 (Metherell, A. F. Ed.), Plenum Press, 1980.

[6] Oppenheim, A. V., and Schafer, R. W. _Digital Signal Processing_, Prentice Hall, Chapter 9, 1975.

ACOUSTIC INVERSE SCATTERING IMAGES FROM SIMULATED HIGHER CONTRAST OBJECTS AND FROM LABORATORY TEST OBJECTS

M. J. Berggren[1], S. A. Johnson[1,2], W. W. Kim[2],
D. T. Borup[2], R. S. Eidens[2], and Y. Zhou[1]

[1]Department of Bioengineering
[2]Department of Electrical Engineering

University of Utah, Salt Lake City, UT 84112

ABSTRACT

Our previously reported acoustic inverse scattering imaging algorithms, based on the exact (not linearized) Helmholtz wave equation, are accurate and robust when applied to scattering data from small or low contrast objects. In order to compare the relative efficacy of our inverse scattering methods with linear approximations such as the Born or Rytov approximations, we have reviewed the limitations of those methods with both theoretical and experimental data. We have augmented our original alternating variable algorithms with Newton-Raphson-like methods which give improved convergence with higher contrast objects (up to 20% contrast in the scattering potential) and/or larger grid sizes. We have also developed a new, fast algorithm for computing forward scattering solutions from two non overlapping right circular cylinderical objects using a Bessel function series expansion. When using our inverse scattering algorithms to reconstruct images from this independently generated data, we find excellent agreement. We demonstrate the ability of our methods to reconstruct separate images of compressibility, absorption, and density for simulated data from a simple breast phantom. We also report on our progress in reconstructing images of larger size, as measured in wavelengths, and in using more realistic tissue simulating models as test objects..
(*Acoustical Imaging 16*, Chicago, Illinois, June, 1987).

INTRODUCTION

We have previously described [1,2] a model for the propagation of an ultrasonic pressure wave $p(\mathbf{x})$ in tissue which includes both compressibility and density fluctuations and how, using the substitution $f(\mathbf{x}) = p(\mathbf{x})/(\rho(\mathbf{x}))^{1/2}$, where ρ is the density, this equation may be transformed to the following two-dimensional formulation of the Lippmann-Schwinger integral equation [3]:

$$f^{(sc)}(\mathbf{x}) \equiv f(\mathbf{x}) - f^{(in)}(\mathbf{x}) = -\int k_o^2 \, \gamma(\mathbf{x}') \, f(\mathbf{x}') \, g(|\mathbf{x} - \mathbf{x}'|) \, d^2\mathbf{x}' \; . \tag{1}$$

Note that the total field, $f(\mathbf{x})$, the scattered field, $f^{(sc)}(\mathbf{x})$, and the incident field, $f^{(in)}(\mathbf{x})$, all depend upon the frequency, ω, and location of the source. The two dimensional outward-going free space Green's function is $g(|\bullet|) = (i/4) \, H_o^{(1)}(k_o|\bullet|)$, where $H_o^{(1)}$ is the zero order Hankel function. Also $\gamma(\mathbf{x})$ is the scattering potential (which in general is frequency dependent) and is given by the expression:

$$\gamma(\mathbf{x}) = 1 - [c_o^2/c^2(\mathbf{x})] + [c_o^2/\omega^2]\, \rho^{1/2}(\mathbf{x})\, \nabla^2[\rho^{-1/2}(\mathbf{x})] - 2\, i\, \{c_o^2/[\omega\, c(\mathbf{x})]\}\, \alpha(\mathbf{x}), \quad (2)$$

where $\alpha(\mathbf{x})$ represents the power absorption. Here $c(\mathbf{x})$ is the inhomogeneous speed of sound. We assume the scattering potential is confined within a known region surrounded by a homogeneous medium (generally water) in which $c(\mathbf{x}) = c_o$ and $\gamma = 0$. Also $k_o = \omega/c_o$ is the wave number of the incident wave. We have described in [4] how Eq. (1) may be transformed into a system of algebraic equations by expanding the product $k_o^2\, \gamma(\mathbf{x}')\, f(\mathbf{x}')$

with a set of basis functions for each frequency, ω, and source location. We have also described [1,5,6] various method for obtaining inverse solutions to these equations from a knowledge of the incident field and measurements of the total field on the boundary.

RELATION TO BORN AND RYTOV APPROXIMATIONS

There has been considerable interest in using the Born or Rytov approximations for diffraction tomography [7,8]. Although there have been a number of investigations concerning the limitations of these methods [9,10], it is of interest for us to examine these limitations independently for confirmation and in order to compare the efficacy of either the Born or Rytov vs. our inverse scattering methods for practical problems in medical imaging or in other areas. There is also a need to further examine imaging using Rytov approximation for the case of large objects with correspondingly small gradients in the eikonal at every point. This condition may hold when either the scattering potential γ is small and/or the gradient in γ is small. Our images (Fig. 2) are as good as any that have been reported for cylinders with sharp boundaries.

We note that when the density fluctuations are zero, the real part of our scattering potential as given by $\gamma_{real}(\mathbf{x}) = 1 - c_o^2/c^2(\mathbf{x}) = 1 - n^2(\mathbf{x})$ where $n(\mathbf{x}) = c_o/c(\mathbf{x})$ is the (real part of the) index of refraction. If we use the notation that $n = 1 + \delta_n$, where δ_n is a small perturbation, then $\gamma_{real} \approx -2\,\delta_n$. It has been demonstrated [10] that the Born approximation is only valid if the phase difference between the incident and actual fields through the object is less than π. If the radius of an object in terms of wavelength is (a/λ), then this phase difference is give by $\Delta\phi = 4\,\pi\,\delta_n\,(a/\lambda)$. The limits of the Rytov approximation are more difficult to apply but can be simply stated as a requirement that the gradient of the eikonal w (complex phase) of the actual field in the object divided by the wavelength be small compared to $2\,\delta_n$, i.e. $k_o^2\,|\gamma| \gg |\nabla w|$. Although we do not obtain the true internal fields in the Rytov approach, it has been suggested that (at least for right circular cylinders) the practical limits of the Rytov approximation are to require that δ_n be less than a few per cent [10]. Still, the Rytov approximation does appear to hold more potential promise for at least semi-quantitative images of tissues than does the Born approximation. These conculsions are clearly illustrated by the Born and Rytov reconstructions shown in Figs. 1 and 2. The data for these simulations was generated from analytic Bessel function expansions solutions for the scattering from circular cylinders with care taken to: (1) use sufficient number of terms in the expansion for accuracy and (2) unwrap the phase correctly after taking the complex logarithm to form the eikonal. Such synthetic data have been shown to match accurately data from real scattering experiments of others [11] and are in accord with the laboratory experiments which we report in this paper. Fig. 1(a) shows the real part of our reconstructions using the Born approximation for cylinders of radius $r = 2\lambda$, 4λ, and 6λ

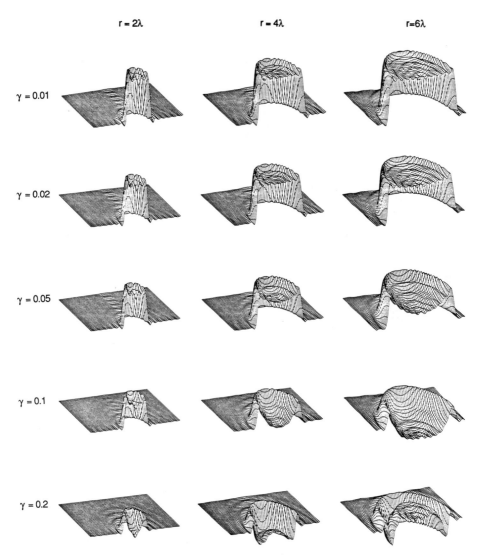

Fig. 1(a)   Reconstructions using the Born Approximation—Real Part. These reconstructions were obtained from simulated scattering data using 64 equispaced detectors and 64 equispaced source angles. The reconstructions were done on a 64x64 grid of pixels each of width λ/4. However for display purposes we only show 64x32 pixel values which effectively cuts the cylinders in half and allows us to view the interior values.

355

r = 2λ   r = 4λ   r=6λ

γ = 0.01

γ = 0.02

γ = 0.05

γ = 0.1

γ = 0.2

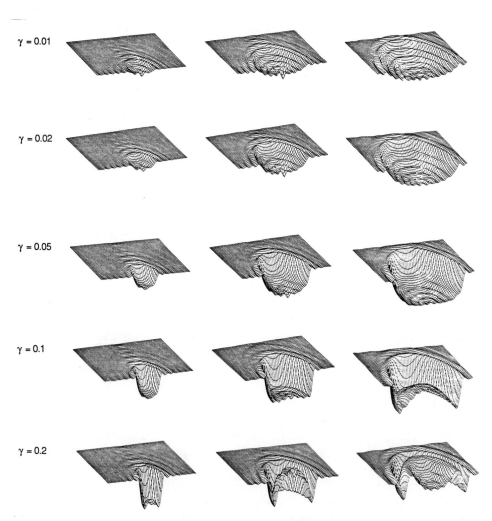

Fig. 1(b)   Reconstructions using the Born approximation—Imaginary Part. Since the
              true scattering potential had no imaginary component but only the real values
              indicated along the left side, these images represent the leakage of energy into
              the imaginary component due to the intrinsic nature of the Born
              approximations.

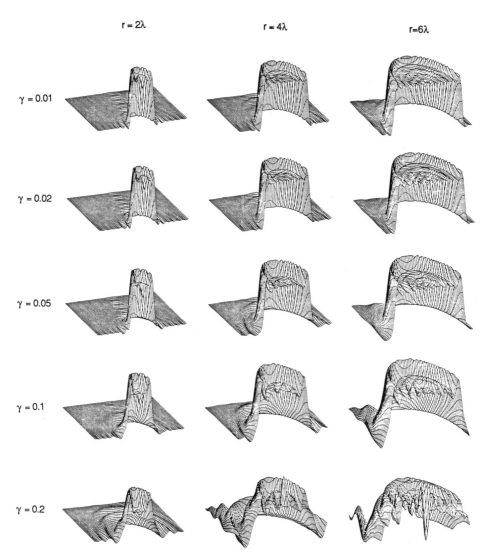

Fig. 2(a)   Reconstructions using the Rytov Approximation—Real Part. These reconstructions were obtained from simulated scattering data using 64 equispaced detectors and 64 equispaced source angles. The reconstructions were done on a 64x64 grid of pixels each of width λ/4. However for display purposes we only show 64x32 pixel values which effectively cuts the cylinders in half and allows us to view the interior values.

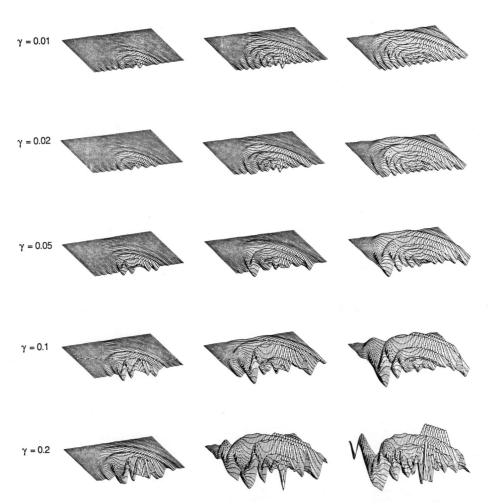

r = 2λ   r = 4λ   r = 6λ

$\gamma = 0.01$

$\gamma = 0.02$

$\gamma = 0.05$

$\gamma = 0.1$

$\gamma = 0.2$

Fig. 2(b)   Reconstructions using the Rytov approximation—Imaginary Part. Since the true scattering potential had no imaginary component but only the real values indicated along the left side, these images represent the leakage of energy into the imaginary component due to the intrinsic nature of the Rytov approximation.

and with γ = .01, .02, .05, 0.1, 0.2 (real only, no imaginary component–i.e., no absorption) and Fig. 1(b) shows the leakage of energy into the imaginary parts of the reconstructions under the Born approximations. Fig. 2 shows the same examples using our implementation of the Rytov approximation.

PRELIMINARY EXPERIMENTAL RESULTS

We have begun a laboratory experimental program to test and verify various diffraction tomography and inverse scattering reconstructions on known objects prior to the construction of prototype clinical scanners. Figure 3 gives a schematic diagram of the experimental arrangement for our initial tests. The incident plane wave is produced by exciting a large 19.6 x 14.6 cm sheet of polyvinylidene floride (PVDF) with 15–20 cycles of a gated 625 kHz sine-wave (λ_o = .238 cm in water). The cylindrical test objects are made of a solution of alcohol in agar [12] and these particular test objects have very little absorption. Note that while individual detector elements are shown in Fig. 3, in actual practice the detector and the PVDF source were held fixed and the object was moved in $\lambda_o/2$ steps. The size of the detector elements are about .16 λ_o wide by 4.2 λ_o high.

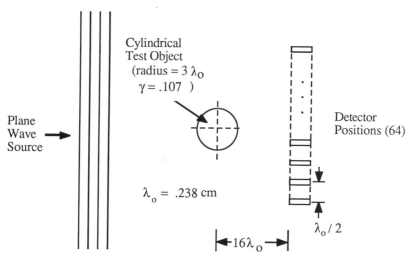

Fig. 3   Schematic diagram of experimental setup.

Figure 4a shows the measured values of the magnitue of the received signal (normalized by the signal recieved in the absence of the scattering object) and Fig. 4b shows a plot of the normalized phase values. The speed of sound in the sample was then measured directly with a time-of-flight measurement along the long axis of the cylinder. This value was then used to generate simulated scattering data with the Bessel function series solutions which have been shown to be very accurate [11], and are shown as solid lines in Fig 4(a) and 4(b). Note that while the agreement between the experimental points and the theoretical curve is good, we definitely expect to improve on this in future work.

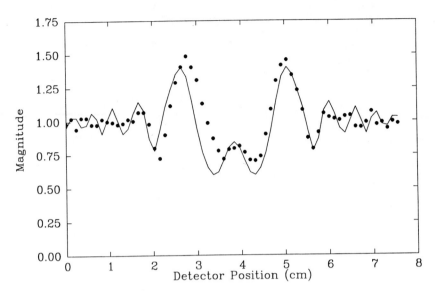

Fig. 4(a)   Magnitude of normalized detector signal (dots) and simulated data (solid line).

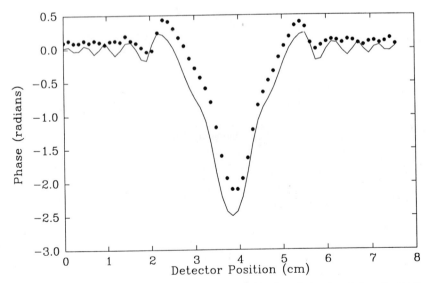

Fig. 4(b)   Phase of normalized detector signal (dots) and simulated data (solid line).

Images were then reconstructed from both the experimental data and the simulated data using the Rytov approximation. Figure 5 shows a isometric projection plot of the reconstruction from the experimental data and Figure 6 is a plot of the reconstruction from the simulated data. These figures illustrates quite clearly that the Rytov approximation is adequate in this case (i.e., $\gamma = .107$, $r = 3\lambda$) for finding the shape of the real part of the scattering potential with a small amount of energy leakage into the imaginary part of the reconstruction. However, the quantitative values of the real part in the center of the object for the reconstructions from either set of data are significantly lower than the actual object. This can be most readily seen in Figure 7 which gives a quantitative plot of the real and imaginary parts of a center line profile of both the reconstruction from real data (dots) and from the simulated data (solid line). This example, along with our previous simulations, clearly illustrates that while the Rytov approximation may give qualitatively correct images, the reconstructions will not be quantitatively correct for objects with significantly large gradients.

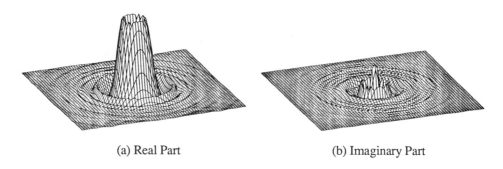

(a) Real Part   (b) Imaginary Part

Fig. 5   Reconstruction using the Rytov approximation on laboratory experimental data for a cylinder with $\gamma = .107$ and $r = 3\lambda$.

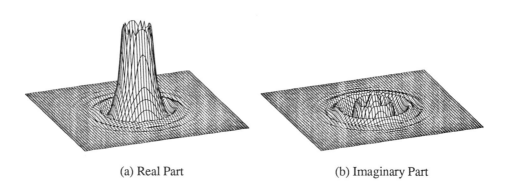

(a) Real Part   (b) Imaginary Part

Fig. 6   Reconstruction using the Rytov approximation on simulated data from the Bessel function series solution for a cylinder with $\gamma = .107$ and $r = 3\lambda$.

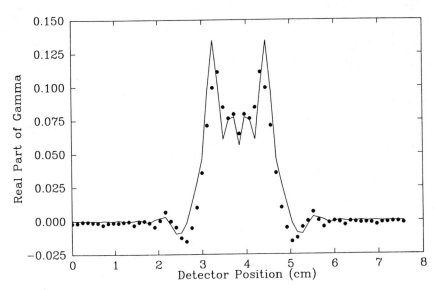

Fig. 7(a)   Center line profile for the real parts of the reconstructions shown in Figs. 5 and 6. The dots correspond to the reconstruction from the experimental data and the solid line corresponds to the reconstruction from simulated data.

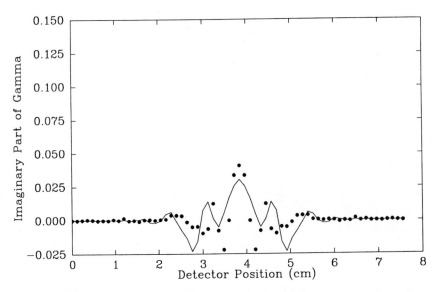

Fig. 7(b)   Center line profile for the imaginary parts of the reconstructions shown in Figs. 5 and 6. The dots correspond to the reconstruction from the experimental data and the solid line corresponds to the reconstruction from simulated data.

NEW METHOD FOR CALCULATING SCATTERING FROM TWO CYLINDERS

We have developed a new method for calculating the scattering from 2 cylinders as an independent method of verifying our algorithms with non-symmetric data. Rather than applying an iterative multiple scattering theory as Azumi [13], we use the well-known analytic solution for one circular cylinder and apply it simultaneously to two cylinders. Consider the geometrical arrangement sketched below where an incident plane wave propagates with a wavelength k_o in a uniform homogeneous medium (such as water) and impinges upon two right circular cylinders with radii of a_1 and a_2 and with wave numbers of k_1 and k_2:

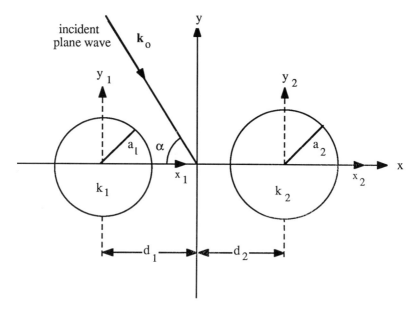

Fig. 8   Geometry for two-cylinder scattering

Let (x, y) be a coordinate system whose origin lies between the two cylinders. The first cylinder lies at $x = -d_1$ and the second cylinder lies at $x = +d_2$. Let (x_1, y_1) be a coordinate system centered on the first cylinder of the left and (x_2, y_2) be a coordinate system centered on the second cylinder. Note that we require $d_1 + d_2 \geq a_1 + a_2$.

From separation of variables we know that there exist the expansions:

$$f_1(\rho_1, \phi_1) = \sum_{n=-\infty}^{\infty} a_n J_n(k_1 \rho_1) e^{in\phi_1} \qquad \text{relative to the } (x_1, y_1) \text{ coordinates} \qquad (3)$$

$$f_2(\rho_2, \phi_2) = \sum_{n=-\infty}^{\infty} b_n J_n(k_2 \rho_2) e^{in\phi_2} \qquad \text{relative to the } (x_2, y_2) \text{ coordinates} \qquad (4)$$

and there also exists the following expansion for the incident plane wave, f^{in} :

$$f^{in}(\rho, \phi) = e^{ik_o \rho \cos(\phi + \alpha)} = \sum_{n=-\infty}^{\infty} i^n e^{in\alpha} J_n(k_o \rho) e^{in\phi} \qquad (5)$$

Inserting f_1 into the integral equation for the scattered field, which can be written as follows:

$$f_1^{sc}(\rho_1,\phi_1) = \frac{i\,(k_1^2-k_o^2)}{4} \int_0^{a_1}\int_0^{2\pi} f_1(\rho',\phi')\,H_0^{(1)}(k_o|\vec{\rho}_1-\vec{\rho}'|)\,\rho'\,d\rho'\,d\phi' \qquad (6)$$

results in the following expression for the scattered field from cylinder 1:

$$f_1^{sc}(\rho_1,\phi_1) = \sum_{n=-\infty}^{\infty} a_n\,c_n\,H_n^{(1)}(k_o\rho_1)\,e^{in\phi_1} \qquad (7)$$

where

$$c_n = \frac{i\,\pi\,k_o\,a_1}{2}\,[\,J_n(k_1 a_1)\,J_n'(k_o a_1) - \frac{k_1}{k_o}\,J_n(k_o a_1)\,J_n'(k_1 a_1)\,]\;. \qquad (8)$$

Similarly for the field scattered by cylinder 2 we obtain the following:

$$f_2^{sc}(\rho_2,\phi_2) = \sum_{n=-\infty}^{\infty} b_n\,d_n\,H_n^{(1)}(k_o\rho_2)\,e^{in\phi_2} \qquad (9)$$

where

$$d_n = \frac{i\,\pi\,k_o\,a_2}{2}\,[\,J_n(k_2 a_2)\,J_n'(k_o a_2) - \frac{k_2}{k_o}\,J_n(k_o a_2)\,J_n'(k_2 a_2)]\;. \qquad (10)$$

(Note that the primes indicate differentiation with respect to the argument.)

In order to solve for a_n and b_n we enforce the continuity of f at the boundaries of the cylinders; i.c., $f_1 = f_1^{sc} + f_2^{sc} + f^{in}$ at $\rho_1 - a_1$ and $f_2 = f_1^{sc} + f_2^{sc} + f^{in}$ at $\rho_2 = a_2$. Note that the continuity of the normal component of the particle velocity is guaranteed by continuity of f and the fact that f^{sc} was derived from the scattering integral equation.

Now we need a means of translating coordinates from (ρ,ϕ) to (ρ_1,ϕ_1) and (ρ_2,ϕ_2). This is provided by Graff's addition theorem for which we will use the following geometry:

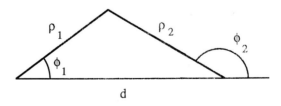

Fig. 9   Geometry for Graff's Addition Theorem

Using Graff's theorem, we can express any Bessel function of one coordinate in Fig. 9 system in terms of a series expansion in the other coordinate system. For example, we may express a Bessel function of $(k_0\rho_2)$ as follows:

$$B_n(k_0\rho_2) e^{in\phi_2} = \begin{cases} \displaystyle\sum_{k=-\infty}^{\infty} B_{k-n}(k_0 d) \, J_k(k_0\rho_1) \, e^{ik\phi_1} & \rho_1 < d \\[3ex] \displaystyle\sum_{k=-\infty}^{\infty} J_{k-n}(k_0 d) \, B_k(k_0\rho_1) \, e^{ik\phi_1} & \rho_1 > d \end{cases} \tag{11}$$

where $B_n = J_n, Y_n, H_n^{(1)}, H_n^{(2)}$, etc.

Upon writing a series expansion for f_1^{sc} in terms of (ρ_2, ϕ_2) and for f_2^{sc} in terms of (ρ_1, ϕ_1), we can then apply the boundary conditions at $\rho_1 = a_1$ and at $\rho_2 = a_2$. By gathering terms with common factors of $\exp(in\phi_1)$ and $\exp(in\phi_2)$, we can then obtain the following expressions:

$$a_n - e_n \sum_{m=-\infty}^{\infty} b_m \, d_m \, H_{n-m}^{(1)}(k_0 d) = e_n \, i^n \, e^{in\alpha} \, e^{ik_0 d_1 \cos(\alpha)} \tag{12}$$

where

$$e_n = \frac{J_n(k_0 a_1)}{J_n(k_1 a_1) - c_n \, H_n^{(1)}(k_0 a_1)} \tag{13}$$

and

$$b_n - f_n \sum_{m=-\infty}^{\infty} a_m \, c_m \, H_{n-m}^{(1)}(k_0 d) = f_n \, i^n \, e^{in\alpha} \, e^{ik_0 d_2 \cos(\alpha)} \tag{14}$$

where

$$f_n = \frac{J_n(k_0 a_2)}{J_n(k_2 a_2) - c_n \, H_n^{(1)}(k_0 a_2)} . \tag{15}$$

We now truncate these series in Eqs. (3) and (4), keeping a sufficient number of terms for an accurate expansion of the fields, as for example:

$$a_m = 0 \quad \text{for} \quad |m| > 2\,|\,k_1 a_1\,|$$
$$b_m = 0 \quad \text{for} \quad |m| > 2\,|\,k_2 a_2\,|. \tag{16}$$

This gives us a linear system of equations (12,14) for a finite set of unknown coefficients a_n and b_n, which may be readily solved by a number of linear methods such as conjugate gradient methods [14]. The resulting algorithm is quite fast.

This new method was used to simulate forward scattering data from which reconstruction were made with our previously described methods. The lower half of Figure 10 shows a digitized version on a 32x32 grid (pixels width = $\lambda/4$) of a test object which consisted of two cylinders each with a radius of $3\lambda/4$, separated by $3\lambda/2$, and with $\gamma = 0.05$ in one and .025 in the other, and the resulting reconstruction is shown in the upper half of the figure. This example not only illustrates our fast new method for simulating the forward scattering from two cylinders, but also gives additional independent verification of the correctness of our direct scattering and inverse scattering reconstruction algorithms.

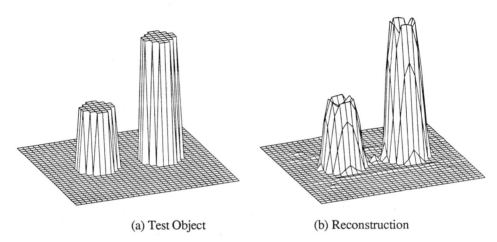

(a) Test Object   (b) Reconstruction

Fig. 10   Digitized version of test object (a) from which scattering data was simulated using our new method for forward scattering from two cylinders. Its reconstruction is given in (b). Both cylinders have a radius of $3\lambda/4$ and were separated by $3\lambda/4$. We set $\gamma = 0.025$ in one and $\gamma = .05$ in the other.

RECENT EXTENSIONS OF OUR INVERSE SCATTERING ALGORITHMS

We have augmented our original alternating variable algorithms with Newton-Raphson like methods which give improved convergence with higher contrast objects and larger grids. We have used $\gamma_{real} = 0.2$ on 32x32 grids with very good results, but this is a considerably higher contrast than we anticipate in normal tissue (for which a typical maximum would be $\gamma_{real} = 0.1$). Fig. 11 shows a reconstruction of a cylinder with $\gamma_{real} = 0.2$ and a radius of $3\lambda/2$. Fig. 12 show a reconstruction on a 64x64 grid from simulated scattering data for a cylinder of radius r = 4λ and $\gamma = 0.1+0i$, which exceed in accuracy the corresponding reconstruction using the Born or Rytov approximations as shown in Figs. 1 and 2. This reconstruction demonstrates quite well our ability to reconstruct objects of this size with the maximum expected variation expected in soft tissue. However, there remains a problem in imaging objects for which the phase difference between the incident and the actual field becomes greater than π. We have recently verified with our algorithms the observations reported by Cavicchi [15] of convergence to an erroneous but stable solution. Methods to insure convergence to unique and correct solutions are now under test.

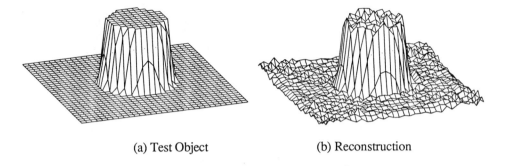

(a) Test Object   (b) Reconstruction

Fig. 11   Reconstruction by inverse scattering from simulated data: (a) test object and
(b) its reconstruction. We show here only the real part of the reconstruction;
the imaginary part was relatively small similar to the results shown in Fig.
12. The test object is a right circular cylinder with $\gamma = 0.2 + 0i$ and a radius
of $3\lambda/2$. The grid has 32x32 pixels and the width of each pixel is $\lambda/4$.

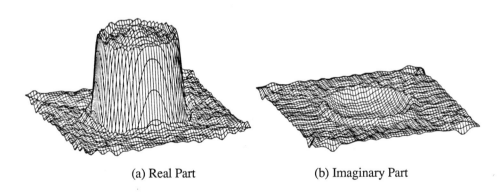

(a) Real Part   (b) Imaginary Part

Fig. 12   Reconstruction by inverse scattering from simulated data: (a) the real part
and (b) the imaginary part. The test object is a right circular cylinder with $\gamma =$
$0.1 + 0i$ and a radius of 4λ. The grid has 64x64 pixels and the width of each
pixel is $\lambda/4$.

We have also enhanced our techniques for simultaneously reconstructing all components of a complex γ as given in Eq. (2). As shown in [16], we may express Eqs. (1) and (2) in terms compressibility (κ), absorption (α), and density (ρ). This follows from the relation $\kappa\rho = 1/c^2$ and the definitions of γ_κ and γ_ρ given in the legend of Fig. 13.

Fig. 13 illustrates the potential advantages of having images of all three components. This figure depicts a simulated breast phantom in which we inserted one tumor (on the left) with a difference in compressibility and absorption and a second tumor (on the right) with only a density variation. The reconstructions of this test object from data equispaced around 360° as shown in the lower half of the figure is excellent and illustrates how tumors might be differentiated by their intrinsic physical properties. Reconstructions of the triad (c, ρ, and α) directly from Eqs. (1) and (2) will be given in [17].

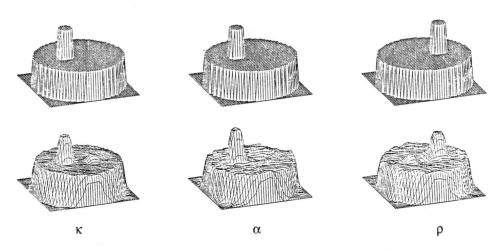

κ $\quad\quad\quad\quad\quad\quad\quad\quad$ α $\quad\quad\quad\quad\quad\quad\quad\quad$ ρ

Fig. 13   Comparison of true and reconstructed values of compressibility (κ), absorption (α) and density (ρ). The top rows of images are respectively from left to right, true $\text{Re}(\gamma_\kappa) = 1 - \kappa/\kappa_o$, 10 times true $\text{Im}(-\gamma_\kappa) \propto \alpha$, and true $\gamma_\rho = 1 - \rho/\rho_o$. The bottom row are respectively, left to right, reconstructed values of γ_κ, α(scaled) and γ_ρ. The image size is 40 x 40 pixels and is smoothed for a spatial resolution of no better than λ_{min}, where λ_{min} corresponds to the wavelength of ω_{max} in water. The pixel separation is $\lambda_{min}/4$. The data results from using two frequencies (ω_{max} and .75 ω_{max}), 16 incident plane waves equispaced around 360°. The true γ_κ and γ_ρ for the breast background are $\gamma_\kappa = .01 - .001i$ and $\gamma_\rho = .01$. In the right tumor $\gamma_\kappa = .02 - .002i$ with γ_ρ unchanged. In the left tumor $\gamma_\rho = .02$ with γ_κ unchanged. About 10 iterations were made. Further improvement is obtained with more iterations, but the results shown are already excellent.

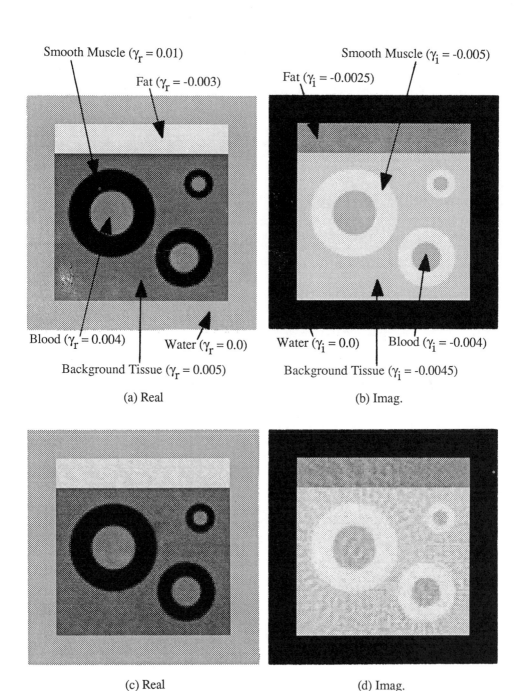

Smooth Muscle ($\gamma_r = 0.01$)

Fat ($\gamma_r = -0.003$)

Smooth Muscle ($\gamma_i = -0.005$)

Fat ($\gamma_i = -0.0025$)

Blood ($\gamma_r = 0.004$)   Water ($\gamma_r = 0.0$)

Background Tissue ($\gamma_r = 0.005$)

(a) Real

Water ($\gamma_i = 0.0$)   Blood ($\gamma_i = -0.004$)

Background Tissue ($\gamma_i = -0.0045$)

(b) Imag.

(c) Real

(d) Imag.

Fig. 14   Shown here are (a) the real components, and (b) the imaginary components of a test object used to simulate arteries in a matrix of soft tissue, muscle, fat and water; and the corresponding components of its reconstruction are shown in (c) and (d). All panels consist of 128x128 pixels each of width $\lambda/4$. Thus each panel is 32 λ wide and the diameter of the arteries are 12 λ, 8 λ and 4 λ. The real and imaginary parts of γ for the different tissues are as shown. They were scaled by a factor of 10 and 2, respectively, from normal tissue values in order to aid the convergence rate. This reconstruction was obtained from simulated scattering data for 114 source angles using 508 equispace detectors around the outer border.

We have begun studies of test object which more closely simulate actual tissue. For example, Fig. 14 shows a gray scale image of a test object (upper half) and its reconstruction (lower half). We have used this test object to simulate arteries in a matrix of soft tissue, muscle, fat, and water. Although typical values of the speed of sound, density and absorption were used in smaller simulations with this test object, for this our largest grid (128x128), we removed density as a factor and reduced the real and imaginary components by a factor of 10 and 2 respectively in order to increase the rate of convergence. Figure 14 is labeled with these scaled down values with the real components shown on the left panels and the imaginary components shown on the right panels. The reconstruction was made using the alternating variable approach [4] and is shown after 5 outer loop iterations. Further details concerning the methods used in reconstruction of the object shown Fig. 14, and applications of constraint methods in the presence of noisy data will be given in [17]. Due to memory limitations on the CRAY-XMP that we are using (about 4 mega-words for any single user) this is the largest sized image we can reconstruct without resorting to disk storage for intermediate values.

SUMMARY

We have demonstrated that our inverse scattering methods give accurate quantitative images of an unknown real scattering potential γ. We have verified the limitations of the Born and Rytov approximation with both numerical simulations and experimental data. We have reconstructed Rytov images from laboratory data and have shown that this data matches that predicted by Bessel function series solutions to the direct scattering problem. Our inverse scattering images are superior to Rytov images reconstructed from similar data generated by the same Bessel function series solutions. We have developed a new, independent method for generating forward scattering data from two cylinders and have used this as a further verification of our algorithms. We have also demonstrated our progress in obtaining reconstructions of higher contrasts and larger test objects and have demonstrated simultaneous images of compressibility (κ), absorption (α), and density (ρ).

ACKNOWLEDGMENTS

We wish to thank Professors Frank Stenger and Calvin H. Wilcox, Department of Mathematics, and Professor James S. Ball, Department of Physics, University of Utah, for their advice and stimulating discussions. This work supported in part by PHS Contract No. N43-CM-57805 and Grant No. 2 R01 CA29728-05 from NCI, and PHS Grant No. 1 R01 HL34995-02 from NHLBI. Computer time was supplied by the University of Utah VLSI facility and by a research grant from Cray Research, Inc., through the San Diego Supercomputer Center.

REFERENCES

1. M. J. Berggren, S. A. Johnson, B. L. Carruth, W. W. Kim, F. Stenger, and P. K. Kuhn, "Ultrasound inverse scattering solutions from transmission and/or reflection data," *SPIE Vol. 671, Physics and Engineering of Computerized Multidimensional Imaging and Processing*, ed. by Nalcioglu, Cho, and Budinger, pp. 114–121 (1986).
2. S. A. Johnson, F. Stenger, C. Wilcox, J. Ball, and M. J. Berggren, "Wave Equations and Inverse Scattering Solutions for Soft Tissue," *Acoustical Imaging 11*, pp. 409-424 (1982).
3. S. A. Johnson and M. L. Tracy, "Inverse Scattering Solutions by a Sinc Basis, Multiple Source, Moment Method -- Part I: Theory", *Ultrasonic Imaging 5*, pp. 361-375 (1983).
4. M. L. Tracy, and S. A. Johnson, "Inverse Scattering Solutions by a Sinc Basis, Multiple Source, Moment Method -- Part II: Numerical Evaluations", *Ultrasonic Imaging 5*, Academic Press, pp. 376-392 (1983).

5. S. A. Johnson, Y. Zhou, M. L. Tracy, M. J. Berggren, and F. Stenger, "Inverse Scattering Solutions by a Sinc Basis, Multiple Source, Moment Method -- Part III: Fast Algorithms", *Ultrasonic Imaging 6*, pp. 103-116 (1984).

6. S. A. Johnson, Y. Zhou, M. L. Tracy, M. J. Berggren, and F. Stenger, "Fast Iterative Algorithms for Inverse Scattering of the Helmholtz and Riccati Wave Equations," *Acoustical Imaging 13,* Plenum Press, pp. 75–87 (1984).

7. R. Mueller, M. Kaveh, and G. Wade, "Reconstructive Tomography and Applications to Ultrasonics", *Proc. IEEE* Vol 67, No.4, pp. 567–587 (1979).

8. M. Kaveh, M. Soumekh, and R. K. Mueller, "A Comparison of Born and Rytov Approximations in Acoustic Tomography", *Acoustical Imaging 11*, pp. 325–335 (1982).

9. M. Kaveh, M. Soumekh, Z. Lu, R. K. Mueller, and J. F. Greenleaf, "Further Results on Diffraction Tomography Using Rytov's Approximation", *Acoustical Imaging 12*, pp. 599–608, (1982).

10. M. Slaney, A. Kak, and L. Larsen, "Limitations of Imaging with First-Order Diffraction Tomography", *IEEE Trans. on MW Theo. and Tech.*, Vol MTT-32, pp. 860–874 (1984).

11. B. S. Robinson and J. F. Greenleaf, "Measurement and Simulation of the Scattering of Ultrasound by Penetrable Cylinders," *Acoustical Imaging 13*, Plenum Press, pp. 163–178 (1984).

12. M. Burlew, E. Madsen, J. Zagzebski, et al, "A New Ultrasound Tissue-Equivalent Material", *Radiology 134*, pp. 517–520 (1980).

13. M. Azimi and A. C. Kak, "Distortion in Diffraction Tomography Caused by Multiple Scattering," *IEEE Trans. on Med. Imag.*, Vol. MI-2, pp. 176–195 (1983).

14. R. D. Fletcher, *Practical Methods of Optimization, Vol.. 1, Unconstrained Optimization*, John Wiley and Sons, New York, (1980).

15. T. J. Cavicchi and W. D. O'Brien, "Tomographic Reconstruction by the Sinc Basis Moment Method", *Acoustical Imaging 16*, Chicago, IL, June, 1987 (this volume).

16. S. A. Johnson and F. Stenger, "Ultrasound Tomography by Galerkin or Moment Methods", in *Selected Topics in Image Science* ed. by O. Nalcioglu and Z. H. Cho, Springer-Verlag, pp. 254–276 (1984).

17. Y. Zhou, S. A. Johnson, and M. J. Berggren, "Constrained Reconstruction of Object Acoustic Parameters From Noisy Ultrasound Scattering Data", submitted to IEEE 1987 Ultrasonics Symposium.

ACOUSTICAL IMAGING BEYOND BORN AND RYTOV

R.J.Wombell and M.A.Fiddy

Department of Physics, King's College London
The Strand, London, WC2R 2LS, U.K.

ABSTRACT

Diffraction tomography, the quantitative imaging of structures from measurements of the scattered field, has recieved much attention over the last five years, the method of inverse scattering being based either on the First Born or the Rytov approximation and there has been much discussion on their relative merits and domains of validity.

In this paper we examine methods to better model the direct scattering problem for structures not satisfying the above conditions. These include the distorted wave Born approximation (DWBA) which models the scattering as a first order perturbtion about a known field distribution within the structure (e.g. A.J.Devaney and M.L.Oristaglio (1983)) and renormalisation methods that make perturbation expansions more uniformly valid (e.g. the multiple scales analysis of H.D.Ladouceur and A.K.Jordan (1985)).

For the distorted-wave method we consider an inversion procedure which uses algorithms previously developed for inversion in the Born or Rytov approximations. This we demonstrate using some numerical examples.

1. INTRODUCTION

Over the past decade a large number of papers have been written on inverse scattering and it's implementation as a technique of imaging, i.e. diffraction tomography (see, for example, Wolf (1969), Adams and Anderson (1982), Devaney (1985)) the key aspect of these papers being that the inversion algorithm is based on casting the integral equation of scattering into a form suitable for inversion through the first Born approximation or, by a suitable transformation, the Rytov approximation. These approximations lead to inversion algorithms based around the Fourier transform and since they can thus be implemented using the FFT they are computationally attractive. Implementation can be either by 'backpropagation' or Fourier plane interpolation of data from semi-circular arcs to a grid of regular points (Kaveh *et al* (1984), Kak (1983)). The numerical consequences of this are easy to study by determining the scattered field data on the basis of assuming the validity of one of these approximations. In practice, of course, their validity for real scatterers depends upon a strong inequality being satisfied. This can be loosely described as ensuring that the change in the amplitude and phase of the scattered field from that of the incident field is negligable (Born) or that the rate of change of the scattered field is slow on the scale of the wavelength of the incident field (Rytov). For this reason then there have been a number of papers evaluating their suitability on the basis of numerical experiments (Pan and Kak (1983), Slaney *et al* (1984)). These have been complemented by some effort to try and explicitly determine approximation errors introduced by the Born and Rytov approximations - notably by Kaveh and his coworkers (Langenberg and Schmitz (1985), Soumekh *et al* (1985)). Some papers have appeared which apply these methods to real data (see, for example, Tomikawa *et al* (1986)).

Unfortunately, it remains a fact that little use has been made of these diffraction tomography procedures (particularly when compared to the sucess of projection tomography) because of the severe restrictions put on the scatterer for the Born or Rytov approximations to be reasonable. Moreover, because of the nature of these approximations, namely that they depend upon a strong inequality holding, there has been a long debate over their relative merits in the various communitities that have tried to use them (see, for example, Fiddy (1986) and references therein).

One way of surmounting such problems is to develop inversion procedures with wider ranges of validity than those based on the Born or Rytov approximations. In the next section we consider two such techniques; renormalisation and distorted-wave methods. In section 3 we describe an inversion procedure based upon a distorted-wave approach and present numerical results from such a procedure in section 4.

2. EXTENDING THE BORN AND RYTOV APPROXIMATIONS

For the scattering of scalar waves ψ by a potential V of compact support on domain D we have

$$(\nabla^2 + k^2)\psi(\underline{r}, k\hat{\underline{r}}_o) = k^2 V(\underline{r})\psi(\underline{r}, k\hat{\underline{r}}_o) \tag{1}$$

where $\hat{\underline{r}}_o$ specifies the incident field direction and where ψ_o is the incident field of fixed k satisfying

$$(\nabla^2 + k^2)\psi_o(\underline{r}, k\hat{\underline{r}}_o) = 0 \tag{2}$$

By using the free-space Greens function (1) maybe written as an integral equation for ψ, which in the asymptotic limit, $r \to \infty$, is

$$\psi(\underline{r}, k\hat{\underline{r}}_o) = \psi_o(\underline{r}, k\hat{\underline{r}}_o) + k^2 \frac{e^{ikr}}{4\pi r} f(k\hat{\underline{r}}, k\hat{\underline{r}}_o) \tag{3}$$

where $f(k\hat{\underline{r}}, k\hat{\underline{r}}_o)$, is the scattering amplitude, given by

$$f(k\hat{\underline{r}}, k\hat{\underline{r}}_o) = \int_D e^{-ik\hat{\underline{r}}\cdot\underline{r}_1} V(\underline{r}_1)\psi(\underline{r}_1, k\hat{\underline{r}}_o)d\underline{r}_1 \tag{4}$$

The first Born approximation (BA) assumes that $\psi \simeq \psi_o$ in D, which from equation (3) implies that the potential is of small dimensions and/or weakly scattering. Within the BA and for plane wave incidence the scattering amplitude is given by

$$f^{BA}(k\hat{\underline{r}}, k\hat{\underline{r}}_o) = \int_D \psi_o(\underline{r}_1, -k\hat{\underline{r}})V(\underline{r}_1)\psi_o(\underline{r}_1, k\underline{r}_o)d\underline{r}_1 = \int_D e^{-ik(\hat{\underline{r}}-\hat{\underline{r}}_o)\cdot\underline{r}_1}V(\underline{r}_1)d\underline{r}_1 \tag{5}$$

This then is the familiar form of the Born approximation and the recovery of V from the data f assuming (5) being inversion under the Born approximation. The condition for the approximation to be valid can be written as

$$|kVa| \ll 1 \tag{6}$$

where a specifies the dimensions of V.

Alternativly, if (1) and (2) are transformed in Ricatti equations by substituting ψ for it's complex phase, i.e. $\psi = e^S$ and $\psi_o = e^{S_o}$ then it is found, again in the asymptotic limit, that

$$\psi(\underline{r}, k\hat{\underline{r}}_o) = \psi_o(\underline{r}, k\hat{\underline{r}}_o)e^{S_s(\underline{r}, k\hat{\underline{r}}_o)} \tag{7}$$

where $S_s = S - S_o$ and is given in the Rytov approximation by

$$S_s(\underline{r}, k\hat{\underline{r}}_o) = \{\frac{k^2 e^{ikr}}{4\pi r\psi_o(\underline{r}, k\hat{\underline{r}}_o)}\}f^{BA}(k\hat{\underline{r}}, k\hat{\underline{r}}_o) \tag{8}$$

under the condition

$$|\nabla S_s| \ll k \tag{9}$$

The necessary inversion to recover V thus becomes again

$$d(\underline{r}, k\hat{\underline{r}}_o) = \int_D \psi_o(\underline{r}_1, -k\hat{\underline{r}})V(\underline{r}_1)\psi_o(\underline{r}_1, k\hat{\underline{r}}_o)d\underline{r}_1 \tag{10}$$

where the data d are now

$$d(\underline{r}, k\hat{\underline{r}}_o) = \{\frac{4\pi r e^{-ikr}}{k^2}\}\psi_o(\underline{r}, k\hat{\underline{r}}_o)\ln\frac{\psi(\underline{r}, k\hat{\underline{r}}_o)}{\psi_o(\underline{r}, k\hat{\underline{r}}_o)} \tag{11}$$

This leads to the same inversion procedure as was required for the Born approximation, but with different data and a very different condition for the approximation to be valid.

To consider how we may usefully go beyond the limitations embodied in equations (6) and (9) we begin by observing that if (3) is regarded as the first two terms of an exponential series and recast it into an exponential then we obtain the Rytov approximation (RA) to the field of equation (7). The technique of recasting into an exponential form is a well known method of renormalistion - the effective summation of secular terms (Nayfeh (1973))- and thus the Rytov approximation is the Born approximation renormalised.

The proceedure of renormalisation has been applied to the direct problem of scattering by scalar waves as described by (1) and (2) for some years. One treatment, due to Tatarski and Gertenstein (1962) lets the scattered field be composed of a regular and random part,i.e. $\psi = <\psi> +\phi$ where $<\phi> = 0$ and then expresses $<\psi>$ in the form of a power series of a 'renormalised' operator which already accounts for multiple scattering. Renormalisation techniques have also recently begun to be applied to the inverse problem. Jaggard and Kim (1985) used a non-linear correction to the one-dimensional Ricatti equation to effect a renormalisation. Ladouceur and Jordan (1985), again working only in one-dimension, used a method based on an exact theory developed by Kay and Moses from the work of Gel'fand, Levitan and Marchenko (Kay and Moses (1982)). Using the method of multiple scales (Nayfeh (1973)) to effect a renormalisation they obtained a rapidly convergent solution equivalent to the second order solutuon of a perturbation treatment of the exact problem - an extension over the Born approximation of nearly a factor of 50 in ka.

While currently limited to one-dimensional treatment, such work indicates how renormalisation might be used to extend diffraction tomography algorithms.

An alternative approach, and one we shall consider in more detail, is the distorted-wave method. The scatterer is expressed as the sum of two parts, $V = V_1 + V_2$ and V_2 is treated as a perturbation to the background V_1, and can be treated in a fashion analogous to either the Born or Rytov approach to the solution of equation (1).

Using a Born treatment (assuming the incident field is a plane wave $\psi_o(\underline{r}, k\hat{\underline{r}}_o) = e^{ik\hat{\underline{r}}_o \cdot \underline{r}}$) it is found that the scattering amplitude in (3) is given by the two-potential formula (Taylor (1972), Newton (1982)),

$$f(k\hat{\underline{r}}, k\hat{\underline{r}}_o) = f_1(k\hat{\underline{r}}, k\hat{\underline{r}}_o) + f_2(k\hat{\underline{r}}, k\hat{\underline{r}}_o) \tag{12}$$

where

$$f_1(k\hat{\underline{r}}, k\hat{\underline{r}}_o) = \int_D e^{-ik\hat{\underline{r}} \cdot \underline{r}_1} V_1(\underline{r}_1)\psi_1(\underline{r}_1, k\hat{\underline{r}}_o)d\underline{r}_1 \tag{13}$$

$$f_2(k\hat{\underline{r}}, k\hat{\underline{r}}_o) = \int_D \psi_1(\underline{r}_1, -k\hat{\underline{r}})V_2(\underline{r}_1)\psi(\underline{r}_1, k\hat{\underline{r}}_o)d\underline{r}_1 \tag{14}$$

and where ψ_1 is the field for the case $V_2 = 0$ which satisfies the equation

$$(\nabla^2 + k^2 - k^2V_1(\underline{r}))\psi_1(\underline{r}, k\hat{\underline{r}}_o) = 0 \tag{15}$$

The distorted-wave Born approximation (DWBA), in an exactly analagous fashion to the BA, replaces the total field ψ by an approximation to the field within D, in this case ψ_1, in (13) to give a DWBA estimate to the scattering amplitude (just as in the BA it is approximated by ψ_o, the incident field)

$$f_2^{DWBA}(k\hat{\underline{r}}, k\hat{\underline{r}}_o) = \int_D \psi_1(\underline{r}_1, -k\hat{\underline{r}})V_2(\underline{r}_1)\psi_1(\underline{r}_1, k\hat{\underline{r}}_o)d\underline{r}_1 \tag{16}$$

and hence, to be valid, places the requirements of the Born approximation on V_2 alone, rather than on the entire potential, i.e.

$$|kV_2a| \ll 1 \tag{17}$$

Under this approximation scattering by V_1 is accounted for exactly (equation (13)) but only to first order in V_2 (equation (14)).

The derivation of the Distorted-Wave Rytov approximation (DWRA) follows in an exactly parallel fashion to that of the RA and it is found that (Wombell and Fiddy (1987b)) again in the asymptotic limit

$$\psi(\underline{r}, k\hat{\underline{r}}_o) = \psi_1(\underline{r}, k\hat{\underline{r}}_o)e^{S_s(\underline{r}, k\hat{\underline{r}}_o)} \tag{18a}$$

where $\psi_1 = e^{S_1}$ and $S_s = S - S_1$

$$S_s(\underline{r}, k\hat{\underline{r}}_o) = \frac{k^2 e^{ikr}}{4\pi r \psi_1(\underline{r}, k\hat{\underline{r}}_o)} \int_D \psi_1(\underline{r}_1, -k\hat{\underline{r}})V_2(\underline{r}_1)\psi_1(\underline{r}_1, k\hat{\underline{r}}_o)d\underline{r}_1 \tag{18b}$$

with the validity of this approximation depending upon the condition

$$|\nabla S_s| \ll |\nabla S_1| \tag{19}$$

Thus both the distorted-wave techniques remove the appropiate Born and Rytov restrictions, replacing them with ones that are very similar but apply to only a perturbation in the potential rather than to the entire potential itself. It should also be noted that the relation between the DWRA and DWBA is the same as that between the RA and BA, namely that the DWRA is the DWBA renormalised in the sense of being recast into exponential form.

Thus there are a number of techniques for modelling the direct problem which apply conditions for validity of considerably wider extent than those of the Born and Rytov approximation - the only problem is to cast them into a form suitable for inversion.

3. DISTORTED-WAVE INVERSION

If we assume that V_1 is known exactly or to a good approximation *a priori* then the reconstruction of V_2 is the inversion of

$$d(\underline{r}, k\underline{\hat{r}}_o) = \int_D \psi_1(\underline{r}_1, -k\underline{\hat{r}})V_2(\underline{r}_1)\psi_1(\underline{r}_1, k\underline{\hat{r}}_o)d\underline{r}_1 \qquad (20)$$

where in the DWBA

$$d(\underline{r}, k\underline{\hat{r}}_o) = f_2(k\underline{\hat{r}}, k\underline{\hat{r}}_o) = f(k\underline{\hat{r}}, k\underline{\hat{r}}_o) - f_1(k\underline{\hat{r}}, k\underline{\hat{r}}_o) \qquad (21a)$$

or for the DWRA it is

$$d(\underline{r}, k\underline{\hat{r}}_o) = \{\frac{4\pi r e^{-ikr}}{k^2}\}\psi_1(\underline{r}, k\underline{\hat{r}}_o)\ln\frac{\psi(\underline{r}, k\underline{\hat{r}}_o)}{\psi_1(\underline{r}, k\underline{\hat{r}}_o)} \qquad (21b)$$

This problem has been addressed by Devaney and Oristaglio (1983) who developed an inversion proceedure based upon an eigenvalue/eigenvector decomposition - but a readily computed inversion algorithm was not forthcoming. Here we propose an inversion algorithm to recover V_2 based upon algorithms already developed for diffraction tomography and valid whenever the DWBA or DWRA are valid. The conditions for their validity suffer from the same kind of loose interpretation as the Born and Rytov approximation - but the new procedures open up a much wider class of potential applications.

We note from (20) that we need to calculate ψ_1 for $\underline{r} \in D$. For simple structures such as spheres or cylinders analytical expressions for the internal field are available - but for a general field this will not be so, and could only be calculated by numerical techniques. However, as we shall see, it is not necessary to know ψ_1 explicitly (Wombell and Fiddy (1987a)).

From equation (4) for the case of $V_2 = 0$ we know

$$f_1(k\underline{\hat{r}}, k\underline{\hat{r}}_o) = \int_D e^{-ik\underline{\hat{r}}\cdot\underline{r}_1}V_1(r_1)\psi_1(\underline{r}_1, k\underline{\hat{r}}_o)d\underline{r}_1$$

but were we to make a first Born assumption and invert the data d regardless of the appropiateness of taking the first Born approximation we would obtain a first Born estimate $\tilde{V}_1$ of V_1 satisfying the equation

$$f_1(k\underline{\hat{r}}, k\underline{\hat{r}}_o) = \int_D e^{-ik\underline{\hat{r}}\cdot\underline{r}_1}\tilde{V}_1(\underline{r}_1)\psi_o(\underline{r}_1, k\underline{\hat{r}}_o)d\underline{r}_1$$

subtracting these two equations we have

$$\int_D e^{ik\underline{\hat{r}}\cdot\underline{r}_1}\left[V_1(\underline{r}_1)\psi_1(\underline{r}_1, k\underline{\hat{r}}_o) - \tilde{V}_1(\underline{r}_1)\psi_o(\underline{r}_1, k\underline{\hat{r}}_o)\right]d\underline{r}_1 = 0 \qquad (22)$$

Assuming that there are no non-scattering scatterers in the potential (see, for example, Kim and Wolf (1986)) which is a reasonable physical assumption then a sufficient but not necessary condition for equation (22) to hold is

$$V_1(\underline{r})\psi_1(\underline{r}, k\underline{\hat{r}}_o) - \tilde{V}_1(\underline{r})\psi_o(\underline{r}, k\underline{\hat{r}}_o) = 0$$

so giving

$$\psi_1(\underline{\hat{r}}, k\underline{r}_o) = \frac{\tilde{V}_1(r)}{V_1(r)}\psi_o(\underline{r}, k\underline{\hat{r}}_o) \qquad (23)$$

From this we can evaluate the field ψ_1 within D, namely the domain of the scatterer V_2. On substituting (23) into equation (20) (where we have already assumed that the incident field is a plane wave) we obtain

$$d(k\hat{r}, k\hat{r}_o) = \int_D e^{-ik(\hat{r}-\hat{r}_o)\cdot r_1} U(r_1) dr_1 \tag{24}$$

where

$$U(r) = \frac{|\tilde{V}_1(r)|^2}{|V_1(r)|^2} V_2(r) \tag{25}$$

The form of the integral in (24) is that of the usual Born inversion procedure required for diffraction tomography. Explicitly, the far-field data are the values of the Fourier transform of U on the boundaries on Ewald spheres, exactly as for the case of far field data in the Born approximation (Wolf (1969)). Thus any reconstruction algorithm based on the first Born approximation can be used to invert the data d, e.g. Fourier interpolation, filtered backpropagation etc, reconstructing U. As V_1 is known a priori, its reconstruction under the first Born approximation $\tilde{V}_1$ can be calculated. If $V_1(r) \neq 0, r \in D$ and we are able to identify the extent of D, i.e. the object support, then using these in (24) we can obtain an estimate of V_2.

4. NUMERICAL RESULTS

Here we demonstrate inversion using the DWBA through a number of numerical experiments. Our object is an infinitely long homogeneous cylinder in a zero background - the scattering amplitude from which can be calculated analytically for a plane wave of perpendicular incidence (van de Hulst (1957)).

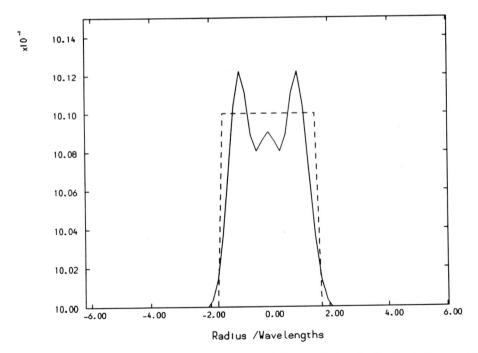

Figure 1: The Born reconstruction of an infinitly long cylinder of refractive index $n = 1.01$ and radius $a = 1.5\lambda$ in a background of refractive index $n_o = 1.0$. For this cylinder $| kVa | = 0.09$ and a fairly good reconstruction of the cylinder is obtained.

378

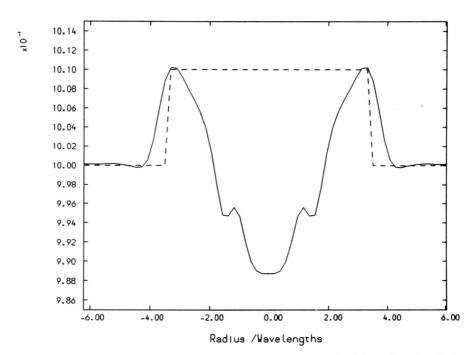

Figure 2: The reconstruction of a cylinder, still with a refractive index of $n = 1.01$, but now with a radius of $a = 3.5\lambda$ so that $|kVa| = 0.2$. The reconstruction is now poor, but recognisable as the Born reconstruction of a cylinder.

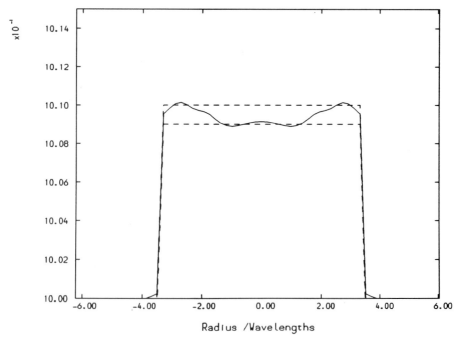

Figure 3: A DWBA reconstruction of the cylinder taking V_1 to be a cylinder of radius 3.5λ and of refractive index 1.009, so that $|kV_2a| = 0.02$. The reconstruction of V_2 is similar to Born reconstructions of a cylinder and the reconstruction of the total potential is considerably improved over that in figure 2.

The reconstruction of a cylinder under the Born approximation when equation (6) is satisfied is shown in figure 1,we obtain a good estimate of the object with the type of distortion typical of Born reconstructions (Zapolowski $et\ al$ (1985)). If we now increase the size of the cylinder so that the Born approximation is no longer valid we obtain a poor estimate as seen in figure 2. To try and obtain a better estimate we use the DWBA. We guess the object is a cylinder, estimate it's magnitude and take this as V_1, actually underestimating it such that V_2 satisfies (17). For V_1 we calculate f_1 and hence obtain $\tilde{V}_1$. Calculating f_2 and reconstructing U using Born approximation proceedures we calculate $\tilde{V}_2$ using equation (25) and hence $\tilde{V} = V_1 + \tilde{V}_2$ as shown in figure 3. As we see the reconstruction compared to figure 2 is much improved.

5. CONCLUSIONS

We have discussed how the limitations of the Born and Rytov approximations might be extended by a variety of techniques, particularly renormalisation and distorted-wave methods.

In detail we have considered a distorted-wave method, and in particular the DWBA and DWRA for direct and inverse scattering. Both of these techniques relax the strict condition for the Born and Rytov approximations from the entire potential to only a perturbation about a known structure. The distorted-wave approach to inverse scattering is more appropiate in many applications than the usual approach. For example - in a medical/industrial diagnostic role the basic form of the object will be known and it will be deviations from this that are frequently required. While the Born or Rytov approximations will estimate V under conditions which are doubtless violated, the DWBA and DWRA will estimate only V_2 under conditions much more likely to be fulfilled. We have also illustrated the DWBA by a number of numerical examples. Experimental work to test the algorithms developed and study their limitations using ultrasound and microwave scattering is currently being undertaken.

ACKNOWLEDGEMENTS

The authors are grateful for financial support from SERC and BP (Sunbury-on-Thames).

REFERENCES

Adams M.F. and Anderson A.P. (1982) Synthetic aperture tomographic (SAT) imaging for microwave diagnostics. Proc. Inst. Electr. Eng. Part 8 **129** pp.83-88.

Azimi M. and Kak A.C. (1983) Distortion in diffraction tomography caused by multiple scattering. IEEE **MI-2** pp.176-195.

Devaney A.J. (1982) Inversion formula for inverse scattering within the Born approximation. Opt. Lett. **7** pp.111-112.

Devaney A.J. (1985) Diffraction Tomography. W.-M. Boerner $et\ al$ (eds.),Inverse Methods in Electromagnetic Imaging, Part 2 pp.1107-1135. D. Reidel Publ. Co.

Devaney A.J. and Oristaglio M.L. (1983) Inversion procedure for inverse scattering within the distorted-wave Born approximataion. Phys. Rev. Letts. **51** pp.237-240.

Fiddy M.A. (1986) Inversion of optical scattered field data. J. Phys. D: Appl. Phys. **19** pp.301-317.

Jaggard D.L. and Kim Y. (1985) Accurate one-dimensional inverse scattering using a non-linear renormalising technique. J. Opt. Soc. Amer. A **2** pp.1922-1930.

Kaveh M., Soumekh M. and Greenleaf J.E. (1984) Signal processing for diffraction tomography. IEEE **SU-31** pp.230-239.

Kay and Moses (1983) Inverse scattering papers: 1955-1963. Math Sci, Brookline, Mass.

Kim K. and Wolf E. (1986) Non-radiating monochromatic sources and their fields. Opt. Comm. **59** pp.1-6.

Ladouceur H.D. and Jordan A.K. (1985) Renormalisation of inverse scattering theory for inhomogeneous dielectrics. J. Opt. Soc. Amer. A **2** pp.1916-1921.

Langenberg K.J. and Schmitz V. (1985) Generalized tomography as a unified approach to linear inverse scattering. Berkhout A.J. *et al* (eds.). Acoustical Imaging Vol. 14. pp.283-294.

Nayfeh M. (1973) Perturbation methods. J. Wiley and sons. Springer-Verlag. New York.

Newton R.G. (1982) Scattering theory of waves and particles. Springer-Verlag, New York.

Pan S.X. and Kak A.C. (1983) A computational study of reconstruction algorithms for diffraction tomography: interpolation versus filtered backpropagation. IEEE **ASSP-31** pp.1262-1275.

Slaney M., Kak A.C. and Larsen L.E. (1984) Limitations of imaging with first order diffraction tomography. IEEE **MTT-32** pp. 860-873.

Soumekh M., Kaveh M. and Soumekh B. Theoretical and numerical results on scattering from soft cylinders with implications for diffraction tomography. Berkhout A.J. *et al* (eds.). Acoustical Imaging Vol. 14 pp.741-748.

Taylor R.G. (1972) Scattering theory. John Wiley and Sons. New York.

Tatarski V.I. and Gertenstein M.E. (1962) Propagation of waves in a medium with strong fluctuations in the refractive index. Soviet Physics. J.E.T.P. **17** pp.458-463.

Tomikawa Y., Iwase Y. Arita K. and Yamada H. (1986) Non-destructive inspection of a wooden pole using ultrasound computed tomogrpahy. **UFFC-33** pp.354-357.

van de Hulst H.C. (1957) Light scattering by small particles. John Wiley and sons. New York.

Wolf E. (1969) Three-dimensional structure determination of semi-transparent objects from holographic data. Opt. Comm. **1** pp.153-156.

Wombell R.J. and Fiddy M.A. (1987a) Inverse scattering within the distorted-wave Born approximation. Submitted to Inverse Problems.

Wombell R.J. and Fiddy M.A. (1987b) A distorted-wave approach to direct and inverse scattering with the Rytov approximation. In preparation.

Zapolowski L., Leeman S. and Fiddy M.A. (1985) Image reconstruction fidelity using the Born and Rytov approximations. Berkhout *et al* (eds.), Acoustical Imaging Vol. 14. pp.295-304.

SURFACE IMAGING VIA WAVE EQUATION INVERSION

Mehrdad Soumekh

Department of Electrical and Computer Engineering
State University of New York at Buffalo
Amherst, NY 14260

ABSTRACT

This paper examines the consequences of the inversion of refelcted waves from an object based on the Helmholtz wave equation model; the primary wave is a multi-frequency source and the object/transmitter/receiver geometry is fixed. We show that a parametric reconstruction in this inverse problem is not possible; this is true for both the Born and Rytov inversions. However, the available reflected data can be used to accurately reconstruct the surface (shape) of objects. Reconstruction results for penetrable and non-penetrable objects are presented.

INTRODUCTION

Inversion of reflected data arises in many imaging problems of diagnostic medicine, remote sensing and industrial inspection. There are two common inversion schemes used in these reflection imaging systems. One scheme, pulse-echo, is based on an object/wave interaction model that neglects diffraction effects (e.g., B-scan) [2]. The second inversion scheme is based on the inhomogeneous wave equation model; this scheme assumes a particular geometry/property for the inhomogeneous object under study (e.g., the object function is a separable function) and/or requires processing the reflected data through computationally intensive iterations (e.g., migration) [11].

With the success of diffraction tomography (inhomogeneous wave equation inversion) in the transmission-mode [1], it was natural to apply similar concepts in the reflection-mode (e.g., [7]). In our perliminary work in reflection diffraction tomography, we utilize a multi-frequency plane wave as the primary wave. Our initial objective was to obtain *parametric* reconstructions from the reflected waves comparable to the ones obtained from the transmitted waves. However, our study indicated that three factors (inversion model, frequency dependence of the object's parameters and lack of low frequency data) prevented us to directly obtain such tomograms.

In spite of these difficulties, the wave equation is the linear (though shift-varying) model that best describes the interaction between the primary wave and the test object. Thus, this model provides a better explanation for the physical phenomenon of scattering in inhomogeneous media than the pulse-echo models. In this paer, we show that by solving the wave equation inversion, based on measurements of the reflected waves, one can accurately reconstruct the surface (boundary) of homogeneous structures that a test object is composed of. This was achieved through defining a synthetic object function; this function was related to not only the object parameters but also the scattered wave. In this case, we show that this synthetic function carries boundary structure information.

INVERSE PROBLEM

Fig. 1 shows the geometry of the imaging system in the spatial domain. The test object is an inhomogeneous isotropic viscoelastic fluid or solid structure embedded in a homogeneous fluid. We denote the speed of sound in this medium by c_o. The incident multi-frequency plane wave is traveling in the positive x-direction. Thus, the temporal Fourier transform of the incident field at frequency ω is $exp(-jkx)$ (without loss of generality, we assume that the amplitude of the field is one at this frequency), where $k = \frac{\omega}{c_o}$ is the wavenumber. The receiver line resides in the surrounding homogeneous fluid, i.e., outside the inhomogeneous object. We assume that the distance of the origin from the receiver line, i.e., OR, is equal to R and the angle between OR and the x-axis is θ. Note that the reflection data corresponds to the case $\theta = \pi$. However, we formulate the problem for an arbitrary θ that may vary from $-\pi$ to π.

The general model used in our inverse problem for both fluid and solid objects is the following scalar differential equation:

$$[\nabla^2 + k^2]\phi_s(x,y,\omega) = e(x,y,\omega) \cdot exp(-jkx) \qquad (1),$$

where ∇^2 is the Laplacian operator, $\phi_s(x,y,\omega)$ is the temporal Fourier transform of the scalar *scattered wave* and $e(x,y,\omega)$ is the source function that is to be reconstructed.

When the object can be modeled as an inhomogeneous fluid, the incident plane wave does not experience mode change inside the object; i.e., shear waves are not generated inside the test object [3,8]. In this case, $\phi_s(x,y,\omega)$, throughout the spatial domain, is the differnce between the total pressure wave with the object present, $\phi(x,y,\omega)$, and the pressure wave with the object removed, i.e., $exp(-jkx)$. Moreover, the total wave satisfies the following operator [8]:

$$[\nabla^2 + \frac{\omega^2}{c_e^2(x,y,\omega)} + \frac{\nabla \rho_e(x,y,\omega)}{\rho_e(x,y,\omega)} \cdot \nabla] \phi(x,y,\omega) = 0 \qquad (2),$$

where c_e, effective speed, and ρ_e, effective density, are frequency dependent complex quantities that represent the distributions of sound speed, sound attenuation and density in the spatial domain. In this case, the source function is defined by the following:

$$e(x,y,\omega) \cdot exp(-jkx) \equiv k^2[1 - \frac{c_o^2}{c_e^2(x,y,\omega)}]\phi(x,y,\omega) + \frac{\nabla \rho_e(x,y,\omega)}{\rho_e(x,y,\omega)} \cdot \nabla \phi(x,y,\omega) \qquad (2a).$$

Note that $e(x,y,\omega)$ is composed of the pressure wave and its derivatives modulated with the variations of effective speed and density in the spatial domain.

The interaction of a sound field with an inhomogeneous soild object can be modeled via the constitutive equation and equation of motion [3,9]. These equations relate strain and stress tensors to the two local Lame' constants. Using these tensor differential equations, Mueller [3] derived the model of (1) in terms of a scalar scattered wave function and a scalar source function. In this model, the scattered wave function, $\phi_s(x,y,\omega)$, is a complex function of the dilatation and the variations of the object's elastic parameters [3]. However, at the location of the receiver array where the elastic parameters' variations are

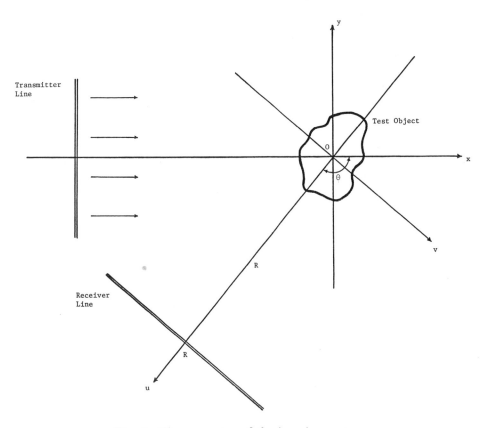

Fig. 1. The geometry of the imaging system.

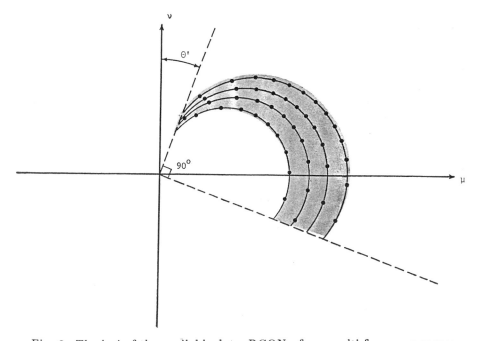

Fig. 2. The loci of the available data, RCONs, for a multi-frequency source.

all equal to zero, $\phi_s(x,y,\omega)$ is equal to the difference between the observed pressure field with the object present and the observed pressure field with the object removed [3]. The source function $e(x,y,\omega)$ is composed of the scattered wave and its partial derivatives modulated with the variations of the object's elastic parameters. One can also derive an expression similar to (2a) for this source function. This requires defining numerous wave functions and object parameters [3,12] and is not presented in our discussion.

The source function for both fluid and solid models possess a common property that is crucial in the formulation of the inverse problem: Due to the fact that the source function is a modulated form of the wave function and the variations of the elastic parameters, the support of $e(x,y,\omega)$ (call it D) is the intersection of the test object's support and the wave fuction's support. We will discuss the properties of $e(x,y,\omega)$ and its relation to the boundary information in the next section.

We denote $E(\mu,\nu,\omega) \equiv F.T._{(x,y)}[e(x,y,\omega)]$, and the spatial Fourier transform of the observed scattered wave with respect to v by $\chi(\beta,\omega)$. Then, it can be shown that

$$E(\mu,\nu,\omega) = 4\pi j\sqrt{k^2 - \beta^2} \cdot exp(j\sqrt{k^2 - \beta^2}R) \cdot \chi(\beta,\omega) \quad for \ |\beta| < k \qquad (3).$$

where

$$\mu(\beta,\omega) = k - \sqrt{k^2 - \beta^2}\,Cos\theta + \beta\,Sin\theta$$
$$\nu(\beta,\omega) = \sqrt{k^2 - \beta^2}\,Sin\theta + \beta\,Cos\theta \qquad (4).$$

Note that for $|\beta| < k$ the pair (μ,ν) trace a line contour in the spatial frequency plane. From (4), it can be seen that this contour, called Reconstruction CONtour (RCON), is in the form of a semicircle. A collection of these contours for different values of k is depicted in Fig. 2.

By employing single scattering assumption, the above results are used in the Born inversion [1]; a similar inversion also exists for the Rytov approximation [1]. We had presented theoretical and numerical results indicating that in the transmission-mode wave equation inversion, the Born approximation performed poorly while the Rytov approximation provided good results [5,6]. However, our study showed that in the reflection diffraction tomography, both the Rytov and Born approximations yielded erroneous results; the Born approximation failed when the test object's size was chosen to be larger than the wavelength of the impinging wave * , and the Rytov error (a term related to the derivatives of the scattered wave) dominated the desired signal on the locus of the available data. The empirical study that led us to these conclusions was similar to the one performed in [5] for the transmission diffraction tomography.

SYNTHETIC OBJECT FUNCTION

Equation (3) provides information regarding a freqeuncy dependent function on a semicircular contour in the spatial frequency domain. In this section, we show that the collection of this form of data on RCONs produces a *synthetic* object function that carries boundary structure information. We define the object function in the spatial frequency domain by

$$F(\mu,\nu) = E(\mu,\nu,\omega) \qquad (5),$$

where the pair (μ,ν) are given in (4). Our purpose is to examine the properties of the object function, $f(x,y)$, in the spatial domain.

---

* The test object used in [7] was a disk of radius one wavelength.

We can assume, in good approximation, that the variations of sound speed and density over the range of the excitation frequencies are negligible. However, the attenuation coefficient varies with the temporal frequency, i.e., ω (or k) [8,10]. In this case, it can be shown that

$$c_e(x,y,\omega) = \frac{c(x,y)}{\sqrt{1 + j\frac{\gamma(x,y)}{\omega\rho(x,y)}}}$$

and

$$\rho_e(x,y,\omega) = \rho(x,y) \cdot [1 + j\frac{\gamma(x,y)}{\omega\rho(x,y)}]$$

where $c(x,y)$, $\gamma(x,y)$ and $\rho(x,y)$ represent the distributions of the sound speed, flow resistance and density, respectively. We denote the indicator function for the test object domain, D, by $i(x,y)$; i.e., $i(x,y) = 1$ for $(x,y) \in D$, and zero otherwise. To simplify our analysis, we assume that the test object is a homogeneous structure. Thus, we have

$$c(x,y) = c_o + (c' - c_o) \cdot i(x,y),$$
$$\rho(x,y) = \rho_o + (\rho' - \rho_o) \cdot i(x,y),$$
$$and \quad \gamma(x,y) = \gamma_o + (\gamma' - \gamma_o) \cdot i(x,y),$$

where c_o, ρ_o and $\gamma_o \approx 0$ (negligible loss in the medium) are the medium parameters, and c', ρ' and γ' are the object parameters. Consider the case of fluid test objects. We assume that the mechanical parameters of the test object vary slightly from the surrounding medium parameters. The total wave is predominately a plane wave traveling in the positive x-direction. Hence, the total wave in D can be approximated by

$$\phi(x,y,\omega) \approx exp(-jk'x) + first \ order \ perturbation \ terms \quad for \ (x,y) \in D \quad (6),$$

where $k' = \frac{\omega}{c'}$. The first order perturbation terms include sound attenuation and scattering effects in D (note that in the Born approximation, the total wave is approximated by the incident wave, i.e., $exp(-jkx)$, plus first order perturbation terms). Substituting (6) in (2a) and neglecting the second order perturbation terms yields

$$e(x,y,\omega) \approx \left[k^2[1 - \frac{c_o^2}{c_e^2(x,y,\omega)}] - j\frac{k'}{\rho_e(x,y,\omega)} \cdot \frac{\partial \rho_e(x,y,\omega)}{\partial x}\right] \cdot exp[-j(k'-k)x] \quad (7),$$

for $(x,y) \in D$, and zero otherwise. Using the expressions for c_e and ρ_e in (7) and after some rearrangements and approximations one obtains

$$e(x,y,\omega) \approx [(\Delta_c k^2 + j\Delta_\gamma k)i(x,y) - j(\Delta_\rho k' + j\Delta_\gamma)\frac{\partial i(x,y)}{\partial x}] \cdot exp[-j(k'-k)x] \quad (8),$$

where $\Delta_c \equiv 1 - (\frac{c_o}{c'})^2$, $\Delta_\rho \equiv ln(\frac{\rho'}{\rho_o})$, and $\Delta_\gamma \equiv \frac{\gamma'-\gamma_o}{c_o\rho_o}$.

Taking the spatial Fourier transform of both sides of (8) and using (5) yields

$$F(\mu,\nu) \approx [\Delta_c k^2 + j\Delta_\gamma k + \Delta_\rho \mu k' + j\Delta_\gamma \mu] \cdot I(\mu + k' - k, \nu) \quad (9).$$

Consider the right side of (9). The term inside [ ] is composed of powers of μ, k and k'. The powers of μ correspond to highpass filtering in the spatial frequency domain. It can be shown that $k = \frac{\mu^2 + \nu^2}{2\mu}$. In this case, powers of k and k' also represent highpass filtering for the reflected data (see Fig. 2). Thus, the term inside $|\ |$ represents a highpass filter in the spatial frequency (wavenumber) domain on I. This results in edge enhancement of the indicator function in the spatial domain and reconstruction of the surface structure of the object. The presence of $\mu + k' - k$ in $I(\cdot)$ on the right side of (9) results in a linear wavenumber compression. If k' is close to k over the range of the excitation frequencies, this form of compression can be assumed to be a shift in the wavenumber domain. Hence, $|f(x,y)|$ is a function that carries information about the surface of the test object.

RECONSTRUCTION FROM ONE-QUADRANT DATA

We now examine the coverage of the data obtained in the spatial frequency domain. We can make the following observations from Fig. 2. The RCONs, theoretically, cover an area equal to one quadrant in the (μ, ν) domain. For $|\theta| \geq 90^o$ (this includes the reflection-mode), the area of the available data is composed of a single set in the (μ, ν) domain. It can be shown that this domain is equivalent to the first quadrant in the (μ, ν) domain when it is rotated by (see Fig. 2)

$$\theta' = \frac{2\theta - \pi}{4} \tag{10}.$$

We can define the function that can be reconstructed from the one-quadrant reflection data by

$$F_1(\mu, \nu) \equiv F(\mu, \nu) \cdot G_\theta(\mu, \nu) \tag{11},$$

where

$$G_\theta(\mu, \nu) \equiv U(\mu Cos\theta' - \nu Sin\theta') \cdot U(\mu Sin\theta' + \nu Cos\theta') \tag{12},$$

and U is the step function; i.e., $U(a) = 1$ for $a \geq 0$ and zero otherwise. In practice, the incident multi-frequency ultrasonic source is a bandpass signal. In addition, the transmitting and receiving transducers have finite bandwidths. Thus, in practice the data can only be obtained over a midband values of k. We define the function that may be reconstructed from this finite range of the available data in the k-domain by

$$F_2(\mu, \nu) \equiv F_1(\mu, \nu) \cdot H_\theta(\mu, \nu) \tag{13},$$

where H_θ is an indicator function that its support varies with θ and the range of the available k values. The support of H_θ in Fig. 2 is indicated by the shaded region.

The effect of variations of θ on G_θ is simply the rotation of the (μ, ν) domain by θ' (see (12)). However, as θ varies, H_θ takes on different shapes. For $\theta = 180^o$, the effect of H_θ on f_1 is similar to bandpass filtering. This reuslts in edge enhancement and loss of uniform structure for $f_1(x,y)$. Although our results have also indicated the loss of uniform structure for $f_1(x,y)$ due to H_θ for other values of θ, the effect of H_θ is too complicated to study and cannot be generalized. In this section, we study the information that might be obtained through the function $f_1(x,y)$.

Consider the case of $\theta = 90^o$, $(\theta' = 0)$. From (11) and (12) we can write

$$F_1(\mu, \nu) = F(\mu, \nu) \cdot U(\mu) \cdot U(\nu) \tag{14}.$$

Taking the inverse spatial Fourier transform of both sides of (14) and using the fact that $Inverse F.T._{(a)}[U(a)] = \frac{\delta(b)}{2} + \frac{j}{2\pi b}$ yields

$$f_1(x,y) = f(x,y) * [\frac{\delta(x)}{2} + \frac{j}{2\pi x}] * [\frac{\delta(y)}{2} + \frac{j}{2\pi y}]$$

$$= \frac{f(x,y)}{4} + jf(x,y) * [\frac{1}{2\pi x} + \frac{1}{2\pi y} - \frac{1}{4\pi x} * \frac{1}{4\pi y}] \qquad (15).$$

We denote the Hilbert transforms of $f(x,y)$ in the x-domain and y-domain by $f_x(x,y)$ and $f_y(x,y)$, respectively. Hence, (15) can be rewritten as follows:

$$f_1(x,y) = .25 f(x,y) + .5j[f_x(x,y) + f_y(x,y)] - .25 f_{xy}(x,y) \qquad (16).$$

To facilitate the study, we assume that the test object is a uniform disk of radius r centered at the origin in the spatial domain; i.e., $f(x,y) = 1$ for $x^2 + y^2 \le r$ and zero otherwise.

The Hilbert transform of a pulse function has a positive spike at the positive edge of the pulse and a negative spike at the negative edge of the pulse (see Fig. 3). For a constant value of x, $f(x,y)$ is a pulse function in the y-domain, and for a constant value of y, $f(x,y)$ is a pulse function in the x-domain. With the help of these facts, it can be shown that at the boundary of the disk, i.e., the locus of $x^2 + y^2 = r^2$, we have

i) $\quad f_x(x,y) \cdot f_y(x,y) > 0 \quad if \quad \dfrac{\partial f(x,y)}{\partial x} \cdot \dfrac{\partial f(x,y)}{\partial y} > 0$

this occurs in the first and third quadrants for a centered disk where the tangent line at the surface of the object has negative slope (see Fig. 4); and

ii) $\quad f_x(x,y) \cdot f_y(x,y) < 0 \quad if \quad \dfrac{\partial f(x,y)}{\partial x} \cdot \dfrac{\partial f(x,y)}{\partial y} < 0$

this occurs in the second and fourth quadrants for a centered disk where the tangent line at the surface of the object has positive slope (see Fig. 4).

In other words, in the first and third quadrants of the spatial domain, $f_x(x,y)$ and $f_y(x,y)$ have same polarities and their sum in (16) is constructive. However, in the second and fourth quadrants of the spatial domain, $f_x(x,y)$ and $f_y(x,y)$ have opposite polarities and their sum in (16) is destructive. The last term in (16), i.e., $f_{xy}(x,y)$, can be shown to produce a combination of negative and positive spikes at the boundary of the object.

From the above results one can conclude that $|f_1(x,y)|$ resembles a smeared version of $f(x,y)$ with enhanced edges in the first and third quadrants when the object function is a unifrom disk centered at the origin. Thus, $|f_1(x,y)|$ is a structure on the disk of radius r with its highest energy around the points $x = y = r/\sqrt{2}$ and $x = y = -r/\sqrt{2}$. In general, the reflections from a "surface element" could be detected at the receiver line at $\theta = 90°$ if the tangent line at that boundary point has negative slope. (This is analogous to the reflection from a flat mirror.) Thus, for any arbitrarily shaped object function that is composed of several "uniform" structures, the first quadrant data in the (μ, ν) domain results in the enhancement of the edges where the first partial derivatives of $f(x,y)$ have same polarities. Also, at the boundary points where the first partial derivatives of $f(x,y)$ have opposite polarities the edges lose some energy.

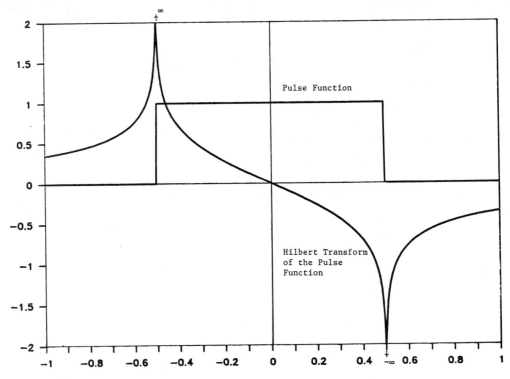

Fig. 3. A pulse function and its Hilbert transform.

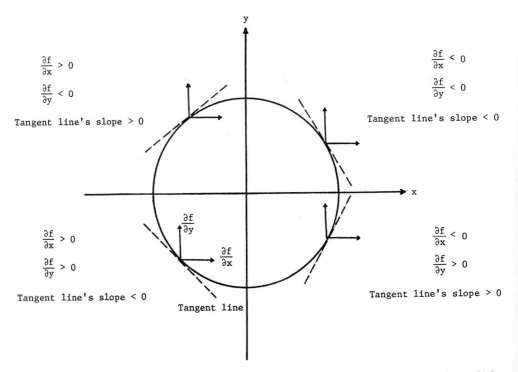

Fig. 4. The partial derivatives and the tangent line at the boundary of a uniform disk.

390

When the first quadrant data in the (μ, ν) domain is rotated by θ', the above principles hold for the $f(x, y)$ rotated by θ'. Thus, we can make the following generalizations. We denote the rotational transformation of the object function by θ' with

$$f_\theta(w, z) \equiv f(x, y)$$

where

$$\begin{bmatrix} x \\ y \end{bmatrix} = \begin{bmatrix} Cos\theta' & Sin\theta' \\ -Sin\theta' & Cos\theta' \end{bmatrix} \cdot \begin{bmatrix} w \\ z \end{bmatrix}$$

Thus, when the receiver line is located at θ, the one-quadrant data produces an image, $|f_1(x, y)|$, with high energy at the surface points where

$$\frac{\partial f_\theta(w, z)}{\partial w} \cdot \frac{\partial f_\theta(w, z)}{\partial z} > 0 \qquad (17).$$

Note that in the above analysis of the one-quadrant data we assumed that $f(x, y)$ is a uniform function. However, as we indicated earlier, $f(x, y)$ is a highpass version of $i(x, y)$ which is a uniform function in D. But both the highpass filter and G_θ represent linear operators and they commute. Hence, $|f_1(x, y)|$ provides information about the surface points that satisfy (17) (due to the one-quadrant data) and are enhanced by highpass filtering (see (9)).

There are two aspects of this imaging problem that are not discussed here due to the lack of space. One problem is associated with the finite width of the receiver aperture; this can be shown to limit the depth of reconstruction. The other problem involves reconstructing the object function from the available unevenly-spaced sampled data; this is done by an interpolation scheme known as Unified Fourier Reconstruction [4].

RESULTS

In this section, we examine the merits of the surface imaging scheme using uniform disks. The parameters used in this study are as follows: the test object is a disk of one centimeter radius, centered at the origin in the spatial domain; the sound speed in the homogeneous medium that the object in which is embedded is equal to $c_o = 150000$ cm/sec; the source has 32 frequencies from .3 MHz (wavelength = .5 cm) to .9 MHz (wavelength = .167 cm); the distance of the origin from the receiver line is equal to R = 5 cm; the size of the receiver line is 32 cm (extended in $[-16, 16]$ in the v-domain) and there are 256 equally-spaced elements on the receiver; the reconstruction is made on a 4 cm x 4 cm (128 pixels x 128 pixels) uniform grid; and the reconstructed image has 94 gray scale levels.

Fig. 5 exemplifies a penetrable object with speed ratio = 1.05, density ratio = .98, and attenuation coefficient=0. Fig. 5a shows the reconstruction for $\theta = 180°$ $(\theta' = 45°)$. Note that the sound waves were capable of penetrating the test object, thus producing reflection at the side of the disk that does not directly face the transmitting transducer. Fig. 8b depicts the reconstructions for $\theta = 90°$ $(\theta' = 0°)$. The next example is for a non-penetrable disk with speed ratio = 3, density ratio = 10, and attenuation coefficient=$.2\omega$ (ω is in MHz), shown in Fig. 6. In this case, the ultrasonic waves cannot penetrate the object. Thus, only the surface of the object that faces the transmitting transducer can be reconstructed.

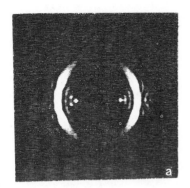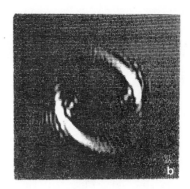

Fig. 5. Reconstructions of a 1 cm radius penetrable disk with speed ratio=1.05, density ratio=.98 and attenuation coefficient=0. : a) $\theta = 180°$, and b) $\theta = 90°$.

Fig. 6. Reconstructions of a 1 cm radius non-penetrable disk with speed ratio=3, density ratio=10, and attenuation coefficient=.2ω : a) $\theta = 180°$, and b) $\theta = 90°$.

REFERENCES

1. M. Kaveh, R.K. Mueller, and R.D. Iverson, "Ultrasonic Tomography Based on Perturbation Solutions of the Wave Equation," Comp. Graph. Image Proces., 1979.

2. A. Macovski, "Ultrasonic Imaging Using Arrays," Proc. IEEE, 67:484, April 1979.

3. R.K. Mueller, "Diffraction Tomography I: The Wave Equation," Ultrasonic Imaging, Vol. 2, pp. 213-222, 1980.

4. M. Soumekh and M. Kaveh, "Image Reconstruction from Frequency Domain Data on Arbitrary Contours," Proc. ICASSP '84.

5. M. Soumekh and M. Kaveh, "A Theoretical Study of Model Approximation Errors in Diffraction Tomography," IEEE Trans. on UFFC, January 1986.

6. M. Kaveh, M. Soumekh and R.K. Mueller, "A Comparison of Born and Rytov Approximations in Acoustic Tomography," Acoustical Imaging, 10, Plenum, 1981.

7. B. Roberts and A. Kak, "Reflection Mode Diffraction Tomography," ULtrasonic Imaging, 7:300, 1985.

8. P.M. Morse and K.V. Ingard, *Theoretical Acoustics*, McGraw-Hill, 1968.

9. B.A. Auld, *Acousitc Fields and Waves in Solids*, Wiley, 1973.

10. C.M. Sehgal and J.F. Greenleaf, "Diffraction of Ultrasound by Soft Tissues: The inhomogeneous Continuous Model," Acoustical Imaging, Vol. 13, Plenum, 1984.

11. E.A. Robinson, "Image Reconstruction in Exploration Geophysics," IEEE Trans. Sonics Ultra., pp. 259-270, July 1984.

12. Ben Soumekh, "Theoretical Background of the Measured Data in a Scanning Laser Acoustic Microscope," M.S. Thesis, University of Minnesota, August 1985.

ACCURACY IN PHASE TOMOGRAPHIC RECONSTRUCTION

A. Markiewicz, J. Berry and H. W. Jones

Engineering Physics Department, Technical University of Nova Scotia, P.O. Box 1000, Halifax, Nova Scotia, B3J 2X4

Abstract

A series of studies have been conducted on numerical and gelatine phantoms. These phantoms were used to obtain data and test reconstruction techniques of 31 x 31 pixels by phase tomograms in which the value of the $\{1 - [c(r)^2/c_0^2]\}$ varied from 0 - 0.1 approximately.

The variation in the accuracy of reconstruction with different reconstruction and data gathering techniques is described in some detail. The difficulties in obtaining satisfactory data for such error assessments is commented upon.

Some conclusions on preferred methods of treating and obtaining data and deriving reconstructions will be presented.

Introduction

We have for some time been concerned with the accurate reconstruction of phase tomograms of soft human tissue based on a 31 x 31 pixel array. It is hoped in the longer term to use such tomograms to infer temperature changes in tissue. In order to explore the possibilities of such a technique we have followed a program of work which has allowed us to investigate the practical and theoretical barriers to the achievement of sufficient accuracy. In earlier papers[1,2] we have described our initial work on this problem. In those papers we set ourselves the task of determining the time of flight of a pulse of ultrasound between two transducers so that the reproducibility of the results over a 30 cm path length was +/- 2 nanoseconds (rms). Clearly, this was a first step. We were then in the position to determine the velocities of sound to an accuracy of 3×10^{-2} m/sec provided that all the usual conditions for such experiments are satisfied particularly the stability of temperature of the sound path involved and the supposition of zero flow velocities and that zero "end effects" applied to the transducers. This accuracy is quite sufficient for our purposes. If we can determine changes in velocity with temperatures to about 1 m/sec or better in the pixel array then we would appear to have put ourselves into a position of determining temperature changes of about 1° C. This we are told is a degree of accuracy which is sufficient for many medical purposes.

Two types of experiments are possible and each has its difficulties. First, we can make phantoms in which the geometry and velocities are sufficiently well known that we can obtain data for reconstruction. Second, we can use numerical phantoms and obtain data for reconstruction. It appears to us that both approaches are necessary. The need for experimental data is the simple need to stay with reality. The need for numerical phantoms arises from reasons of economy, the precision which can be obtained (in some respects) and the generation of an understanding of the propagation process. We shall describe our endeavors with both types of phantoms.

The problems of reconstruction are well understood physically. Unfortunately, this means that the basic processes can be described in mathematical outline by equations which cannot, in general, be solved quickly and with precision. The reason for this situation has been the subject of comment elsewhere[3,4]. The fact that we are obliged, in tomography, to assign a value which characterizes the velocity for the whole pixel, over which the velocity usually is varying, creates our first problem. No progress is possible unless the choice of the characteristic pixel quantity is judicious to the degree which is required. This is a problem which has exercised us considerably and we will present some discussion on the point. Given that we can find a value for the velocity in each pixel which reasonably describes the situation we must accept that we are obliged to obtain progressively more accurate values by some scheme of successive approximations. Such a scheme was proposed in an earlier paper[2]. We can demonstrate that initial solutions may not be particularly accurate. Once we have obtained a first approximation for the value in the tomogram we face a new problem. We have a set of values for velocity which are spatially fixed on a set of x-y coordinates. We have to adjust progressively the whole set to obtain a new set which provides the most accurate fit to the data. There are a variety of ways of obtaining such a result presupposing we can beg such questions as the uniqueness of the solution. Put another way, utility and speed in achieving a practical result is the only factor which is of concern to us. We shall discuss some of the approaches to obtaining solutions which might satisfy our requirements. A difficulty which always remains arises from the empirical nature of our approach. We are obtaining solutions for particular geometries. It might be that other geometries do not lend themselves to the methods we are developing, consequently we cannot claim that we are developing a universally applicable approach, perhaps not even to the circumstances that we are concerned with. Only experience with a wide range of real specimens will demonstrate the general validity of this work.

Finally, we should state the basic physical factors which guide our thinking. It was noted in the previous paper[2] that the quantity:

$$(1 - (c^2(r)/c_\emptyset^2))$$

(where cø is the background sound velocity and c(r) the velocity
at the position r) is significant in obtaining solutions of the
wave equation. We postulated that this had, typically, for our
circumstances a value of about Ø.1 (maximum). If we concern
ourselves with ray tracing techniques then the deviation of the
ray by refraction is given by:

$$\delta\theta = (\mu - 1)\tan\theta \qquad\qquad\qquad (1)$$

where μ= refractive index, $\delta\theta$ is the deviation of the ray and θ
the angle of incidence. Given that μ is in the region of 1.Ø5
it follows that

$$\delta\theta = Ø.Ø5 \tan\theta \quad ,\text{for example:}$$

Table 1

θ	Øo	1Øo	3Øo	6Øo	8Øo
$\delta\theta$	Øo	Ø.ØØ9o	Ø.Ø3o	Ø.Ø9o	Ø.28o

If the velocity of the sound is varying with a spatial
coordinate then the variant of equation (1) is required which
leads to curved paths with essentially the same deviation. It
hardly needs emphasizing that the deviations are small and this
fact is of significance in our later considerations.

STUDIES WITH REAL PHANTOMS

The phantom shown in Fig. 1 was used for our studies.
Typically the error in the tomographic reconstruction, supposing
measured values for the water and the tomogram, are as shown in
Fig. 2. On an earlier occasion when we reported measurements we
observed that the gelatine parts of the phantom appeared to have
become softened by their immersion in the water. This factor
was recognized from "double humped" error distribution which was
observed for the pixel region relating to the gelatine edge, see
Fig. 3. If we assume that the edge softening caused the
velocity in the gelatine to be reduced by 5 m/sec then the error
distribution becomes that shown in Fig. 4. Attempts were made
to measure the velocity of sound in the edge region but, it was

Figure 1 Gelatine phantom

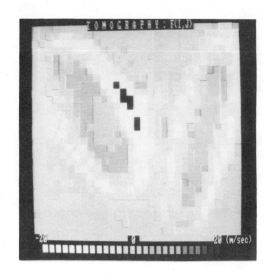

Figure 2. Reconstruction error

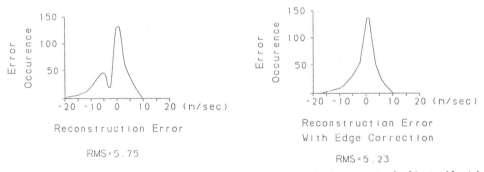

Figure 3 Error distribution   Figure 4 Corrected distribution

difficult to obtain the required accuracy because of the
problems of dimensional stability of the specimens. This led us
to a repeat of these experiments. New and accurately machined
molds for the gelatine were made and the velocity of sound in
the gelatine was determined as accurately as possible in several
parts of the sample of the gelatine. The gelatine phantom was
assembled and its geometry determined by means of a measuring
microscope. A bath of distilled water which was temperature
stabilized by immersion in a larger tank of water was used with
the gelatine to complete the phantom. First, however, a set of
tomographic data was collected for the distilled water alone.
It was shown that the error in the reconstruction was less than
0.3 m/sec (rms) based on comparison with direct velocity
measurements. Next the gelatine was immersed in the water and
the measurements repeated. Directly after the data had been
obtained the gelatine was removed from the water and allowed to
dry. The most difficult problem which remained was the
determination of the registration of the gelatine parts of the
phantom with the calculated pixel position which was used for
the error analysis. By trial and error both phantoms were
brought into close registration. This allowed an error analysis
to be made. Table 2 shows the errors observed for both
phantoms.

Table 2

Original Phantom	first reconstruction	first correction	second correction
total error	5.75 m/s	5.2 m/s	5.3 m/s
water only	4.65 m/s	4.37 m/s	4.46 m/s
center of gelatine and water	4.74 m/s	4.80 m/s	4.84 m/s
Edges	8.20 m/s	6.95 m/s	7.12 m/s
"Precision" Phantom total error	3.3 m/s	3.0 m/s	

No discrimination in error distribution (water, edge or center.)

It is to be observed that in the original phantom there was an overall error in the initial error determination which is approximately twice that which was determined for the second phantom. Given the errors due to the velocity variation at the edges of the gelatine in the original phantom, the data from the second phantom does not give results which are remarkably better. We are, of course, approaching something like the practical limits of accuracy for the determination of the velocity of sound both in the water and the gelatine. Our estimate for the worst case errors for these values, assuming well mixed gelatine and very little movement of the water is given below.

  i) velocity error in water   0.5 m/sec
  ii) velocity error in gelatine 2.0 m/sec

These errors arise from various causes, primarily the determining of the "end effects" in the transmitting and receiving transducers. Great care was taken to maintain a constant temperature both spatially and temporally. There was a stirring effect in the water due to the movement of the transducers (at a velocity of about 1.4 mm/sec). Because the geometry of the phantoms was well defined makes it unlikely that there was a large error on this account; an estimate of the error arising from this cause is +/- 1.5 m/sec maximum. Such errors would be variable over the tomogram.

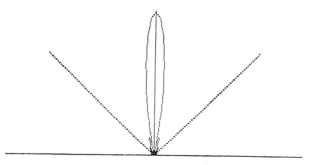

Figure 5. Beam pattern

Taking account of these errors and the fact that the assumption of planar data gathering (our transducers have a beam pattern which is cylindrically symmetric (as shown in Fig. 5) is not completely true. It appears that we cannot expect much better determination than that which has been obtained. Unfortunately, the diffractive correction (see first approximation, Table 2) is very small and it can hardly be said that it makes a significant difference to the results. Finally, it should be noted that our interest is in the changes of velocity which indicate temperature changes. We shall be reporting data from such experiments at a later date.

Numerical phantoms

The limitation of experimental phantoms naturally disposes us to the use of numerical phantoms, if only because it might be possible to explore the effects of diffraction and refraction a little more definitely because, in principle, we can reduce the inherent errors. The simplest phantom to use is one with maximum symmetry, consequently we chose that shown in Fig. 6. It might be objected that a phantom with some lack of symmetry would be more useful in some respects and this is a point of view which we accept. However, the economy and simplicity of the chosen phantom were important to us given the constraints upon our work.

The difficulties which arise in these phantoms is that of obtaining tomographic data which accurately accounts for diffraction, scatterings and to a lesser degree refraction. The problems with diffraction and scattering arise from the difficulties from obtaining accurate solutions to the wave equation in circumstances where the wave front may be curved and the apertures of the transducers are defined by the analytical functions which somewhat lack reality.

Initially we obtained sets of "straight line" data i.e. the sort of information which x-rays would provide. We used this data to investigate two factors. One was the importance to the accuracy of reconstruction of the number of sampling intervals (along R- see Fig. 6). In principle this requirement needs to be satisfied if we are to recognize all spatial frequency information in the tomogram.

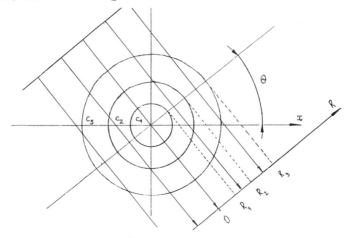

Figure 6. Numerical phantom

Second, it is convenient to use the time of flight, which in
practice is a histogram because of the receiver's aperture, to
provide backprojection using the most convenient reconstruction
intervals. These intervals can be obtained by linear or cubic
spline interpolation. Table 3 shows the backprojection
reconstruction error as a function of the number of angular
projections obtained and interpolation method for a sampling
interval ΔR which satisfies the Nyquist requirements.

Table 3

ΔR = 2.83 mm (according to Nyquist theory)

No. of projections	Backprojection only	Backprojection & linear int.	Backp. & cubic spline inter.
12.	6.04	8.57	3.7
24.	4.45	5.36	1.73
28.	4.21	4.92	1.43
32.	3.72	4.34	1.25
36.	3.55	4.04	1.03
40.	3.59	3.88	1.01
44.	3.46	3.64	0.96
48.*	3.24	3.55	0.09

* according to Nyquist theory (number of projections).

Table 4 shows similar data as function of sampling interval(ΔR).

Table 4

48 projections

	Backprojection average	Back & linear average	Back & cubic average
ΔR = 2.83	3.24	3.55	0.90
ΔR = 4.40	3.86	3.73	1.30
ΔR = 5.66	4.40	3.98	1.82

Further studies with this phantom concern the accuracy of
reconstruction of the image with diffractive and refractive
corrections. A variety of studies were undertaken and limited
comment is given on the results. Fig. 7 shows the error
distribution attained from refractive studies. In this case we
determined the "experimental" data by tracing rays and taking
account of the refraction which occurred between the transmitter
and receiver. We then reconstructed by a variety of methods.
First, by simple backprojection, then the use of the Rytov
approximation and finally by progressive correction (which will
be described later). The errors for the simple backprojection
are somewhat worse than those for the experimental phantom. The
errors reveal very clearly an edge effect. This effect arises
from the design of the phantom. The Fourier series for the
velocity change in the phantom contain components which cannot
be recognized by the reconstruction which is band-width limited
in its frequency response to $n\pi/x$ where n is half the pixel
array number and x is the pixel dimension. It is possible to

reduce these errors by choosing a phantom in which the velocity
variation has fewer components at the unwanted higher
frequencies. However, all artificial phantoms real and
numerical suffer from this difficulty to a lesser or greater
degree. It might be that it is appropriate to use a spatial
frequency filter to treat the original phantom i.e. to remove
the unwanted components and then to obtain an error
distribution. However, such procedure is effectively changing
the original data in a way in which is of doubtful utility.

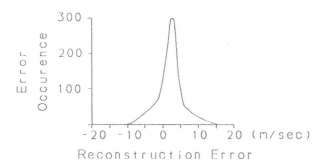

Figure 7. Refractive error distribution

It is not possible in this publication to list in detail
all the effects which have been observed. Some general
conclusion can be stated however. We find that when the best
account of the various factors are considered we obtain an error
which is marginally worse than those obtained with the
experiments with the real phantom. We also find that the
refractive and diffractive treatments reduced the observed
errors by marginal amounts.

We must conclude from these studies that it is difficult to
design experiments which, at least for pixel numbers of about
1000, provide a means of studying and reducing error in our
circumstances to less than about 2m/s. We notice that the
errors in the phantom which had the edge softened by soaking in
water were marginally worse than those for the phantom in which
the velocity was constant and very considerable care had been
taken to obtain the best accuracy with the equipment at our
disposal. This suggests to us that the gradation of velocity at
the edge of the phantom had its effect in reducing the observed
errors. In the lack of better evidence we must speculate that
the effect of velocity reduction was to reduce the spatial
frequency components in the phantom. A rough check on this was
done by finding the Fourier series for the error tomogram (Fig.
8) and reconstructing it by omitting frequencies above the
reconstruction bandwidth. Some reduction in the edge errors is
observed.

A Commentary on Reconstruction Methods

It can be supposed that any relatively quick iterative
routine which reduces the observed errors is of itself
justified. A method with two variants appear to us to be of
possible use.

Figure 8. DFT through error tomogram

First, we can obtain the initial reconstruction by the use of some convenient technique (for example the Rytov approximation or backprojection). At this point we have an array as shown in Fig. 9. We can fit to this array a two-dimensional Fourier series. The matrix of constants from this series contain the full information of the reconstruction up to the spatial frequency limit of the tomogram. We can use this to generate constant velocity contours within the reconstruction. The contours, Fig. 10, allow us to trace rays through the reconstruction and to build up sets of tomographic data to compare with the original experimental data. By progressive approximation the difference between the two sets of data can be reduced. This is a fairly substantial calculation and unfortunately takes longer than we would wish.

Second, we can obtain the same results less accurately by interpolation between several adjacent velocity values. We are not yet able to comment on the time saving and compare it with the loss of accuracy arising from this technique.

A difficulty remains with this technique--it is only applicable to phantoms or specimens in which the constant velocity contours enclose a reasonable number of pixels. An object with many small areas of differing velocities cannot be expected to provide accurate velocity contours.

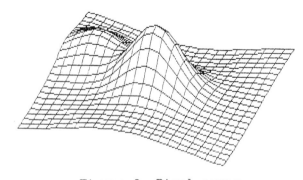

Figure 9. Pixel array

Conclusion

A discussion with some illustration by examples of the limits of the accuracy of reconstruction which can be obtained from experimental and real phantoms is presented. It is shown that for phase or velocity tomograms of about 1000 pixels, with velocities in the range of 1500 m/s, the limiting errors appear

403

Figure 10. Contours

to be about 2m/s on average. The reason for the existence of
these errors can be attributed either to experimental error or,
in the case of numerical phantoms, artifacts arising from the
bandwidth limitations in the systems under study. A method of
reconstruction using ray tracing is proposed.

ACKNOWLEDGEMENTS

The authors wish to acknowledge the National Sciences and
Engineering Research Council of Canada, Esso Resources Canada
and the Nova Scotia Cancer Treatment and Research Foundation for
their support of this work.

REFERENCES

1.   H. W. Jones, M. Mieszkowski, J. Berry, "Phase
      Tomography,Velocity and Temperature Measurement",
      Acoustical Imaging, Vol. 14., 1985, pp. 617-628,

2.   J. Berry, H. W. Jones, M. Mieszkowski,"Ultrasonic Phase
      Tomography for Medical Applications".,pp 79-89,
      Acoustical Imaging Vol. 15 ,1986.

3.   R. K. Mueller, M. Kaveh and G. Wade, "Reconstructive
      tomography and Applications of Ultrasonics", Proc. IEEE
      67:567 (1979).

4.   H. W. Jones, "Reconstruction of Phase and Amplitude
      Tomograms in Media with Significant Acoustic
      Refraction", Third Spring School on Acoustics and
      Acousto-Optics, Gdansk, Poland, 1986.

A SUBOPTIMAL TOMOGRAPHY RECONSTRUCTION TECHNIQUE

Serge Mensah and Jean-Pierre Lefebvre

C.N.R.S. L.M.A.
31 Chemin Joseph Aiguier
13402 Marseille Cedex 9
France

INTRODUCTION

This note describes a new suboptimal ultrasonic reflection mode tomography technique. This method is the last step towards our goal : quantitative images of acoustic impedance.

We begin with presenting the algorithm of the quantitative reconstruction technique in reflection mode. That method, using wide band signals and spatial scanning, is based on the technique of filtered backprojections that has now become usual.

Furthermore, we apply to the recorded data a deconvolution which eliminates all the perturbations inherent in physical systems. The algorithm relies on a stochastic model, and uses a priori information to widen the spectral band of the received response from the medium.

In the reconstruction algorithm, projections are filtered before being back-projected. As the preprocessed pulses tend to Diracs, and since the convolution of Dirac with any functions resembles the function itself, we had better acting in the spatial field and replace the pulses by the deblurring filter function. The latter is an experimental one which has been developed with a view to simplicity.

At last we present our original tomography system which has been built by our staff in order to test every possible scanning. Its precision allows us to work on the RF signal up to 10 MHz.

ACOUSTIC IMPEDANCE TOMOGRAPHY

Introduction

The present researches in acoustic imaging are focussed on the quantification of the image. For echography is only a qualitative technique, lacking any physical interpretation. Going further i.e. giving each pixel a value of the parameter characterizing the medium, implies raising the inverse problem in physical terms. We have to find a realistic and synthetic formulation, which as far as possible, reduces the number of parameters in order to lead to a practicable inversion. By focussing our attention on the reflection mode, we succeed in reducing to one parameter, the acoustical impedance. Then, we are able to solve the linear inverse problem, by applying algorithms similar to those used in X rays tomodensitometry.

Modelization

The theory which is presented here, is based on the assumption that we are working with plane wave, and that the following hypotheses are verified :

- The medium of propagation is slightly inhomogeneous.

- The measurements of the scattered field are made far from the scattering object.

- The dispersion absorption phenomena are negligeable compared with scattering phenomena.

To begin with, the medium may be described only by its density and compressibility fluctuations.

The acoustic pressure propagation equation is :

$$-\frac{1}{c^2}\frac{\partial^2 P}{\partial t^2} + \rho \, div\left(\frac{1}{\rho}\overrightarrow{grad}\,P\right) = 0 \tag{1}$$

The target is placed in a reference medium of known characteristics ρ_0, c_0.

Let $\alpha = \dfrac{c^2 - c_0^2}{c^2}$ the square velocity fluctuations.

$\varsigma = \log\dfrac{z}{z_0}$ the logarithmic impedance $(z = \rho c)$ fluctuations.

We have in the case of weak inhomogeneities $(\alpha \sim \varsigma \sim 10^{-2})$ at the first order of the limited development.

$$-\frac{1}{c_0^2}\frac{\partial^2 P}{\partial t^2} + \Delta P = -\frac{\alpha}{c_0^2}\frac{\partial^2 P}{\partial t^2} - \frac{1}{2}\overrightarrow{grad}\,\alpha\,\overrightarrow{grad}\,P + \overrightarrow{grad}\,\varsigma\,\overrightarrow{grad}\,P \tag{2}$$

This is the modelization adopted, it is perfectly suitable to biological media. It allows us to describe the forward problem in order to solve the inverse problem.

Forward problem

So density and compressibility fluctuations are responsible for ultrasonic scattering. One shows (see LEFEBVRE /1/) that the two parameters work together in such a way that pure transmission is caused solely by velocity fluctuations and pure reflection by impedance fluctuations. For small parameters fluctuations, the scattered field from a plane wave is within the far field first order Born approximation.

$$P_d^{(1)}{}_\infty (\vec{x}, \omega) = \frac{P_0\, e^{-ik|\vec{x}|}}{4\pi |\vec{x}|}\, h^{(1)}(\vec{n_0}, \vec{n}, \omega) \tag{3}$$

Where P_0, ω, $k = \omega/C_0$, $\vec{n_0}$, are the amplitude, angular frequency, wave number and the unit vector in the direction of propagation, and $h(\vec{n_0}, \vec{n}, \omega)$ the transfer-function of the medium for (far-field) scattering.

The first order BORN approximation of $h(\vec{n_0}, \vec{n}, \omega)$ is :

$$h^{(1)}(\vec{n_0}, \vec{n}, \omega) = -k^2 \left[\left(1 + \vec{n_0}\vec{n}\right) \hat{\alpha}(\vec{K}) + \left(1 - \vec{n_0}\vec{n}\right) \hat{\zeta}(K) \right] \quad \vec{K} = k(\vec{n_0} - \vec{n}) \tag{4}$$

where $\hat{\alpha}(K)$ and $\hat{\zeta}(K)$ are the space Fourier transforms of α and ζ.

Thus, the (far-field) scattering transfer function is a combination of the space Fourier transforms of the two parameters $\alpha(x)$ (velocity fluctuations) and $\zeta(x)$ (impedance fluctuations).

But near back-scattering $(\vec{n} = -\vec{n_0})$ and pure transmission $(\vec{n} = \vec{n_0})$ only one parameter is significant.

In reflection
$$h^{(1)}(\vec{n_0}, \vec{n}, \omega) \approx -k^2(1 - \vec{n_0}\vec{n})\zeta(K) \approx -2k^2 \hat{\zeta}(K) \tag{5}$$

In transmission
$$h^{(1)}(\vec{n_0}, \vec{n}, \omega) \approx -k^2(1 + \vec{n_0}\vec{n})\hat{\alpha}(K) \approx -2k^2 \hat{\alpha}(K) \tag{6}$$

The error made in this approximation is less than 1 % within $\pm 12°$ about $(\vec{n} = \pm\vec{n_0})$.

Our main objective is impedance reconstruction (acoustic impedance tomography) from wide band reflection measurements, that is the reason why we shall adopt the following representation :

$$P_d(\vec{x}, \omega)_\infty = -2k^2 \frac{e^{-ik|\vec{x}|}}{4\pi |\vec{x}|}\, \hat{\zeta}(K) \tag{7}$$

The inverse problem

The previous formula allows us, by using the tridimensional Fourier transform to find the object $\zeta = \mathrm{Log}(Z/Z_0)$. However, in order to reduce the acquisition and computational requisements, thanks to an acoustic focussing and the use of the slice projection theorem, we replace the problem by a unidimensional one.

Then, let the scattering angle $\theta = (\vec{n_0}, \vec{n})$ be constant, by carrying out frequency scanning between and, and spacial scanning of the incidence angle from $-\pi$ to π, we fill in the Fourier plane with successive radii inscribed in a corona centered at the origin and of radii :

$$\chi_{min} = 2K_{min}.\sin\theta, \qquad \chi_{max} = 2K_{max}.\sin\theta$$

After several calculations, we obtain the solution of the inverse problem (for derivation see LEFEBVRE /1/).

$$\zeta(x,y) = \frac{1}{2\pi} \int_0^\pi d\varphi \int_{-\chi_{max}}^{\chi_{min}} d\chi \, |\chi| \, \tilde{P}_\varphi(\theta,\chi) \, e^{i\chi\xi}$$
(8)

$P_\varphi(\theta,\xi)$ slice of the bidimensional Fourier transform of the object ζ along the direction $\vec{O\xi}$, is the projection of ζ on $\vec{O\xi}$ (projection slice theorem).

So, the complete reconstruction procedure of the bidimensional profile of the logarithmic impedance $\zeta(x,y) = Log(Z/Z_0)$ is :

1. Calculus in the frequency plane of the projection :

$$\tilde{P}_\varphi(\chi) = -\frac{h(\varphi,\omega)}{\chi^2}$$

2. Construction of the filtered backprojection :

$$\Pi_\varphi(\xi,\eta) = P_\varphi(\xi) \underset{\xi}{*} FT^{-1}(|\chi|) = P_\varphi(\xi) \underset{\xi}{*} H(\xi)$$

3. Sum of the filtered backprojections :

$$\zeta(x,y) = \frac{1}{2} \int_0^\pi \Pi_\varphi(\xi,\eta) \, d\varphi$$

This algorithm allows us to reconstruct the slice of the object from its projections . The theory is ideal in so far as $h(\varphi,\omega)$ is not the true response of the medium, but only the measured response obtained from the acquisition system, and inevitably filtered by it. So, in order to reduce the distance between the theorical solution and reality, we apply a deconvolution on the recorded data. The principle of this algorithm may be summed up in the following terms :
A priori information to widen the response band of the echoes recorded.

DECONVOLUTION

Resolution improvement using tomography can never be achieved if we do not take into account the dynamic characteristics of the transducer(s) and measurements errors.

The deconvolution theory, based on a stochastic approach, deals with linear cases, the point spread function of which is shift invariant.

All the information introduced is concentrated in the first and second moments of the distribution of random processes $\underline{X}$ (state vector) and $\underline{b}$ (noise vector).

$$E\{\underline{X}\} = \underline{X}_o, \qquad E\{(\underline{X}-\underline{X}_o).(\underline{X}-\underline{X}_o)^t\} = R_{XX} = P_0$$

$$E\{\underline{b}\} = \underline{O}, \qquad E\{\underline{bb}^t\} = R_{bb}, \qquad E\{\underline{b}\underline{X}^t\} = 0$$

In order to reduce the dimension of the state vector, the convolution model is replaced by a state-space realization of minimal order. The L.M.V. estimator is recursively computed using Chandrasekhar type equations applied on a degenerate case of Kalman filter.

The algorithm of the deconvolution which has been developped by DEMOMENT can be find in /2/.

The intended results from the deconvolution are :

- an improvement of the axial resolution. One tenth of the wave length (i.e. at 10 MHz the precision is 0.015mm in water)

- time saving process. After a deconvolution, the interface echoes appear as Dirac functions. Filtering such responses, amounts to replacing the convolution operation (which is often done by using D.F.T.) by the filter function itself.

THE DEBLURRING FILTER

The deblurring filter introduced by the theory is written as $h(\varphi) = FT^{-1}(|\chi|)$. Generally, filters ar developped in the Fourier space, and are band limited. That is why a time window is applied on the filter, in order to reduce leakage effects caused by sequence truncation at the end points.

The figure 1 represents the shape of the filters, with a view to simplicity, we adopt the one developed on three points by Mehran MOSHFEGHI. /3/.

It is an empirical three point filter, which has the essential features of a positive main lobe, with negative sidelobes on either side. This positive main lobe, gives the intensity contribution of the point reflector.

The negative sidelobes contribute to reducing most of the blurring near and around the point object.

The simplicity of that numerical filter leads to an important time saving process all the more so as we just have to replace Dirac functions (obtained by deconvolution) with this three points support function.

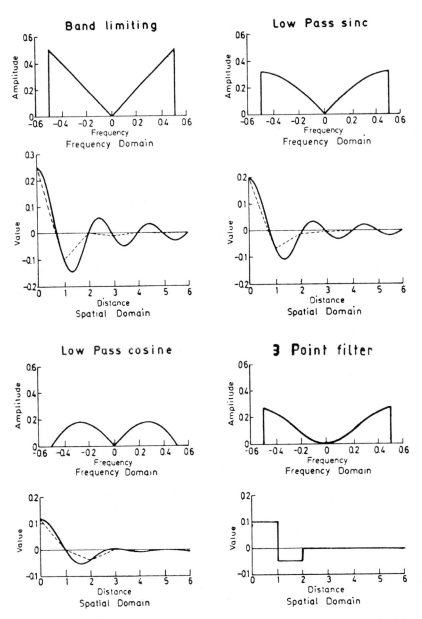

Fig. 1.   The frequency and spatial domain representations of the deblurring filters.

THE MECHANICAL SECTION OF THE TOMOGRAPHY SYSTEM

The dimensions of the mechanical section are 2.5 meters by 1.5 meters with a height of 2.5 meters. The mechanical section is constructed about a strong fram which bears a 1 meter diameter rotatable arm.

The arm is driven in rotation by a strong stepping motor via a screw-gear, attaining an accuracy of one one hundredth of degree.

Bellow this arm are fixed two carriages which can move along the arm or transverse to it with an increment of 0,0075 mm. This latter is useful for a linear scanning.
Further, the two carriages carry transmetters or receivers, rotatable in increments of 0,03 degree.

The target hangs from an adjustable rotatable support, and can be stepped by 0,01 degree units.

Vertical motion is obtained by a lead-screw which will later be numerically controlled.

Electronics

The overall precision is such that it will be possible to work with radiofrequency signals (in both amplitude ad phase) up to 10 MHz. The whole mechanical system is controlled by a PC Olivetti M 24 SP microcomputer.

The electro-acoustic system is composed of wide-band Panametrix transducers driven by a high energy 60 V-50/100/250 ns pulse generator-receiver (for back scattering experiments) especially built by H.HERMENT (IDCV-INSERM) for interfacing with a Biomation 8100 analogue-to-digital converter for high frequency (100 MHz) sampling and signal capture. All the operations (transducers positionning, signal emission and capture, processing and imaging) are controlled by the microcomputer.

Planed system utilisation

The system permits classical echography : linear, angular or coumpound (summation of images from various angles) scanning.

It will also allow testing classical velocity tomography from time of flight measurements (transmitter and receiver facing each others). With the same geometry, it will also allow attenuation imaging.

The system, further, is capable of testing velocity imaging with wide-band or narrow-band transmission measurements. For wide-band pulsed measurements, only one scattering angle will be necessary. For narrow-band measurements, linear scanning of the scattering pattern will be necessary. Multi-Frequency imaging will also be possible.

Finally, the system will be used to test diffraction tomography of the impedance parameter (acoustic impedance tomography) from wide band back-scattering measurements.

Presently, only backscattering measurements are made (the same transducer acting as transmitter and receiver) Multistatic reflection measurements are planed in the future by adding a second transverse arm in addition to the one supporting the transmitter.

Then, seismic imaging measurements will also be possible using seismic models.

CONCLUSION

A wide band Ultrasonic reflection image technique has been dealt with in detail, that uses aperture synthesis for filling in the Fourier plane and reconstructing bi-dimensional quantitative images.

In the present state, we do not yet exploit the phase information and only develop a high resolution qualitative tomography. But this stage is compulsory if we want to have an actual phase detection.

The deblurring filter that we use, is a simple empirical three point filter directly applied on the pre-processed data which are deconvolved.

The deconvolution permits to suppress the filter function of the data acquisition channel, and to obtain wide band pulses.

At last, in order to test our and all other techniques, we have built an original tomography rig, the overal precision of which enables us to work on the R.F. signal up to 10 MHz. First results have been obtain in NDT but images in vitro experiments may as well be tempted in order to obtain high resolution ultrasonic tomography images.

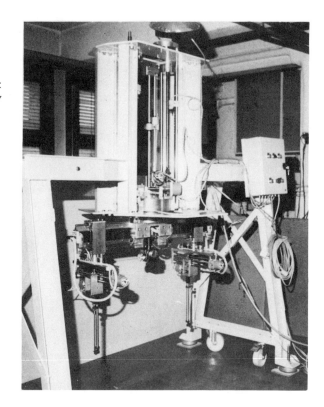

Fig. 2. The ultrasonic diffraction tomography system.

Fig. 3. Section of a carbon composite. The (0.05mm) fissure appears twice, since the celerity in the material is twice as much as in water.

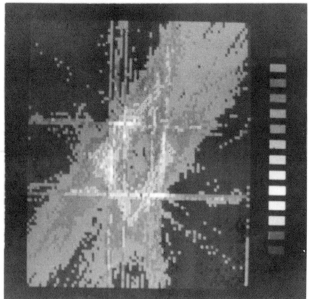

REFERENCES

[1] LEFEBVRE J.P. - La tomographie d'impédance acoustique - Traitement du signal - Volume 2. n° 2 - 1985.

[2] DEMOMENT G., REYNAUD R. and SEGALEN - Estimation sous-optimale rapide pour la deconvolution en temps réel. 1983 9 Colloque du GRETSI Proceedings p. 205 - 210.

[3] DEMOMENT G., REYNAUD R. and HERMENT A. - Fast minimum variance deconvolution. LSS Intern. Report n° 061 - 1983.

[4] MEHRAN MOSHFEGHI - IEEE Transaction on Ultrasonics - Ultrasound Reflection Mode Tomography Using Fan-Shaped-Beam Insonification. Vol. n° 3, May 1986.

[5] HO Y.C. and LEE R.C.K.- A Bayesian approach to problems in stochastic estimation and control - IEEE Transaction Automatic Control. AC-9 p. 333 - 339. 1964.

[6] FRANKLIN J.N. - Well-posed stochastic extensions of ill-posed linear problems. J. Math. Analy. Appl. 31. p. 682 - 716. 1970.

A FAST RECONSTRUCTION ALGORITHM FOR DIFFRACTION TOMOGRAPHY

Hui-Liang Xue, and Yu Wei

Department of Biomedical Engineering
Nanjing Institute of Technology
Nanjing, China

ABSTRACT

A fast reconstruction algorithm for diffraction tomography is derived. The proposed algorithm increases the reconstruction speed of the filtered backpropagation algorithm to be comparable with that of straight line mode reconstruction. Equal image quality is proved remained by simulation of phantom.

1. Introduction

The filtered backpropagation algorithm proposed by A.J. Devaney(1) for diffraction tomography is a promising reconstruction algorithm within Born or Rytov approximation but requires vast amount of computation. To save the computation time, Zhang Ming et al.(2) introduced the Radon transform method which reduces the multiplication times from $N^3 Log_2 N$ of the filtered backpropagation algorithm to N^3. Based on this method, a very fast reconstruction algorithm for diffraction CT is derived in this paper with the speed being comparable with that of straight line mode reconstruction(e.g., X-CT). Simulation of phantom demonstrates equal image quality.

2. A fast reconstruction algorithm

The Radon transform of $f(x,y)$ in Cartesian coordinate(see Fig.1) is given by

$$Rf(l, \theta) = \int_l f(x,y) dl$$
$$= (\frac{1}{\sin\theta}) \cdot \int_{-\infty}^{\infty} f(x, -x ctg\theta + \frac{l}{\sin\theta}) dx \qquad (1)$$

By the filtered backpropagation algorithm(1), we obtain

$$f(x,y) = \frac{1}{4\pi^2} \int_0^{2\pi} d\phi \int_{-k_o}^{k_o} r(k,\phi) \cdot |k| \cdot e^{i(\gamma-k_o)(\eta-l_o)} e^{ik\xi} dk \qquad (2)$$

where $\gamma = \sqrt{k_o^2 - k^2}$. Defining the following functions

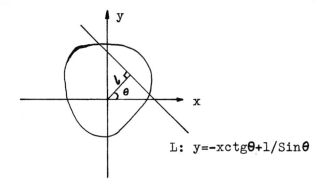

L: y=-xctgθ+1/Sinθ

Fig.1. Radon transform of an object

$$\begin{cases} tg\,\beta(k) = \dfrac{k_0 - \gamma}{k} \\[2mm] N(k) = \sqrt{(\gamma - k_0)^2 + k^2} = \sqrt{2k_0(k_0 - \sqrt{k_0^2 - k^2})} \end{cases} \qquad (3)$$

If $\theta \neq 0$, there is

$$[Rf](l,\theta) = \frac{1}{4\pi^2 Sin\theta} \int_{-\infty}^{\infty}\!\int_{0}^{2\pi}\!\int_{-k_0}^{k_0} \Gamma(k,\phi)|k|\cdot exp\{i(\gamma - k_0)[x\,cos\phi +$$

$$(-x\,ctg\,\theta + \frac{l}{Sin\theta})Sin\,\phi - l_0]\}\cdot exp\{ik[x\,Sin\phi - (-x\,ctg\,\theta + \frac{l}{Sin\theta})cos\phi]\}\,dk\,d\phi\,dx$$

$$= \frac{1}{4\pi^2 Sin\theta} \int_{0}^{2\pi}\!\int_{-k_0}^{k_0} \Gamma(k,\phi)exp\left[-ilN(k)\frac{cos(\phi - \beta(k))}{Sin\theta} - i(\gamma - k_0)l_0\right]\cdot|k|\cdot$$

$$\int_{-\infty}^{\infty} exp\left[ix\cdot N(k)\,Sin\,(\theta + \beta(k) + \frac{\pi}{2} - \phi)/Sin\theta\right]dx\,dk\,d\phi$$

$$= \frac{1}{2\pi Sin\theta} \int_{0}^{2\pi}\!\int_{-k_0}^{k_0} \Gamma(k,\phi)\cdot|k|\cdot exp\left[-il\,N(k)\frac{cos(\phi - \beta(k))}{Sin\theta} - i(\gamma - k_0)l_0\right]\cdot$$

$$\delta\left[N(k)\cdot Sin\,(\theta + \beta(k) - \phi + \frac{\pi}{2})/Sin\theta\right]dk\,d\phi$$

Obviously, $\phi = \theta + \beta(k) + \frac{\pi}{2} + m\pi = \phi_0 + m\pi$. m should be two successive integers to meet the need of $0 \leq \phi \leq 2\pi$. Taking m and m+1 as values of m gives rise to the expression

$$[Rf](l,\theta) = \frac{1}{2\pi}\int_{-k_0}^{k_0}\left[\Gamma(k,\phi_0 + m\pi)\,exp[il\,N(k)(-1)^m] + \right.$$

$$\left.\Gamma(k,\phi_0 + m\pi + \pi)exp[-il\,N(k)(-1)^m]\right]c(k)\,dk \qquad (4)$$

where
$$c(k) = \frac{|k|\cdot e^{-i(\gamma - k_0)l_0}}{N(k)}$$

In the case of $\theta = 0$, a similar formula can be derived:

$$[Rf](L,0) = \frac{1}{2\pi}\int_{-k_0}^{k_0}\left[\Gamma(k,\beta(k)+\tfrac{\pi}{2})e^{iLN(k)} + \right.$$
$$\left. \Gamma(k,\beta(k)+\tfrac{3}{2}\pi)e^{-iLN(k)}\right]c(k)dk \qquad (5)$$

Therefore, we can always take the form of Eq.(4) as the executive formula.

Define k_p which satisfies $\quad \theta+\beta(k_p)+\tfrac{\pi}{2}=\pi \qquad (6)$

It is clear that $\quad 0 \leqslant \theta \leqslant \pi, \quad -\tfrac{\pi}{4}\leqslant\beta(k)\leqslant\tfrac{\pi}{4}, \quad \tfrac{\pi}{4}\leqslant\phi_0\leqslant\tfrac{7}{4}\pi.$

So, we have

If $\quad k \leqslant k_p,$ then $\quad \phi_0 \leqslant \pi, \quad m = 0$

If $\quad k > k_p,$ then $\quad \phi_0 > \pi, \quad m = -1$

Eq.(4) can be rewritten as

$$[Rf](L,\theta) = \frac{1}{2\pi}\int_{-k_0}^{k_p} c(k)\left[\Gamma(k,\phi_0)e^{iLN(k)} + \Gamma(k,\phi_0+\pi)e^{-iLN(k)}\right]dk$$

$$+\frac{1}{2\pi}\int_{k_p}^{k_0} c(k)\left[\Gamma(k,\phi_0-\pi)e^{-iLN(k)} + \Gamma(k,\phi_0)e^{iLN(k)}\right]dk$$

$$= \frac{1}{2\pi}\int_{-k_0}^{k_0} c(k)\left[\Gamma(k,\phi_0)e^{iLN(k)} + \Gamma(k,\phi_0+\pi)e^{-iLN(k)}\right]dk$$

$$= \frac{1}{2\pi}\int_{-k_0}^{k_0} c(k)\left[\Gamma_A e^{iLN(k)Sign(k)} + \Gamma_B e^{-iLN(k)Sign(k)}\right]dk \qquad (7)$$

where
$$\Gamma_A = \begin{cases} \Gamma(k,\phi_0), & k \geqslant 0 \\ \Gamma(k,\phi_0+\pi), & k < 0 \end{cases} \qquad (8)$$

$$\Gamma_B = \begin{cases} \Gamma(k,\phi_0+\pi), & k \geqslant 0 \\ \Gamma(k,\phi_0), & k < 0 \end{cases} \qquad (9)$$

$$Sign(k) = \begin{cases} 1, & k \geqslant 0 \\ -1, & k < 0 \end{cases} \qquad (10)$$

Let $\quad k = k'\sqrt{1-k'^2/4k_0^2} \qquad (11)$

It follows that $N(k)Sign(k) = k' \qquad (12)$

and $\quad \dfrac{dk}{dk'} = \dfrac{1-k'^2/2k_0^2}{\sqrt{1-k'^2/4k_0^2}} \qquad (13)$

Substituting Eqs.(11) and (12) and (13) into Eq.(7), we have

$$[Rf](l,\theta) = \frac{1}{2\pi} \int_{-\sqrt{2}\,k_0}^{\sqrt{2}\,k_0} \left[\Gamma_A e^{ilk'} + \Gamma_B e^{-ilk'} \right].$$

$$\frac{|k'|\sqrt{1 - k'^2/4k_0^2}}{|k'|} \cdot \frac{1 - k'^2/2k_0^2}{\sqrt{1 - k'^2/4k_0^2}}\, e^{-i(\gamma - k_0)l_0}\, dk'$$

$$= \frac{1}{2\pi} \int_{-\sqrt{2}\,k_0}^{\sqrt{2}\,k_0} \left[\Gamma_A e^{ilk'} + \Gamma_B e^{-ilk'} \right] c(k')\, dk' \qquad (14)$$

where

$$c(k') = \left(1 - \frac{k'^2}{2k_0^2} \right) e^{-i(\gamma - k_0)l_0} \qquad (15)$$

Eq.(14) is the executive formula to calculate the equivalent Radon values of scattering field in straight line mode.

The whole course of image reconstruction by the proposed algorithm consists of two steps. The first step is to figure out the equivalent Radon values of scattering field in straight line mode. As shown above, this step needs 2N FFT's for an NxN image reconstructed from N diffracted projections. The second step is the inverse Radon transform. It can be completed by the filtered backprojection algorithm which is a fast operation. The elapsed time of the first step only amounts to a small part of the whole reconstruction period. Therefore the new algorithm proposed in this paper increases the speed of the filtered backpropagation algorithm greatly to be comparable with that of straight line mode reconstruction.

3. Simulation Study

A lot of simulation results have been obtained by study of the proposed algorithm on Shepp-Logan phantom. Fig.2 and Fig.3 are two of them. By the simulation results, we can see that the speed of the proposed algorithm is increased greatly to be comparable with that of straight line mode reconstruction.

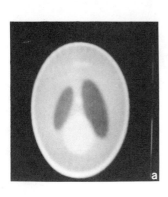 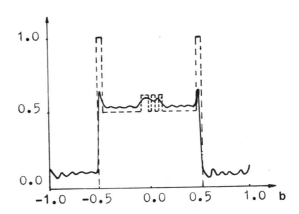

Fig.2 (a) A 128x128 reconstruction obtained with the proposed method from 64 projections and 128 samples per projection. The CPU processing time on a vax-11/730 is 4 min. (b) A numerical comparison of the true and reconstructed values on the line y=-0.605 through the phantom.

418

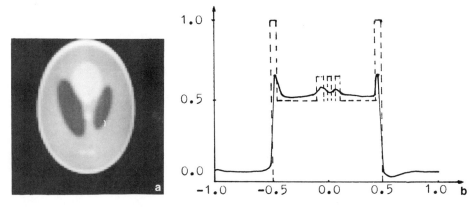

Fig.3 (a) A 128x128 reconstruction obtained with the proposed
method from 64 projections and 128 samples per projection.
The CPU processing time on a vax-11/730 is 6 min. (b) A
numerical comparison of the true and reconstructed values on
the line y=-0.605 through the phantom.

Reference

1. A.J. Devaney
   A filtered backpropagation algorithm for diffraction images
   Ultrasonic Imaging, vol.4, pp336-350, 1982

2. Zhangming, Chai Zhen-Ming & Wei Yu
   A new reconstruction algorithm based on Radon transform
   for diffraction tomography
   Conference of proceedings of radio engineering, Budapest,
   October, 1986

A COMPUTATIONAL STUDY OF RECONSTRUCTION ALGORITHMS FOR SYNTHETIC APERTURE DIFFRACTION TOMOGRAPHY : INTERPOLATION VERSUS INTERPOLATION-FREE

Jian-Yu Lu

Department of Biomedical Engineering
Nanjing Institute of Technology
Nanjing, China

INTRODUCTION

X-ray computerized tomograpy (X-CT) has been a great success in medical diagnoses [1-7]. Its finders, Hounsfield (English) and Cormack (American), had got Nobel Prize In physiology and medicine in 1979. But, X-CT is harmful to human body because large dosage of the X-ray is used. In addition, the equipment of X-CT is very expensive. In order to overcome the disadvantages of X-CT, people developed the ultrasonic computerized tomography (U-CT) [8]. Compared to X-ray, the wavelength of the ultrasound is much longer and, thus, the diffraction effects of the ultrasound are remarkable. In order to solve the diffraction problem of the ultrasound in U-CT, people developed the concept of ultrasonic diffraction computerized tomography (DUCT) [9-10], and, after then, many reconstruction algorithms for DUCT have been developed and further studied [11-18].

The concept of DUCT is directly based on the inversion of wave equation under the weak scattering assumption (i.e. the scattered wave is much smaller than the incident wave). Conventional ultrasonic diffraction computerized tomography uses the method analogous to that used in commercial X-CT. A plane-wave is insonified on an object to be imaged and is rotated 360° around the object. For each position of the rotation, a diffraction projection will be obtained (usually, we call such a projection a view). Thus, as the incident wave rotates, many projections will be obtained. From these projections, the images can be reconstructed. The advantage of the conventional DUCT is that high quality images can be obtained by relatively less projections and less points contained in each projection. This is because relatively more information of the spectrum of the object can be obtained by using this measuring geometry. But, the conventional DUCT requires its measuring system rotating 360° around the object, and is not applicable when obstacles exist in the path of the rotation. In addition, it requires a large scale plane-wave insonification, which will be difficult if it is used in practical medical imaging.

In order to overcome the disadvantages of the conventional DUCT, D.Nahamoo et al. [16] put forward a new type of DUCT — the synthetic aperture diffraction computerized tomography (SADCT). In SADCT, for a good image reconstruction, the measuring system is only needed to rotate around the object once and, in principle, any kind of insonifications can be used. Furthermore, in reference [15], D.Nahamoo et al. developed an interpolation-free reconstruction algorithm (IFRA) for SADCT and performed its computer simulation study. But, because this algorithm containes a space-variant filter, it requires large amount of computations. For an NxN image reconstructed by NxN diffracted data, this algorithm requires approximatly $O(N^3 + N^2 \log_2 N)$ complex multiplications. Besides, even if for a rather big value of N, such as, N=128, the reconstructed image is still not satisfied. In order to reduce the number of the complex mutiplications and to improve the image reconstructions, we used Fourier-domain interpolation reconstruction algorithms (FDIRAs) for the reconstruction of SADCT, and performed a detailed computational study of these algorithms. For an NxN image reconstructed by NxN diffracted data, the FDIRAs require only approximately $O(N^2 \log_2 N)$ complex multiplications, and, the larger the N is, the more the computation is saved. From the results of the computational study of SADCT, one can see that the quality of the images reconstructed by FDIRAs is better than that reconstructed by IFRA. In addition, in FDIRAs, the nearest-neighbor interpolation reconstruction algorithm, in general, will give better results than the bilinear interpolation reconstruction algorithm. But, this is not the case for the conventional DUCT, where the Fourier-doamin bilinear interpolation reconstruction algorithm gives better results than the nearest-neighbor interpolation reconstruction algorithm. This conclusion was demonstrated in reference [11] and has been re-proved by us in this paper.

In the computational study of FDIRAs, we discovered that better reconstructions would be obtained when the shift of the coordinate of the object was performed before the Fourier-domain interpolations, and, we also discovered that the accuracy of the interpolations of the points near the boundaries of the Fourier-domain coverage areas A or B (see Fig.2) had a great influence on the reconstructions.

In this paper, all the reconstructed images will be quantitatively evaluated by distance criteria as well as reconstructed values on a line through three smallest ellipses of the phantom used in our computer simulation.

This paper is organized as follows : First, we will state simply the basic principles of SADCT and obtain two diffraction projection formulas. Next, we will introduce FDIRAs and derive the relationships between curvilinear and rectangular coordinates. Then, we will give the results of the computer simulation and the comparisons among these results. Finally, we will make a brief summary of this paper.

BASIC PRINCIPLES FOR SADCT

Wave Equation

In this paper, we consider only the two-dimensional case,

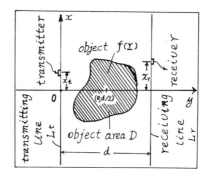

Fig.1 *The datum acqusition geometry of the SADCT*

i.e., we assume that the object is not varied with the axis z (the axis z is normal to the x-y plane shown in Fig.1). The coordinates x_t and x_r in Fig.1 represent the positions of the transmitter and the receiver on the transmitting line L_t and the receiving line L_r, respectively, and the center of the object is sited at the point (0,d/2). The transmitter can be moved to N positions on the transmitting line L_t, and for each transmitter position, the receiver can also be moved to N positions on the receiving line L_r. Thus, by using this datum acquisition system, NxN diffracted data will be obtained.

We now confine our discussion to soft bio-tissues. We assume that the object is immersed in the surrounding homogeneous medium (such as water), and the ultrasonic field in the area between the two lines L_t and L_r is governed approximatly by the inhomogeneous Helmholtz equation

$$(\nabla^2 + k_o^2)u(\underline{r}) = -f(\underline{r})u(\underline{r}) \tag{1}$$

where ∇^2 is the Laplacian operator; $u(\underline{r})$ represents the total complex wave field at the point $\underline{r}=(x,y)$; k_o is the wave number of the surrounding homogeneous medium; and $f(\underline{r})$ is the object function. $f(\underline{r})$ is related to the distribution of a refrative index $n(\underline{r})$ by

$$f(\underline{r}) = \begin{cases} k_o^2[n^2(\underline{r})-1] & , \ \underline{r} \in D \\ 0 & , \ \text{otherwise} \end{cases} \tag{2}$$

and $f(\underline{r})$ will be a real function if the attenuation of the object is not considered. (In the following, we will not consider the attenuation of the object).

Solution of Helmholtz Equation and Born Approximation

The total wave field $u(\underline{r})$ in Eq.(1) can be written as the sum of the incident field $u_i(\underline{r})$ and the scattered field $u_s(\underline{r})$ [16]

$$u(\underline{r}) = u_i(\underline{r}) + u_s(\underline{r}) \tag{3}$$

where $u_s(\underline{r})$ is given by [15]

$$u_s(\underline{r}) = \int_D f(\underline{r}_o)u(\underline{r}_o)G(\underline{r}|\underline{r}_o)d\underline{r}_o \tag{4}$$

423

and D is the area shown in Fig.1; $G(\underline{r}|\underline{r}_o)$ is the Green function associated with the datum acquisition system. Here, we assume $G(\underline{r}|\underline{r}_o)$ is a two-dimensional free-space Green function, and can be expressed as [20]

$$G(\underline{r}|\underline{r}_o) = \frac{j}{4} \, H_o(k_o|\underline{r}-\underline{r}_o|) \tag{5}$$

where H_o is the zero order Hankel function with the first kind; and $|\underline{r}-\underline{r}_o|$ is the distance between the field point $\underline{r}=(x,y)$ and the source point $\underline{r}_o=(x_o,y_o)$.

We further assume that the Born approximation is held, i.e., the weak scattering assumption is satisfied

$$|u_s(\underline{r})| \ll |u_i(\underline{r})| \tag{6}$$

Under the condition of the Born approximation, the total field $u(\underline{r})$ in Eq.(4) may be simply replaced by the incident field $u_i(\underline{r})$

$$u_s(\underline{r}) = \frac{j}{4} \int_D f(\underline{r}_o)u_i(\underline{r}_o)H_o(|\underline{r}-\underline{r}_o|)d\underline{r}_o \tag{7}$$

Thus, from Eq.(7), we obtain the weak scattering solution of the Helmholtz equation (1).

Diffraction Projection Formulas for SADCT

Let $u_i(\underline{r};x_t)$ represent the incident field at the point $\underline{r}$ with the transmitter located at the point $(x_t,0)$, by using the angular spectrum expansion [19,21], and through simple derivations [15], one obtains

$$u_i(r;x_t) = \frac{1}{2\pi} \int_{-\infty}^{\infty} A_t(k_x)e^{-jk_x x_t} e^{j\underline{K}\cdot\underline{r}} dk_x \tag{8}$$

where $A_t(k_x)$ is the Fourier transform of the function $u_i(x,0;0)$ which is the incident field on the transmitting line L_t while the transmitter is placed at the origin of the coordinate shown in Fig.1

$$A_t(k_x) = \int_{-\infty}^{\infty} u_i(x,0;0)e^{-jk_x x} dx \tag{9}$$

where $\underline{K}=(k_x,k_y)$, and

$$k_y = \sqrt{k_o^2-k_x^2} \tag{10}$$

H_o in Eq.(7) can also be expanded by the angular spectrum expansion [21]

$$H_o(k_o|\underline{r}-\underline{r}_o|) = \frac{1}{\pi} \int_{-\infty}^{\infty} \frac{e^{j\underline{T}\cdot(\underline{r}-\underline{r}_o)}}{t_y} dt_x \tag{11}$$

424

where $\underline{T}=(t_x,t_y)$, and

$$t_y=\sqrt{k_o^2-t_x^2} \qquad (12)$$

Substituting Eqs.(8) and (11) into Eq.(7), and using the notation $u_s(x_r;x_t)$ to represent the received scattered field at the point $\underline{r}=(x_r,d)$, one obtains

$$u_s(x_r;x_t)=\frac{1}{(2\pi)^2}\int\int_{-\infty}^{\infty}\frac{je^{jt_yd}}{2t_y}A_t(k_x)[\int_D f(\underline{r}_o)$$

$$e^{-j(\underline{T}-\underline{K})\cdot\underline{r}_o}d\underline{r}_o]e^{j(t_xx_r-k_xx_t)}dt_xdk_x \qquad (13)$$

Let $U_s(t_x;k_x)$ be the Fourier transform of $u_s(x_r;-x_t)$, one obtains

$$U_s(t_x;k_x)=\frac{je^{jt_yd}}{2t_y}A_t(k_x)\int_D f(\underline{r}_o)e^{-j(\underline{T}-\underline{K})\cdot\underline{r}_o}d\underline{r}_o \qquad (14)$$

If we take the filtering properties of the receiver into account and define $P_{sa}(t_x;k_x)$ as the Fourier transform of the scattered field received, we have

$$P_{sa}(t_x;k_x)=\frac{je^{jt_yd}}{2t_y}A_r(t_x)A_t(k_x)\int_D f(\underline{r}_o)e^{-j(\underline{T}-\underline{K})\cdot\underline{r}_o}d\underline{r}_o \qquad (15)$$

where $A_r(t_x)$ is the filter function of the receiver. The integral on the right hand side of Eq.(15) represent the Fourier transform of $f(\underline{r})$ evaluated on the curvilinear coordinate : $\{\underline{T}-\underline{K}; \ |t_x|\leqslant k_o,|k_x|\leqslant k_o\}$, and is defined as $F(\underline{T}-\underline{K})$. From Eq.(15), we obtain the diffraction projection formula

$$F(\underline{T}-\underline{K})=-\frac{2jt_ye^{-jt_yd}}{A_r(t_x)A_t(k_x)}P_{sa}(t_x;k_x) \qquad (16)$$

In order to make the Fourier space covered sufficiently by the data measured, we must rotate the measuring system or the objet $90°$ around the point $(0,d/2)$ once. According to the similar derivations given above, and defining the notation $P_{sb}(t_x;k_x)$ as the Fourier transform of the scattered field measured after the $90°$ rotation, we obtain another diffraction projection formula

$$F[Q^{-1}(\underline{T}-\underline{K})]=-\frac{2jt_ye^{-jt_yd}}{A_r(t_x)A_t(k_x)}P_{sb}(t_x;k_x) \qquad (17)$$

where Q is a $90°$ rotation matrix, and Q^{-1} is its inversion

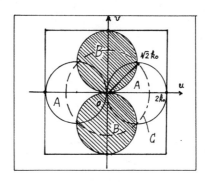

Fig.2 The Fourier-domain coverages. The area A and B (hatched) are the coverage before and after the 90° rotation of the measuring system, respectively

$$Q = \begin{bmatrix} 0 & , & -1 \\ 1 & , & 0 \end{bmatrix} \qquad (18)$$

From Eqs.(16) and (17) and by using the inverse Fourier transform, the object function $f(\underline{r})$ can be reconstructed.

FOURIER-DOMAIN INTERPOLATION RECONSTRUCTION ALGORITHMS

Fourier-Domain Coverages

Suppose that u-v is a rectangular coordinate on the Fourier space and $\underline{W}=\underline{T}-\underline{K}$, where $\underline{W}=(u,v)$ represents a vector at the point (u,v), and according to Eqs.(10) and (12), one obtains

$$\begin{cases} u=t_x-k_x \\ v=t_y-k_y \end{cases} \qquad (19)$$

After the 90° rotation of the measuring system, if we define $\underline{W}=Q^{-1}(\underline{T}-\underline{K})$, from Eqs.(10), (12) and (18), we obtain

$$\begin{cases} u= t_y-k_y \\ v=-t_x+k_x \end{cases} \qquad (20)$$

where $|t_x|\leqslant k_o$ and $|k_x|\leqslant k_o$. For the case $|t_x|>k_o$ or $|k_x|>k_o$, the scattered wave is not a propagation wave, but really an attenuated field. If the receiving line L_r is placed several wavelengths, say, ten wavelengths away from the object, the effects of the attenuated field is negligible. By the way, we will state that the Eqs.(16) and (17) are held only when the conditions $|t_x|\leqslant k_o$ and $|k_z|\leqslant k_o$ are satisfied.

The Fourier-domain coverage areas A and B can be obtained from Eqs.(19) and (20), respectively, as shown in Fig.2

Relationships Between Curvilinear and Rectangular Coordinates

In order to perform the Fourier-domain interpolations, we

must find the relationships between the curvilinear and the rectangular coordinates. From Eqs.(12) and (10), we know that $t_y \geqslant 0$ and $k_y \geqslant 0$, and we obtain

$$t_x^2 + t_y^2 = k_o^2 \qquad (21)$$

$$k_x^2 + k_y^2 = k_o^2 \qquad (22)$$

From Eq.(19), we obtain

$$\left\{ \begin{array}{l} (u-t_x)^2 + (v-t_y)^2 = k_o^2 \qquad (23) \\[2mm] (u+k_x)^2 + (v+k_y)^2 = k_o^2 \qquad (24) \end{array} \right.$$

Eq.(23) represents the circles centered at (t_x, t_y), with radius of k_o, on the u-v plane. The trace of the center of these circles (t_x, t_y) is also a circle but centered at origin, with radius k_o, on the u-v plane, defined by Eq.(21). Because $t_y \geqslant 0$, Eq.(21) represents only the super-half of the circle. Similarly, Eq.(24) represents the circles centered at $(-k_x, -k_y)$, with radius of k_o, on the u-v plane. The trace of the center of these circles $(-k_x, -k_y)$ is on the lower-half of the circle defined by Eq.(22). All the points (u,v) which satisfy Eqs.(23) and (24) simultaneously will form a set which will cover the area A shown in Fig.2. From Eqs.(21) to (24), we will obtain the relationships between a point (t_x, k_x) on the curvilinear coordinate and a point (u,v) on the rectangular coordinate (See Fig.3).

If a point (u,v) belongs to the first or the third quarter of the coordinate u-v, one obtains

$$\left\{ \begin{array}{l} t_x = \dfrac{1}{2}(u-q1) \\[4mm] k_x = -\dfrac{1}{2}(u+q1) \end{array} \right. \qquad (25)$$

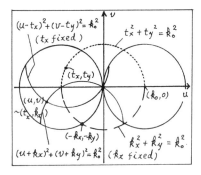

Fig.3 The relationships between the curvilinear and the rectangular coordinates

where

$$q1 = |v| \cdot \sqrt{\frac{4k_o^2}{u^2 + v^2} - 1} \qquad (26)$$

If a point (u,v) belongs to the second or the fourth quarter of the coordinate $u-v$, one obtains

$$\begin{cases} t_x = \dfrac{1}{2}(u+q1) \\[3mm] k_x = -\dfrac{1}{2}(u-q1) \end{cases} \qquad (27)$$

From Eq.(20), one obtains

$$\begin{cases} (u-t_y)^2 + (v+t_x)^2 = k_o^2 & (28) \\[3mm] (u+k_y)^2 + (v-k_x)^2 = k_o^2 & (29) \end{cases}$$

Similarly, we will obtain the relationships between a point (t_x, k_x) on the curvilinear coordinate and a point (u,v) on the rectangular coordinate in area B (the area B is formed by a set of the points (u,v)s which satisfy the Eqs.28 and 29 simultaneously as shown in Fig.2). If a point (u,v) belongs to the first or the third quarter of the coordinate $u-v$, one obtains

$$\begin{cases} t_x = -\dfrac{1}{2}(v-q2) \\[3mm] k_x = \dfrac{1}{2}(v+q2) \end{cases} \qquad (30)$$

where

$$q2 = |u| \cdot \sqrt{\frac{4k_o^2}{u^2 + v^2} - 1} \qquad (31)$$

If a point (u,v) belongs to the second or the fourth quarter of the coordinate $u-v$, one obtains

$$\begin{cases} t_x = -\dfrac{1}{2}(v+q2) \\[3mm] k_x = \dfrac{1}{2}(v-q2) \end{cases} \qquad (32)$$

Bilinear Interpolation and Nearest-Neighbor Interpolation

In order to reconstruct the object function $f(\underline{r})$ from its Fourier transform $F(\underline{W})$ using IFFT, we must know the values of $F(\underline{W})$ on rectanglar grids. To arrive this, we use two commonly used interpolation methods : the bilinear interpolation and the nearest-neighbor interpolation. The bilinear interpolation formula is given by

$$P_S(t_x;k_x) = \sum_{i=1}^{N_{t_x}} \sum_{j=1}^{N_{k_x}} P_S(t_{x_i};k_{x_j})\, h1(t_x-t_{x_i})\, h2(k_x-k_{x_j}) \quad (33)$$

where $(t_{x_i};k_{x_j})$ are the discrete points on the t_x-k_x plane, on which the values of the function $P_S(\underline{T}\text{-}\underline{K})$ are known. N_{t_x} and N_{k_x} are the numbers of the discrete points of t_x and k_x respectively, and h1 and h2 are given by (here, $\Delta t_x = \Delta k_x = $ const.)

$$h1(t_x-t_{x_i}) = \begin{cases} 1- \dfrac{|t_x-t_{x_i}|}{\Delta t_x} & , \ |t_x-t_{x_i}| \leqslant \Delta t_x = t_{x_{i+1}}-t_{x_i} \\ 0 & , \ \text{otherwise} \end{cases}$$

$$\begin{cases} & (34) \\ h2(k_x-k_{x_j}) = \begin{cases} 1- \dfrac{|k_x-k_{x_j}|}{\Delta k_x} & , \ |k_x-k_{x_j}| \leqslant \Delta k_x = k_{x_{j+1}}-k_{x_j} \\ 0 & , \ \text{otherwise} \end{cases} \end{cases}$$

The procedures of the bilinear interpolation are given below. Given a grid point (u,v) on the rectangular coordinate, if it belongs to the area A, we will calculate the corresponding point (t_x,k_x) on the curvilinear coordinate according to Eqs.(25) or (27), and then calculate $P_{SA}(t_x;k_x)$ using Eq.(33), finally, we will find the Fourier transform of the object function $F(\underline{W})$ from the Eq.(16); If the point (u,v) belongs to the area B, we will calculate (t_x,k_x) according to Eqs.(30) and (32), and then, using Eqs.(33) and (17), we will obtain the value of $F(\underline{W})$ at the point (u,v); If the point (u,v) belongs to A∩B, we will average these two interpolation resultes and assign it to $F(\underline{W})$ at this point. (The interpolations of points near the boundaries of the areas A or B (see Fig.2) will be specially considered in next section.)

The procedures of the nearest-neighbor interpolation are the same as those of the bilinear interpolation except that after the point (t_x,k_x) is calculated, the value of the Fourier transform of the object function on the point $(t_{x_{i_0}},k_{x_{j_0}})$ (which is the nearest-neighbor of the point (t_x,k_x) among the points (t_{x_i},k_{x_j}) (i=1,2,...,N_{t_x}, j=1,2,...,N_{k_x}) on which the Fourier transform of the object function are known) is taken as the value of the Fourier transform of the object function on the point (t_x,k_x).

In order to diminish the Gibbs oscillation, we apply a two-dimensional blackman window $b(\underline{W})$ to the function $F(\underline{W})$ prior to IFFT

$$b(\underline{W}) = \begin{cases} 0.42-0.5\cos\dfrac{2\pi r}{2\sqrt{2}k_o}+0.08\cos\dfrac{4\pi r}{2\sqrt{2}k_o} & , \ |\underline{W}| \leqslant \sqrt{2}k_o \\ & \quad (35) \\ 0 & , \ |\underline{W}| > \sqrt{2}k_o \end{cases}$$

where

$$r = \sqrt{u^2+v^2+\sqrt{2}k_o} \quad (35')$$

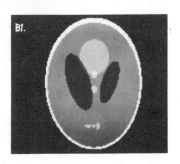

Fig.4 *The 128x128 head phantom image used in our computer simulation*

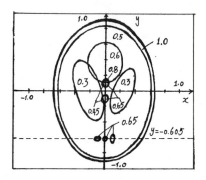

Fig.5 *The gray level asignment of the phantom. The dashed line through the phantom is the line y:-0.605*

If we take the 2-D IFFT of the function $F(\underline{W})b(\underline{W})$, we will obtain the low-pass filtered version of the object function $f(\underline{r})$.

COMPUTER SIMULATION RESULTS OF THE FOURIER-DOMAIN INTERPOLATION AND INTERPOLATION-FREE RECONSTRUCTION ALGORITHMS

The Phantom Used in Our Computer Simulation

The phantom used in our computer simulation was the same as that used by Shepp and Logan [22], but had the gray level assignment changed to those used by Devaney [13] and Pan and Kak [11]. The image of the phantom and the gray level assignment of the phantom are shown in Figs.4 and 5, respectively.

Definition of the Distance Criteria

For the convenience of the quantitative comparision of the reconstructed images and the phantom, we have adopted the distance criteria defined in [7] and defined the average adjusted version of them. The definitions of these distance criteria are given by

$$d1 = \sqrt{\frac{\sum\limits_{i=1}^{128}\sum\limits_{j=1}^{128}(r_{ij}-p_{ij})^2}{\sum\limits_{i=1}^{128}\sum\limits_{j=1}^{128}(p_{ij}-\overline{p})^2}} \tag{36}$$

$$r1 = \frac{\sum\limits_{i=1}^{128}\sum\limits_{j=1}^{128}|r_{ij}-p_{ij}|}{\sum\limits_{i=1}^{128}\sum\limits_{j=1}^{128}|p_{ij}|} \tag{37}$$

$$e1 = \max_{\substack{1<I<64\\1<J<64}} |r_{IJ}-p_{IJ}| \tag{38}$$

where

$$r_{IJ} = \frac{1}{4} (r_{2I,2J} + r_{2I+1,2J} + r_{2I,2J+1} + r_{2I+1,2J+1}) \qquad (39)$$

$$P_{IJ} = \frac{1}{4} (P_{2I,2J} + P_{2I+1,2J} + P_{2I,2J+1} + P_{2I+1,2J+1}) \qquad (40)$$

r_{ij} and p_{ij} in the above equations are pixel values of the reconstructed images and the phantom for the ith row and the jth column respectively, while $\bar{r}$ and $\bar{p}$ are the average values of the reconstructed images and the phantom respectively

$$\bar{r} = \frac{1}{128 \times 128} \sum_{i=1}^{128} \sum_{j=1}^{128} r_{ij} \qquad (41)$$

$$\bar{p} = \frac{1}{128 \times 128} \sum_{i=1}^{128} \sum_{j=1}^{128} p_{ij} \qquad (42)$$

If the reconstructed image is the phantom, the distance critera d1, r1 and e1 will be all equal to zero. Therefore, the smaller the distance critera are, the better the reconstructed images will be. The three distance criteria reflect the different nature of the errors of the reconstructed images. d1 is sensitive to the individual large errors of the reconstructed image; r1 is sensitive to the accumulation of the small errors; while e1 indicates the maximum error of the elements of the reconstructed images, which is important for the quantitative image reconstructions (for obtaining the distance criterion e1, we have chosen the average of every four pixel values as the value of one element of a reconstructed image, the element represents part of the tissues and is usually of the size of several pixels).

If the images reconstructed have their averages different from the average of the phantom, we will use the average adjusted version of the distance criteria defined above. The average adjusted version of the distance criteria are defined and calculated as follows. First, we calculate the averages $\bar{r}$ and $\bar{p}$ of the reconstructed image and the phantom according to Eqs.(41) and (42), respectively, and then we add the difference between $\bar{p}$ and $\bar{r}$ to every pixel of the reconstructed image, and finally, we re-calculate the distance critera using Eqs.(36), (37) and (38) and define the newly calculated average adjusted distance critera as d2, r2 and e2, respectively. From the definition of the average adjusted distance criteria, one may see that the average adjusted distances get rid of the factor of the shift of the averages of the reconstructed images and, therefore, they are more closely connected to the qualities of the reconstructed images shown on a monitor (because the brightness of the images shown on the monitor can be adjusted arbitrarily to make the images looked better). In this paper, we will using following eight criteria d1, r1, e1, d2, r2, e2, max and min for all the images reconstructed. The notations max and min represent the maximun and the minimun values of the reconstructed images before the

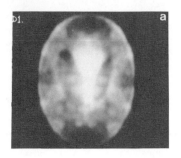

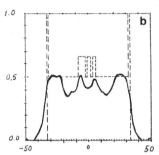

Fig.6 (a) The reconstructed 128x128 image with N=64 for the SADCT (Bilinear) (b) Numerical comparison on the line y=-0.605 (See Fig.5)

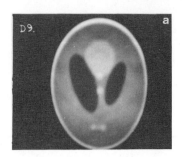

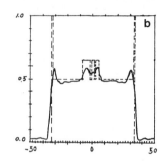

Fig.8 (a) The reconstructed 128x128 image with N=256 for the SADCT (Bilinear) (b) Numerical comparison on the line y=-0.605 (See Fig.5)

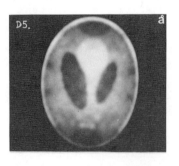

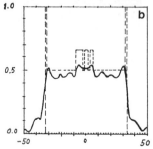

Fig.7 (a) The reconstructed 128x128 image with N=128 for the SADCT (Bilinear) (b) Numerical comparison on the line y=-0.605 (See Fig.5)

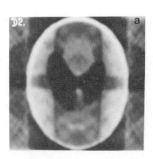

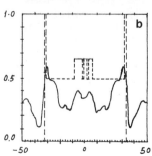

Fig.9 (a) The reconstructed 128x128 image with N=64 for the SADCT (Nearest-neighbor) (b) Numerical comparison on the line y=-0.605 (See Fig.5)

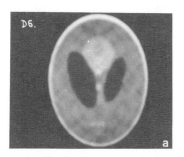

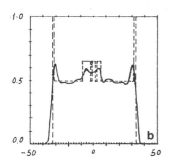

Fig.10 (a) The reconstructed 128x128 image with N=128 for the SADCT (Nearest-neighbor) (b) Numerical comparison on the line y=-0.605 (See Fig.5)

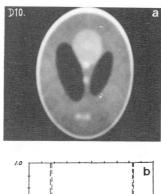

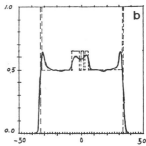

Fig.11 (a) The reconstructed 128x128 image with N=256 for the SADCT (Nearest-neighbor) (b) Numerical comparison on the line y=-0.605 (See Fig.5)

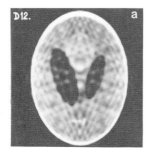

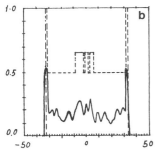

Fig.12 (a) The reconstructed 128x128 image with N=128 for the SADCT (Interpolation-free) (b) Numerical comparison on the line y=-0.605 (See Fig.5)

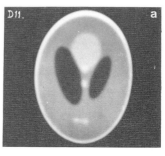

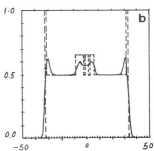

Fig.13 (a) The 128x128 image reconstructed from the windowed 128x128 Fourier-domain grids (delta=4k0/128) (b) numerical comparison on the line y=-0.605 (See Fig.5)

433

Table 1. The distances of the images reconstructed by the FDIRAs and the IFRA for the SADCT

Fig.	N	Interpolation	d_1	r_1	e_1	d_2	r_2	e_2	max	min
6	64	Bilinear	0.6355	0.4781	0.7959	0.6355	0.4781	0.7958	1.090	-0.2864
9	64	Nearest-Neighbor	0.9571	0.9100	0.6181	0.9570	0.9105	0.6139	0.8422	-0.2703
7	128	Bilinear	0.4659	0.3179	0.6089	0.4659	0.3179	0.6087	0.8364	-0.1203
10	128	Nearest-Neighbor	0.3367	0.1634	0.4617	0.3367	0.1634	0.4616	0.9202	-4.976×10^{-2}
8	256	Bilinear	0.3608	0.1943	0.5224	0.3608	0.1943	0.5223	0.8310	-4.943×10^{-2}
11	256	Nearest-Neighbor	0.3310	0.1421	0.4661	0.3310	0.1422	0.4659	0.9125	-2.593×10^{-2}
12	128	Interpolation-free	1.041	1.010	0.8365	0.5091	0.4224	0.6020	0.7724	-0.4149
13	/	Direct-calculation	0.3288	0.1228	0.4586	0.3288	0.1231	0.4584	0.9209	2.268×10^{-4}

adjustment of the averages of these images.

The Comparison of FDIRAs and IFRA

Figs.6(a), 7(a) and 8(a) show the 128x128 images reconstructed by the Fourier-domain bilinear interpolation reconstruction algorithm with N=64, 128 and 256, respectively; while Figs.9(a), 10(a) and 11(a) are the same as Figs.6(a), 7(a) and 8(a) respectively, except that they were reconstructed by the nearest-neighbor interpolation reconstruction algorithm. Fig.12(a) shows the 128x128 image reconstructed by IFRA, and Fig.13(a) is the 128x128 image reconstructed by direct calculation of the Fourier transform of the phantom on the rectangular grids. The sampling interval of Fig.13(a) on the Fourier-domain was taken as delta=$4k_o$/128. Fig.13(a) is the best image which can be obtained by DUCT with the use of 2-D blackman window of radius $\sqrt{2}k_o$ and this sampling interval. Figs.6(b) to 13(b) are the comparisons of the reconstructed values (real lines) and the real values (dashed lines) on the line y=-0.605 (see Fig.5) corresponding to Figs.6(a) to 13(a), respectively.

From Figs.7, 10 and 12, we can see that FDIRAs give better image reconstructions than IFRA with N=128 (for SADCT, the reconstructions will not be good if N is less than 128, this can be seen from Figs.6 and 9). From the comparison of Figs.8, 11, and Fig.9 in reference [15], the same conclusion will be obtained. In addition, we can see that the reconstructions obtained by the nearest-neighbor interpolation algorithm are better than those obtained by the bilinear interpolation, except for the case of N=64. These conclusions can also be seen clearly from table 1 (Table 1 shows the comparison of the distances of the reconstructed images from Figs.6 to 13). By the way, we shall state that the quality of the image of Fig.11 (N=256) is very close to its limit reconstruction

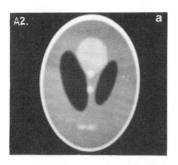

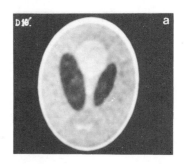

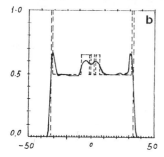

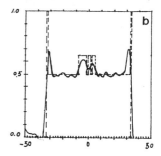

Fig.14 (a) The reconstructed 128x128 image with 360° rotated plane-wave insonification (Bilinear) (b) Numerical comparison on the line y=-0.605 (See Fig.5)

Fig.15 (a) The reconstructed 128x128 image with 360° rotated plane-wave insonification (Nearest-neighbor) (b) Numerical comparison on the line y=-0.605 (See Fig.5)

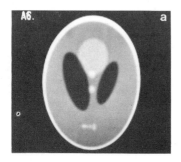

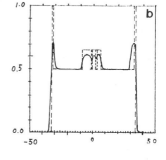

Fig.16 (a) The 128x128 image reconstructed from the windowed 128x128 Fourier-domain grids (delta=2√2k0/128) (b) numerical comparison on the line y=-0.605 (See Fig.5)

Fig.13, and this indicates that the information obtained from the 256x256 diffracted data for the SADCT are sufficient and the errors caused by the reconstruction algorithm itself is very small, provided that the weak scattering assumption is satisfied and the data obtained by the measuring system are sufficiently accurate.

For the conventional DUCT, the conclusion that the nearest-neighbor interpolation algorithm is better than the bilinear interpolation algorithm is not true. Fig.14 and 15 are the reconstructed images obtained by the conventional DUCT, and Fig.16 is the image reconstructed by the direct

Table 2. The distances of the images reconstructed by the bilinear and the nearest-neighbor interpolation reconstruction algorithms for the 360° rotated plane-wave insonified DUCT

Fig.	Interpolation	d_1	r_1	e_1	d_2	r_2	e_2	max	min
14	Bilinear	0.2910	0.1209	0.4100	0.2910	0.1209	0.4098	0.9436	-3.127×10^{-2}
15	Nearest-neighbor	0.3105	0.1895	0.4298	0.3105	0.1897	0.4292	0.9926	-8.233×10^{-2}
16	Direct calculation	0.2725	9.151×10^{-2}	0.3789	0.2725	9.186×10^{-2}	0.3788	0.9861	-1.880×10^{-4}

calculation of the Fourier transform of the object function on the rectangular grids. The conditions of the reconstruction of the image in Fig.16 are the same as those in Fig.13 except for delta=$2\sqrt{2}k_o/128$ (the radius of the blackman window is kept the same). The distances of the reconstructed images of Fig.14 to 16 are shown in Table 2.

From Fig.14, Fig.15 and Table 2, we can see that the image reconstructed by the bilinear interpolation algorithm is superior than that reconstructed by the nearest-neighbor interpolation. By the way, we can see that the image in Fig.16 is superior than that in Fig.13. This is because the image in Fig.16 contains more grid points in Fourier space than the image in Fig.13.

Figs.17, 18 and 19 are the moduli of the 128x128 spectra of the object obtained by the bilinear interpolation, the nearest-neighbor interpolation (N=128) and the direct calculation, respectively. They are corresponding to Figs.7, 10 and 13, respectively. In order to show the details of the spectra, we assigned those values which were greater than 2 to 2 in the spectra (the maximan value in these spectra was about 20 and occured at the center of these spectra). From these spectra, we can see that in the case of SADCT, the spectrum obtained by the nearest-neighbor interpolation is more accurate than that obtained by the bilinear interpolation, especially in the overlaped regions of the Fourier coverage areas A and B. This is why for SADCT, the nearest-neighbor interpolaiton gives better reconstructions.

Fig.17   The 128x128 spectrum image obtained by the bilinear interpolation (N=128). (before windowing)

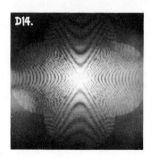

Fig.18   The 128x128 spectrum image obtained by the nearest-neighbor interpolation (N=128). (before windowing)

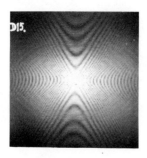

Fig.19 (a) The 128x128 spectrum image obtained by direct calculations (Before windowing)

Shift of Coordinate before and after Interpolations

Because we take the coordinate system shown in Fig.1, the center of the object is located at point (0,d/2). Therefore, the coordinate of the object must be shifted in Fourier domain before the 2-D IFFT is performed (the shift of the coordinate of the object in the Fourier domain is fulfiled simply by multiplying the spectrum of the object with a phase factor).

In the computer simulation, we discovered that the order of the shift of the coordinate and the interpolations had a great influence on the reconstructed images. The shift of the coordinate before the interpolations was better than the shift of the coordinate after the interpolations. Figs.20 and 21 show the images reconstructed by the bilinear and the nearest-neighbor interpolation algorithms, respectively, with the shift of the coordinate after the interpolations. Table 3 gives the comparison of the distances of these images and the images shown in Figs.7 and 10.

From Fig.20 we can see that there are noticeable unsymmetrics in the reconstructed image as compared with Fig.7, and, from Fig.21 we can see that artifacts are increased in the reconstructed image as compared with Fig.10. In Table 3, we will see that the distances for Figs.20 and 21 are increased greatly. These results indicate that the qualities of the reconstructed images will be degraded if the shift of the coordinate is performed after the interpolations.

With and Without Special Considerations of the Points Near the Boundaries

In the computer simulation, we also discovered that for SADCT, the accuracy of the interpolations of the points near the boundaries in the Fourier-domain coverage areas A or B (see Fig.2) had a great influence on the reconstructed images. This can be illustrated by Fig.22.

From Fig.22, we can see that the distributions of the discrete points on which the Fourier transform of the object is known are highly uneven. Near the axes of the rectangular coordinate, the points distribute densely, while in the areas near the boundaries of the Fourier-domain coverage areas A or B, the points distribute sparsely. Therefore, the interpola-

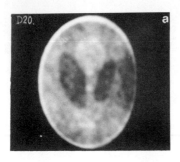

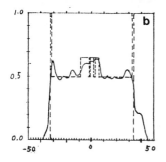

Fig.20 (a) The reconstructed 128x128 image with N=128 for the SADCT. The co-ordinate was shifted after the bilinear interpolation. (b) Numerical comparison on the line y=-0.605 (See Fig.5)

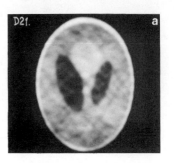

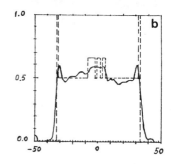

Fig.21 (a) The reconstructed 128x128 image with N=128 for the SADCT. The co-ordinate was shifted after the nearest-neighbor interpolation. (b) Numerical comparison on the line y=-0.605 (See Fig.5)

tions of the points near the boundaries will be less accurate than those near the axes.

The special considerations of the points near the boundaries are as follows. Rather than interpolate the points near the boundaries by averaging two interpolation results obtained from the areas A and B respectively as described in previous section, we interpolate the points near the boundary of the area A by the data in the area B, and vice versa. This will produce better results because a point near the boundary of one area will be the point near the axis in another (because the high values of the spectrum are concentrated in the low-

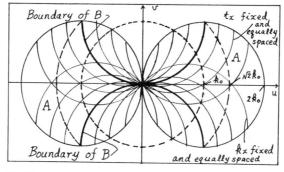

Fig.22 The distributions of the known values on the curvilinear coordinate for the SADCT

438

Table 3. The distances of the images reconstructed by the FDIRAs for the SADCT (coordinate shift after the interpolations)

Fig.	N	Specifications	d1	r1	e1	d2	r2	e2	max	min
20	128	Coordinate shifted after bilinear interpolation	0.4990	0.3591	0.7446	0.4990	0.3592	0.7445	0.8414	-0.2446
7	128	Coordinate shifted before bilinear interpolation	0.4659	0.3179	0.6089	0.4659	0.3179	0.6087	0.8364	-0.1203
21	128	Coordinate shifted after nearest-neighbor interpolation	0.3596	0.2017	0.5084	0.3596	0.2017	0.5082	0.9795	-9.540×10^{-2}
10	128	Coordinate shifted before nearest-neighbor interpolation	0.3367	0.1634	0.4617	0.3367	0.1634	0.4616	0.9202	-4.976×10^{-2}

frequency region, we just consider mainly those points in the region of lower frequency). In practical implementation of the interpolations of the points near the boundaries, we must decide which points are those near the boundaries. We solved this problem by experiments. As a result, we interpolated the points near the boundary of one area by the data in another when the conditions $|t_x| \geqslant k_o/32$ or $|k_x| \geqslant k_o/32$ were satisfied. Figs.23 and 24 are the reconstructed images obtained by the Fourier-domain bilinear and nearest-neighbor interpolations respectively, and with no specical considerations of the points near the boundaries (i.e., the interpolations of a point near the boundary of one area is performed by the data in the same area, then, if the point is in the intersections of two areas, the results of the interpolations obtained from these two areas are averaged).

Table 4. The distances of the images reconstructed by the FDIRAs for the SADCT (no considerations of the vicinity points of the boundaries)

Fig.	N	Specifications	d1	r1	e1	d2	r2	e2	max	min
7	128	Bilinear interpolation with the considerations of the vicinities of the boundaries	0.4659	0.3179	0.6089	0.4659	0.3179	0.6087	0.8364	-0.1203
23	128	Bilinear interpolation with no considerations of the vicinities of the boundaries	0.4935	0.3287	0.6660	0.4935	0.3288	0.6659	0.7883	-0.1204
10	128	Nearest-neighbor interpolation with the considerations of the vicinities of the boundries	0.3367	0.1634	0.4617	0.3367	0.1634	0.4616	0.9209	-4.976×10^{-2}
24	128	Nearest-neighbor interpolation with no considerations of the vicinities of the boundaries	0.3740	0.2151	0.5416	0.3740	0.2152	0.5415	0.9120	-0.1095

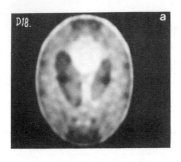

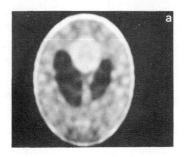

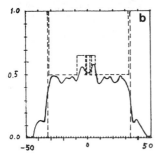

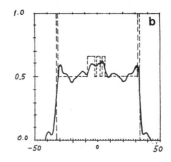

Fig.23 (a) The reconstructed 128x128 image with N=128 for the SADCT (bilinear). No special considerations of those points in the vicinities of the boundaries (b) Numerical comparison on the line y=-0.605 (See Fig.5)

Fig.24 (a) The reconstructed 128x128 image with N=128 for the SADCT (nearest-neighbor). No special considerations of those points in the vicinities of the boundaries (b) Numerical comparison on the line y=-0.605 (See Fig.5)

From Figs.23 and 24, we can see that the reconstructed images degraded greatly as compared with figs.7 and 10 respectively (Fig.7 and 10 are the images reconstructed with the special considerations of the points near the boundaries of the areas A and B using the method described above). In Table 4, we can see that the values of the distances for Figs.23 and 24 are increased remarkably.

For the results obtained by the shift of coordinate after the interpolations and, at the same time, with no special considerations of the points near the boundaries, the readers may refer to reference [18].

SUMMARY

In this paper, we have made a detailed study of FDIRAs for SADCT, and, the results of the study has been compared with those obtained by IFRA developed by Dr. D.Nahamoo et al..

The major conclusion of our computer simulation study is in the following :

(1) The number of the complex multiplications has been greatly reduced from the order $O(N^3+N^2\log_2N)$ for IFRA derived by Dr. D.Nahamoo at el. to $O(N^2\log_2N)$ for FDIRAs for an NxN image reconstructed from NxN diffracted data. (For N=128, VAX-11/730 CPU processing time were 2.61 min (bilinear) and 2.57 min (nearest-neighbor) for FDIRAs and 19.28 min for IFRA).

(2) FDIRAs gives better reconstructions than IFRA.

(3) For FDIRAs of SADCT, the nearest-neighbor interpolation reconstruction algorithm will give better results than the bilinear interpolation, but, it is not the case in the conventional DUCT.

(4) For FDIRAs of SADCT, the shift of the coordinate of the object must be performed before the Fourier-domain interpolations, otherwise, the quality of the reconstructed images will be degraded.

(5) The accuracy of the interpolations of the points near the boundaries of the Fourier-domain coverage areas has a great influence on the reconstructions. Therefore, to ensure a good image reconstruction, the interpolations of these points must be specially considered.

(6) In this paper, all the reconstructed images have been evaluated by the distance criteria, and these criteria are well coincidence with the quality of the reconstructed images.

(7) The relationships between the curvilinear coordinate and the rectangular coordinate for SADCT are not so straight-forward as compared with the conventional DUCT. It depends on which quarter of the Cartesian coordinate and which area of the Fourier-domain coverages a point (u,v) on the rectangular coordinate belongs to.

(8) For SADCT with $N=256$, the Fourier transform of the diffracted data will provide sufficient information of the Fourier transform of the object, and the nearest-neighbor interpolation reconstruction algorithm itself will cause little distortion and will produce a reconstruction close to the limit reconstruction (see Fig.13).

(9) The disadvantage of SADCT is that the distribution of the points on which the Fourier transform of the object are known are highly uneven as compared with the conventional DUCT [11], and, thus, more diffracted data are required for SADCT to obtain the same reconstruction quality as the conventional DUCT.

ACKNOWLEDGEMENT

The author is gratitude to Prof. Yu Wei, the author's doctoral supervisor, in Department of Biomedical Engineering, Nanjing Institue of Technology (who is now the president of the Institute) for her many suggestions in this work, such as, to adopt the distance critera for the quantitative evaluations of the reconstructed images. The author is also gratitude to Miss Li Lin, a graduated student in the Pharmacy University of China, for her help in drawing many figures in this paper.

REFERENCES

1. L. Axel, P.H. Arger, and R.A. Zimmerman, "Applications of Computerized Tomography to Diagnostic Radiology", <u>Proceedings of IEEE</u>, Vol. 71, No. 3, March 1983, pp. 293-297.
2. D.P. Boyd and M.J. Lipton, "Cardiac Computerized Tomography", <u>Proceedings of IEEE</u>, Vol. 71, No. 3, March 1983, pp. 308-319.
3. R.A. Robb, E.A. Hoffman, L.J. Sinak, L.D. Harris, and E.L. Ritman, "High-Speed Three-Dimensional X-Ray Computed Tomography", <u>Proceedings of IEEE</u>, Vol. 71, No. 3, March 1983, pp. 308-319.
4. H.T. Bates, K.L. Garden, and T.M. Peters, "Overview of Computerized Tomography with Emphasis on Future Developments", <u>Proceedings of IEEE</u>, Vol. 71, No. 3, March 1983, pp. 356-372.
5. J.K. Udupa, "Display of 3D Information in Discrete 3D Scenes Produced by Computerized Tomography", <u>Proceedings of IEEE</u>, Vol. 71, No. 3, March 1983, pp. 420-431.
6. P. Bloch and J.K. Udupa, "Application of Computerized Tomography to Radiation Therapy and Surgical Planning", <u>Proceedings of IEEE</u>, Vol. 71, No. 3, March 1983, pp. 351-355.
7. G.T. Herman, "Image Reconstruction from Projections", Academic Press, 1980.
8. J.F. Greenleaf, "Computerized Tomography with Ultrasound", <u>Proceedings of IEEE</u>, Vol. 71, No. 3, March 1983, pp. 330-337.
9. R.K. Mueller, M. Kaveh, and G. Wade, "Reconstructive Tomography and Applications to Ultrasonics", <u>Proceedings of IEEE</u>, Vol. 67, No. 4, 1979, pp. 567-587.
10. R.K. Mueller, M. Kaveh, and R.D. Iverson, "A New Approach to Acoustic Tomography Using Diffraction Techniques", <u>Acoustical Imaging</u>, A.F. Metherell, ed., Vol. 8, 1980, pp. 615-629.
11. S.X. Pan, and A.C. Kak, "A Computational Study of Reconstruction Algorithms for Diffraction Tomography: Interpolation Versus Filtered-Backpropagation", <u>IEEE Transactions on Acoustical Speech Signal Processing</u>, Vol. ASSP-31, 1983, pp. 1262-1276.
12. Cong-Qing Lan, Gail T. Flesher, and Glen Wade, "Plane-Scanning Reflection-Diffraction Tomography", <u>IEEE Transactions on Sonics and Ultrasonics</u>, Vol. SU-32, No. 4, July 1985, pp. 562-565.
13. A.J. Devaney, "A Filtered-Backpropagation Algorithm for Diffraction Tomography", <u>Ultrasonic Imaging</u>, Vol. 4, 1982, pp. 336-350.
14. A.J. Devaney, "A Computer Simulation Study of Diffraction Tomography", <u>IEEE Transactions on Biomedical Engineering</u>, Vol. BME-30, No. 7, July 1983, pp. 377-386.
15. D. Nahamoo, S.X. Pan, and A.C. Kak, "Synthetic Aperture Diffraction Tomography and Its Interpolation-Free Computer Implementation", <u>IEEE Transactions on Sonics and Ultrasonics</u>, Vol. SU-31, No. 4, July 1984, pp. 218-229.
16. D. Nahamoo and A.C. Kak, "Ultrasonic Diffraction Imaging", Tech. Rep. TR-EE-82-20, School of Electrical Engineering, Purdue University.
17. D.J. Vezzetti and S.ठ. Aks, "Reconstruction from Scattering Data: Analysis and Improvements of the Inverse Born Approximation", <u>Ultrasonic Imaging</u>, Vol. 1, 1980, pp. 335-345.
18. Jian-Yu Lu, "A Study of the Fourier-Domain Reconstruction Algorithms for Synthetic Aperture Diffraction Tomography", <u>Proceedings of China-Japan Joint Conference on Ultrasonics</u>, May 11-14, 1987, Nanjing, China, Edited by the Organizing Committee.
19. J.W. Goodman, "Introduction to Fourier Optics", <u>McGraw-Hill</u>, New York 1968, Chapter 3.
20. P.M. Morse and K.V. Ingard, "Theoretical Acoustics", <u>McGraw-Hill</u>, 1968, Chapter 8.

21. P.M. Morse and H. Feshbach, "Methods of Theoretical Physics", McGraw-Hill, 1953, p. 823.
22. L.A. Shepp and B.F. Logan, "The Fourier Reconstruction of a Head Phantom", IEEE Transactions on Nuclear Science, Vol. NS-21, 1974, pp. 21-43.

THE RANDOM PHASE TRANSDUCER : A NEW TOOL FOR SCATTERING MEDIUM

M. Fink, F. Cancre, D. Beudon and C. Soufflet

Groupe de Physique des Solides de l'E.N.S., Université Paris
VII - Tour 23 - 2, place Jussieu
75251 Paris Cedex 05 - France

ABSTRACT

In pulse echo measurement, interference between scatterers observed
with a coherent transducer, leads to scalloping in spectral estimation
(attenuation measurement) and to speckle noise in envelope detection
(imaging).

In order to get rid of these effects, Miller has proposed phase
insensitive transducers (CdS transducer and array of point like detectors). These transducers have shown promising results, but suffer from a
lack of sensitivity.

We have developed a new approach to control the degree of spatial
coherence of ultrasonic transducers.

Random Phase (R.P.) transducers consist in one or two rotating random
phase screens (R.P.S.) located in front of a coherent transducer. The
ultrasonic beam generated by the transducer is then transmited and
received through the moving R.P.S.. Spectral analysis and envelope
detection are processed by averaging the data on a set of locations of the
R.P.S. Optimal choice of the location of the R.P.S. will be discussed in
order to get spectral smoothing and speckle reduction. Experimental
results will shown the strong efficiency of the R.P. transducer in
attenuation measurement of tissue like phantom and in low constrast
specular reflector detection.

INTRODUCTION

Ultrasonic backscatter appears to represent a promising mode for
clinical tissue characterization and for quantitative imaging. Spectral
analysis and envelope detection are the commonly used tools in backscatter pulse echo measurement. However in biological tissues, interference among echoes from randomly situated scatterers within the resolution volume leads to statistical uncertainties in spectral estimation and
to speckle noise in envelope detection. These noises are linked to the
spatial coherent behavior of piezoelectric transducers : the complex
summation at the transducer face is assumed to be linear.

In spectral estimation, within each spectral element [narrow band
filtering], the spectral amplitude is a random variable whose values vary
from one scan line to the next. Within each narrow band, the ultrasound

spectral amplitude results from the coherent accumulation of random narrow band echoes from scatterers within the resolution cell. In this case the envelope exhibits Rayleigh statistics and the square of the envelope, which corresponds to the power amplitude obeys chi-square statistics. Measured power spectra exhibit a stochastic spectral rippling about the mean spectral shape. Ultrasound attenuation estimation from spectral measurement [narrow band or spectral centroid techniques] is thus strongly effected by this random noise, [1,2]. The incertainties associated with attenuation measurement can be reduced by averaging the running power spectra from a set of uncorrelated A lines. Usually an average of 128 A lines is performed and is obtained from a large tissue volume. A set of uncorrelated A lines is obtained by shifting the transducer laterally between each digitized echographic line. The shift must be of the order of the transducer beam width ($\simeq$ 2 mm). Therefore a volume of some cm^3 is needed in order to get accurate value of the attenuation.

On the other hand, in ultrasound imaging, <u>envelope detection</u> of the broad band pulse echo signal leads to the same kind of speckle noise. This phenomenon degrades the inherent target detectability. It appears as a random mottle superimposed on an image : It strongly reduces the perceived resolution and degrades the minimum detectable contrast level. The first and second - order statistics of speckle can be predicted by calculating the coherent echo sum of many such targets for Rayleigh statistics,[3,4]. These predictions have been experimentally verified by several groups,[5]. The commonly used figure of merit in speckle texture analysis is the ratio of the mean gray scale level (μ) to the standard deviation σ. For a Rayleigh probability function μ/σ equals 1.91. In general the size of speckles on 2D images is related to the bandwidth (in the axial direction) and to the beam width (in the lateral direction).

Many techniques have been proposed in order to reduce the speckle in envelope detected images, including multi frequency compounding,[6,7], spatial compounding,[8,9] and phase insensitive transducing,[10,11].

1 - Compounding

In the first two techniques, the speckle is reduced by forming an image which is the incoherent average of images with differing speckle patterns after envelope detection. Such images may be acquired by changing the spectrum of the acoustical pulse in transmit or in receive mode (frequency compounding), or by varying the angle from which a target is imaged (spatial compounding). The efficiency of these technics depends on the magnitude of speckle pattern change caused by the compounding process. It may be described by the speckle decorrelation between detected A lines or between ultrasonic images.

In <u>frequency compounding</u> the speckle correlation versus the center frequency has been studied by Trahey,[6] and it shows, that for a 2 Mhz bandwidth an improvement of the signal to noise ratio μ/σ of 2 may be obtained by averaging 5 partially correlated images with and by varying the center frequencies. However the speckle correlation decreases slowly and needs practically a shift of 500 KHz to ensure a correlation coefficient of 0.2. Besides the μ/σ increase of 2 is obtained at the expense of the axial resolution which is reduced by a factor of 6.5.

The success of <u>spatial compounding</u> in speckle reduction is a result of the statistical independance of speckle patterns arising from a fixed target region viewed from different angles. Optimisation of speckle reduction in phased array has been done. Trahey,[8] has studied the rate of decrease of speckle pattern decorrelation coefficients with lateral aperture translation. Division of the avalaible aperture into overlapping subapertures resulted in an increase of μ/σ. Independant speckle patterns are obtained with a translation of approximately 40 percent of the aperture. For example an increase of √3.2 of μ/σ is obtained at the expense of a factor 2 loss in lateral resolution.

2 - Phase insensitive transducers

Spectral and envelope speckle are linked to the coherent behavior of the piezoelectric transducer. There is a linear summation of the acoustical pressure over the whole transducer front face. In order to get rid of these effects, Miller has proposed phase insensitive transducers,[10,11]. He notes that phase cancellation effects (destructive interferences) at an extended aperture piezoelectric receiving transducer can compromise pulse-echo measurement. In the first approach he used acoustoelectric receivers (CdS) which suffer from a lack of sensitivity. In a recent work,[11], he investigated the application of a two-dimensional pseudo-array of point-like receivers for the phase-insensitive characterization of the scattered field. Phase insensitive measurements were obtained by summing the power spectra from all elements of the pseudo array. He shows an enhancement of the accuracy of spectral measurement. An increase of the signal to noise ratio of √6 has been obtained. It must be noted that the pseudo-array may be considered as the limiting case of the spatial compounding technique, where each sub-aperture is reduced to a single array element. In this case a strong loss of resolution is expected, due to the small dimension of the sub-apertures.

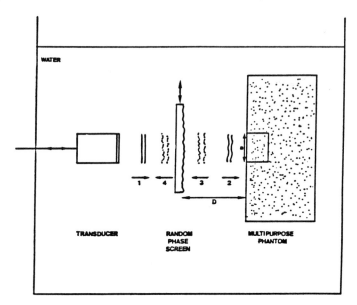

Figure 1 - Experimental procedure

I - THE RANDOM PHASE TRANSDUCER

In order to reduce the spectral and the envelope speckle noise, with a limited loss in resolution, we have developed a new approach yielding spatial incoherence. To control the degree of spatial coherence of US transducers, we place a moving random phase screen in front of a coherent transducer (Fig. 1). The ultrasonic beam generated by the transducer is then transmited and received through the moving random phase screen. Envelope detection and spectral analysis are then processed for each position of the Random phase screen, and averaging of these data is then performed on a set of locations of the Random phase screen.

The random phase screen is constructed using a material whose acoustic impedance is well matched to water, but whose sound velocity is quite different. Different rubbers used to build acoustic lenses may be chosen. The phase screen thickness varies randomly, in order to introduce a spatial random delay law, which corresponds to phase shift up to 2π at the center frequency. In monochromatic mode such a thin phase screen can be considered a transmittance of the form $e^{i\phi}$, where ϕ is random. The effects of such a moving diffuser has been studied in coherent optics, and it has been shown that reduction of the speckle noise can be achieved in optical imaging.

II - THEORETICAL APPROACH

In a first approach, in order to understand the behavior of such a random phase screen in pulse-echo mode, we shall neglect the Random phase screen effects in the transmit mode. We shall assume that the scattering medium can be represented by a random distribution of point targets located in the insonified region. We consider the case where this region has a finite dimension with a microscopic correlated microstructure.
In this case the scattering strengh of the medium can be considered as a complex random function $s(x)$. Where x is the position and refer to two dimensional coordinates. In the case of plane wave insonification. If the microstructure is uncorellated e.g randomly dispersed fine particles, then the autocorrelation function of the process $s(x)$ can be written :

$$R_s(x_2,x_1) = 0(x_2)\ 0^*(x_1)\ \delta(x_2-x_1)$$

where $0(x)$ is the aperture function of the insonified field.
If the insonified region is limited by a circular aperture of radius a

$$0(x) = circl\ (r/a)$$

The diffracted wave field originating from this random aperture propagates as a random field whose phase and amplitude fluctuate in space.

Usually, in coherent mode, this random wave field is directly intercepted by the transducer front face, and the resulting RF signal is corrupted by the speckle noise.

- The complex diffracted field observed on the random phase plane, located at a distance z from the scatterers is the Fresnel transform of $s(x)$. It can be written :

$$U(x) = s(x)\ \otimes\ \exp\ (j\alpha x^2)$$

where $\otimes$ is the usual convolution, and $\alpha = \pi/\lambda z$.

- The autocorrelation of the complex field observed in the plane z is then

$$R_U(x_2,x_1) = R_s(x_2,x_1)\ \overset{1}{\otimes}\ \exp\ (-j\alpha x_1^2)\ \overset{2}{\otimes}\ \exp(j\alpha x_2^2)$$

Using the definition of $R_s(x_2, x_1)$ we found :

$$R_U(x_2, x_1) = |0(x_1)|^2 \exp(-j\alpha\Delta x^2) \overset{\Delta x}{\circledast} \exp(j\alpha\Delta x^2)$$

with $\Delta x = x_2 - x_1$.
It is nothing else than the Fourier transform of $|0(x)|^2$ taken at the spatial frequency $F_x = \dfrac{\Delta x}{\lambda z}$

If $0(x) = circl(r/a)$. We found the point spread function of a finite aperture lens of focal length z.
Thus the spatial correlation scale or "speckle size" of the scattered wave field is then proportionnal to $\lambda z/a$.

- This random scattered field $U(x)$ will transmit through the random phase screen of transmittance $t(x)$, and the new random resulting field $U(x) t(x)$ will diffract up to the transducer front face.
If the correlation length of the random phase screen is well adapted to the correlation length of the field $U(x)$ a displacement of the phase screen will give decorrelated data.
    Taking into account the speckle size formula, it may be showed that the scattered wave field from a random medium located after the focus of the transducer will diffract a random wave field, whose grain size at the focal plane is constant and equal to $\lambda F/D$ where D is the transducer diameter and F the focal length. In our case, the aperture number (F/D) is equal to 5, with $\lambda = .3$mm, which corresponds to a speckle size of approximatively 1.5 mm. The most efficient random screen, located in the focal plane, must have an equivalent coherence length.

III - EXPERIMENTAL RESULTS

    The data were collected with 5 MHz focused transducer ($\emptyset$ = 19 mm or 13 mm and F = 90mm). The phantoms used contained graphite scatterers distributed throughout gelatin from Doctor Zabzeski, a water immersed fine-grained sponge and an echobloc phantom made of two types of foam. The data was obtained by digitizing the RF echoes bakscattered from the phantom at a rate of 32 MHz using a 8 bits Lecroy digitizer.
    The random phase screen is made of rubber and has a coherence length of approximatively 1 mm. During our acquisition sequence, the random phase screen is laterally shifted in front of the transducer. Depending on the speckle size and on the coherence length of the random phase screen, the set of resulting RF signals collected are partially or totally uncorre-lated.
    In the first experiments made on gelatin phantoms, 400 RF data are collected for different positions of the random phase screen. The lateral displacement is incremented by steps of .25 mm between each acquisition. The RF lines are envelope detected or then averaged. The same procedure may be done for spectral estimation. In a moving window the power spectrum is eveluated and then averaged for a set of random phase screen positions.
    The experimental results presented concerning envelope detection are the decorrelation efficiency, the SNR improvement, the influence of the position of the random phase screen and in the echobloc phantom, the detectability of the interface between two foams.

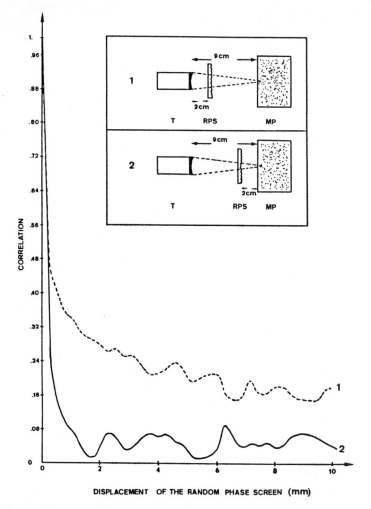

Figure 2 - Correlation coefficient versus RPS displacement

With the gelatin phantom and the ∅ 19 mm focused transducer, we have plotted the correlation coefficient [Fig. 2] between detected A lines versus the lateral translation of the random phase screen. The front face of the phantom was located near the focus of the transducer, and the treated data corresponded to a 20 µs window. Special attention was given to avoid multiple reflections between the phantom and the random phase screen. The two plotted curves correspond to 2 different locations of the random phase screen. The correlation coefficient decreases more quickly when the screen is located near the focus of the transducer. This is related to the fact that the beam intercepted by the screen is smaller, and then a short displacement of the screen decorrelates the data.

Taking into account the uncertainties on the correlation coefficient, we observed that a lateral shift of .5 mm gives a correlation coefficient of approximately .1.

450

SNR vs Number
of lines averaged

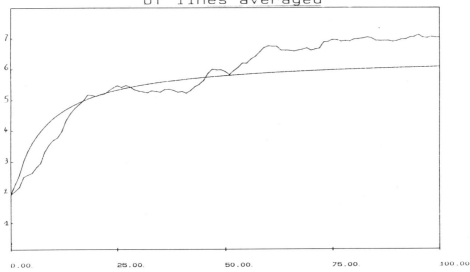

Figure 3

We then presented the calculated SNR versus number of averaged detected lines [Fig. 3]. In this case each A lines corresponds to a lateral shift of .5 mm, the random phase screen being located 2 cm in front of the transducer focus.

We observed on this figure an improvement of the SNR with the number of averaged lines. The optimal number of averaged A lines for this example is 60 with an expected speckle μ/σ increase up to a value of 7. This increase corresponds, compared to 1.91, to a relative increase greater than 3.5. It corresponds to an effective number of 12 completely decorrelated A lines obtained by averaging 60 partially correlated A lines. Increasing the number of A lines beyond 60 is not efficient.

These results are consistent with the formula which gives the increase in speckle μ/σ as a result of the averaging of N partially correlated lines.

It can be calculated using :

$$\sigma^2_{av} = \frac{\sigma^2}{N^2} + 2 \sum_{k=1}^{N} \sum_{j=k+1}^{N} \rho(X_k, X_j) \frac{\sigma^2}{N^2}$$

Where σ^2 is the variance of each line and $\rho(X_k, X_j)$ the correlation coefficient of line X_k and X_j.
Corresponding to a constant mean correlation coefficient of .1, a fit of the SNR is plotted on figure 3.

Single
Envelope

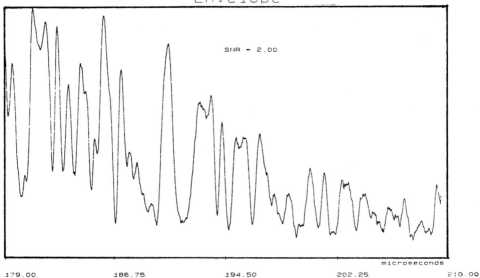

SNR = 2.00

microseconds

179.00   186.75   194.50   202.25   210.00

Average of
Envelopes

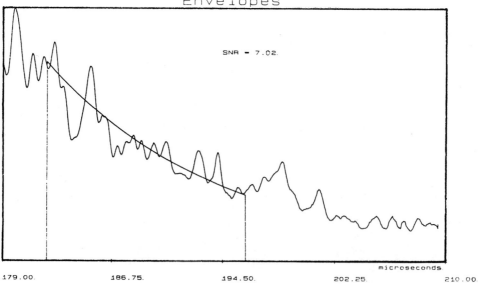

SNR = 7.02

microseconds

179.00   186.75   194.50   202.25   210.00

Figure 4

Figure 4 shows the improvement on the envelope detected A line through the random phase screen. The top curve corresponds to one single A line. The bottom curve is the average of 60 single A lines.

In order to evaluate the defocusing effect introduced by the random phase screen on the directivity pattern of the transducer, we have compared the directivity patterns of the focused transducer without the random phase screen and with the screen located at 2 cm and at 7 cm away from the focus.

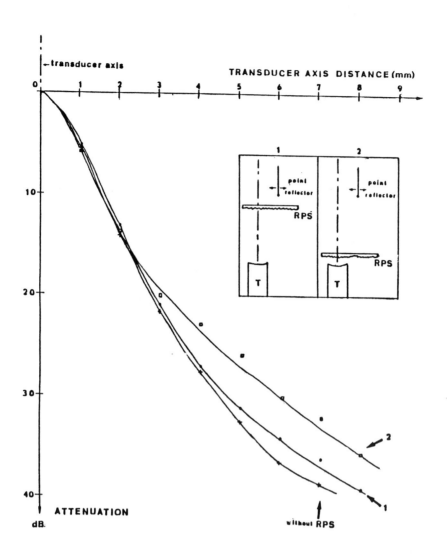

DIRECTIVITY PATTERN

Figure 5

Figure 5 shows that the - 6 db and the - 20 db lateral resolution is not affected by the presence of the random phase screen. However the random phase screen introduces divergent waves which reduce the - 40 db lateral resolution. This result shows that the improvement of the SNR is obtained, without an important loss of resolution.

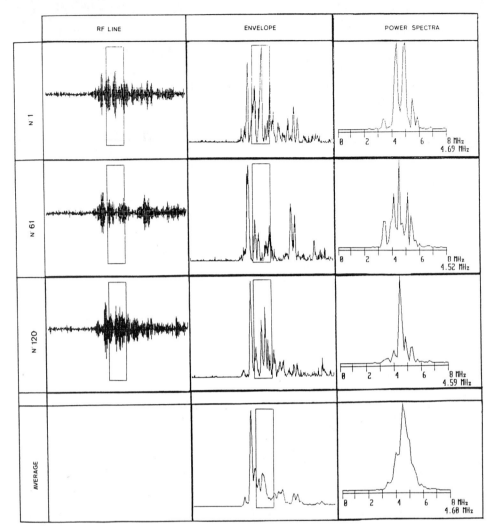

Figure 6

Figure 6 presents some results obtained on a fine grain sponge. 3 particular A lines are presented corresponding to 3 positions of the random phase screen. The complete envelopes and the power spectra calculated in the plotted window are presented for these 3 lines. The averaged envelope and power spectrum are plotted on the bottom of the drawing. It can be observed that the power spectrum noise is strongly reduced, and do not present the classical "rippling".

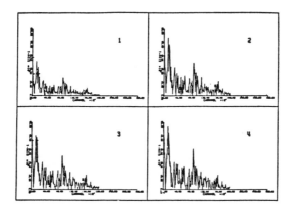

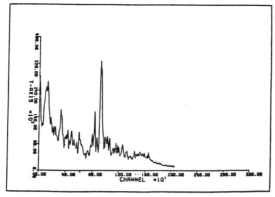

Figure 7 - Top curves : Envelope detected A lines from a 2 foams phantom
Bottom curve : Average of 60 detected A lines

Figure 7 shows the data collected from an echobloc phantom, made of two different foams. The first plots correspond to four single detected A lines, and the bottom curve to the average of 60 detected A lines. It may be checked that specular reflector are strongly enhanced in this process compared to the speckle noise which is greatly reduced.

CONCLUSION

Random phase screens moving in the beam of a coherent transducer can strongly reduce the speckle noise and the uncertainties in spectral estimation. Optimal choice of the location of the random phase screen is important in order to obtain a good decorrelation coefficient. An increase of the SNR of 3.5 has been obtained using this technique, with focused transducer. New geometries using two random phase screens are being studied and may give greater SNR.

REFERENCES

1. R.Kuc, Bounds on estimating the acoustic attenuation of small tissue regions from reflected ultrasound, Proc. IEEE. 73:1159 (1985).

2. M. Fink, F. Hottier and J.F. Cardoso, Ultrasonic signal processing for in vivo attenuation measurements : short time Fourier Analysis, Ultrasonic Imaging. 5:117 (1983).

3. C.B. Burckhardt, Speckle in ultrasound B-mode scans, IEEE Trans. Sonics Ultrasonics. SU-25:1 (1978).

4. R.F. Wagner, S.W. Smith, J.M. Sandrik and H. Lopez, Statistics of speckle in ultrasound B-scans, IEEE Trans. Sonics Ultrasonics. 30:156-163 (1983).

5. R.F. Wagner, M.F. Insana and S.W. Smith, Fundamental correlation lengths of coherent speckle in medical ultrasonic images, IEEE Trans. Ultrasonics, Ferroelectrics and Frequency control. (to be published).

6. P.A. Magnin, O.T. Von Ramm and F.L. Thurstone, Frequency compounding for speckle contrast reduction in phased array images, Ultrasonic Imaging. 4:267 (1982).

7. G.E. Trahey, J.W. Allison, S.W. Smith and O.T. Von Ramm, A quantitative approach to speckle reduction via frequency compounding, Ultrasonic Imaging. 8:151 (1986).

8. D.P. Shattuck and O.T. Von Ramm, Compound scanning with a phased array, Ultrasonic Imaging. 4:93 (1982).

9. G.E. Trakey, S.W. Smith and O.T. Von Ramm, Speckle pattern correlation with lateral aperture translation : experimental results and implications for spatial compounding, IEEE trans. Ultrasonics, Ferroelectrics and Frequency control. 33:257 (1986).

10. L.J. Busse and J.G. Miller, Detection of Spatially non uniform Ultrasonic Radiation with phase sensitive and phase Insensitive Receivers, J. Acoust. Soc. Am.. 70:1377 (1981).

11. P.H. Johnston and J.G. Miller, A comparaison of backscatter measured by phase sensitive and phase insensitive detection, Proc. IEEE Ultrasonics symposium. 827 (1985).

CALIBRATION OF ULTRASONIC TRANSDUCERS BY TIME DECONVOLUTION OF THE DIFFRACTION EFFECTS

D. Cassereau, and D. Guyomar*

Etudes et Productions Schlumberger
26 rue de la Cavée
92140 Clamart, France

ABSTRACT

Due to diffraction effects, a wavefront launched by a planar transducer becomes distorted. In the impulse domain, these effects are represented mathematically by a temporal filter. The signal observed from a transducer working in the emitting/receiving mode results from a time convolution
of the diffraction filter with the acousto-electrical response of the
transducer. This transducer response is not directly observable experimentally since a calibration can not be achieved without a propagation fluid.
To overcome this problem, a method, based on a time-deconvolution of the
radiation filter, is proposed. This method leads to an absolute calibration
of the transducer impulse response and explains the results of experimental
observations in a simple way. The comparisons between theory and experiment
show an excellent agreement. The proposed method enables the prediction of
the output signal at different observation distances. The proposed concept,
applied to annular transducers, gives a simple explanation to the vibration
behavior of the transducer external rings.

I. INTRODUCTION

A lot of interest has been given lately to the computation of the field
radiated from pulsed sources. Several techniques have been developped to
solve the direct problem corresponding to a description of the diffracted
pattern in terms of the transducer surface displacement [1-8]. We presently
adress the problem of a pulsed source insonifying a planar reflector.

To observe a transducer response, the probe has to be immerged in a
propagating medium in front of a reflector. Consequently for an adequate
transducer characterization, the influence of the front medium and the
reflector must be considered. The perturbations are due to absorption and
diffraction in the propagating medium and acoustical impedance and geometry
of the reflector. In the following a lossless propagation medium and a hard
plane reflector will be considered.
By integrating the pulsed diffracted field over the surface of a finite
receiver, T.L. Rhyne [9] has adressed a closed form solution to the radiation coupling of a disk. Time and spectral characteristics of this function
have been detailled and he showed that the transfer function associated to
the diffraction corresponds to a high pass filter.

---

* Currently with Thomson-Sintra A.S.M., Chemin des Travails, 06801 Câgnes
  sur Mer, France.

In this paper we first use the concept of radiation coupling to explain the results of classic experiments such as the enhancement of high frequency components with the transducer/reflector distance.
In a second part a method is proposed to remove the diffraction effects perturbing the transducer response in the common plane transducer/plane reflector configuration. The method presented in this paper to take away the radiative effects is based on a numerical time deconvolution of the radiation coupling function. Although the deconvolution is usually a delicate operation it is shown that the radiation coupling function is always an inversible filter in this case. Therefore a straightforward deconvolution scheme can be applied to the observed time signal to get the specific transducer response.
The deconvolved waveform allows comparisons with existing transducer response model or/and an estimation of the radiation coupling gain influence on the transducer behavior.

The concept of radiation coupling function is also extended to annular arrays resulting in an easy interpretation of the most external annuli transient response. Although the external annuli response exhibits significative differences from centered ones, the deconvolved waveforms look similar. This confirms the fact that the discrepancies appearing on the time signals are mainly due to the diffraction effects.
Comparisons between predicted and observed waveforms are given for disk and annular sources.

II. BASIC THEORY

A common way to calibrate a transducer, in terms of shape and amplitude, is to have a probe working in the emitting/receiving mode insonifying a planar reflector (Fig. 1). The time response of the acoustic chain can be written as a combinaison of the acousto-electrical response of the transducer, the impulse response of the firing/receiving electronics and the radiation coupling function.

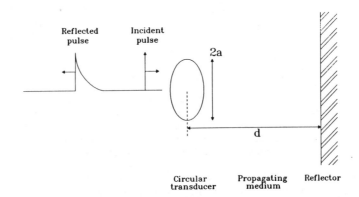

Fig.1 : Geometrical configuration. The two pulses symbolize the emitted and reflected waves. The distortion of the initial pulse is clearly visible.

It can be shown, for linear propagation, that these successive effects can be expressed in terms of iterated convolutions. Therefore the output signal W(t) is written as follows :

$$W(t) = F(t) \overset{*}{\underset{t}{}} H(t) \overset{*}{\underset{t}{}} D(t)$$ \hfill (1)

with :
  - F(t) is the response characterising the firing/receiving filter.
  - H(t) is the acousto-electrical response of the transducer.
  - D(t) is the radiation coupling function.

For calibration purposes, the quantity of interest is the transducer response H(t). The determination of this function is quite important if comparisons must be made with theoretical predictions or numerical results. Because the firing/receiving electronics has usually a broad bandwidth (Dirac like behavior in time), or because this function is easily modeled, the response of the filters is not separated from the transducer one :

$$W(t) = F(t) \overset{*}{\underset{t}{}} H(t) \overset{*}{\underset{t}{}} D(t) = P(t) \overset{*}{\underset{t}{}} D(t) \tag{2}$$

with

$$P(t) = F(t) \overset{*}{\underset{t}{}} H(t) \tag{3}$$

The output signal appears now as a simple convolution between P(t), the transducer response in its electronics environment and D(t) corresponding to the diffraction effects. It is worth noting that P(t) depends only upon the transducer system electrical characteristics while D(t) depends only on the geometrical characteristics of the probe.
The main goal of this paper is to remove the effects of D(t) from the experimental waveform W(t) to get a numerical expression of P(t).

The function D(t) has been described by T.L. Rhyne [9] and other authors [10] in terms of velocity potential and pressure for a transducer vibrating in a piston mode. A closed form solution has been adressed only for a planar disk, however its extension to annular arrays is straightforward. Both radiation functions are presented in the appendix and graphic representations are given in Fig. 2 and Fig. 3.
It can be shown that the radiation coupling function D(t) is a function of the transducer diameter and the distance probe/reflector (See appendix).

The coupling function starts always on a discontinuity corresponding to the first wave arrivals, then decreases to reach zero on a finite time period. This time period increases with the transducer diameter and decreases with an increasing transducer/reflector distance (Fig. 2).
The preceding description in terms of linear system theory leads to an easy interpretation of the following experimental results :

Required distance for optimal experimental calibration

It is well known that to calibrate a transducer and get a good evaluation of its response, the probe must be located as close as possible from the reflector. The radiation coupling function provides a simple explanation to this empirical way of calibrating. Observing that the slope on the decreasing part of the potential coupling function is low, that implies a small negative part after the Dirac on the pressure response (Fig. 4). The diffraction effects in terms of pressure have a Dirac like behavior and consequently do not perturb the transducer response. If the transducer/reflector distance is larger then the negative part of the pressure coupling function is not negligeable anymore and has a strong contribution to the signal output.

Enhancement of high frequencies with the distance probe/reflector

As it can be seen on Fig. 4, the pressure radiative function shows two discontinuities of opposite signs. The amplitude of the negative peak

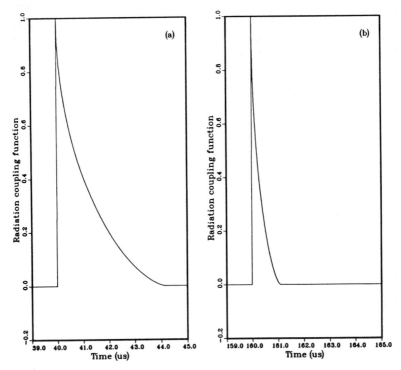

Fig.2 : Radiation coupling function of a disk
(a=14 mm) in terms of the velocity potential :
a) d=30 mm – b) d=120 mm.

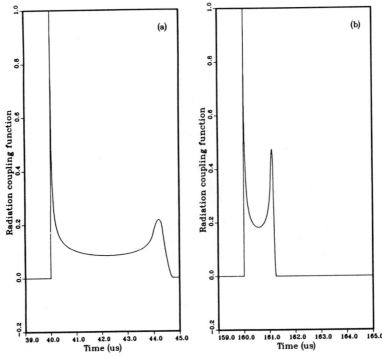

Fig.3 : Radiation coupling function of an annulus
in terms of the velocity potential (a_1=13.7 mm,
a_2=15 mm) : a) d=30 mm – b) d=120 mm.

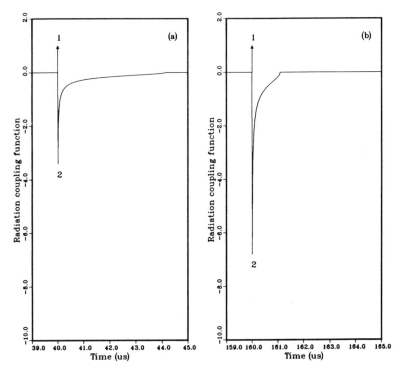

Fig.4 : Pressure radiative function of a disk
(a=14 mm) : a) d=30 mm - b) d=120 mm.

increases with the transducer/reflector distance. It is enough to derivate
a signal to convolve it with a Dirac derivative. Therefore the convolution
of the pressure radiation function with the transducer response corresponds
to something close to a time differentiation of the latter when the trans-
ducer/reflector distance is large. That implies an enhancement of the high
frequency components of the signal as it is visible on Fig. 5.

III. COMPUTATION OF THE SPECIFIC TRANSDUCER RESPONSE

Starting with the convolution equation :

$$W(t) = P(t) \overset{*}{\underset{t}{}} D(t)$$

(where $\overset{*}{\underset{t}{}}$ is the convolution operator over the time variable t), the process
to obtain P(t) consists in removing the effects of the linear filter D(t)
from the output signal W(t). Mathematically it corresponds to a deconvolution or a convolution with the inverse convolution filter $D^{-1}(t)$.
The deconvolution is a delicate operation unless the filter exhibits
specific properties. As it is proved later the filter D(t) is inversible,
therefore resulting in an exact deconvolution. Writing an explicit
expression for W(t) leads to :

$$W(t) = \int_{-\infty}^{+\infty} P(\tau)D(t - \tau)d\tau \qquad (4)$$

Noting that P(t) and D(t) are causal functions, Eq. (4) can be written as :

$$W(t) = \int_{0}^{t} P(\tau)D(t - \tau)d\tau \qquad (5)$$

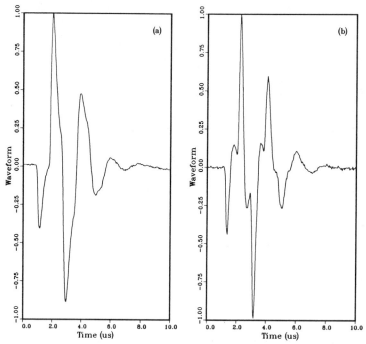

Fig.5 : Experimental waveform of a circular
transducer (a=14 mm) : a) d=25 mm – b) d=225 mm.

A discrete form of the preceding equation can be expressed at each step
of time T by :

$$W_n = \sum_{i=1}^{n} P_i D_{n-i+1} \qquad (6)$$

with n=E(t/T) (where E(x) returns the integer part of the real argument x).
The subscript i corresponds to the function taken at t=iT : W_i =W(iT).
Assuming a waveform sampled on N points and writing the discrete integral
equation at each step of time leads to the linear system :

$$[W] = [D][P] \qquad (7)$$

with

$$[D] = \begin{pmatrix} D_1 & 0 & \cdots & 0 \\ D_2 & D_1 & \cdots & 0 \\ \vdots & \vdots & \ddots & \vdots \\ D_N & D_{N-1} & \cdots & D_1 \end{pmatrix} \qquad (8)$$

$$[W] = \begin{pmatrix} W_1 \\ W_2 \\ \vdots \\ W_N \end{pmatrix}, \qquad [P] = \begin{pmatrix} P_1 \\ P_2 \\ \vdots \\ P_N \end{pmatrix} \qquad (9)$$

The [D] matrix is triangular and its first diagonal corresponds to the
value D_A of D(t) on the discontinuity. The determinant of a triangular
matrix is the product of the diagonal elements. Elementary computation
shows that the determinant equates D_A^N . Since D_A never vanishes, the
determinant never cancels. Therefore from algebra theorems, the matrix [D]
is inversible. As a consequence the specific transducer response [P] is

462

obtained by multiplying $[W]$ and $[D]^{-1}$:

$$[P] = [D]^{-1}[W] \qquad (10)$$

IV. NUMERICAL RESULTS

To apply the proposed method on real waveforms, experiments have been run. The different steps of the process can be summarize as follows :
- Digitization and storage of the output transducer signal observed at a distance d.
- Evaluation of the pressure coupling function at this distance from Rhyne formulation.
- Numerical deconvolution of the pressure coupling function from the output signal.

These experiments have been repeated for various distances d (25 mm and 225 mm). The output signal exhibits significant differences. However the deconvolved waveforms, obtained from different output signals, look similar. This similarity could be expected since the transducer response P(t) depends only on the acousto-electrical characteristics of the probe. The deconvolved waveforms correspond to the real transducer response. Fig. 6 represents the impulse response of the transducer obtained by numerical deconvolution of the diffraction effects at d=25 mm (Fig. 6a) and d=225 mm (Fig. 6b). The obtained impulse response is very close to the observed waveform at 25 mm.
Although the experimental waveforms in these two cases look different, the two deconvolutions lead to the same numerical determination of the impulse response of the transducer. This result confirms the fact that the signal distortion is due to the diffraction effects in the propagating medium.

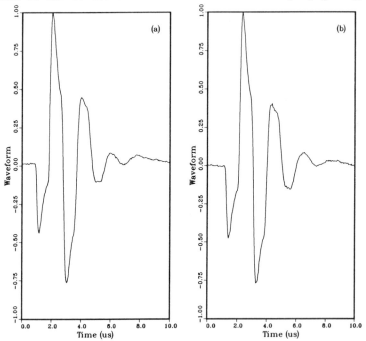

Fig.6 : Transducer response obtained by deconvolution of the waveform at : a) d=25 mm - b) d=225 mm.

V. PREDICTION OF THE TRANSDUCER WAVEFORM

Once the transducer response is known, a forward convolution enables the prediction of the output waveform for any other geometrical configu-

ration. To avoid self compensating errors, the prediction of the output signal at 225 mm was obtained by convolution of the pressure coupling function computed at 225 mm with the transducer response obtained by deconvolution of the experimental waveform digitized at 25 mm.
The same scheme was applied to evaluate the waveform at 25 mm from the deconvolved waveform at 225 mm.
Fig. 7 represents the comparison between the experimental (continuous curves) and the predicted waveforms (dashed curves) at 25 mm (Fig. 7a) and 225 mm (Fig. 7b). In both cases we obtain an excellent agreement.

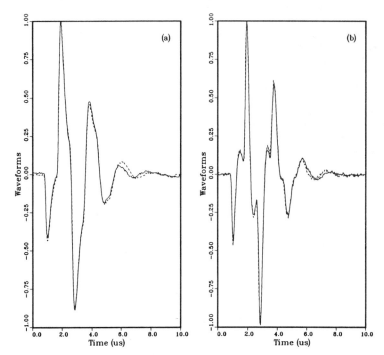

Fig.7 : Comparison between experimental and predicted waveforms at : a) d=25 mm – b) d=225 mm.

VI. APPLICATION TO ANNULAR ARRAYS

It has been observed experimentally that the external annuli response differs appreciably from the central one. Once the concept of radiation coupling function is generalized to include annular geometries, it leads to an explanation of this experimental result.
T.L. Rhyne [9] has adressed a closed form solution to the radiation coupling function in terms of the velocity potential for two plane disks (emitter and receiver) of same radius a. In the appendix, the radiation coupling function is first extended to a plane emitting disk of radius a and a plane receiving disk of radius b. The corresponding closed form solution is simply obtained by integration of the pulsed diffracted field over the receiving surface.
Looking at this expression, we can make the two following remarks :
  - Interchanging a and b does not change the expression of the radiation coupling function.
  - Assuming b=a we rediscover the Rhyne formulation.

Considering the linearity in emission and in reception, it becomes possible to express the radiation coupling function of an annulus in terms of the generalized radiation coupling function of two different disks. The final expression given in Eq. (13) shows a cross coupling term between the internal and external radii of the annulus.

The effect of this cross coupling term is illustrated in Fig. 3 (representing the radiation coupling function in velocity potential) and in Fig. 8 (representing the radiation coupling function in pressure). By time convolution with the impulse response of the transducer, this cross coupling term introduces a second echo separated from the first one. For a great distance between the transducer and the reflector, this separation do not appear clearly since the two different echos overlap in the time domain.

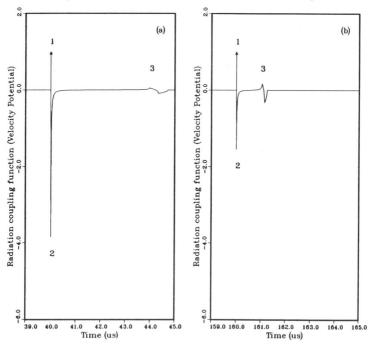

Fig.8 : Pressure radiative function of an annulus
(a_1=13.7 mm, a_2=15 mm) : a) d=30 mm − b) d=120 mm.

Fig. 9 represents at 50 mm (Fig. 9a) and 70 mm (Fig. 9b) the comparison between the recorded (continuous curves) and the predicted (dashed curves) waveforms. The theoretical prediction leads to a correct determination of the time behavior in the two cases, but the peak relative amplitudes are not retrieved exactly. This figure illustrates the differences between the central and external annuli responses and the effects of annular geometry.

VII. CONCLUSION

A method based on a numerical deconvolution of the diffraction effects from the transducer output signal was presented. The method enables the determination of the real acousto−electrical response of the transducer. It corresponds to a calibration of the transducer response.
The concept of radiative function has been extensively used to interpret experimental observations. This notion is quite powerful since it allows a separation of the output signal in two terms : the acousto−electrical response of the probe and a second term that depends only upon the geometrical configuration of the system.
The deconvolution method has been applied to annular arrays to explain the external annuli behavior. It has been shown that differences between central and external annulus responses are mainly due to diffraction effects.

APPENDIX : THE RADIATION COUPLING FUNCTION

We first consider an emitting and a receiving disk of radii a and b.

465

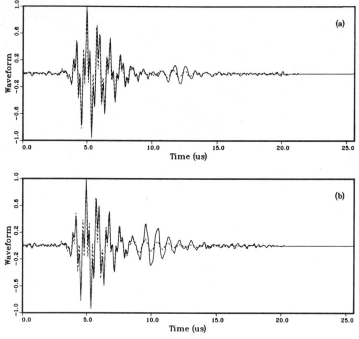

Fig.9 : Comparison between experimental and predicted
waveforms (a_1=13.7 mm, a_2=15 mm) at :
a) d=50 mm - b) d=70 mm.

The surfaces of the disks are parallel and distant of d. By integrating
the potential field radiated from the emitter over the receiving surface,
it can be shown that the potential radiation coupling function $H_v(a,b,d,t)$
is given by :

$$H_v(a,b,d,t) = \begin{cases} 0 & \text{if } t < t_1 \\ c\pi \left[\text{Min}(a,b)\right]^2 & \text{if } t_1 \leq t \leq t_2 \\ ca^2 \cos^{-1}\left(\dfrac{c^2 t^2 - d^2 + a^2 - b^2}{2a\sqrt{c^2 t^2 - d^2}}\right) + \\ cb^2 \cos^{-1}\left(\dfrac{c^2 t^2 - d^2 + b^2 - a^2}{2b\sqrt{c^2 t^2 - d^2}}\right) + & \text{if } t_2 < t \leq t_3 \\ \dfrac{c^3}{2}\sqrt{(t^2 - t_2^2)(t_3^2 - t^2)} \\ 0 & \text{If } t > t_3 \end{cases} \quad (11)$$

with the following notations :

$$t_1 = \frac{d}{c} \qquad t_2 = \frac{\sqrt{d^2 + (a-b)^2}}{c} \qquad t_3 = \frac{\sqrt{d^2 + (a+b)^2}}{c}$$

Interchanging a and b in this expression does not change the radiation
coupling function. This property expresses the reciprocity relations of
the considered system :

$$H_v(a,b,d,t) = H_v(b,a,d,t) \quad (12)$$

Considering now that the receiver is identical to the emitter, we obtain
from Eq. (12) the Rhyne formulation [9] for the radiation coupling function
in terms of the velocity potential.

We can now use the preceding results to obtain the closed form solution to

466

the radiation coupling function of an annulus of internal and external radii a_1 and a_2. The pulsed diffracted field from this annulus (in the emitting mode) can be written as the difference of the field radiated from a disk of radius a_2 and a_1. The same argument can be naturally applied to the integration of the diffracted field over the receiving surface (by linearity). Finally we obtain the following expression :

$$H_{v/\text{ann}}(a_1, a_2, d, t) = H_v(a_2, a_2, d, t) + H_v(a_1, a_1, d, t) - 2H_v(a_1, a_2, d, t) \qquad (14)$$

In this expression appears a cross coupling term that takes the ring geometry into account. This cross coupling term allows an easy interperation of the particuliar behavior of the most external annuli.

REFERENCES

1   G.R. Harris, "Review of transient field theory for a baffled planar piston", J. Acoust. Soc. Am. 70 (1), 1981, pp. 10–19.
2   P.R. Stepanishen, J. Acoust. Soc. Am. 49, 1971, pp. 283–292.
3   P.R. Stepanishen, J. Acoust. Soc. Am. 49, 1971, pp. 1629–1838.
4   G.R. Harris, "Transient field of a baffled piston having an arbitrary vibration amplitude distribution", J. Acoust. Soc. Am. 70, 1981, pp. 186–204.
5   P.R. Stepanishen, "Experimental verification of the impulse response method to evaluate transient acoustic fields", J. Acoust. Soc. Am. 63 (6), 1981, pp. 1610–1617.
6   D. Guyomar and J. Powers, "Propagation of transient acoustic waves in lossy and lossless media", Acoust. Imag. Vol. 14, 1985, Plenum Press New-York.
7   D. Guyomar and J. Powers, "Transient fields radiated by curved surfaces : Application to focusing", J. Acoust. Soc. Am. 76 (5), 1984, pp. 1564–1572.
8   D. Guyomar and J. Powers, "Transient radiation from axially symmetric sources", soumis et révisé, J. Acoust. Soc. Am.
9   T.L. Rhyne, "Radiation coupling of a disk to a plane and back or a disk to a disk : An exact solution", J. Acoust. Soc. Am. 61, 1977, pp. 318–324.
10   D. Guyomar, "Théorie et méthodes de la diffraction impulsionnelle", Thèse de Doctorat d'Etat, Janvier 1986, Université Paris VII.

DETERMINATION OF THE IMPULSE DIFFRACTION

OF AN OBSTACLE BY RAY MODELING

D. Cassereau, and D. Guyomar*

Etudes et Productions Schlumberger
26 rue de la Cavée
92140 Clamart, France

ABSTRACT

A method is proposed to model the output signal of an ultrasonic
transducer after reflection of the pulsed radiated wave on a reflector of
arbitrary geometry. This method, based on an impulse ray model, corresponds
to a discrete approach of the Kirchhoff-radiation integral. It allows prediction of the radiation coupling function obtained by T.L. Rhyne [1] in the
simple case of a planar circular transducer and a planar reflector of infinite acoustical impedance. It also permits an explanation of the spectral
distortion of the echo due to the presence of an obstacle. The developed
concept provides an easy understanding of the reflected waveform shape.
After a presentation of the principles of the method and its implications,
we present comparisons between theoretical and experimental results obtained for different reflector shapes (cylinder, inclined plane). In all these
cases, we can not obtain a closed form solution for the radiation coupling
function. The ray model leads to a numerical evaluation of this function.

I. INTRODUCTION

Due to the diffraction effects, the wavefront launched by a planar
transducer becomes distorted. In the impulse domain these distortion
effects can be mathematically represented by a temporal filter called "the
radiation coupling function" [1].
This radiation coupling function is obtained assuming a Dirac time behavior
of the normal velocity over the ceramic surface and integrating the pulsed
diffracted field over the receiver surface. In this case we obtain the
radiation coupling function in terms of the velocity potential. Since our
transducer is sensible to the acoustic pressure applied on its surface, the
evaluation of the pressure radiation coupling function is obtained by time
differentiation of the velocity potential radiative function.

Considering the linearity of the propagation equation, the transient
response of the transducer can be expressed as the time convolution of the
radiation coupling function in pressure with the impulse response of the
transducer [2]. This convolution equation shows a complete separation of
two filtering processes appearing on the transducer response. The impulse

----

* Currently with Thomson-Sintra A.S.M., Chemin des Travails, 06801 Câgnes
sur Mer, France.

response of the transducer depends strictly upon the acousto-electrical properties of the transducer while the radiation coupling function depends only on the geometrical configuration emitter/reflector/receiver.

The radiation coupling function contains all the informations about the geometry of the system. Since the signal perturbations are due to absorption and diffraction in the propagating medium and acoustical impedance of the reflector, it appears clearly that the radiation coupling function is a very important parameter in order to predict spectral or/and temporal distortions appearing on the waveform. T.L. Rhyne [1] has adressed a closed form solution to the radiation coupling of a planar circular transducer working in the emitting/receiving mode and a hard plane infinite reflector parallel to the transducer surface. For an arbitrary reflector geometry, a closed form solution can not be found.
In this paper we present a numerical method to obtain the radiation coupling function for any geometrical configuration emitter/reflector/receiver. This method is based on an impulse ray modeling, corresponding to a discrete approach of the Kirchhoff-radiation integral.
We first present the method and examine the convergence of the numerical process. A comparison is presented between the results obtained with the proposed method and the Rhyne solution [1].
Then the numerical process is applied to several geometrical configurations to get the pressure radiative function. The time convolution of the acousto-electrical response of the transducer leads to the prediction of the transducer output signal and explains qualitatively the signal distortion. Comparisons between experimental and theoretical waveforms are also given.

In the following a lossless propagation medium and a hard reflector will be considered.

II. THE IMPULSE RAY MODEL. PRINCIPLES AND JUSTIFICATION

We use a time domain approach of the diffraction that leads to a separation in time of the different waves contributing to the radiation coupling function. In particuliar the impulse ray modeling and the concept of transit time associated to the propagation of each ray allow an immediate geometrical interpretation of the finite time interval where the radiation coupling function is defined.

The basic principle of the ray model is close to the one commonly used in geometrical optics : an impulse spherical wave is described by "acoustical rays" propagating along straight lines. The following parameters are associated to each ray :
   - The emitting point of spatial coordinates x_o, y_o and z_o.
   - The direction of propagation given by u_x, u_y and u_z.
   - The amplitude A.
   - The emission time t_o.

Each "acoustical ray" is reflected from the infinite impedance reflector and propagated back to the transducer. We only consider the reflected rays intersecting the receiver surface. For these rays it is easy to evaluate the arrival time considering the geometrical path between the emitting and the receiving points.

It is well known that the amplitude of a spherical propagating wave varies as the inverse of the radius. This variation of the amplitude is a consequence of the conservation of the total energy carried by the spherical wave.
Fig. 1 represents an emitting point E and a plane reflector. The dashed arrows correspond to the acoustical rays propagating from E in direction of

470

the reflector sampling points. The circles symbolize the spherical wave emitted by E for different times. It is clearly visible that the number of acoustical rays modeling the spherical wave does not vary as the time increases. Therefore we do not consider any amplitude variation along a ray. The local density of rays decreases as time increases but the total energy is conserved.

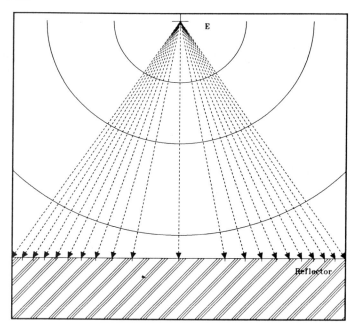

Fig.1 : The ray model compared to the theoretical wave. The number of rays intersecting the impulse spherical wave does not vary as time goes by.

We consider sampling points on the emitting transducer and on the reflector, each ray having the same amplitude A=1.
In order to justify this model we consider an emitting point E (whose coordinates are $x_e=0$, $y_e=0$ and $z_e=0$), an infinite hard reflector at the altitude z=d and an infinite receiver in the plane z=0. The diffracted field in the receiving plane is observed after reflection. By reciprocity, this problem is equivalent to that of an infinite receiver in the plane z=2d without reflector [1].

For the computation of the radiation coupling function, a time interval $[t_{min}, t_{max}]$ and the number N of data points must be defined. The numerical process consists in the evaluation, at each step of time t_i, of the number of acoustic rays contributing to the diffracted field in the time interval $[t_i, t_i+\delta t[$, δt being a fixed step :

$$\delta t = \frac{t_{\max} - t_{\min}}{N} \tag{1}$$

We now try to define the spatial sampling versus the time sampling. It is easy to verify that the rays contributing to the radiation coupling function in the time interval $[t_i, t_{i+1}[$ correspond to the spatial interval $[r_i, r_{i+1}[$ on the reflector. Elementary algebra leads to :

$$\begin{cases} 2r_i = \sqrt{c^2 t_i^2 - 4d^2} \\ 2r_{i+1} = \sqrt{c^2 t_{i+1}^2 - 4d^2} \end{cases} \tag{2}$$

Writing

$$t_i = \frac{2d}{c} + (i-1)\delta t \tag{3}$$

Eq. (2) becomes :

$$r_i = \frac{1}{2}\sqrt{c^2\delta t^2(i-1)^2 + 4c\delta td(i-1)} \tag{4}$$

The preceding relation gives the spatial sampling once the time sampling
δt is fixed. It is worth noting that the spatial sampling is non-linear.

We now restrict the receiving plane to a finite surface. The radiation
coupling function is given by an integration of the field generated by the
emitting surface over itself. The sampling on the emitting transducer
corresponds to a discrete approach of the radiation integral : this sampling takes into account the finite dimensions of the emitter. The geometry
effects of the receiver are introduced by the numerical process that keeps
only the rays impinging on the receiver surface.
These arguments show that the radiation coupling function can be concep-
tually retrieved by the model. The problem is now to extend these results
to any geometry of the reflector.

The sampling we have to consider on the reflector in the cylindrical
coordinates (r,φ) is given by :

$$\varphi_n = (n-1)\frac{2\pi}{N_\varphi} \tag{5}$$

$$r_n = \sqrt{B^2(n-1)^2 + 4dB(n-1)} \tag{6}$$

with N_φ number of sampling points for the angular variable φ (B constant).
Considering the spherical angular coordinates (θ,φ) with the origin given
by E, Eq. (6) can be expressed as :

$$\cos\theta_n = \frac{2d}{\sqrt{r_n^2 + 4d^2}} = \frac{2d}{B(n-1) + 2d} \tag{7}$$

Finally, φ varying from 0 to 2π and θ from 0 to θ_{max}, N_φ and N_θ being the
number of sampling points for the φ and θ variables, we have the following
expressions :

$$\begin{cases} \varphi_n = \frac{2\pi}{N_\varphi}(n-1) \\ \theta_n = \cos^{-1}\left[1 + \frac{n-1}{N_\theta-1}\frac{1-\cos\theta_{max}}{\cos\theta_{max}}\right]^{-1} \end{cases} \tag{8}$$

For smooth surfaces, these expressions are independent of the geometry of
the reflector. All the simulation programs are based on Eq. (8).

III. THE RAY MODEL IN THE PLANE TRANSDUCER/PLANE REFLECTOR CONFIGURATION

The model is tested on the plane transducer/plane reflector configuration to retrieve the radiation coupling function [1].
We consider a plane circular transducer of radius a=14 mm working in the
piston mode and a plane infinite reflector. Fig. 2 illustrates the considered geometrical configuration. The distance between the tranducer and
the reflector along the vertical line is d=50 mm and the angle of the plane

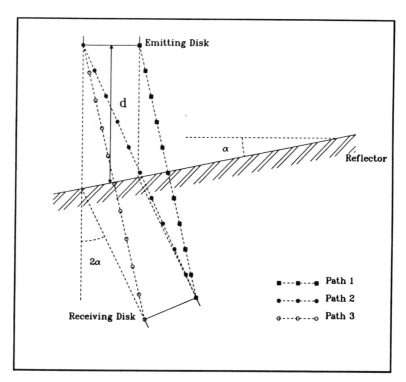

Fig.2 : Geometrical configuration. The emitting disk
is distant of d from the reflector tilted of an angle
α. The receiving disk results from a projection of the
emitter through the plane reflector.

reflector is α (the reference α=0 corresponding to a plane parallel to the
transducer front layer). Since the reflector has an infinite acoustical
impedance, this problem of a transducer working in the emitting/receiving
mode is equivalent to an emitting transducer and a receiver obtained by
geometrical symmetry of the emitter through the reflector.

In the particuliar case α=0, T.L. Rhyne [1] has adressed a closed form
solution . Fig. 3a presents the comparison between this closed form solution
and the numerical result obtained by the ray model. For values of α different from zero, a closed form solution can not be found. It results from the
reciprocity principle that it is possible to evaluate the radiation coupling
function by numerical integration of the diffracted field over the receiving
surface. Fig. 3b presents, for α=5°, the comparison between the solution
obtained by direct integration and the ray model. In both cases we obtain
an excellent agreement.
The continuous and dashed curves correspond to the integration of the radiated field and the ray model respectively. This results validates our
approach.

The radiation coupling function (in velocity potential or pressure) is defined on a finite time interval $[t_1,t_2]$. The values of t_1 and t_2 depend
naturally on the geometrical configuration. The ray model allows a very
easy explanation of this finite time interval in terms of geometrical path.
Indeed the value of t_1 corresponds to the shortest path (Path 1 on Fig. 2)
and the value of t_2 to the longest path (Path 2 or Path 3 on Fig. 2). Path
2 is larger than Path 3 for $α<α_{max}$ while Path 3 is greater than Path 2 for
$α>α_{max}$. In the particuliar case α=0, the time interval $[t_1,t_2]$ is given by
the following equations (c is the sound speed of the propagation medium) :

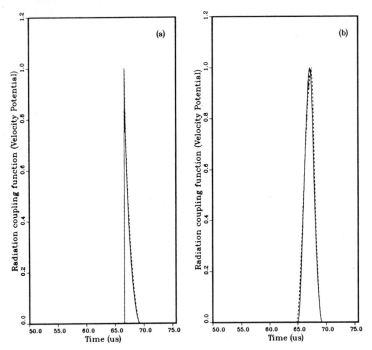

Fig.3 : The radiation coupling function for a plane
tilted reflector (a=14 mm, d=50 mm) :
a) $\alpha=0$ ° – b) $\alpha=5$ °.

$$\begin{cases} t_1 = \dfrac{2d}{c} \\[2ex] t_2 = \dfrac{2\sqrt{d^2 + a^2}}{c} \end{cases} \tag{9}$$

These two values are immediately obtained from the geometrical interpretation of the ray model considering the path length of Path 1 and Path 2
(corresponding respectively to t_1 and t_2).
We can easily verify that t_1 is a decreasing function of the α variable.
Indeed t_1 varies as :

$$t_1 = \frac{2}{c}\left(d\cos\alpha - a\sin\alpha\right) \tag{10}$$

(the condition dcosα–asinα>0 is necessary since the reflector can not
intersect the surface of the emitting transducer for mechanical reasons).
In order to obtain the expression of t_2 as a function of α, we determine the
longest path of Path 2 and Path 3 (Fig. 2). Elementary geometry leads to :

$$t_2 = \begin{cases} \dfrac{2\cos\alpha}{c}\sqrt{d^2+a^2} & \text{if } \alpha < \dfrac{1}{2}\tan^{-1}\left(\dfrac{a}{d}\right) \\[3ex] \dfrac{2}{c}\left(d\cos\alpha + a\sin\alpha\right) & \text{if } \alpha > \dfrac{1}{2}\tan^{-1}\left(\dfrac{a}{d}\right) \end{cases} \tag{11}$$

Fig. 4 represents the variation of t_1 and t_2 with the angular variable α.
The dashed curve corresponds to the time interval length $t_2 - t_1$. The first
arrival time t_1 decreases with α while the time interval length t_2-t_1 is an
increasing function of α.

474

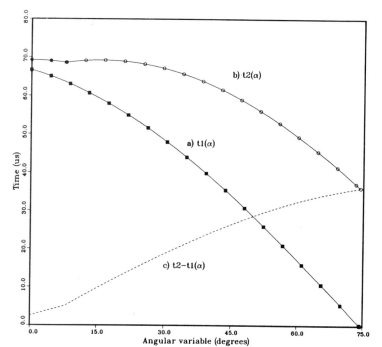

Fig. 4 : Time support $[t_1, t_2]$ of the radiation
coupling function for a disk (a=14 mm, d=50 mm).

The impulse ray model is a statistical and numerical process converging on the radiation coupling function (in velocity potential) of the
system emitter/reflector/receiver. Since the amplitude of the numerical
result depends upon the number of sampling points over the spherical angular variables θ and φ, the amplitude of the radiation coupling function is
obtained up to a constant.

Fig. 5 and 6 represent the radiation coupling function in velocity potential
and in pressure for different values of the angle α (a=14 mm and d=50 mm).
The time window is $[50 \, \mu s, 75.6 \, \mu s]$, the curves have been normalized to one.
For α=0, the pressure coupling function shows a Dirac distribution at $t=t_1$.
For graphic clearness this distribution has been represented as a vertical
arrow. These two figures illustrate very well the length variation of the
time interval $[t_1, t_2]$ with the angle α. As α increases, the time interval
gets larger and the curves look smoother, therefore enhancing the low frequency components of the output transducer signal. The spurious peaks
appearing on Fig. 5h and Fig. 6f correspond to numerical noise. The time
convolution of the radiation coupling function with the acousto-electrical
transducer response reduces the effects of the noise.

Fig. 7 represents the experimental echos recorded from a plane circular
transducer of radius a=14 mm at a distance d=50 mm and a hard plane reflector for different values of the angle α. This figure illustrates the signal
distortion due to the reflector tilting.
The experimental waveform observed for α=0 has been used for calibration of
the transducer [2]. Once the impulse response of the transducer is known, a
time convolution with the radiation coupling function gives the output
signal for any α configuration. Fig. 8 represents the comparison between the
experimental (continous curves) and the predicted (dashed curves) waveforms for different values of α. This figure shows an excellent agreement
between the numerical model and the experimental results. The differences
observed at 6 degrees are due to the finite acoustical impedance of the
reflector.

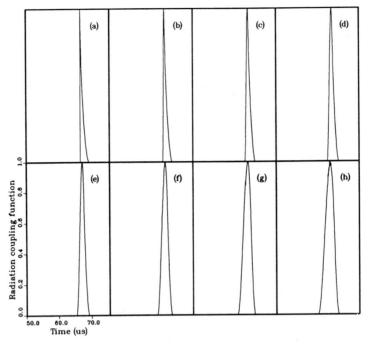

Fig.5 : Radiation coupling function (a=14 mm, d=50 mm)
for different values of the angle α :
a) α=0 ° – b) α =1 ° – c) α =2 ° – d) α =3 °
e) α =4 ° – f) α =6 ° – g) α =8 ° – h) α =10 °

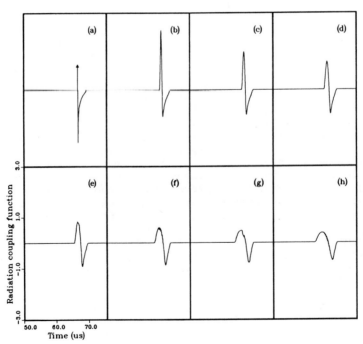

Fig.6 : Pressure coupling function (a=14 mm, d=50 mm)
for different values of the angle α :
a) α =0 ° – b) α =1 ° – c) α =2 ° – d) α =3 °
e) α =4 ° – f) α =6 ° – g) α =8 ° – h) α =10 °

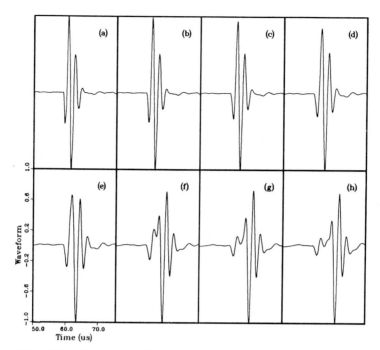

Fig. 7 : Experimental waveforms recorded with a plane
disk (a=14 mm, d=50 mm) for different values of the
angle α : a) α =0 ° – b) α =1 ° – c) α =2 ° – d) α =3 °
e) α =4 ° – f) α =6 ° – g) α = 8 ° – h) α =10 °

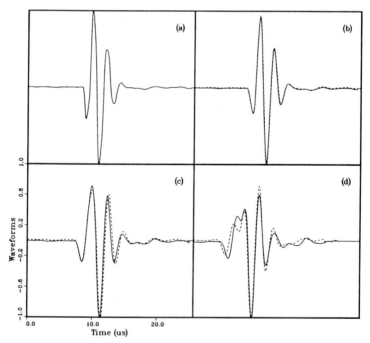

Fig.8 : Comparison between experimental and predicted
waveforms (a=14 mm, d=50 mm) for different values of the
angle α :
a) α =0 ° – b) α =2 ° – c) α =4 ° – d) α =6 °

477

IV. THE RAY MODEL IN OTHER GEOMETRICAL CONFIGURATIONS

The ray model is now applied on a cylindrical reflector. The determination of the radiation coupling function in velocity potential or in pressure permits the prediction of the time behavior of the reflected echo. The influence of the curvature of the reflector surface and the echo distortion are studied.
The appendix presents the numerical treatment to be done for each ray considering a specular reflexion. We determine the intersection point P of each acoustical ray with the reflector surface and the corresponding normal direction.

Assuming a plane circular transducer of radius a=14 mm, located in the plane z=0, the origin of the spatial coordinates being the center of the transducer, the geometrical parameters introduced in the numerical process for a cylindrical reflector are :
   – The spatial coordinates of the center of the cylinder : X_c, Y_c and Z_c.
   – The radius of the cylinder : R.
   – The distance d between the plane containing the transducer and the tangent plane to the reflector surface parallel to the transducer : $d=Z_c-R$.

The ray model has been run for a cylindrical reflector in the different geometrical configurations described in Table 1.
Fig. 9 represents the radiation coupling functions in velocity potential (continous curves) and in pressure (dashed curves) for a cylinder. The eccentering parameter e mentionned on these curves corresponds to the value of X_c.

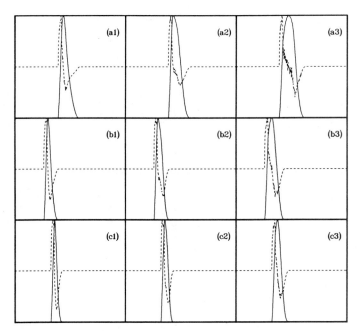

Fig.9 : Radiation coupling function for a hard
cylindrical reflector :
a) d=30 mm – b) d=50 mm – c) d=70 mm
1) e= 0 mm – 2) e= 5 mm – 3) e=10 mm

The following remarks can be made :

- For large values of Z_c, the effects of curvature are very low and the curves look similar to the radiation coupling function obtained in a planar configuration.
- The radiation coupling function in pressure shows two contributions of opposite sign and similar amplitude. As a consequence, the time convolution with the impulse response of a high frequency tranducer leads to a separation of two different echos. This separation can not be observed with a low frequency transducer, but the resulting signal distortion is important.

Table 1 : Numerical parameters of the ray model. The values of X_c, Y_c, Z_c and R are given in millimeters.

Fig.	X_c	Y_c	Z_c	R
a_1	0	0	80	50
a_2	5	0	80	50
a_3	10	0	80	50
b_1	0	0	100	50
b_2	5	0	100	50
b_3	10	0	100	50
c_1	0	0	120	50
c_2	5	0	120	50
c_3	10	0	120	50

V. CONCLUSION

A numerical model was presented here to obtain the radiation coupling function in arbitrary geometrical configuration emitter/reflector/receiver. The model considers a non-penetrable reflector of smooth profile. This model is quite powerful since it permits the numerical determination of the diffraction effects for any smooth surfaces. The prediction of the observed signal in such configurations becomes possible and the signal distortion receives a simple explanation. The proposed model can be extended to include focused transducers and an amplitude variation over the emitting surface.

APPENDIX : THE RAY MODELING FOR ARBITRARY GEOMETRY

This appendix presents the extension of the ray modeling to arbitrary reflector geometry. We consider an emitting point E on the transducer with spatial coordinates (x_e, y_e, z_e) and an acoustical ray direction given by the two angular variables θ and φ. The reflector surface is given by the equation $z=f(x,y)$.
The straight line corresponding to the propagation of the ray is given by the equations :

$$\begin{cases} x = x_e + \lambda \sin \theta \cos \varphi \\ y = y_e + \lambda \sin \theta \sin \varphi \\ z = z_e + \lambda \cos \theta \end{cases} \qquad (12)$$

where λ is the parameter describing the line.

The first step is to determine the spatial coordinates of the intersection point $P(x_p, y_p, z_p)$ of the straight line with the reflector surface. This determination leads to the following equation :

$$z_e + \lambda \cos \theta = f(x_e + \lambda \sin \theta \cos \varphi, y_e + \lambda \sin \theta \sin \varphi) \qquad (13)$$

the unknown variable being λ. The solution of this transcendental equation is usually obtained numerically.

The tangent plane to the reflector at P is given by the two vectors :

$$\vec{T_1} = \begin{pmatrix} 1 \\ 0 \\ \dfrac{\partial f}{\partial x}(x_p, y_p) \end{pmatrix} \qquad \vec{T_2} = \begin{pmatrix} 0 \\ 1 \\ \dfrac{\partial f}{\partial y}(x_p, y_p) \end{pmatrix} \qquad (14)$$

and the normal vector is given by the vector product :

$$\vec{N} = \vec{T_1} \otimes \vec{T_2} = \begin{pmatrix} -\dfrac{\partial f}{\partial x}(x_p, y_p) \\ -\dfrac{\partial f}{\partial y}(x_p, y_p) \\ 1 \end{pmatrix} \qquad \vec{N'} = \dfrac{\vec{N}}{\| \vec{N} \|} \qquad (15)$$

The direction of the incident ray is given by $\overrightarrow{U} = \overrightarrow{EP}$.

Considering that the ray is reflected according the Snell laws, the directions of the incident and reflected rays can be written respectively :

$$\begin{cases} \vec{U} = (\vec{U}.\vec{N'})\vec{N'} + [\vec{U} - (\vec{U}.\vec{N'})\vec{N'}] & \text{Incident ray} \\ \vec{U'} = -(\vec{U}.\vec{N'})\vec{N'} + [\vec{U} - (\vec{U}.\vec{N'})\vec{N'}] = \vec{U} - 2(\vec{U}.\vec{N'})\vec{N'} & \text{Reflected ray} \end{cases} \qquad (16)$$

We now introduce the following notation :

$$A = \| \vec{N} \| = \sqrt{1 + \left(\dfrac{\partial f}{\partial x}(x_p, y_p) \right)^2 + \left(\dfrac{\partial f}{\partial y}(x_p, y_p) \right)^2} \qquad (17)$$

From Eq. (16) we obtain the expression of the different components of the propagation vector $\overrightarrow{U'}$:

$$\begin{cases} U'_x = (x_p - x_e) - \dfrac{2}{A}\dfrac{\partial f}{\partial x}(x_p, y_p)\Bigg[(x_p - x_e)\dfrac{\partial f}{\partial x}(x_p, y_p)+ \\ \qquad\qquad\qquad (y_p - y_e)\dfrac{\partial f}{\partial y}(x_p, y_p) - (z_p - z_e)\Bigg] \\ U'_y = (y_p - y_e) - \dfrac{2}{A}\dfrac{\partial f}{\partial y}(x_p, y_p)\Bigg[(x_p - x_e)\dfrac{\partial f}{\partial x}(x_p, y_p)+ \\ \qquad\qquad\qquad (y_p - y_e)\dfrac{\partial f}{\partial y}(x_p, y_p) - (z_p - z_e)\Bigg] \\ U'_z = (z_p - z_e)+ \\ \qquad \dfrac{2}{A}\Bigg[(x_p - x_e)\dfrac{\partial f}{\partial x}(x_p, y_p) + (y_p - y_e)\dfrac{\partial f}{\partial y}(x_p, y_p) - (z_p - z_e)\Bigg] \end{cases} \qquad (18)$$

As the only quantity of interest is the direction of the reflected ray, we can multiply each component of $\overrightarrow{U'}$ with A without changing the result. We then obtain Eq. (19).
The last step is to determine the receiving point $R(x_r, y_r, z_r)$ and the arrival time of the ray. If the condition $U'_z < 0$ is verified (this condition corresponds to a reflected ray in the direction of the plane z=0 containing the receiver), we obtain Eq. (20).

$$
\begin{cases}
U'_x = (x_p - x_e)\left[1 - \dfrac{\partial f}{\partial x}(x_p, y_p)^2 + \dfrac{\partial f}{\partial y}(x_p, y_p)^2\right] - \\[2mm]
\qquad 2(y_p - y_e)\dfrac{\partial f}{\partial x}(x_p, y_p)\dfrac{\partial f}{\partial y}(x_p, y_p) + 2(z_p - z_e)\dfrac{\partial f}{\partial x}(x_p, y_p) \\[4mm]
U'_y = (y_p - y_e)\left[1 + \dfrac{\partial f}{\partial x}(x_p, y_p)^2 - \dfrac{\partial f}{\partial y}(x_p, y_p)^2\right] - \\[2mm]
\qquad 2(x_p - x_e)\dfrac{\partial f}{\partial x}(x_p, y_p)\dfrac{\partial f}{\partial y}(x_p, y_p) + 2(z_p - z_e)\dfrac{\partial f}{\partial y}(x_p, y_p) \\[4mm]
U'_z = 2(x_p - x_e)\dfrac{\partial f}{\partial x}(x_p, y_p) + 2(y_p - y_e)\dfrac{\partial f}{\partial y}(x_p, y_p) - \\[2mm]
\qquad (z_p - z_e)\left[1 - \dfrac{\partial f}{\partial x}(x_p, y_p)^2 - \dfrac{\partial f}{\partial y}(x_p, y_p)^2\right]
\end{cases}
\tag{19}
$$

$$
\begin{cases}
x_r = x_p - z_p\dfrac{U'_x}{U'_z} \\[3mm]
y_r = y_p - z_p\dfrac{U'_y}{U'_z} \\[3mm]
z_r = 0
\end{cases}
\tag{20}
$$

The reflected ray intercepts the transducer (radius a) if the relation $x_r^2 + y_r^2 \leqslant a^2$ is satisfied. In such case the arrival time t of the ray is given by :

$$
t = \frac{1}{c}\sqrt{(x_p - x_e)^2 + (y_p - y_e)^2 + (z_p - z_e)^2} + \frac{1}{c}\sqrt{(x_r - x_p)^2 + (y_r - y_p)^2 + (z_r - z_p)^2}
\tag{21}
$$

The model does not include secondary diffraction. This approximation is realistic if the spatial derivatives of the reflector profile are small.

REFERENCES

1   T.L. Rhyne, "Radiation coupling of a disk to a plane and back or a disk to a disk : An exact solution", J. Acoust. Soc. Am. 61, 1977, pp. 318-324.

2   D. Cassereau and D. Guyomar, "Calibration of ultrasonic transducers by time deconvolution of the diffraction effects", Acoust. Imag., Vol. 16, Ed. Dr. Lawrence W. Kessler, Plenum Press, New-York, 1987.

THERMAL WAVE IMAGING IN ANISOTROPIC MEDIA

M. Vaez Iravani and M. Nikoonahad

Philips Laboratories
North American Philips Corporation
345 Scarborough Road, Briarcliff Manor, New York 10510, USA

1. INTRODUCTION

Thermal wave and photoacoustic techniques rely on the excitation of the specimen by a modulated laser, electron, or X-ray beam {1}. These techniques have led to new forms of imaging and microscopy for NDE and material characterization. The motivation for using such techniques is that the sources of image contrast are primarily different from those responsible for contrast formation in optical, electron or acoustic microscopy {2}. One is often probing the thermal properties of the specimen (eg thermal conductivity). When a sample is excited by a laser beam, the light is partially absorbed by the sample, leading to a temperature rise. When the light is modulated, there is an ac component of temperature which is superimposed on a dc temperature. It is this ac temperature variation which, over a short range, behaves as a wave- a *thermal wave*. Various ways of detecting this wave provide the essence of almost all forms of photothermal techniques.

Since in all these techniques one relies on the transformation of one type of energy into another, the contrast mechanisms and quantitative imaging in this field are somewhat more complicated than situations where only one form of energy is involved. Furthermore, in many cases, the candidate samples (eg graphite based composites) for thermal wave imaging exhibit a large thermal anisotropy, which makes the contrast issues even more complex. Also, owing to the finite expansivity of the medium, a thermal wave is almost invariably accompanied by an acoustic wave. In general we are, therefore, concerned with coupled thermo-acoustic waves. It is clear that understanding the image contrast and quantitative imaging requires a formalism which adequately accounts for all these processes. Reviewing the literature indicates that the one-dimensional thermo-acoustic equations have been analyzed for a wide variety of cases {3-6}. One dimensional models often provide immediate design answers for a system. They are particularly useful for thermoacoustic phenomena in liquids where the acoustic wave can be described by a scalar function, and we do not have any thermal anisotropy. However, they fail to provide analyses for: (i) diffraction and therefore thermal beam formation in the specimen and (ii) propagation in anisotropic media. The existing 3D techniques for thermo-acoustic problems are generally based on numerical analysis which can be computationally intensive {7-9} and not easily applicable to the anisotropic case.

A major part of this work was carried out when the authors were with the Department of Electronic and Electrical Engineering, University College London.

The aim of this paper is to give a 3D analysis based on a *transfer function formalism* using a 2D FFT. This technique is simple, mostly analytical, and retains the physics of the problems at every stage of its development. The question we wish to address is: Given an ac energy distribution at the input plane, figure 1, what is the resulting temperature distribution at an output plane, after propagation through the specimen? The input energy can be a "given" temperature distribution - a thermal aperture - or, as is generally the case, can be the profile of a laser or electron beam.

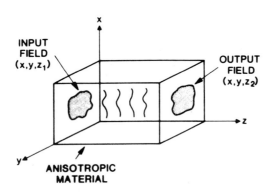

Figure 1. The basic geometry of the system.

Our approach is basically an extension of Fourier optics {10} and has been reported for circularly symmetric isotropic thermal wave problems{11}. We first decompose the input field into a spectrum of primary waves and then propagate each one of these waves through the sample. The propagation imposes an appropriate phase and attenuation on each one of these waves. At the output plane we perform an inverse Fourier transform to find the field. In section 2 we develop the thermo-acoustic equations for a general solid and arrive at two coupled equations for the acoustic waves in the presence of thermal coupling, and thermal waves with acoustic coupling. In the same section, we give a formulation for a pure thermal wave (ie ignoring the coupling) in an anisotropic medium and derive a transfer function for such a medium. Section 3 deals with a number of simulated results for thermal wave propagation in graphite. Conclusions are presented in section 4.

2. THEORY

The thermo-acoustic coupling is primarily due to a time dependent thermal expansion of the medium. As far as the acoustic waves are concerned, this dynamic expansion gives rise to a stress field which acts as a source, leading to a source term in the acoustic wave equation. As far as the thermal waves are concerned, on the other hand, the dynamic stress gives rise to a temperature change which provides an additional source term in the thermal conduction equation. In this section, we first derive the coupled thermo-acoustic equations and then derive solutions for the decoupled thermal waves in anisotropic solids.

2.1 Thermo-acoustic Waves in Solids

For a temperature variation ϕ, in a solid whose expansivity is β, the temperature induced strain, S, is given by:

$$S \equiv \nabla_s u = \beta \phi, \tag{1}$$

where ∇_s is a matrix differential operator and u is the particle displacement field {12}. The expansivity β is given in reduced notation as $\beta = \beta_i$, for $i = 1$ to 6. For a mass density ρ_m, the body force F, associated with this expansion can be related to the temperature as:

$$\nabla_s F = \rho_m \beta \frac{\partial^2 \phi}{\partial t^2}. \tag{2}$$

We now substitute this body force in the acoustic field equation, {12}. We have:

$$\nabla_s \{\nabla . c : \frac{\partial S}{\partial t}\} = \rho_m \frac{\partial^3 S}{\partial t^3} - \rho_m \beta \frac{\partial^3 \phi}{\partial t^3}, \tag{3}$$

where c is the stiffness tensor. The next step is to derive the heat conduction equation in the presence of an acoustic coupling. From the conservation of energy we have {13}:

$$\rho_m C \frac{\partial \phi}{\partial t} + \left\{ \frac{\partial \xi_z}{\partial x} + \frac{\partial \xi_y}{\partial y} + \frac{\partial \xi_z}{\partial z} \right\} + s + A = 0, \tag{4}$$

where s is the source term due to the external input, C is the specific heat capacity, ξ_z, ξ_y, ξ_z are thermal fluxes, which for a conductivity matrix K, are defined by: $\xi \equiv -K \times (\nabla \phi)$ and the additional heat due to acoustic strain in the sample is given by: $A = \phi c \beta \partial S / \partial t$, {14}. By substituting this in eq. (4) and rearranging and also by integrating both sides of eq. (3) with respect to time, we arrive at the coupled thermo-acoustic equations in a general solid, given by:

$$\nabla_s \{\nabla . c : S\} - \rho_m \frac{\partial^2 S}{\partial t^2} = -\rho_m \beta \frac{\partial^2 \phi}{\partial t^2}, \tag{5a}$$

$$\nabla . \{K \times (\nabla \phi)\} - \rho_m C \frac{\partial \phi}{\partial t} = s + \phi \{c \beta\} \frac{\partial S}{\partial t}. \tag{5b}$$

where $\nabla \cdot$ is the divergence operator.

2.2 Thermal Wave Imaging in Anisotropic Media

Photothermal imaging of surface and subsurface features involves a number of processes, which can be classified into three major categories: wave generation and propagation, wave-matter interaction, and finally detection. Of these, it is the first which needs special attention in the case of anisotropic media; the remaining two steps can be expected to have much in common with those in the more established isotropic imaging. It is for this reason that in the remainder of this paper we shall concentrate on the generation and propagation of thermal waves in anisotropic media. We shall further confine our study to those cases where acoustic coupling can be ignored.

In the absence of any acoustic wave, we are concerned with the decoupled version of eq. (5b), which can be expanded into

$$K_{11} \frac{\partial^2 \phi}{\partial x^2} + K_{22} \frac{\partial^2 \phi}{\partial y^2} + K_{33} \frac{\partial^2 \phi}{\partial z^2} + (K_{12} + K_{21}) \frac{\partial^2 \phi}{\partial x \partial y} + (K_{13} + K_{31}) \frac{\partial^2 \phi}{\partial x \partial z} + (K_{23} + K_{32}) \frac{\partial^2 \phi}{\partial y \partial z}$$

$$= \rho_m C \frac{\partial \phi}{\partial t} + \eta e^{-\alpha z} s(x, y). \tag{6}$$

Here, $K_{ij}(i,j=1,2,3)$ are the elements of the conductivity matrix, $\mathbf{K}$, and $s(x,y)$ defines the source function, which is assumed to attenuate as $exp(-\alpha z)$ as it propagates into the sample; η is a normalization factor.

We assume an $exp(j\omega t)$ time dependence for the source and temperature, and directly perform a 2D Fourier transform on eq. (6) to obtain:

$$a\frac{\partial^2\Phi}{\partial z^2}+b\frac{\partial\Phi}{\partial z}+c\Phi=\eta e^{-\alpha z}S(f_x,f_y),\tag{7}$$

where $\Phi(f_x,f_y,z)$ and $S(f_x,f_y)$ are the Fourier transforms of $\phi(x,y,z)$ and $s(x,y)$, respectively. The parameters a, b, and c are given by:

$$a=K_{33};\tag{8a}$$

$$b=-2j\pi\{(K_{13}+K_{31})f_x+(K_{23}+K_{32})f_y\};\tag{8b}$$

$$c=-4\pi^2\{K_{11}f_x{}^2+K_{22}f_y{}^2+(K_{12}+K_{21})f_xf_y\}-j\omega\rho_m C.\tag{8c}$$

A solution to eq. (7) can be written as:

$$\Phi(f_x,f_y,z)=A(f_x,f_y)e^{(mz)}+\eta pS(f_x,f_y)e^{(-\alpha z)},\tag{9}$$

where

$$m_{1,2}=\{-b\pm(b^2-4ac)^{\frac{1}{2}}\}(2a)^{-1},\tag{10}$$

and p is a constant which depends on the shape of the source. The subscripts 1,2 in eq. (10) denote propagation in opposing z directions. Before applying this solution to a specific example, we can gain much useful insight by examining the parameters of eq. (7): a is mainly responsible for the propagation along z, whereas a finite b will result in beam steering due to asymmetric propagation. The preferential spreading of the temperature is determined by c.

Let us now consider the case depicted in fig. 2, where a modulated Gaussian energy source is focused onto the surface of a thermally uniaxial material, with $K_{11}=K_{22}$. Clearly, a simple transformation performed on $\mathbf{K}$, the principal conductivity matrix, directly gives the elements of the conductivity ellipsoid, as defined by the left hand side of eq. (6):

$$K'_{ij}=T_{ik}T_{jl}K_{kl},\tag{11a}$$

with

$$\mathbf{T}=\begin{pmatrix} cos\theta & 0 & sin\theta \\ 0 & 1 & 0 \\ -sin\theta & 0 & cos\theta \end{pmatrix}.\tag{11b}$$

The non-zero elements of the new matrix are:

$$K'_{11}=K_{11}cos^2\theta+K_{33}sin^2\theta;\tag{12a}$$

$$K'_{13}=K'_{31}=(K_{33}-K_{11})cos\theta sin\theta;\tag{12b}$$

$$K'_{33}=K_{33}cos^2\theta+K_{11}sin^2\theta;\tag{12c}$$

$$K'_{22}=K_{11}=K_{22}.\tag{12d}$$

By substituting these parameters into eq. (8), the terms a, b, and c, and subsequently the propagation constant, m, are determined. Let the Gaussian source term be given by:

$$s(x,y)=exp\{-(x^2+y^2)/W_1{}^2\},\tag{13}$$

where W_1 is the half waist of the beam. The normalized Fourier transform of this function is also a Gaussian, given by:

$$S(f_x,f_y)=exp\{-(f_x{}^2+f_y{}^2)/W_2{}^2\},\tag{14}$$

where $W_2=(\pi W_1)^{-1}$ The constant p can easily be shown to be

486

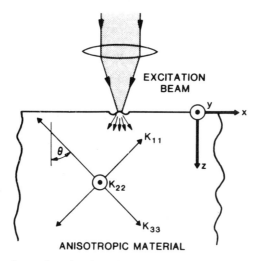

EXCITATION
BEAM

K_{11}

θ

K_{22}

K_{33}

ANISOTROPIC MATERIAL

Figure 2. Photothermal excitation of a uniaxial material with a focused beam.

$$p = \pi W_1^2 (a\alpha^2 - b\alpha + c)^{-1}. \tag{15}$$

To determine A in eq. (9), the boundary conditions are invoked which, since the sample is assumed to be in contact with a vacuum, require only that the flux should vanish at $z=0Z$:

$$\left\{ -K'_{13}\frac{\partial\phi}{\partial x} - K'_{23}\frac{\partial\phi}{\partial y} - K'_{33}\frac{\partial\phi}{\partial z} \right\}_{z=0} = 0, \tag{16}$$

which results in:

$$A(f_x, f_y) = \eta p \left(\frac{2\alpha a + b}{2m_1 a - b} \right) S(f_x, f_y). \tag{17}$$

Hence, all the relevant terms in our solution, eq. (9), have been determined. It is particularly significant to note that despite the fact that the excitation source is assumed to be symmetric, owing to anisotropy, the resulting temperature distribution at any z level is, in general, asymmetric. It should, in addition, be remembered that although for clarity only a simple example has been given, the technique presented here is quite general, and can be applied to many practical cases. For example, a multi-layered structure can be dealt with using this method, in conjunction with a scattering matrix analysis, in an analogous way to that already shown in isotropic media {11}.

3. SIMULATED RESULTS AND DISCUSSION

We shall now apply our basic formulations for uniaxial materials to the specific case of graphite with the following parameters {15}: $\rho_m = 2.25 \times 10^3 Kgm^{-3}$; $C = 714 JKg^{-1}K^{-1}$; $K_{22} = K_{11} = 250\,Wm^{-1}K^{-1}$; $K_{33} = 80\,Wm^{-1}K^{-1}$. In all the following cases, the sample is assumed to be semi-infinite in contact, at $z=0$, with a vacuum. The first case to be considered is that of a source-free propagation. For this case the aperture function is assumed to be Gaussian with a waist of 4 μm, at a modulating frequency of 5 kHz, as shown in fig. 3. Here, θ is taken to be 45°. Anisotropy-induced field asymmetry effects are clearly seen in figures 4(a), and 5(a), where the temperature distribution is plotted at $z=3$, and $z=6\mu m$, respectively. We note firstly that there is a greater spreading of the field along the y axis as compared with that along the x axis. This is easily understood in conjunction with fig. 2, where the greater overall conductivity is seen to coincide with the y direction. The distribution along the $x-$axis shows a shift in the position of the maximum

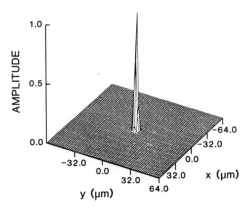

Figure 3. Gaussian thermal aperture with a waist of $4\mu m$.

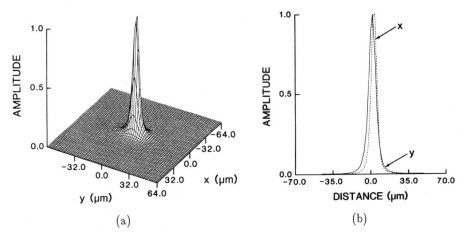

(a)   (b)

Figure 4. a) Temperature distribution in a material with $\theta=45^{\circ}$ at $z=3\mu m$; b) distribution along x and y directions, showing preferential spreading, and beam steering.

temperature, as well as a preferential spread in the $+x$ direction. These related effects are again associated with a greater net conductivity along $+x$, which is evident from fig. 2. Such effects are more clearly shown in the two-dimensional profiles of temperature distribution along x and y, through the point of maximum temperature. The position of the peak temperature, as one progresses into the sample, moves further and further away from the point vertically below the center of the aperture. This suggests that there is a preferred "channel" for thermal waves, resulting in beam steering.

It is important to note that the occurrence of beam steering is distinct from asymmetric spreading of temperature along x and y. This point is illustrated in figures 6 and 7, where the distribution is plotted at $z=6\mu m$, for θ of 0°, and 90°, respectively. In the former, one observes a circularly symmetric distribution, representative of the planar-symmetric nature of the problem. In the latter case, the distribution is elliptic, with the

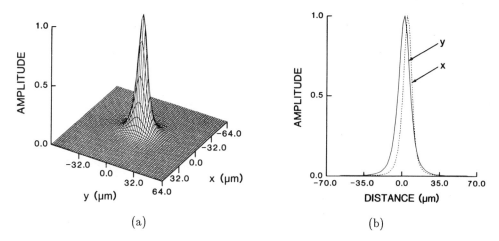

(a)   (b)

Figure 5. As in fig. 4, with $z=6\mu m$.

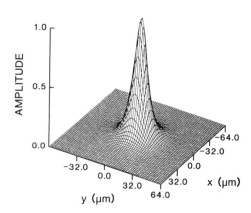

Figure 6. As in fig. 4, with $\theta=0^{\circ}$ and $z=6\mu m$, showing planar field symmetry.

major axis along y, and minor axis along x, which is due to greater conductivity along the $y-$axis. There is no beam steering in either of these situations, since the principal conductivity axes coincide with the coordinate system. Finally, fig. 8 illustrates the effect of varying θ where, as expected, both asymmetric spreading, and beam steering are found to be dependent on θ. Next, we wish to study the effect of anisotropy on thermal wave propagation in the presence of an optical Gaussian source. The sample is a semi-infinite material in contact with a vacuum, as in the source-free case. The source is assumed to have a waist of $4\mu m$, and is modulated at 20 kHz. The resulting tempera-ture distributions at $z=0$, $z=3$ and $z=6\mu m$, are respectively shown in figures 9 through 11. Again, for greater clarity, the two-dimensional plots through maximum temperature along the x and y directions are also shown.

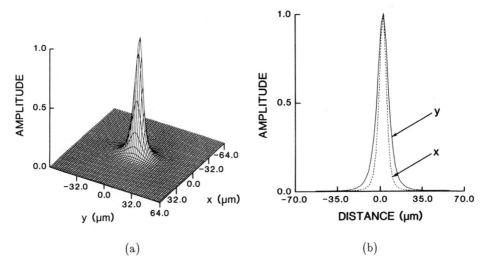

(a)   (b)

Figure 7. As in fig. 4, with $\theta=90°$; elliptical temperature distribution is evident.

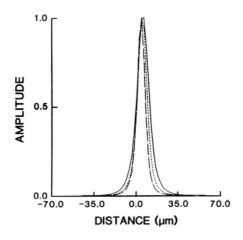

Figure 8. Effect of θ on temperature distribution at $z=6$, and frequency of 5 kHz for $\theta=30°$ (solid line), $\theta=45°$ (chained line), and $\theta=60°$ (chain-dot line).

We can see that, although the source is circularly symmetric, the anisotropy has resulted in a distorted thermal aperture at the surface, fig. 9. That is, even before any propagation has taken place, the surface temperature distribution is elongated in a preferred direction. However, the peak temperature still coincides with the peak of the source, as would be expected. Propagation into the sample is found to accentuate the asymmetric distribution even further, as well as to result in beam steering, which amounts to 3 and $6\mu m$, at $z=3$ and $z=6\mu m$, respectively.

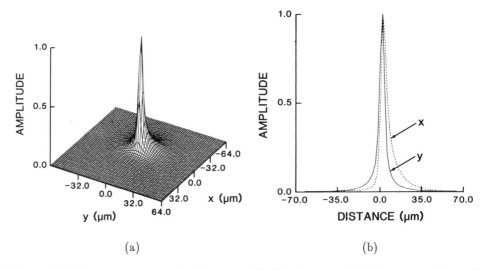

(a)   (b)

Figure 9. Surface temperature distribution at 20 kHz for a semi-infinite material $(\theta=45°)$, irradiated by a Gaussian source with $W_1=2\mu m$.

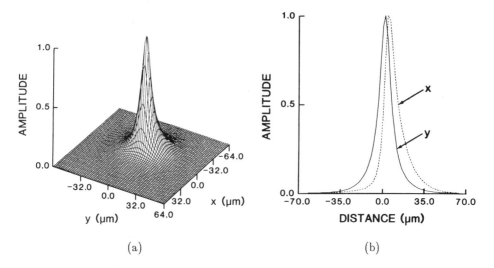

(a)   (b)

Figure 10. As in fig. 8, with $z=3\mu m$.

4. CONCLUSIONS

We have derived the coupled thermo-acoustic equations, describing the generation and propagation of waves in solids. We have analyzed thermal waves in anisotropic media in the absence of any acoustic coupling, and applied the formulation to the particular case of thermally uniaxial materials. Numerical examples for propagation in graphite predict anisotropy-induced field distortion, as well as beam steering. Such effects can play a significant role in image formation in thermal wave imaging systems, especially in sub-surface photothermal microscopy. It thus becomes important to remove field distortion effects through a deconvolution process, so as to restore a one-to-one image/object relationship.

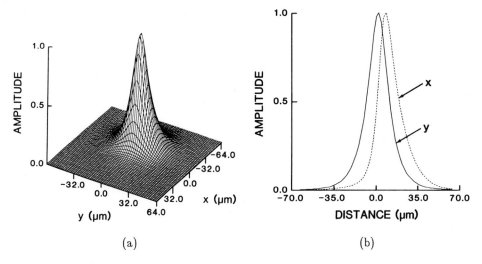

(a)   (b)

Figure 11. As in fig. 8, with $z=6\mu m$.

REFERENCES

1.   See IEEE Trans. Ultra. Ferr. Freq. Cont., Vol. UFFC-33, edited by D. Hutchins and A.C. Tam, (1986)

2.   E.A. Ash, (editor), "Scanned Image Microscopy", (Academic Press, London, 1980)

3.   A. Rosencwaig and A. Gersho, Jour. Appl. Phys. 47, 64 (1976)

4.   F.A. McDonald and G.C. Wetsel, Jour. Appl. Phys. 49, 2313 (1978)

5.   J.C. Murphy and L.C. Aamodt, Jour. Appl. Phys. 51, 4580 (1980)

6.   G.C. Wetsel, Jr., IEEE Trans. Ultra. Ferr. Freq. Cont., Vol. UFFC-33, 450 (1986)

7.   Y.H. Wong, in Scanned Image Microscopy, edited by E.A. Ash, (Academic Press, London, 1980) p. 247

8.   W.B. Jackson, N.M. Amer, A. Boccara, and D. Fournier, Appl. Opt. 20, 1333 (1981)

9.   C.R. Petts and H.K. Wickramasinghe, Proc. 1981 IEEE Ultrasonics Symposium, edited by B.R. McAvoy, (IEEE, New York, 1981), p.832

10   J.W. Goodman,"Introduction to Fourier Optics", (McGraw-Hill, New York, 1968)

11.   M. Vaez Iravani and H.K. Wickramasinghe, Jour. Appl. Phys. 58, 122 (1985)

12.   B.A. Auld, "Acoustic Fields and Waves in Solids", Vol. 1, (Wiley, New York, 1973)

13.   H.S. Carslaw and J.C. Jaeger, "Conduction of heat in solids", 2nd ed., (Oxford University Press, London, 1959)

14.   W.P. Mason, "Crystal Physics of Interaction Processes", (Academic Press, New York, 1966)

15.   American Institute of Physics Handbook, 3rd ed., (McGraw Hill, New York, 1982)

CHARACTERIZATION OF HETEROGENEOUS MATERIALS USING

TRANSMISSION PHOTOACOUSTIC MICROSCOPY

T. Ahmed, J.W. Monzyk, K.W. Johnson, G. Yang, and S. Goderya

Department of Physics
Southern Illinois University
Carbondale, IL 62901

INTRODUCTION

Photoacoustic microscopy, in which modulated light is absorbed by an opaque material and converted into a thermal wave, can be used as an in-situ technique for determining the thermal diffusivity of the material.[1-5] If a microscope is used to image the light onto a heterogeneous sample containing different components embedded in a matrix, the thermal diffusivities of components as small as fifty micrometers can be probed. This paper illustrates how photoacoustic microscopy can be used to determine the diffusivities of carbon-carbon composite materials.

THEORY

Figure 1 illustrates the process by which the back-surface photoacoustic signal is produced. A modulated beam of light is focused on the front surface of the material. In a material such as a carbon-carbon composite, a substantial fraction of the light is absorbed and converted into a diffusive heat wave. The wave propagates to the back (non-illuminated) surface of the sample, the amplitude decaying exponentially with distance, as the drawing illustrates. If the thickness of the sample is comparable to the wavelength of the thermal wave, part of the heat reaches the back surface and effuses into the surrounding air. The air is contained within a closed chamber, so the periodic heat flux produces a change in pressure that is measured with a microphone. The electrical signal from the microphone is the photoacoustic signal.

The time required for the thermal wave to diffuse through the sample depends on the thermal diffusivity and thickness of the sample, as well as the modulation frequency of the incident light. Thus, a phase difference exists between the thermal wave generated at the front surface and that at the back surface. This "back-surface" phase Φ is given by[6-12]

$$\tan \Phi = \frac{(1 + b)e^{x} + (1 - b)e^{-x}}{(1 + b)e^{x} - (1 - b)e^{-x}} \tan x \tag{1}$$

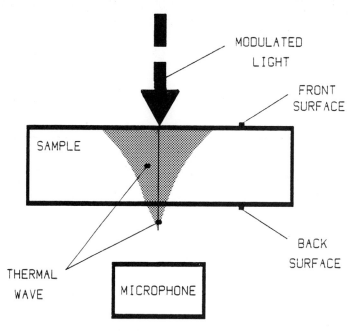

Fig. 1.   Generation of the back-surface photoacoustic signal.

In this expression b is the ratio of the effusivity of air to the effusivity of the sample, and $x = (\pi f/f_c)^{1/2}$ where:

f = modulation frequency of the light

f_c = characteristic frequency of the sample; $f_c = \alpha/L^2$

α = thermal diffusivity of the sample

L = sample thickness

Figure 2 shows a plot of the back surface phase as a function of $(f/f_c)^{1/2}$, calculated from Equation 1, for the case where b = 0, a situation that approximates a carbon composite/air interface. When $(f/f_c)^{1/2} > 1$, the curve becomes a straight line, and Equation 1 reduces to

$$\Phi = \left(\frac{\pi f}{f_c}\right)^{1/2} = \left(\frac{\pi L^2}{\alpha}\right)^{1/2} f^{1/2} \qquad (2)$$

The slope S of Equation 2 is

$$S = \left(\frac{\pi L^2}{\alpha}\right)^{1/2} \qquad or \qquad \alpha = \frac{\pi L^2}{s^2} \qquad (3)$$

Equation 3 states that the thermal diffusivity α of the sample can be measured from a knowledge of the sample thickness L and the slope S of the linear part of the phase curve.

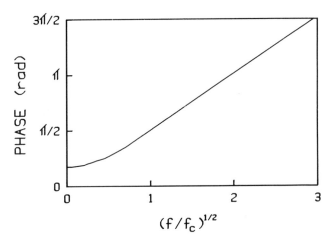

Fig. 2. The calculated phase of the back-surface photoacoustic
signal is plotted as a function of the modulation
frequency f. The characteristic frequency is f_c.

EXPERIMENTAL

The main features of the photoacoustic microscope are illustrated in
Figure 3. The light source is a Xenon arc lamp, and the beam is modulated
by an mechanical chopper. The modulated light beam is directed into a
microscope and focused on the surface of the sample under study. The
sample, along with the microscope, is placed in a sealed cell. The
microphone senses the pressure changes that arise within the cell and
outputs its signal to a preamplifier. The signal is sent from the
preamplifier to a lock-in amplifier that measures the phase of the
photoacoustic signal relative to that of a reference signal derived from the
chopper. The phase and frequency information are relayed to a computer for
analyses.

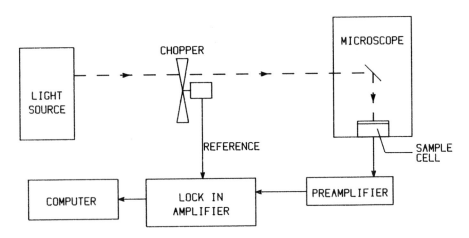

Fig. 3. The experimental layout of a photoacoustic microscope.

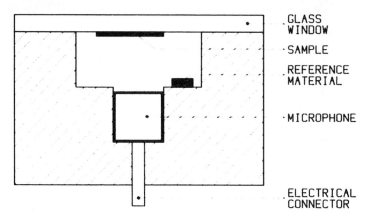

Fig. 4. The sample cell.

A cross-sectional view of the photoacoustic cell is shown in Figure 4. The cell body is a machined aluminum block. The microphone is placed in the center of the cell body. The cylindrical space above the microphone contains the sample and a reference material. A one-millimeter-thick glass slide seals the chamber while permitting the entrance of the light beam.

The sample was bonded to the interior surface of the glass slide with a small amount of glue. This bonding was necessary for two reasons. First, the sample thickness was reduced to the desired value, typically 0.5 mm, by polishing the free surface, and the glass slide served as a mechanical support during this process. Second, the glass slide reduced the flexing of the sample caused by the gradient of the thermal wave.

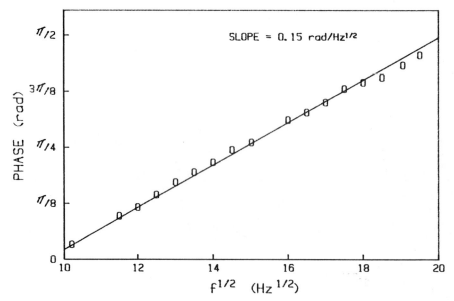

Fig. 5. Back-surface phase for graphite as a function of modulation frequency.

Table 1. Thermal diffusivities of graphite, glass, and zinc.

SAMPLE	THERMAL DIFFUSIVITY (cm²/sec)	
	Photoacoustic Data	Known Value
Graphite	0.91	0.89 - 1.18
Glass	0.009	0.005 - 0.009
Zinc	0.42	0.43

To eliminate the phase shift caused by the system (cell, microphone, and electronics) a piece of optically thick and thermally thick glassy carbon was placed inside the cell chamber to serve as a reference material. Theoretically[1], the phase of the front surface photoacoustic signal from such a material is known to be independent of frequency. Therefore, any frequency-dependent phase changes that arise from the reference material are due to the system, and these changes can be subtracted from the sample phase measurements to give the corrected back-surface phase.

RESULTS

Figure 5 shows the experimental results for a graphite sample. The corrected phase of the back-surface photoacoustic signal is plotted as a function of the square root of the frequency. The straight line is a least squares fit to the data points. The thermal diffusivity was found from the slope of this line and the thickness of the sample. Table 1 lists the thermal diffusivity for graphite, as well as for some other common materials. For comparison, the diffusivities of these materials, as measured by other techniques, are also presented. The results compare favorably, but photoacoustic microscopy has the advantage of being able to measure the diffusivity of very small samples in-situ.

Carbon-carbon composites are fiber-reinforced synthetic materials. The individual polyacrylonitrile fibers are approximately 8 micrometers in diameter. Thousands of these fibers are placed together to form a bundle. Bundles are arranged in a pattern and the spaces between them filled with pitch. Heating binds the materials together and partially graphitizes the fibers.

Carbon-carbon fiber composites were chosen for study, because of their heterogeneous nature. The sizes of the bundles (each several hundred micrometers in diameter) and the regions of the pitch matrix were larger than the diameter of the light beam. Therefore, the heat generated at the top surface remained within a bundle as the heat diffused through it; the heat did not diffuse laterally into the matrix material that surrounded each bundle. The thickness of the bundles was such that the samples could be polished to a thickness so that a single bundle lying in the plane of the sample was exposed to both the top and bottom surfaces.

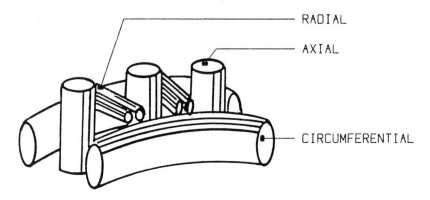

Fig. 6.   A three-dimensional weave for a carbon-carbon
          composite material.

      The samples used in this study were taken from a piece of a cylindrical
billet of composite material.   Figure 6 shows the weave pattern employed in
producing the billet.[13] When the billets were manufactured, the axial and
radial fibers were placed under tension before heating, but the
circumferential fibers were not.

      Figure 7 shows the three types of cuts that were used in producing the
various samples. The surface of each cut had three distinct regions of
interest. These regions consisted of one set of bundles lying parallel to
the surface, one set of bundles running perpendicular to the surface, and a
region of matrix material. Photoacoustic data were on the three regions of
each cut.

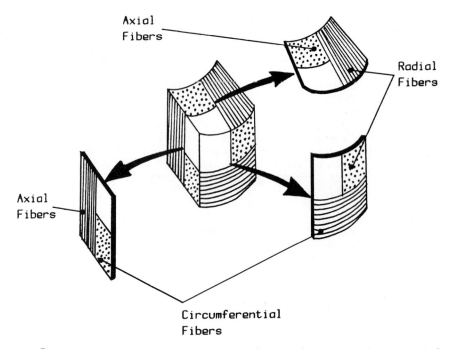

Fig. 7.   Three types of cuts from a carbon-carbon composite material.

Table 2. Thermal diffusivities of the constituents of a
carbon-carbon composite material.

| Fibers | THERMAL DIFFUSIVITY (cm²/sec) | | Anisotropy Ratio |
	Perpendicular to Fibers	Parallel to Fibers	
Circ.	0.6 ± 0.2	3.8 ± 0.5	6
Radial	1.0 ± 0.2	7.2 ± 0.8	7
Axial	1.6 ± 0.3	10.1 ± 0.9	6

Table 2 shows the thermal diffusivities of the three types of bundles
(circumferential, radial, and axial), measured perpendicular to the fibers
and parallel to the fibers. The results are in agreement with diffusivities
obtained by other methods.[14-17] The anisotropic nature of the thermal
diffusivity is evident from the table. The thermal diffusivity measured
parallel to the fibers is always higher than that measured perpendicular to
the fibers, the anisotropy ratio of the diffusivities being about 6:1. The
thermal diffusivity measured on the circumferential bundle is the lowest of
the three bundle directions. This is true for the directions both parallel
and perpendicular to the fibers. The differences in the diffusivities are
attributable to the fact that, in manufacture, the circumferential bundles
are not held under tension, while the radial and axial bundles are kept
under tension.

CONCLUSIONS

The thermal diffusivities of the components of a 3-D carbon-carbon
reinforced composite were measured using transmission photoacoustic
microscopy. Thermal diffusivities were measured in directions parallel and
perpendicular to the fiber reinforcement for the axial, radial, and
circumferential fiber bundles. The anisotropy of the heat flow in the three
orthogonal directions was measured.

REFERENCES

1. A. Rosencwaig, "Photoacoustics and Photoacoustic Spectroscopy," Chem.
    Anal. 57, John Wiley and Sons, New York, N.Y. (1980).
2. A. C. Tam, Rev. Mod. Phys., 58:381 (1986).
3. R. L. Thomas, L. D. Favro, and P. K. Kuo, Can. J. Phys, 64:1234 (1986).
4. P. Charpentier, F. Lepoutre, and L. Bertrand, J. Appl. Phys.,
    53:608 (1982).
5. R. T. Swimm, Appl. Phys. Lett., 42:955 (1983).
6. T. Ahmed, "In Situ Measurements of the Thermal Properties of
    Heterogeneous Materials Using Transmission Photoacoustic Microscopy,"
    Ph.D. Thesis (Unpublished), Southern Illinois University at
    Carbondale (1987).

7. A. Rosencwaig and A. Gersho, J. Appl. Phys., 47:64 (1976).
8. F. A. McDonald and G. C. Wetsel, Jr., J. Appl. Phys., 49:2313 (1978).
9. H. S. Bennett and R. A. Forman, Appl. Opt., 20:911 (1977).
10. L. C. Aamodt and J. C. Murphy, J. Appl. Phys., 49:3036 (1978).
11. P. Korpiun and B. Buchner, Appl. Phys. B, 30:121 (1983).
12. H. C. Chow, J. Appl. Phys., 51:4053 (1980).
13. W. L. Lachman, J. A. Crawford, and L. E. McAllister, "Proc. Inter. Conf.
    on Composite Materials" (Metallurgical Society of AIME), B. Nortor,
    R. Signorelli, K. Street and L. Phillips, eds., New York, N.Y.
    (1978).
14. R. E. Taylor, J. Jortner, and H. Groot, Carbon, 23:215 (1985).
15. D. L. Balageas and A. M. Luc, AIAA Journal, 24:109 (1986).
16. D. L. Balageas, High Temp. - High Press., 16:199 (1984).
17. R. E. Taylor, and B. H. Kelsic, J. Heat Transfer, 162:108 (1986).

NEW TECHNIQUES IN DIFFERENTIAL PHASE CONTRAST

SCANNING ACOUSTIC MICROSCOPY

M. Nikoonahad

Philips Laboratories
North American Philips Corporation
345 Scarborough Road, Briarcliff Manor, New York 10510, USA

1. INTRODUCTION

Differential phase techniques in scanning acoustic microscopy (SAM) rely on phase comparison between signals arriving from two adjacent regions of the object {1-3}. At a given frequency, to maintain a high imaging resolution, ideally one would like these two regions to be as close to each other as possible - one requires *two adjacent foci*. This way, one records the phase gradients caused by the object, as compared to another form of imaging in SAM where a differential image is obtained by digitally subtracting two successive absolute images, {4}. While a reflection differential phase system easily lends itself to imaging topography, a transmission system is particularly useful for imaging velocity variations in biological media such as tissue.

Smith and Wickramasinghe have reported a transmission system with a split transducer geometry in which they achieve the required separation between the spots by defocusing the microscope {1}. In another technique reported by the same authors, the transducer is excited by the fundamental and its third harmonic {2}. This leads to two concentric foci, one of which is 9 times larger in area than the other. In this system, one records the phase difference between one point (small spot) and an *average local reference* which arrives from the large spot. Another differential phase technique in SAM relys on time gating, narrowband filtering and phase comparison between the specularly reflected signal and the leaky Rayleigh wave {3,5}. This is realized by defocusing a reflection microscope and the lateral resolution is basically determined by the Rayleigh wave path length on the specimen. These techniques have led to notable results in accurate determination of Rayleigh velocity in solids {3,5}. It is clear that, although in all these techniques one displays some kind of differential phase signal, one never makes a phase comparison between signals from *two adjacent foci*. The aim of this paper is to demonstrate alternative schemes in which (i) the phase comparison is made between two true foci; (ii) the separation between the foci can be designed to be any specific value, depending on the application; (iii) in principle, no defocusing is required; (iv) one operates at a single frequency; (v) the microscope can be set up in both transmission and reflection, and (vi) does not rely on the generation of leaky Rayleigh waves.

The basic operation of the system resides on a two beam lens shown in figure 1. The back of the lens rod embodies two tilted surfaces with two transducers. The tilt angle, lens rod and the transducer widths are designed so that the axis of each beam passes

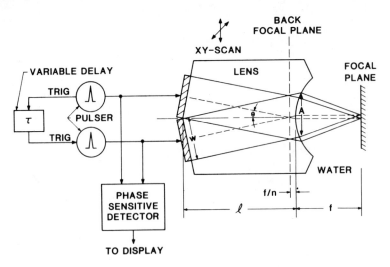

Figure 1. The basic elements of the system for reflection.

through the back focal point of the lens, leading to two off-axis foci. The operation of the two beam lens is analogous to that of the Nomarski objective in optical phase contrast microscopy. The phase difference between the signals arriving from these two foci is then measured in a phase sensitive detector (PSD). The PSD consists of two fast comparators to clip the signals in order to remove any amplitude variation, a balanced mixer, and a low-pass-filter. The lens design based on geometrical acoustics is given in section 2. There are two modes of operation for reflection which are also discussed in section 2. In transmission, a conventional lens acts as transmitter and the two beam lens serves as the receiver. A 10 MHz microscope has been constructed which operates both in reflection and in transmission. In section 3 a range of results obtained from surface topography in solids and velocity variation in soft samples is presented. Section 4 deals with conclusions.

2. LENS DESIGN AND BASIC THEORY

For a velocity ratio n across the lens, the back focal length is n times shorter than the front focal length. It is then clear that for a separation Δx between the spots, from figure 1 we have,

$$\theta \approx \tan^{-1}\left[(n-1)\Delta x/2R\right] \tag{1}$$

Where we have used the definition: $f \equiv nR/(n-1)$, for the paraxial focal length for a lens of radius R. Once the tilt angle is determined, in order for the central ray to pass through the back focal plane we must have: $w=2(l-f/n)\sin\theta$, where w is the transducer diameter. Furthermore, the effective aperture, responsible for resolution is $A=w/\cos\theta \approx 2R\sin\alpha$ where α is half opening angle of the lens. For a given α, and once θ is determined from equation (1), we must have:

$$l \approx \frac{R}{(n-1)}\left[\frac{2R\sin\alpha}{\Delta x}+1\right] \tag{2}$$

For a typical value of 0.8 for $\sin\alpha$ and a given Δx, it is clear that l can turn out to be relatively long. We, therefore, have to aim for the largest n and smallest R to achieve an acceptable l.

502

With this design, when the object is at focus, the energy emitted from each transducer after reflection comes back to itself. However, when the object is slightly defocused, we expect to get cross coupling. The signals collected by the two transducers can, in general, be written in the form:

$$S_1(t)=s_{11}\cos(\omega_o t+\phi_{11})+s_{12}\cos(\omega_o t+\phi_{12}+\phi_o) \qquad (3a)$$

and

$$S_2(t)=s_{21}\cos(\omega_o t+\phi_{21})+s_{22}\cos(\omega_o t+\phi_{22}+\phi_o) \qquad (3b)$$

Where s_{ij} and ϕ_{ij} are the amplitude and phase of the signal emitted from jth and scattered into the ith transducer. When we are at focus, the variation of s_{11} with transverse distance peaks at $+\Delta x/2$; s_{22} peaks at $-\Delta x/2$ and $s_{12}=s_{21}\approx 0$. We then adjust ϕ_o by means of the variable transmit delay, figure 1, so that the output of the PSD is $\sin(\phi_{11}-\phi_{22})$ which ensures small-signal-linearity and maximum sensitivity. We name this mode of operation Mode I. In Mode II operation we defocus the lens slightly and use only one pulser and make use of the cross term component for phase reference. In this mode we have $s_{11}\approx s_{21}$, $s_{22}=s_{12}=0$ and we introduce a $\pi/2$ phase shift in one of the arms in reception. We hence measure $\sin(\phi_{11}-\phi_{21})$. One could, of course, interchange 1's and 2's in the subscripts. In this mode of operation, s_{11} peaks at $\Delta x/2$ from the axis and s_{21} peaks approximately on the axis. This is because s_{21} is effectively the point-spread-function of the first transducer blurred by that of the second transducer.

Figure 2(a) shows ray tracing through a two-beam aluminum lens focusing into water. It is clear that the central ray, which has passed through the back focal point of the lens, after passing through the lens, travels parallel to the axis and, after reflection, all the energy returns to the lower transducer. In this case Δx is 190 μm. Figure 2(b) shows the same situation with the object closer to the lens by 150 μm. The reflected rays follow a different path and some of these rays reach the upper transducer, figure 2(c). We see that with this defocusing distance only a relatively small portion of the rays reach the upper transducer. We therefore expect: (i) The cross terms amplitude to be smaller and (ii) the beam profile corresponding to the cross term signal to be wider than that due to s_{11}.

3. EXPERIMENTAL RESULTS

Based on the design given at the end of the last section, a two beam lens was fabricated and two PZT4 10 MHz transducers were bonded to the sloped surfaces. At 10 MHz with this lens we have: $\Delta x=1.3\lambda$ and we are aware that this lens design is better suited for a lower frequency operation. Figure 3 shows the beam profile of the lens which was measured by scanning the lens over a 100 μm steel wire stretched between two points in water. We see that the separation between the spots is 195 μm which is in good agreement with that predicted by ray tracing. In order to measure the cross term component the lens was defocused by 1.5 λ into $-z$. We observed that, when at focus, the cross terms were only 15 dB down and in order to operate in Mode I the microscope had to be defocused into $+z$ by approximately 1 λ in order to further reduce the cross term amplitudes. This is believed to be due to diffraction effects in the lens rod which are not predicted by ray tracing. Section 3.1 gives results obtained with the reflection set up and our transmission results are presented in section 3.2.

3.1 Reflection Experiments

The differential nature of the microscope can be readily demonstrated by scanning over an edge. Figure 4 shows a series of line scans obtained from a 10 μm ($\lambda/15$) step in

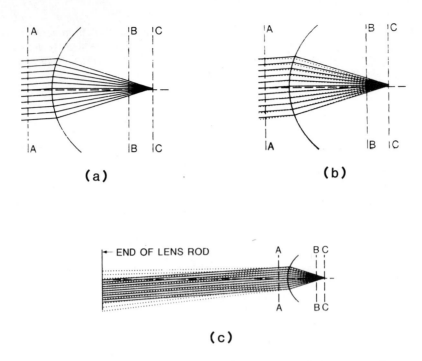

(a) **(b)**

(c)

Figure 2. Ray tracing through an aluminum/water lens. We have: $R=5$ mm; $n=4.2$; $l=32$ mm; $w=3.7$ mm ; $\theta=3.5°$. Solid lines are incident rays and dotted lines are reflected rays. AA is through the back focal point, BB is through the center of the lens and CC is through the object plane. (a) Object at the paraxial focal plane; (b) object 150 μm defocused toward the lens and (c) as (b) but rays shown through the lens rod.

stainless steel using Mode I. It is clear that the signal is differential. The measured SNR for this step was 50 dB in 300 Hz bandwidth. Given that we are measuring $\sin\Delta\phi$, we expect a minimum detectable phase of 0.1° for the microscope with its present electronics. By reducing the bandwidth to 1 Hz we gain another 25 dB in SNR and the expected minimum detectable phase is 0.008°. This means that we should be able to detect height changes down to 16 Å. Figure 5 shows an image obtained from letters inscribed on brass. The inscription depth and width were 25 μm and 0.5 mm respectively. The two spots were along the horizontal scan direction and as expected the image is differential along this direction only. Furthermore, we see that there are two signals for each character width. These correspond to the edge of each letter. In all the following grey scale images the grey background represents zero differential phase signal; white is a positive signal and black represents a negative signals. Figure 6 shows an image obtained from a dime using Mode II. It is clear that the image is differential and we are imaging surface topography. Figure 7(a) shows the surface of a ZnSe crystal used in solid state visible lasers. This crystal was grown at Philips Laboratories by a vertical Bridgman technique {6}. The growth process has led to a number of parallel twins which, after the crystal was polished and etched, have resulted in height changes on the surface. The vertical parallel lines in figure 7(a) are due to this topography variation. A surface profiler was used to scan the object along AA. Figure 7(b) shows three line scans (the separation between scans is 115 μm) along AA which was obtained from the microscope. The two way phase change of 2° corresponds to a height change of 0.4 μm which we measured with the profiler.

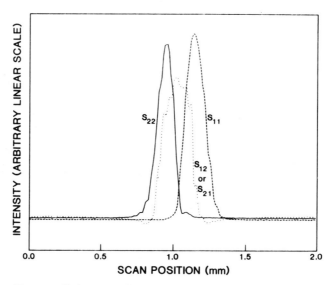

Figure 3. Point-spread-function of the two beam lens.

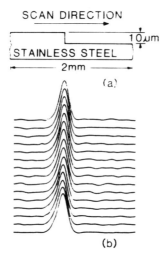

Figure 4. (a) Schematic of the test object; (b) image obtained using Mode I.

3.2 Transmission Experiments

The transmission set up of the microscope is illustrated in figure 8. The two beam lens, used in receiver, probes two adjacent points of the focal plane field distribution of the transmitter lens and it is the phase difference between signals from these two points which provides the contrast. The variable delay line used in one of the arms is used for calibration purposes and normally we set this delay so that, as in reflection, we measure $\sin\triangle\phi$. The motivation for our transmission work is to image velocity variation in tissue specimens. A knowledge of magnitude and the spatial scale of velocity variation in

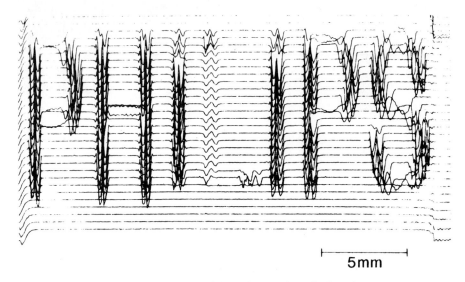

5mm

Figure 5. Image obtained from letters inscribed on brass.

5mm

Figure 6. Image obtained from a coin.

tissue is important both for diagnostic ultrasound and for disease classification. In medical ultrasound, the variation of velocity in tissue can lead to a random phase noise in the wavefront of the echo, resulting in the degradation of the focus, {7}. In tissue characterization, there is evidence that the velocity in diseased tissue can be different from that in healthy tissue {8} and recently there has been a great deal of interest in velocity measurement in tissue {9,10}. As far as high resolution techniques are concerned, both SAM {11} and the SLAM {12} have been used for such velocity measurements.

Given its high imaging resolution and phase sensitivity, a differential phase transmission microscope, figure 8, is particularly suitable for such applications. We have paid particular attention to the design of our sample holders to minimize the thickness variations in such soft samples. Figure 9 shows an image from a tissue-like, but velocity homogeneous, 1 mm thick gel phantom. The semi-circle is the edge of the sample holder. We see that within the sample we do not have any discernible contrast which indicates that the thickness was uniform. The feature indicated by an arrow is an air bubble under the

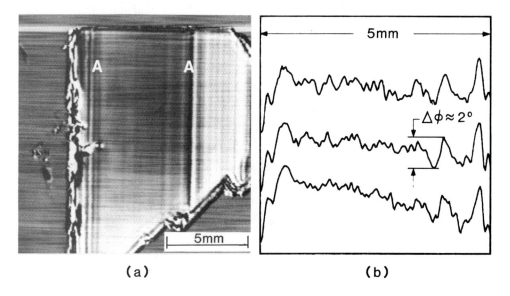

(a)   (b)

Figure 7. (a) Mode II image obtained from twins in a ZnSe crystal. (b) Three line scans between AA.

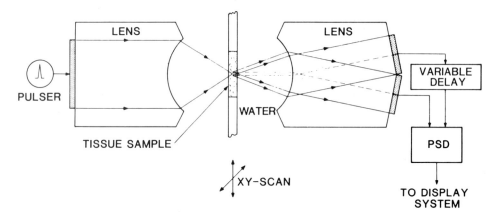

Figure 8. The transmission differential phase microscope.

sample which we observed after we recorded the image. Figure 10 shows a velocity inhomogeneous sample made of gel and agar swirled into each other using the same sample holder. The velocity in the gel matrix and the agar are 1.494±0.005 mm/μs and 1.520±0.002 mm/μs, {13}. The contrast in the image, figure 10, is almost solely due to this small velocity variation. Figure 11 shows an image obtained from a 2 mm thick slice of ham. A large degree of velocity variation is evident and a close inspection of the data indicated that in the visually uniform parts of the sample the differential phase signal varied by ±10°.

These results show that small changes in velocity can be imaged with such a microscope. Quantitative determination of velocity variation primarily requires: (i) calibration of the phase detector and (ii) an angular spectrum analysis of the two beam transmission system.

Figure 9. Transmission image of a tissue-like but velocity homogeneous sample.

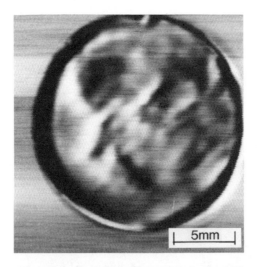

Figure 10. Transmission image of a tissue-like and velocity inhomogeneous sample.

4. CONCLUSIONS

We have shown that with a tilted rod geometry it is possible to achieve two adjacent foci, leading to new forms of differential phase contrast imaging in SAM. Using such a lens, both reflection and transmission experiments were described and illustrated by means of images obtained at 10 MHz. We described the experiments at a relatively low frequency to demonstrate the basic concepts. Extension to higher frequencies is straight-forward.

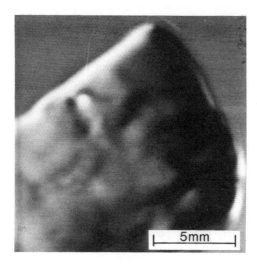

Figure 11. Transmission image obtained from a 2 mm thick slice of ham.

ACKNOWLEDGEMENTS

The author would like to thank Drs. R. Bhargava and B. Singer for their support during this study. Many thanks go to Professor R. Waag of Rochester University, New York for providing the gel phantoms and to Dr. M. Shone for providing the ZnSe sample. Discussions with Drs. M. Vaez Iravani and A. Shaulov and help from Ms. M. Rosar with the display system are greatly appreciated. Finally thanks are due to R. Wassmann and his crew for their excellent technical support.

REFERENCES

1. I. R. Smith and H. K. Wickramasinghe, "Differential Phase Contrast in The Acoustic Microscope", Electronics Letters, 18(2), pp 92-94, 1982.

2. I. R. Smith and H. K. Wickramasinghe, "Dichromatic Differential Phase Contrast Microscopy", IEEE Trans. Vol. SU(29), 6, 1982.

3. K. K. Liang, S. D. Bennett, B. T. Khuri-Yakub and G. S. Kino, "Precision Measurement of Rayleigh Wave Velocity Perturbation", Appl. Phys. Lett, 41(12), pp 1124-1126, 1982.

4. S. D. Bennett and E. A. Ash, "Differential Imaging with the Acoustic Microscope", IEEE Trans. , Vol SU(28), 2, pp 59-64, 1981.

5. K. K. Liang, S. D. Bennett, B. T. Khuri-Yakub and G. S. Kino, "Precise Phase Measurement with the Acoustic Microscope", IEEE Trans. Vol. SU(32), 2, 1985.

6. M. F. Shone, Philips Laboratories, Private Communication.

7. M. Nikoonahad, "Synthetic Focused Image Reconstruction in the Presence of a Finite Delay Noise", Proc. of IEEE Ultrasonic Symposium, pp 819 - 824, 1986.

8. J. F. Greenleaf, S. A. Johnson, R. C. Bahn and and B. Rajagopalan, "Quantitative Cross-sectional Imaging of Ultrasound Parameters", Proc. of IEEE Ultrasonics Symposium, pp 989-995, 1977.

9.   D. E. Robinson, F. Chen and L. S. Wilson, "Measurement of Velocity of Propagation from Ultrasonic Pulse-echo Data", Ultrasound in Med. & Biol., 8(4), pp413-420, 1982.

10. J. Ophir, "Estimation of Speed of Ultrasound Propagation in Biological Tissues: A Beam-Tracking Method", IEEE Trans. on Ultrasonics, Ferroelectrics and Frequency Control, UFFC33(4), pp 359-368, 1986.

11. D. A. Sinclair and I. R. Smith, "Tissue Characterization using Acoustic Microscopy", in Acoustical Imaging, Vol 12, E. A. Ash and C. R. Hill Eds. New York, Plenum, 1982.

12. P. A. Embree, K. M. U. Tervola, S. G. Foster and W. D. O'Brien Jr. "Spatial Distribution of the Speed of Sound in Biological Materials with the Scanning Laser Acoustic Microscope", IEEE Trans. on Sonics and Ultrasonics, SU(32), 2, pp 341-350, 1985.

13. R. C. Waag, University of Rochester, New York, private communication.

A 100 MHZ PVDF ULTRASOUND MICROSCOPE WITH BIOLOGICAL APPLICATIONS

M.D. Sherar and F.S. Foster

Physics Division, Ontario Cancer Institute, and Department
of Medical Biophysics, University of Toronto, 500 Sherbourne
Street, Toronto, Ontario, Canada M4X 1K9

ABSTRACT

A scanning ultrasound microscope has been constructed using a
spherically focused Poly(Vinylidene fluoride) (PVDF) transducer. The
system operates in the frequency range 50 MHz to 110 MHz (a consequence of
the high bandwidth of PVDF) and has a corresponding lateral resolution
limit of 17.5 μm. The system is designed to make two types of
image: Attenuation images of thin specimens using a quartz flat as a
reflector, and also dark field, backscatter images (C-scans) of
cross-sectional planes within specimens up to 4 mm in diameter. A detailed
analysis of the properties of PVDF over the frequency range 5 MHz to 170
MHz is presented and the design of a transducer which meets the bandwidth
and sensitivity requirements for backscatter imaging is described. Axial
resolution of 28 μm is demonstrated while signals from a viable multicell
tumour spheroid showed a scatter signal to noise ratio of approximately
40 dB.

INTRODUCTION

Ultrasound microscopy uses high frequency ultrasound waves to probe
the elastic structure of specimens on a microscopic scale. The ultrasound
wave interacts with the elastic properties of the specimen leading to high
contrast images without the need for staining as is required in optical
microscopy. Since the introduction of practical methods of ultrasound
microscopy in 1974 (1), numerous systems based on transmission and low
temperature reflection methods have been used to give high contrast images
of biological specimens including cells (2), chromosomes (3) and bacteria
(4) down to a resolution of 200 Å. However, in transmission and low
temperature reflection modes, it is not possible to distinguish structure
as a function of depth, limiting these techniques to the imaging of thin
biological sections.

The unique advantage of ultrasound for microscopy over other forms of
radiation is its ability to penetrate specimens without causing damage and
this leads to the possibility of imaging tomographic planes, at depth, in
living specimens. Using this property of ultrasound, several methods have
been developed for subsurface imaging in solid samples including
tomography (5), reflection microscopy (6), and pulse compression
subsurface microscopy (7). One of the most promising methods for

subsurface imaging of tomographic planes in biological specimens is dark field (backscatter) ultrasound microscopy, a technique which involves scanning the specimen with a highly focused ultrasound beam and detecting the ultrasound scattered from within the specimen using the same transducer. Structure can be determined as a function of depth because the scatter signal is detected at a time corresponding to the flight time of the ultrasound pulse to the depth of interest and back to the transducer. This technique has had considerable success in medical ultrasound imaging at low frequencies (1-10 MHz) (8) and is the basis of most clinical ultrasound scanners. We are attempting to apply a particular mode of backscatter imaging, referred to as a C-scan (8), to ultrasound microscopy of planes at depth in biological specimens. In a C-scan, the detected scatter signal is sampled at a time corresponding to a plane of interest in the specimen. A two dimensional image is formed by scanning the beam across the plane and measuring the scatter amplitude at each point in the scan. In this paper, the ultrasound microscope is described with particular reference to the transducer design.

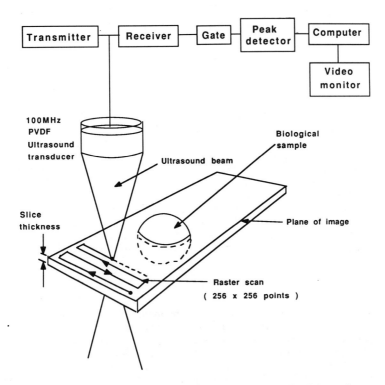

Figure 1. Schematic diagram of the dark field ultrasound microscope. A short, 100 MHz radiofrequency pulse excites a spherically focused PVDF transducer. The ultrasound beam is brought to a focus in the specimen and the resulting scatter is detected by the same transducer. C-scans of cross sectional planes in the specimen are made.

The ultrasound microscope we have developed is shown schematically in Figure 1. A short, 130 volt, 100 MHz electrical pulse is generated by a Matec model 310 Gated Amplifier (Warwick, Rhode Island). The pulse is transformed by the transducer into a short, 100 MHz ultrasound pulse which is transmitted through the water coupling medium to a focus in the specimen. The backscattered ultrasound, detected by the same transducer, is amplified and demodulated by an HP model 423B Coaxial Crystal Detector. The amplitude of the demodulated signal is measured by an Avtech model AVS-101 Peak Detector (Ottawa, Canada) at a time corresponding to scatter from the focal plane of the transducer and the measured value is digitized and stored in a computer. A C-scan is made by moving the focus of the transducer to the depth of interest and scanning the specimen in a raster fashion under the beam. The motion is accomplished by piezoelectric "inchworm" positioners (Burleigh Instruments, Rochester, New York) with an absolute accuracy of $\pm$ 1 μm.

TRANSDUCER MATERIAL: POLY(VINYLIDENE FLUORIDE)

At the heart of the system is the transducer. It must have both high bandwidth to enable the production of short pulses as well as good sensitivity to detect the small scatter signals from biological samples. We have employed the piezoelectric polymer Poly(Vinylidene Fluoride) (PVDF) as the transducer material to satisfy these requirements. PVDF is a semicrystalline polymer consisting of long chain molecules each of approximately 2000 CH_2CF_2 repeat units (9). The semicrystalline state is characterized by 10^{-8}m thick lamella crystals embedded in an amorphous matrix. A single molecule folds back and forth many times within a crystal, such that each fold is about 40 repeat units in length. A ferroelectric state is achieved by first stretching the material which aligns the crystals normal to the stretch direction and then orientating the F_2H_2 dipoles at an elevated temperature under a high electric field. In this state, PVDF thin films can generate and detect the high frequency ultrasound required for microscopy.

The important parameters of PVDF which affect the performance of the transducer are the electromechanical coupling coefficient (k_T), the dielectric permittivity (ϵ_r), the dielectric loss tangent (tan δ_e) and the mechanical loss tangent (tan δ_m). We have measured ϵ_r and tan δ_e in the frequency range 70 MHz to 170 MHz and tan δ_m in the range 5 MHz to 110 MHz and applied the results to the Krimholtz-Leedom-Matthei (KLM) transducer model (10) to determine k_T. The dielectric permittivity was determined by measuring the capacitance of a 9 μm PVDF thin film clamped in air, using an HP model 4191A Impedance Analyzer. The results, plotted in Figure 2a, are extrapolated through the piezoelectric resonance and show the magnitude of the permittivity to be decreasing over the measured frequency range, consistent with a dielectric relaxation in PVDF at about 5 MHz (11). The dielectric loss tangent, shown in Figure 2b, is approximately constant over the measured frequency range. The mechanical loss tangent, shown in Figure 3, was determined by measuring the amplitude of a continuous wave source of ultrasound through a PVDF film, 117 μm in thickness. Reverberations within the film were taken into account. The results, in reasonable agreement with those of Leung and Yung (12), show tan δ_m to be decreasing over the measured frequency range again consistent with the relaxation in PVDF at 5 MHz.

The measured values of ϵ_r, tan δ_e and tan δ_m were then applied to impedance measurements of a 9 μm thick PVDF film, near the fundamental piezoelectric resonance at 117 MHz, to calculate k_T using the KLM model. The calculated value of k_T was 0.13, in good agreement with the results of Bui et al (13), using a similar method. The validity of the KLM model calculation was confirmed by using all the measured parameters to predict the impedance of the 9 μm PVDF film over a wide frequency range and comparing the prediction to the experimental impedance measurements as shown in Figure 4. The effect of mechanical and dielectric losses on the resonance are shown in Figure 5, where the theoretical impedance has been calculated without including the losses. It is apparent that the

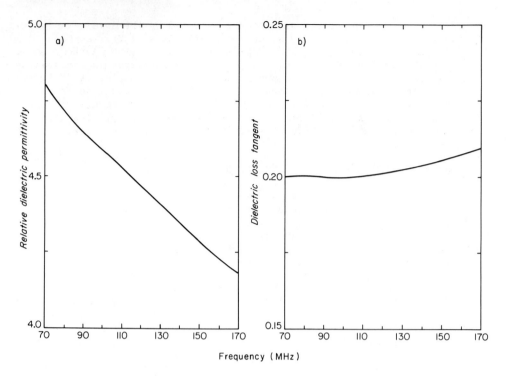

Figure 2. Frequency dependence of   a)   dielectric   permittivity   and   b)
dielectric   loss   tangent for 9 μm PVDF film   (from   impedance
measurements).

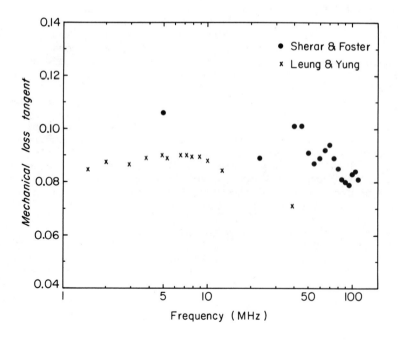

Figure 3.   Frequency dependence of mechanical loss tangent
of PVDF.

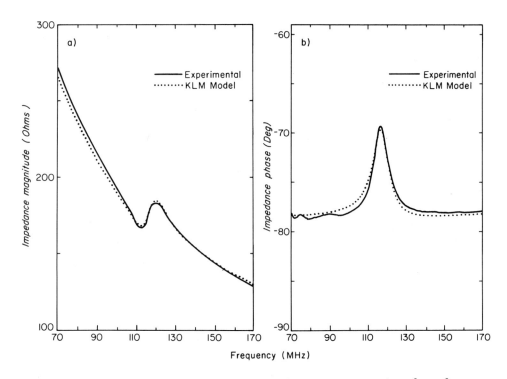

Figure 4. Comparison of experimental impedance measurements of a 9 μm
PVDF film clamped in air with theoretical predictions using
the measured quantities, ϵ_r, tan δ_e, tan δ_m and k_T in the KLM
model.

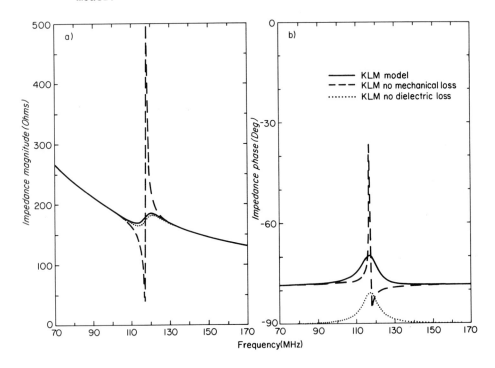

Figure 5. Effect of mechanical and dielectric loss on the
piezoelectric resonance in PVDF. Calculated from KLM model.

mechanical loss has much more effect on diminishing the piezoelectric resonance in PVDF than does the dielectric loss. The bandwidth of the material can be found from the full width at half maximum of the resonance which was measured to be 10 MHz for this sample of PVDF in air. Loading both the front and back of the PVDF film dramatically increases this bandwidth.

The results show that the high mechanical loss gives PVDF both a high bandwidth and a poor sensitivity. However, when incorporated in a transducer, PVDF has a number of other properties which compensate for its poor sensitivity. The mechanical flexibility of PVDF film allows the construction of spherically focused transducers without the need for lenses which incur losses. As well, the acoustic impedance of PVDF is well matched to water which leads to efficient transmission of ultrasound to and from the transducer. Finally, the maximum field strength that can be applied to PVDF without depolarization occuring is about 100 times that which can be applied to conventional ceramic materials (14).

TRANSDUCER

The transducer in our system is shown schematically in Figure 6. A 9 μm thick PVDF film is bonded to a spherically-shaped conductive epoxy backing, which serves as one electrode. The transducer transmits through the front electrode, which is a thin layer of gold, directly into the water coupling medium. The focal length of the transducer is 4 mm, while its diameter is 3 mm, giving an aperture of f/1.33. The PVDF film resonates at a frequency

$$\nu_O = c/4d = 58.5 \text{ MHz}$$

where c is the speed of sound in PVDF (2100 m/s) and d is the film thickness. However, because of the high bandwidth of PVDF, the transducer can be operated at a much higher frequency whilst retaining adequate sensitivity. We have tuned the transducer to 100 MHz for improved resolution using a parallel, single stub, coaxial cable tuner.

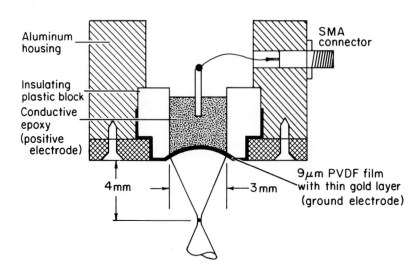

Figure 6. Schematic diagram of spherically focused PVDF transducer.

The sensitivity of the transducer was determined by measuring its two way insertion loss, defined as the ratio of the power received from the transducer after the ultrasound beam has been perfectly reflected, and delivered to a 50 Ω load (P_R), to the incident power available to a 50 Ω load (P_T). Thus, in decibels

$$I = 10 \log(P_R/P_T)$$

The two way insertion loss of the transducer was measured to be 52 dB once the loss due to attenuation in the water (15.3 dB) and the reflection loss at the water-quartz interface (2.0 dB) had been subtracted. As mentioned previously, the poor insertion loss of the PVDF transducer is compensated by its high depolarization field strength which allows the use of high voltage excitation pulses.

The lateral resolution of the microscope is determined by the two way full width at half maximum ($FWHM_{PE}$) of the beam at the focus which is given by standard diffraction theory

$$FWHM_{PE} = (c/\bar{\nu}) \, (f\text{-number}) = \bar{\lambda} \, (f\text{-number})$$

where $\bar{\nu}$ is the average frequency of the ultrasound pulse, c is the speed of sound in the specimen, and $\bar{\lambda}$ is the average wavelength (15). This gives a theoretical resolution of 19.6 μm. The resolution was determined experimentally by scanning the beam across an 8 μm diameter glass fibre and measuring the reflected signal amplitude as a function of position relative to the fibre as shown in Figure 7. The $FWHM_{PE}$ is 17.5 μm, in good agreement with the theoretical resolution.

The slice thickness of the tomographic images is determined by the length of the ultrasound pulse (Δt) and is given by

$$\Delta z = c\Delta t/2$$

where Δz is the slice thickness and c is the speed of sound in the specimen. The pulse reflected from a quartz flat is shown in Figure 8a and has a 6 dB width of 37 ns. This corresponds to a slice thickness of 28 μm. A typical signal received due to scattering within a biological specimen is shown in Figure 8b. In this case the specimen is a spherical aggregation of tumour cells referred to as a spheroid (16). The scatter signal delineates structure as a function of depth through the center of the spheroid and rises to a maximum at a point corresponding to the focal plane of the transducer. The signal amplitude is measured at a time corresponding to scattering at the focus of the beam and a C-scan could be made by scanning the specimen in a raster fashion under the beam. To image a plane at a different depth, the sampling time would be kept constant and the focus moved to the desired depth in the sample. The amplitude of the scatter signal from the spheroid demonstrates a dynamic range of approximately 40 dB (Figure 8b), where the dynamic range is defined as the ratio of the peak scatter signal to the noise. We have begun to apply this ultrasound microscope to the study of the internal structure of spheroids the results of which will be published shortly.

CONCLUSION

A dark field 100 MHz ultrasound microscope based on an f/1.33 spherical PVDF transducer has been developed for biological imaging with a diffraction limited lateral resolution of 17.5 μm. PVDF was chosen for its high bandwidth and adequate sensitivity for dark field microscopy. The PVDF parameters affecting these properties were measured and applied to the KLM transducer model which was able to accurately predict the impedance of a 9 μm thick PVDF film clamped in air, over a wide frequency

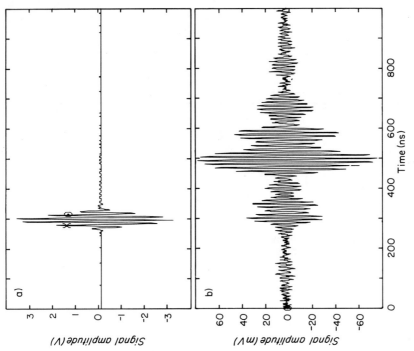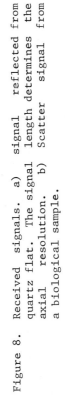

Figure 8. Received signals. a) signal reflected from quartz flat. The signal length determines the axial resolution. b) Scatter signal from a biological sample.

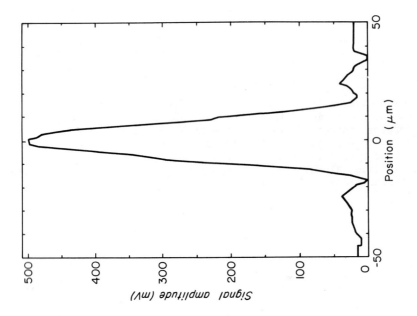

Figure 7. Experimental line response of ultrasound microscope. Full width at half maximum is 17.5 µm.

range. Using the KLM model, the mechanical loss tangent was found to be responsible for the high bandwidth of the material. Consequently the PVDF transducer is able to transmit short 100 MHz pulses resulting in 28 μm axial resolution. The 52 dB insertion loss of the transducer is partially compensated by a large depolarization field strength allowing the use of high voltage excitation pulses. The scatter signal from a biological sample demonstrated approximately 40 dB of dynamic range. We have begun to apply the microscope to the study of the spheroid tumour model system and it is hoped that the microscope will be used in other studies where it can play a unique role as a non-invasive method of sub-surface imaging in biological specimens.

REFERENCES

1. Lemons, R.A. and Quate, C.F., Acoustic microscope - scannning version, Appl. Phys. Lett. **24**:163 (1974).

2. Lemons, R.A. and Quate, C.F., Advances in mechanically scanned acoustic microscopy, Ultrason. Symp. Proc. IEEE Cat. No. 74 CHO 896-ISU:41 (1974).

3. Rugar, D., Heiserman, J., Minden, S. and Quate, C.F., Acoustic microscopy of human metaphase chromosomes, J. Microsc. **120**: 193 (1980).

4. Foster, J.S. and Rugar, D., Low temperature acoustic microscopy, IEEE Trans. Sonics Ultrason. **SU-32**:139 (1985).

5. Lin, Z., Lee, H., Wade, G. and Schueller, C.F., Computer-assisted tomographic acoustic microscopy for subsurface imaging, Acoustical Imaging, vol 13, ed. Kaveh, M., Mueller, R.K. and Greenleaf, J.F. (New York: Plenum 1984) p.91.

6. Jipson, V.B., Acoustic microscopy of interior planes, Appl. Phys. Lett. **35**:385 (1979).

7. Nikoonahad, M., Guangqi, Y. and Ash, E.A., Pulse compression acoustic microscopy using SAW filters, IEEE Trans. Sonics Ultrason. **SU-32**:152 (1985).

8. Hill, C.R., Pulse echo imaging and measurement, in Physical Principles of Medical Ultrasonics ed. Hill, C.R. (Ellis Horwood 1986) p.278.

9. Broadhurst, M.G. and Davis, G.T., Piezo- and pyroelectric properties, in Electrets, Topics in Applied Physics, Vol 33 (Springer-Verlag, Berlin, 1980) p.285.

10. Krimholtz, R., Leedom, D.A. and Matthei, G.L., New equivalent circuits for elementary piezoelectric transducers, Electron. Lett (GB) **6**:338 (1970).

11. Sasabe, H., Saito, S., Asahina, M. and Kakutani, H., Dielectric relaxations in Poly(Vinylidene Fluoride), Journal of Polymer Science, Part A-2, **7**:1405 (1969).

12. Leung, W.P. and Yung, K.K., Internal losses in polyvinylidene fluoride (PVF_2) ultrasonic transducers, J. Appl. Phys. **50**:8031 (1979).

13. Bui, L.N., Shaw, H.J. and Zitelli, L.T., Study of acoustic wave resonance in piezoelectric PVF_2 film, IEEE Trans. Sonics Ultrason. **SU-24**(5):331 (1977).

14. Linvill, J.G., PVF_2 models, measurements and devices, Ferroelectrics **28**:291 (1980).

15. Hunt, J.W., Arditi, M. and Foster, F.S., Ultrasound transducers for pulse-echo medical imaging, IEEE Trans. Biomed. Eng. **BME-30**:453 (1983).

16. Sutherland, R.M., Carlsson, J., Durand, R. and Yuhas, J., Spheroids in cancer research, Cancer Res. **41**:2980 (1981).

FERROELECTRIC POLYMER TRANSDUCERS FOR HIGH RESOLUTION

SCANNING ACOUSTIC MICROSCOPY

H. Ohigashi, K. Koyama, S. Takahashi, Y. Wada,
Y. Maida,[*] R. Suganuma,[*] and T. Jindo[*]

Department of Polymer Materials Engineering, Yamagata University
Jonan 4-3-16, Yonezawa 992, Japan
[*] Honda Electronics Co., Ltd., 20 Oyamazuka, Ohiwa-cho
Toyohashi 441-31, Japan

1. INTRODUCTION

Acoustic microscopy have increasingly become of use for visualization of microstructure and for characterization of acoustic properties of materials[1,2]. Since scanning acoustic microscope (SAM) was first introduced by Lemons and Quate[3], almost all SAM images and V(z) curves have been obtained by using Quate type transducers. This transducer comprises a planar transducer deposited or bonded on the rear surface of a sapphire buffer rod whose front surface is shaped into a concave sphere (acoustic lens) to transmit a focusing acoustic beam in coupling liquid, e.g., water.

Recent studies on thin films of ferroelectric polymers have proved that they are usable for generation and detection of acoustic waves of hundreds MHz and that they are applicable to concave transducers for acoustic microscopy[4,5]. The polymer concave transducer, which is constructed with a concave spherical cavity and a thin film of piezoelectric polymer bonded or coated on the concave surface is expected to have the following advantages as compared to a conventional Quate type transducer; (i) suppression of the reflection loss at interface between the transducer and the coupling liquid because of their good impedance matching, (ii) a large signal to noise ratio due to the absence of spurious echoes in the buffer rod, due to the absence of scattered acoustic waves outside the concave surface, and due to the absence of any transverse waves (iii) the absence of absorption and diffraction losses in the rod lens, and (iv) the broader frequency band characteristics owing to the good impedance matching.

The first trial of concave transducers for SAM by Chubachi and coworker[6,7] has been extended recently by Labreche at al.[8]: SAM images at 30-210 MHz were obtained using a concave transducer composed of a rather thicker film (9 or 28 μm) of poly(vinylidene fluoride) (PVDF) which operates necessarily in high harmonic resonance mode[8].

Ferroelectric copolymer of vinylidene fluoride and trifluoroethylene, P(VDF-TrFE), has a larger electromechanical coupling factor ($k_t \approxeq 0.3$) than PVDF ($k_t \approxeq 0.2$) when the molar ratio of VDF and TrFE is adequately selected[9]. Ultrasonic transducers of P(VDF-TrFE) are successfully applied to medical imaging and non-destructive testing[10-14]. Additional and important advantage of P(VDF-TrFE) is that, in contrast to PVDF, the copolymer does not need drawing to transform the non-piezoelectric phase (α-phase) to the ferroelectric phase (β-phase): P(VDF-TrFE) crystallizes into the ferroelectric phase directly from the melt or solution, and thin films with strong piezoelectricity are easily obtained by coating the solution on substrates, and successively by annealing and poling[9].

In a previous paper[15], we have reported preliminary results on P(VDF-TrFE) concave transducers for SAM. The present paper describes P(VDF-TrFE) concave transducers developed for use in SAM imaging in the extended frequency range from 20-500MHz. High resolution SAM images of polymer materials and V(z) curves obtained with these transducers will be demonstrated.

2. PROPERTIES OF P(VDF-TrFE) THIN FILMS

The piezoelectric and ferroelectric properties of P(VDF-TrFE) films are strongly dependent on the molar ratio of VDF (x) and, TrFE (1-x) and on preparation conditions[9]. The copolymers with $0.65 \leqslant x \leqslant 0.82$ have the largest value of k_t among any other piezoelectric polymers studied so far. The ferroelectric coercive field is 40-50 MV/m. This value is about two order of magnitude larger than inorganic ferroelectrics, implying that a large electrical power can be poured from a source into the film without deterioration of transducing performance.

Although dielectric constant and elastic constants are dependent on temperature, the value of k_t is almost constant over the wide temperature range from 77K up to a temperature just below the Curie point, as shown in Fig. 1.[16]. Therefore, we may predict from the temperature-frequency superposition principle that the P(VDF-TrFE) film works efficiently at higher frequencies, and furthermore, it also operates at lower temperatures, i.e., at liquid helium temperature.

Recent studies on the piezoelectricity of thin P(VDF-TrFE) films (0.06-5 μm) showed that (1) the value of k_t determined by the frequency dependence of the admittance of a piezoelectric free resonator of 5 μm-thick film is 0.26, a value comparable to or larger than those of conventional inorganic thin film transducer material (see Table 1)[4] and (2) the ferroelectric remanent polarization is about $0.1C/m^2$ and is independent of

Table I. Electromechanical Properties of Piezoelectric Thin Films
for Very High Frequency Ultrasonic Transducers (refs. 4,9)

	k_t	$\varepsilon^S/\varepsilon_o$	e_{33} (C/m^2)	ρ $(10^3 kg/m^3)$	v_3 (km/s)	Z_a $(10^6 kg/m^2 \cdot s)$
P(VDF-TrFE)						
5 μm, 250MHz	0.26	4.7	0.17	1.90	2.60	4.9
37 μm, 30 MHz	0.29	4.5	0.18	1.90	2.37	4.49
ZnO	0.28	8.8	1.14	5.65	6.40	36.4
AlN	0,20	8.5	1.03	3.26	10.40	33.4
CdS	0.15	9.5	0.44	4.82	4.50	21.7

thickness of the film down to 0.06 μm, though the electric field required for
polarization reversal becomes higher for the film thinner than 0.3 μm[17].
Since the electromechanical factor k_t is proportional to remanent
polarization[9], a film of P(VDF-TrFE) as thin as 0.06 μm is expected still to
be an efficient high frequency transducer working with the fundamental
resonating mode in a GHz range.

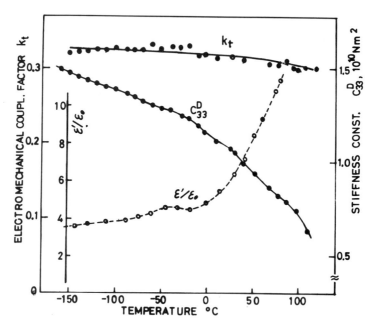

Fig. 1. Temperature dependence of coupling factor k_t, dielectric constant
$\varepsilon^S/\varepsilon_o$, and elastic stiffness constant c^D for P(VDF-TrFE) (x=0.75).

3. CHARACTERISTICS OF P(VDF-TrFE) CONCAVE TRANSDUCERS FOR SAM

A concave transducer developed in the present study is shown in Figs. 2a and 2b. The transducer is composed of a copper rod with a finely polished concave spherical surface, a thin film of P(VDF-TrFE) and a grounded Au electrode. The metal rod serves as the counter electrode and also as a backing load. The procedure for preparing the film on the concave surface was as follows. A dimethylformamide solution of P(VDF-TrFE) (x = 0.75) was spread over the concave surface using a spin coating technique. The film was annealed at 145°C to enhance crystallization[9]. The Au electrode is then deposited by ion sputtering and the film was poled at room temperature under a low frequency AC field applied across the metal rod and the Au electrode. The peak field was 100MV/m, the field sufficiently higher than the coercive field (40MV/m). The thickness of the coated film was controlled by the concentration of the solution and the revolution rate of the spinner. After these procedures, we have fabricated transducers having various center frequencies (50-300MHz), focal lengths (the radii of concave spheres) (3.0 to 0.32mm), and aperture angles ($2\theta = 30\text{-}120°$).

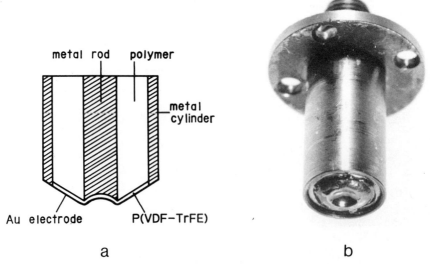

a   b

Fig. 2.   (a) Schematic configuration of a P(VDF-TrFE) concave spherical transducer and (b) photograph of the transducer.

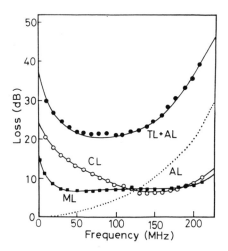

Fig. 3.   Conversion loss (CL), matching loss (ML), transducer loss (TL)
          and absorption loss (AL) in water for a transducer comprising
          a 3 μm-thick P(VDF-TrFE) film. The theoretical values are
          plotted by lines and experimental data by open circles, closed
          circles and closed squares. (from Ref. 4)

Figure 3 shows the frequency dependence of the conversion loss (CL),
transducer loss (TL) and electrical matching loss (ML) reported by Kimura and
one of the present authors[4] for a transducer of a 3 μm-thick P(VDF-TrFE)
film coated on a planer surface of a Cu substrate. The active area was $9mm^2$.
This transducer had essentially the same structure as a concave spherical
transducer and was fabricated in order to collect the fundamental data on the
performance of high frequency polymer transducers. The losses ML, CL and TL
are introduced to evaluate the performance of the transducer quantitatively,
and are defined by $-10\log\,(P_t/P_o)$, $-10\log\,(P_f/P_t)$, and $-10\log\,(P_f/P_o)$,
respectively, where P_o, P_t and P_f are the maximum electrical power available
from a source, the electric input power, and the mechanical output power,
respectively. The experimental results (Fig. 3) are quite in good agreement
with the theoretical ones predicted using Mason's equivalent circuit and the
material constants listed in Table 1. In the Mason's equivalent circuit, the
mechanical and internal losses in the P(VDF-TrFE) film were taken into
account. The minimum value of the conversion loss is 6.5 dB at 150 MHz.

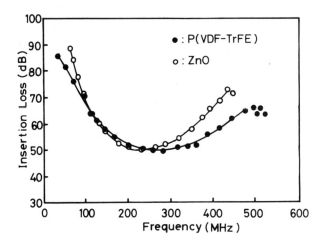

Fig. 4. Two-way insertion loss for a P(VDF-TrFE) concave spherical
transducer and a ZnO transducer with a acoustic sapphire lens.
Contribution of absorption loss in water is excluded for both
transducers.

Figure 4 shows the two-way insertion loss of a concave spherical
transducer (R = 0.32mm, aperture angle = $120°$), consisting of a thinner
(~ 1.5 μm) P(VDF-TrFE) film, and, for comparison, the insertion loss of a
conventional ZnO transducer with a sapphire lens. The sapphire lens has an
antireflection layer. The data on Fig. 4 are corrected for the acoustic loss
in water. The P(VDF-TrFE) concave transducer has a considerably wider

bandwidth than the ZnO transducer, and has the minimum insertion loss
comparable to the loss for the ZnO transducer. This transducer works with
relatively high sensitivity over a wide frequency range covering 20MHz–
500MHz. Low acoustic impedance of the piezoelectric polymer material is
responsible for the high sensitivity and the broader band width
characteristics. The wide band characteristics as well as the high
sensitivity are quite useful for SAM imaging.

The lateral resolution of a concave spherical transducer is essentially
better than that of a planar transducer with an acoustic lens as pointed out
by Li et al.[18]. Deviation from a perfect spherical surface due to
processing error would deteriorate the field distribution. The field
distribution measured for a P(VDF-TrFE) concave transducer, however, was
found to be consistent with that predicted from Fraunhofer diffraction
theory, indicating that there was no distortion in the concave surface within
experimental error.

The dynamic range of the acoustic signal, which is defined as the ratio of the maximum unsaturate signal to noise level, is an important quantity in an imaging system. One of conspicuous advantages of the concave spherical transducer is that it exhibits much wider dynamic range than the conventional ZnO transducer with a sapphire lens. The dynamic range of the concave transducer was found to be wider by 15 dB than that of the latter transducer (typically 50 dB vs 35 dB). In the lens system, the scattered acoustic field outside the aperture of the acoustic lens and reverberation of the acoustic waves in the lens make the dynamic range narrow, while in the concave spherical transducer no scattered field occurs.

4. SAM IMAGE OBTAINED WITH CONCAVE TRANSDUCERS

SAM images of various materials were taken with P(VDF-TrFE) concave transducers using SAM systems HMS-300 and HMS-400 (products of Honda Electronics Co., Ltd.) to test the feasibility.

Figures 5a and 5b compare SAM images of an integrated circuit observed at 150MHz in water with a polymer transducer and a conventional ZnO transducer. (Both the transducers have approximately the same focal lengths, aperture angles and center frequencies.) The image obtained with the concave transducer (Fig. 6a) is more fine as compared to that obtained with the ZnO transducer. This is shown more definitely on the bottom of each figure, in which the spatial distribution of the intensity of the reflection acoustic signals is displayed along the scanning direction.

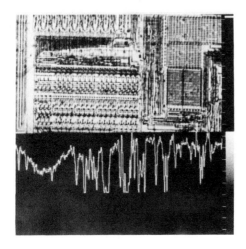 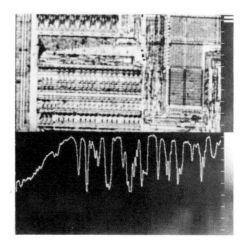

Fig. 5. SAM images of an integrated circuit observed with a P(VDF-TrFE) concave transducer (left) and with a ZnO transducer (right).

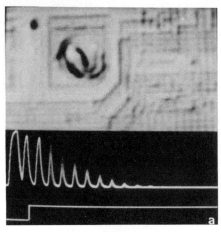

80 MHz

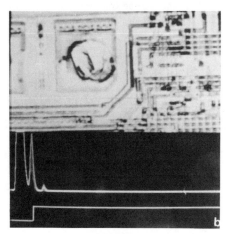

200 MHz

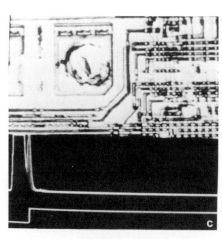

300 MHz

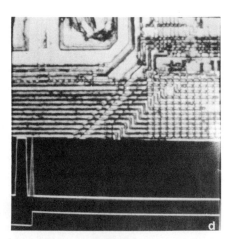

460 MHz

Fig. 6.   SAM images of an integrated circuit observed with a P(VDF-TrFE) concave transducer at different frequencies.

Figures 6a, 6b, 6c and 6d show SAM images of an integrated circuit observed at different frequencies using the same concave transducer having the center frequency around 300MHz, the radius of curvature of 0.32mm and the aperture angle of 120°. The transducer used here was the same as used in Fig. 4. The corresponding reflection signal with respect to time domain (A mode signal) is displayed on the bottom of each figure. These figures

indicate that the present concave transducer operates in a very wide
frequency range. It is also recognized in these figures that the transducer
has a high signal-to-noise ratio, and that higher resolution is attainable at
higher frequency.

Scanning acoustic microscope has been used mainly in characterization of
metals, ceramics and biological materials[19-22]. Despite lack of
accumulation of data, SAM is also useful to observe surface and inner
structures of polymer materials. Figure 7 reveals inner structure of a glass
fiber reinforced plastic plates observed with a P(VDF-TrFE) concave
transducer at 150MHz. Figure 8 shows a 100MHz SAM image of imhomogeneous
structure in a plastic sheet formed by injection molding. As can be seen in
the A mode signal displayed on the bottom of Fig. 8, the plastic plate has
multi-layer structure. Since an acoustic pulse of short duration can be
generated using a P(VDF-TrFE) concave transducer[4,15], high resolution
acoustical tomographic (B-mode) images will be also obtainable with polymer
concave transducers.

Fig. 7. A SAM image of glass fibers embedded in a polymer plate
observed with a P(VDF-TrFE) concave transducer at 150MHz.

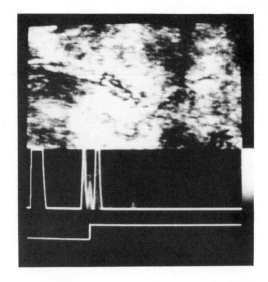

Fig. 8. A SAM image of a inner layer of a polymer sheet formed
by injection molding (100MHz).

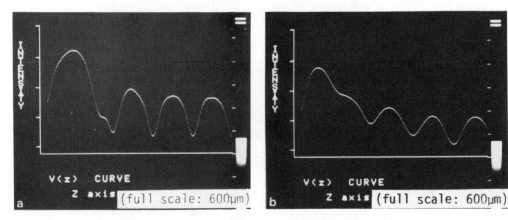

Fig. 9. V(z) curves of a sapphire plate detected at 150 MHz by
(a) a P(VDF-TrFE) concave transducer and
(b) a conventional ZnO transducer.

5. V(z) CURVES

From V(z) curves arising from interference of leaky surface waves with reflected bulk acoustical waves, informations concerning acoustic velocity, and absorption in the objective materials can be obtained[20,23-25]. In this study, we found that V(z) curves are detectable using the P(VDF-TrFE) concave transducer. Figures 9a and 9b show the V(z) curves of a sapphire plate detected at 150MHz by a conventional ZnO transducer and the P(VDF-TrFE) concave transducer, respectively. From these curves, we find that the polymer concave spherical transducers are also essentially suitable for detection of V(z) curves. The analysis of V(z) curve with a spherical concave transducer is now in progress.

6. CONCLUSION

Concave spherical transducers operating in the frequency range from 20 to 500MHz have been developed for a high resolution scanning acoustic microscope using thin films of a ferroelectric copolymer of vinylidene fluoride and trifluoroethylene, P(VDF-TrFE). These transducers were made by coating a dimethylformamide solution on concave metal rod substrates. The transducers have radius of curvature (focal length) of 3.0-0.32mm and the aperture angle (2θ) of 30°-120°.

As compared to conventional ZnO transducers having acoustic lenses (Quate type transducer), the present transducers exhibited the following advantages:
  (1) Broader bandwidth characteristics owing to good acoustic impedance matching between P(VDF-TrFE) and coupling liquid (water).
  (2) Wider dynamic range due to the absence of the acoustic lens which would cause scattering, diffraction, or reverberation of acoustic waves.
  (3) Higher resolution probably due to the absence of spherical abberation and due to wider dynamic range or higher signal-to-noise ratio.
The insertion loss of the polymer concave transducers are comparable to that of ZnO transducers with sapphire lenses having anti-reflection layers.

P(VDF-TrFE) concave transducers have been proved to be applicable to obtaining clear SAM images of various materials including polymers. V(z) signals arising from leaky surface waves were also detectable with the present transducers.

Copolymer films as thin as 0.06 m still retains ferroelectricity even at very low temperatures, and therefore, such thin films are expected to have approximately the same magnitude of k_t as that of thicker films. Thus we may develop effective concave transducers operating at GHz frequencies in liquid nitrogen or helium using P(VDF-TrFE) as transducing material. The work to develop transducers for SAM imaging at GHz frequencies is now in progress.

ACKNOWLEDGMENTS

The authors are greatly indebted to Daikin Kogyo Co., Ltd. for supplying P(VDF-TrFE) copolymers. They are grateful to K. Kimura for her great contribution in early stage of the present work. This work was partly supported by a Grant-in-Aid from the Ministry of Education, Science and Culture of Japan.

REFERENCES

1. R. A. Lemons and C. F. Quate, Acoustic Microscopy, in: "Physical Acoustics" vol.14, Eds. W. P. Mason and R. N. Thurston, pp. 1-92, Academic Press, London. (1979).
2. A. Briggs, "An introduction to Scanning Acoustic Microscopy", Oxford University Press, Oxford (1985). For review of recent developments, see papers in the special issue on acoustic microscopy, IEEE Trans. Sonics Ultrason. SU-32, [2] (1985).
3. R. A. Lemons and C. F. Quate, Acoustic microscope-scanning version, Appl. Phys. Lett. 24:163 (1974).
4. K. Kimura and H. Ohigashi, Generation of very high-frequency ultrasonic waves using thin films of vinylidene fluoride-trifluoroethylene copolymer, J. Appl. Phys. 61:4749 (1987).
5. K. Kimura and H. Ohigashi, Ferroelectric properties of poly(vinylidene-fluoride-trifluoroethylene) copolymer thin films, Appl. Phys. Lett., 43:834 (1983).
6. N. Chubachi and T. Sannomiya, Confocal pair of concave transducers made of PVF_2 piezoelectric films, Jpn. J. Appl. Phys. 16:2259 (1977).
7. N. Chubachi and T. Sannomiya, Composite resonator using PVF_2 film and its application: concave transducer for focusing radiation of VHF ultrasonic waves, 1977 IEEE Ultrasonic Symposium Proceedings, p. 119 (1977).
8. A. Labreche, A. Beausejour, M. Castonguay, and J. D. N. Cheecke, Scanning acoustic microscopy using PVDF concave lenses, Electron Lett. 21:990 (1985).
9. K. Koga and H. Ohigashi, Piezoelectricity and related properties of vinylidene fluoride and trifluoroethylene copolymers, J. Appl. Phys. 56:2142 (1986).
10. H. Ohigashi, K. Koga, M. Suzuki, T. Nakanishi, K. Kimura, and N. Hashimoto, Piezoelectric and ferroelectric properties of P(VDF-TrFE) copolymers and their application to ultrasonic transducers, Ferroelectrics, 60:263 (1984).
11. K. Kimura, N. Hashimoto, and H. Ohigashi, Performance of a linear array transducer of vinylidene fluoride and trifluoroethylene copolymer, IEEE Trans. Sonics Ultrason. SU-32:566 (1985).
12. K. Sakaguchi, T. Sato, K. Koyama, S. Yamamizu, S. Ikeda, and Y. Wada, Wide-band multi-layer ultrasonic transducers made of piezoelectric films of vinylidene fluoride-trifluoroethylene copolymer, Jpn. J. Appl. Phys., 25, Suppl. 25-1:91 (1985).

13. T. Sato, K. Koyama, S. Ikeda, and Y. Wada, Short pulse response of ultrasonic transducers made of piezoelectric copolymer films of vinylidene fluoride-trifluoroethylene, Jpn. J. Appl. Phys. 26, Suppl. 26-1:180 (1987).

14. S. Tsuchiya, T. Sato, K. Koyama, S. Ikeda, and Y. Wada, Application of piezoelectric film of vinylidene fluoride-trifluoroethylene copolymer to highly sensitive miniature hydrophone, Jpn. J. Appl. Phys. 26, Suppl. 26-1:183 (1987).

15. H. Ohigashi, K. Koyama, S. Takahashi, K. Kimura, Y. Maida, and Y. Wada, High-resolution scanning acoustic microscope using a thin film transducer of P(VDF-TrFE), in: "Ultrasonic Technology 1987" (Proceedings of the Toyohashi International Conference on Ultrasonic Technology), Ed. K. Toda, p. 63, Myu, Tokyo (1987).

16. K. Koga and H. Ohigashi, unpublished.

17. K. Kimura and H. Ohigashi, Polarization behavior in vinylidene fluoride-trifluoroethylene copolymer thin films, Jpn. J. Appl. Phys., 25:383 (1986).

18. D. J. Li, G. L. Chen, and K. Q. Zhang, A new transducer-focusing system for acoustic microscope--glass-metal based concave spherical transducer, 1984 IEEE Ultrasonics Symposium Proeedings, p. 567 (1984).

19. J. S. Foster and D. Rugar, Low-temperature acoustic microscopy, IEEE Trans. Sonics Ultrason. SU-32:139 (1985).

20. J. Kushibuki and N. Chubachi, Material characterization by line-focus-beam acoustic microscope, IEEE Trans. Sonics Ultrason. SU-32:189 (1985).

21. K. Yamanaka, Y. Enomoto, and Y. Tsuya, Acoustic microscopy of ceramic surfaces, IEEE Trans. Sonics Ultrason. SU-32:331 (1985).

22. D. A. Sinclair, I. R. Smith, and S. D. Bennett, Elastic constants measurement with a digital acoustic microscope, IEEE Trans. Sonics Ultrason. 31:271 (1984).

23. C. F. Quate, A. Atalar, and H. K. Wickramashinghe, Phase imaging in reflection with the acoustic microscope, Appl. Phys. Lett. 31:791 (1977).

24. W. Parmond and H. L. Bertoni, Ray interpretation of the material signature in the acoustic microscope, Electron. Lett. 15:684 (1979).

25. A. Atalar, A physical model for acoustic signatures, J. Appl. Phys., 50:8237 (1979).

MULTIMEDIA HOLOGRAPHIC IMAGE RECONSTRUCTION

IN A SCANNING LASER ACOUSTIC MICROSCOPE

B. Y. Yu, M. G. Oravecz, and L.W. Kessler
Sonoscan, Inc.
530 East Green Street
Bensenville, IL 60106

R. A. Roberts
Argonne National Laboratories
9700 South Cass Avenue
Argonne, IL   60439

ABSTRACT

    Scanning Laser Acoustic Microscopy (SLAM) is a high resolution
imaging real-time technique, useful for nondestructive evaluation of
solid materials and biological tissues. In SLAM operation, the sample
surface may be parallel to the detecting plane (coverslip) or at an
angle. Owing to the diffraction and the shadowgraphic nature of SLAM,
when a defect inside the sample is a great distance from the detecting
plane, it may be difficult to determine the size and characteristics of
the flaw. Furthermore, when the sample surface and detecting plane are
not parallel, the scattered wave needs to transmit through a nonuniformly
thick water couplant-layer. A method of adjusting the complex wavefield
at the detection plane and at the sample surface prior to conventional
backpropagation is presented here with experimental results. The
reconstructed image demonstrates that this method overcomes the
restrictions of SLAM due to diffraction and the shadowgraphic nature of
the images.

INTRODUCTION

    Scanning Laser Acoustic Microscopy is capable of providing high
quality acoustic micrographs of samples and is useful in the study of
biological tissues, integrated circuits, ceramic capacitors, etc.[1,2]
The technique utilizes a continuous plane wave of ultrasound which
transmits through a sample and causes a dynamic ripple proportional to
the amplitude of the sound field on a plastic mirror surface which acts
as the detection plane. The mirror, which is placed on the top of the
sample is called a coverslip. The acoustic image is produced by means of
a fast scanning laser beam operating at conventional TV rates of 30
images per second.[3] The laser beam reflected from the detection plane
passes through a knife-edge detector and photodiode causing any
homogeneities of the sample to be revealed.

In many applications SLAM technology could be further enhanced by solving two problems. First, the images produced are two dimensional shadowgraphic views of three dimensional objects, similar to x-radiography. Second, when applications involve imaging structures comparable in size to the acoustic wavelength used, overlapping and diffraction cause difficulty in evaluating the exact type and size of a defect from the amplitude image. These difficulties can be reduced by choosing different angles of insonification to minimize overlapping features and using holographic backpropagation to overcome diffraction effects.

SLAM uses a knife-edge optical demodulation method to produce the amplitude acoustic image of the ultrasound wave field.[4,5] A quadrature receiver is needed for holographic reconstruction and it is used by SLAM to obtain both amplitude and phase information at each point of the soundfield. This is accomplished by multiplying the image signal with a coherent electronic reference and then repeating the operation with another reference signal, which is shifted 90 degree with respect to the first, thereby representing the real and imaginary parts of the complex sound field.

In SLAM operation, there are two angles of insonification that are typically used with respect to sample placement, each having its own advantage. In a "10 degree SLAM image", the sample is placed parallel to the coverslip but at a 10 degree angle with respect to the insonification. This causes refraction of the ultrasound wave in the sample and results in shear wave and longitudinal wave image possibilities. In the "0 degree SLAM image", the sample is placed parallel to the transducer and a compressional wave image is produced. These conditions are illustrated in Figure 1.

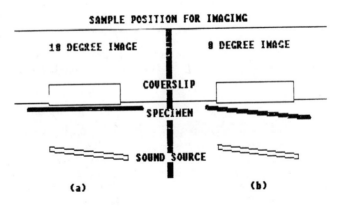

FIGURE 1 - (a) 10 degree SLAM image. The surface of the sample is parallel to the surface of the coverslip and the transducer is tilted at 10 degrees.

(b) 0 degree SLAM image. The surface of the sample is parallel to the surface of the transducer.

SLAM micrographs are shadowgraphs of all structures encountered by the path of ultrasound wave in the sample. If the sample structure has overlapping features of interest, the acoustic image may be hard to analyze. A defect diffraction pattern is observed if the flaw position in the sample is not close enough to the detecting plane or if the flaw size is comparable to the size of the acoustic wavelength. Usually a 0 degree operation mode will have a diffraction pattern that is easier to comprehend, however, if the observed acoustic image is in the far field of the flaw, it may be difficult to interpret the flaw size and determine whether it is an inclusion or void. This problem can be solved by using a modified holographic reconstruction technique.

MODIFIED HOLOGRAPHIC RECONSTRUCTION

In SLAM, the ultrasound wave is transmitted from the transducer into water, scattered by the structure of the specimen and propagated through water to the coverslip. To do holographic reconstruction, we need to backpropagate the complex wavefield from the receiving plane through the water and into the specimen.

If the structure of the specimen is planar, then according to Fourier optics, the propagation phenomenon can be regarded as a linear filter with a finite spatial frequency bandwidth (when the evanescent wave is neglected). The backward propagation equation can be written as:[6]

$$PD(x,y,z)=F^{-1}\{U(k_x,k_y)\cdot H(k_x,k_y)\}$$

$$H(k_x,k_y)=\begin{cases} \exp(iz\sqrt{k^2-k_x^2-k_y^2}) & k_x^2+k_y^2 < k^2 \\ 0 & \text{otherwise} \end{cases} \quad (1)$$

where $U(k_x,k_y)$ is the spatial frequency representation of the spatial spectrum $P(x,y,\theta)$ at the receiving plane, PD (x,y,z) is the wavefield of the reconstructed plane, and z is the distance between the receiving plane and the reconstructed plane. When z is positive, forward propagation is performed and when z is negative, it is backward propagation. Refer to Figure 2.

For a 10 degree SLAM image having backward propagation, the backpropagation from the coverslip, through the water and to the sample surface has been discussed in[7]. The subsurface which we are interested in here is inside the specimen. Because the wavelength in water is different from that of the specimen, and because the specimen may have multiple interfaces when processing the image reconstruction, first the wave field must be backpropagated from the receiving plane to the sample surface, then through the interfaces of the sample, and then to the subsurface plane of interest. In each step the proper wavelength parameters must be employed. The transfer function can now be rewritten as:

$$H2(k_x,k_y)=\begin{cases} \prod_i \exp(iz_i\sqrt{k_i^2-k_x^2-k_y^2}) & k_x^2+k_y^2 < \min(k_i^2) \\ 0 & \text{otherwise} \end{cases} \quad (2)$$

where z_i,k_i are the depth and wave number of the medium, and when i=0 the medium is water.

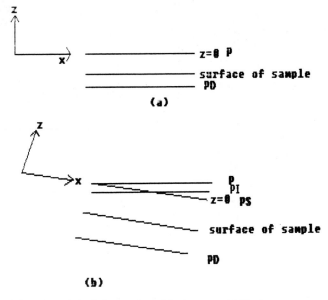

(a)

(b)

FIGURE 2 - (a) Scheme for 10 degree SLAM image
reconstruction. P is the receiving
plane and PD is the reconstructed
plane. The x axis of coordinate is on
the receiving plane P.

(b) Scheme for 0 degree SLAM image
reconstruction. P is the receiving
plane and PS // PD // surface of
sample. Plane P1 is parallel to plane
P. The x axis of coordinate is on
plane PS and the y axis is the line of
interception of plane P and PS. The
angle between P and PS is 0.

Since there is a wedge shaped space between the coverslip and the
sample surface in the 0 degree mode, we need to perform an additional
calculation to obtain the wavefield of the plane parallel to the sample
surface before we use Equation (2). Theoretically, we need to backward
propagate the wavefield from the receiving plane P to plane P1 which is
parallel to plane P by the distance x sin θ, extracting the line of
interception of plane P1 and PS. Plane PS is parallel to the sample
surface and intercepts P at the y axis (Figure 2b), and then we map the
line onto plane PS, and so on. This means that we can write the backward
propagation equation as follows:

$$PS(x,y)=F^{-1}\left\{H(k_x,k_y,x'\sin\theta)U(k_x,k_y)\right\} \quad \delta(x-x') \qquad (3)$$

Once the wavefield of plane PS is reconstructed, we can use Equation
(2) to backward propagate the wavefield inside the sample. This method
is accurate, however, it is very time consuming. A 256 x 240 image needs
256 operations of backward propagation, inverse 2D-FFT, extracting the
intercepted line, and mapping it onto the spatial spectrum of plane PS.
In addition, we need two separate 480k memory blocks to store the 256 x
240 wavefields of planes P and PS during the plane adjustment processing.

538

In order to decrease the computation time and the required memory space, an approximation method can be used. Assuming the longest distance of backward propagation inside the water wedge is comparable to the wavelength in water, and assuming the velocity in the sample is much greater than that of the water, then the refractive angle from the sample surface into water is almost perpendicular to the sample surface. In other words, the phase delay by the water wedge is approximately proportional to the thickness of the water wedge at each point. Thus, we can treat the water wedge as a thin water lens.[6] Because the propagating direction of the ultrasonic wave in water is almost perpendicular to the sample surface, the total phase delay in passing through water wedge can be written as:

$$\emptyset(x,y)=k_w \, X\tan\theta \qquad\qquad (4)$$

where k_w is wavenumber in water and θ is the angle between the receiving plane and the sample surface. We can then incorporate the phase delay into the received complex wavefield to get the wavefield of the plane which is parallel to the sample surface. The relation can be represented as:

$$PS(x,y,o)=P(x,y,z)H1(x,y)$$

$$\qquad\qquad (5)$$

$$H1(x,y)=\exp(-ik_w \, X\tan\theta)$$

This method only requires the multiplaction of the phase delay by the received wavefield to do the adjustment, so it is much faster than using Equation (3). Furthermore it only needs one 480K block memory to implement the algorithm for a 256 x 240 image. The block diagram of the image reconstruction for a 0 degree SLAM image is shown in Figure 3. Once the phase delay is corrected and the 2D-FFT is implemented, we can backpropagate spatial frequency data to the plane of interest.

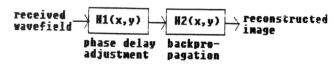

FIGURE 3 - The block diagram of image reconstruction for 0 degree SLAM image.

EXPERIMENTAL RESULTS

In our experiment, the acoustic image is gridded into 256 x 240 pixels and each pixel is digitized into 6 bits. The ultrasound source for the system operates at 105.52 MHz. The field of view is 3.45mm by 2.64mm. The test sample is ceramic with an inclusion at one surface. The thickness of the sample is 3.97mm and the compressional wave velocity inside the sample is 10,300 m/sec. The velocity of ultrasound in the water is 1500 m/sec. The angle between the surface of the sample and the receiving plane is 9.25 degrees. There is a random phase error between the data for real and imaginary components due to the frame grabber not acquiring the two images simultaneously. Specifically, at the frequency of 105.52 MHz, the sampling rate is 0.165 wavelength so the possible largest phase shift error is about 60 degrees. This error is corrected before processing.

Figure 4 shows an optical image of the sample with an inclusion which intersects one surface. Figure 5 shows a 0 degree amplitude SLAM image with the inclusion at the bottom, i.e. farthest away from the coverslip. Here, we can see only diffraction rings in the image and the shape of the defect is not recognizable. Figure 6 shows the reconstructed image at z = 4.0mm without any approximation (Equation 3). Figure 7 shows the reconstructed image at z = 4.0mm by using the thin lens approximation (Equation 5). Comparing both reconstructed images, we see that they are almost the same except that Figure 6 has a little sharper contrast. From the experiment and data, the thin lens approximation appears to be reasonably good for holographic reconstruction of 0 degree SLAM images. For the multilayer samples, tomography reconstruction is needed to separate out the various planes. This is now in process.

FIGURE 4 – The optical image of a ceramic
sample with an inclusion which
intersects the surface. The
thickness of the sample is 3.97mm.

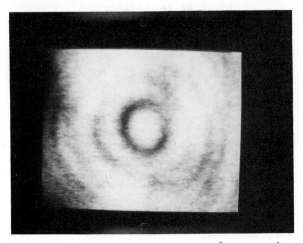

FIGURE 5 – The amplitude image of a ceramic
sample by placing the inclusion
at the bottom and then rotating
the sample by 180 degrees from
Figure 4.

540

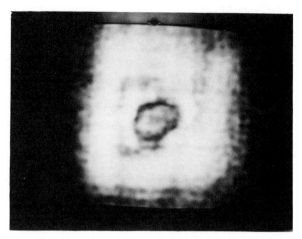

FIGURE 6 - The reconstructed image at z =
4.0mm without using the thin lens
approximation.

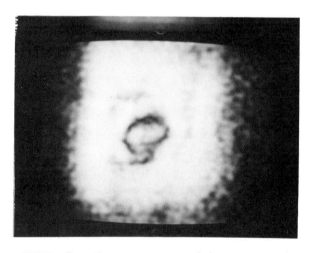

FIGURE 7 - The reconstructed image at z =
4.0mm by using the thin lens
approximation.

REFERENCES

{1} L. W. Kessler and D. E. Yuhas, "Acoustic Microscopy - 1979", Proceedings of IEEE, invited manuscript, Vol. 67, No. 4, April 1979, pp. 526-536.
{2} L. W. Kessler and D. E. Yuhas, "Principles and Analytical Capabilities of the Scanning Laser Acoustic Microscope (SLAM)", SEM/1978, Vol. 1, pp. 555.
{3} L. W. Kessler, "Image with Dynamic-Ripple Diffraction", Acoustical Imaging, G. Wade, Ed., New York, Plenum, 1976, Ch. 10, pp. 229-239.
{4} Z. C. Lin, H. Lee, and G. Wade, "Scanning Tomographic Acoustic Microscope: A Review", IEEE Transactions on Sonics and Ultrasonics, Vol. SU-32, May 1985, pp. 168-180.
{5} R. K. Mueller and R. L. Rylander, "New Demodulation Scheme For Laser Scanned-Acoustic Imaging Systems", Journal of the Optical Society of America, Vol. 69, March 1979, pp. 407-412.
{6} J. W. Goodman, "Introduction to Fourier Optics", McGraw-Hill, 1968, p. 53-54, p. 77-78.
{7} Z. C. Lin, H. Lee, G. Wade, M. G. Oravecz, L. W. Kessler, "Holographic Image Reconstruction in Scanning Laser Acoustic Microscopy", IEEE Transactions on UFFC, Vol. UFFC-34, No. 3, May 1987, pp. 193-300.

Acknowledgement: This work is partially supported by the National Science Foundation under an SBIR contract.

DATA ACQUISITION FOR

SCANNING TOMOGRAPHIC ACOUSTIC MICROSCOPY

A. Meyyappan and G. Wade

Department of Electrical and Computer Engineering
University of California
Santa Barbara, CA 93106

ABSTRACT

Acoustic microscopes are valuable in non-destructive evaluation because of their ability to provide high-resolution images of microscopic structure in small objects. When such a microscope operates in the transmission mode, the micrographs are simply two-dimensional shadowgraphs of three-dimensional objects and the resultant images are frequently difficult to comprehend because of diffraction and overlapping. This is especially true in the case of objects of substantial thickness with complex structures. We have developed a scanning tomographic acoustic microscope (STAM) to overcome these problems.

We have proposed two different rotation schemes to obtain projections for reconstructing the tomograms. The first involves rotating the transducer and the second, rotating the object. To avoid phase errors, the distance between the centers of the transducer and the object should be kept constant, or at least accurately known, throughout the rotations.

In this paper, we examine the stringent geometrical requirement for these schemes. We show, by computer simulation, that small misplacements of the order of a fraction of a wavelength are capable of destroying the image. We therefore propose a third approach which eliminates this problem since it does not require rotating or moving either the transducer or the object.

INTRODUCTION

Acoustic microscopy is an outstanding example of acoustic imaging. It uses ultrasound in the range of hundreds of megahertz and gives images of internal structures of optically opaque specimens. When an acoustic microscope such as the scanning laser acoustic microscope (SLAM) operates in the transmission mode it produces micrographs which are two-dimensional shadowgraphs of all the three-dimensional structure encountered by the acoustic wave passing through the specimen. A schematic diagram of SLAM is shown in Fig. 1. Because of the missing third dimension, SLAM micrographs are frequently difficult to comprehend, especially for structurally complex specimens of substantial thickness.

In order to overcome these difficulties a scanning tomographic acoustic microscope (STAM) is being developed. STAM uses a suitably modified SLAM to acquire the data and employs digital signal processing to reconstruct tomograms of the internal structure of microscopic specimens. [1,2]

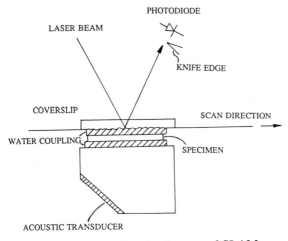

Fig. 1.   Schematic diagram of SLAM.

Computer simulations have shown that STAM is inherently capable of producing high-quality tomograms. In practice, however, many difficulties are encountered in acquiring good projection data. We have successfully modified SLAM to provide suitable data for holographic imaging (that is, reconstructing a tomogram from a single projection) and have produced high-quality, high-resolution subsurface images in this fashion. [3-5]

To obtain tomograms we need data from several projections. These data can be gathered by changing the angular direction of the insonifying waves with respect to the specimen. Two different kinds of rotation have been proposed to accomplish this. [6] The first is to rotate the transducer; the second, to rotate the specimen. Both approaches are prone to introducing phase errors in the acquired data. We have studied the effects of these errors by computer simulations. In this paper we describe the two data acquistion schemes and report the simulation results. We then delineate an arrangement which requires no rotation or movement of the various elements in the system and hence reduces the likelihood of introducing phase errors in the data.

ROTATION SCHEMES FOR DATA ACQUISITION

We use a holographic approach called back-and-forth propagation for reconstructing the tomograms. [7] As described in earlier papers, [1,2] a suitably modified SLAM can provide the data. For tomographic reconstruction, many projections are ordinarily needed. We can systematically rotate either the transducer or the specimen and thus obtain a number of projections in sequence.

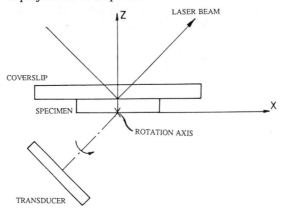

Fig. 2.   Scheme involving transducer rotation.

Transducer Rotation

Fig. 2 depicts the scheme involving transducer rotation. The transducer is rotated about the y-axis which points into the page from the center of the bottom surface of the specimen. The distance between the specimen and the transducer must be kept constant or at least known to avoid phase errors. The transfer function for the knife-edge detection indicated in Fig. 1 is antisymmetric [8] and rotating the transducer may result in operating at angles where the acquired signal is weak and the signal-to-noise ratio low.

We have used computer simulations to investigate the problem of transducer stability. We assumed we could compensate for the detrimental effects of the antisymmetric knife-edge transfer function and thereby obtain a flat response. We also assumed that the data were noise-free and that evenescent waves were negligible.

A three-dimensional object with planar structure was the simulated specimen. The geometry is shown in Fig. 3. We assumed it to be attenuation-free except for two thin layers ten wavelengths apart. The top layer was assumed to be eight wavelengths away from the receiving plane. The layers were assumed to contain binary attenuation patterns with the regions of greatest opacity being fifty percent transparent and the other regions, one hundred percent transparent to the acoustic waves. Fig. 4 shows the patterns corresponding to these regions for the two layers.

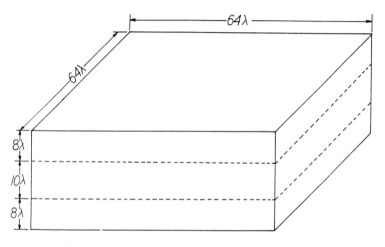

Fig. 3.   Geometry of the simulated specimen.

We simulated nine projections generated by assuming that the incident angle of the ultrasonic plane wave inside the specimen was increased in uniform angular increments from -40^0 to $+40^0$. The intensity images computed for each of these projections are simulations of ordinary SLAM images for this object. One such image is shown in Fig. 5. We see many image ambiguities because of overlapping structure from the two layers and diffraction of the wave traveling through the specimen.

The projections were then processed using an algorithm for tomographic reconstruction based on back-and-forth propagation.[2] Fig. 6 shows the reconstructed tomograms for the two layers when there was no error in the assumed position of the transducer for the different projections. Fig. 7 shows the reconstructed tomograms when the assumed angular position of the transducer was in random error for each projection by up to $\pm 1^0$. These tomograms are quite comparable to those in Fig. 6.

Fig. 8 shows the result when the assumed spatial distance between the transducer and the specimen was in random error by up to $\pm \frac{\lambda}{2}$. The patterns displayed in these reconstructions are barely recognizable. Fig. 9 shows the result when the two types of

Fig. 4. Simulated patterns of the two layers of interest within the specimen.

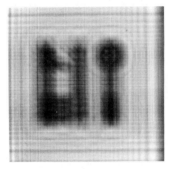

Fig. 5. A simulated intensity (i.e., SLAM) image.

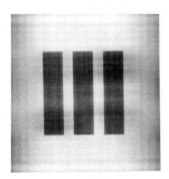

Fig. 6. Simulated images obtained from nine projections with transducer rotation, assuming no error in transducer postion.

errors (angular and spatial) are combined. Such a combination of errors is likely to occur in practical situations. The patterns in the resulting tomograms are, again, barely recognizable as the object patterns.

Fig. 7. Simulated images obtained from nine projections with transducer rotation, assuming random errors in the angular postion of the transducer by up to $\pm 1^0$.

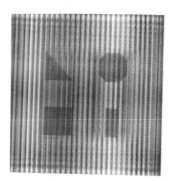 

Fig. 8. Simulated images obtained from nine projections with transducer rotation, assuming random errors in the spatial position of the transducer by up to $\pm\frac{1}{2}\lambda$.

Fig. 9. Simulated images obtained from nine projections with transducer rotation, assuming random errors in both angular and spatial position of the transducer.

547

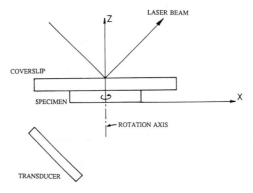

Fig. 10.   Scheme involving specimen rotation.

Specimen Rotation

Fig. 10 shows the scheme involving specimen rotation about the z-axis. Since the transducer is held stationary, the incident angle can be chosen to have optimum response, thus avoiding low signal-to-noise ratios. Nevertheless, to prevent phase errors, the specimen should be maintained precisely in its original plane.

We used simulations to check the consequences of this problem in keeping the specimen in the right position. The transducer was assumed to be held stationary with an angle of incidence of -40^0 inside the specimen. The specimen was assumed to be rotated circularly with a constant angular increment through 360^0 to produce nine projections.

Figs. 11 through 14 show the results. Fig. 11 displays the reconstructed tomograms when we assume the specimen was held flat in its plane and the rotation was exactly 40^0 with no error. Fig. 12 shows the result when the angle of rotation was assumed to be in random error by up to $\pm 1^0$. Fig. 13 shows the result when the specimen was moved up or down from its original plane by a randomly chosen distance of up to $\pm\frac{\lambda}{2}$. Fig. 14 depicts the reconstructed images when both of these types of errors were present simultaneously.

These results are very similar,but perhaps marginally superior, to those for transducer rotation. We can clearly observe that knowing the position of the transducer and the specimen for every projection is critical in obtaining good tomograms of the layers of interest.

One way to avoid these errors is to keep both the transducer and the specimen fixed. This would obviously eliminate any errors produced by the type of motion we have just studied.

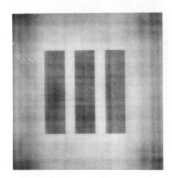 

Fig. 11.   Simulated images obtained from nine projections with specimen rotation, assuming no error in positioning the specimen.

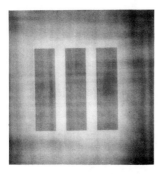

Fig. 12.   Simulated images obtained from nine projections with specimen rotation, assuming random errors in the angular position of the specimen by up to $\pm1^0$.

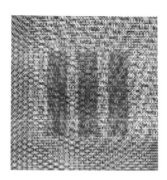

Fig. 13.   Simulated images obtained from nine projections with specimen rotation, assuming random errors in the spatial position of the specimen by up to $\pm\frac{1}{2}\lambda$.

Fig. 14.   Simulated images obtained from nine projections with specimen rotation, assuming random errors in both angular and spatial position of the specimen.

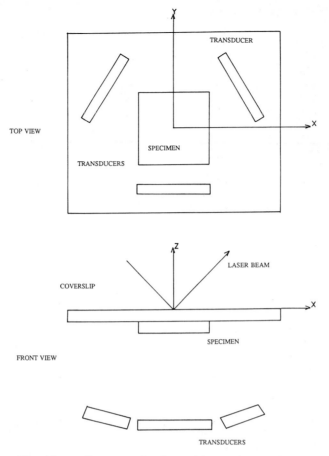

Fig. 15.   Geometry for the multi-transducer scheme.

A SCHEME WITH MULTIPLE TRANSDUCERS

To avoid the above problems we have devised a scheme involving a multiplicity of fixed transducers. Fig. 15 shows the geometry with three such transducers. For this scheme we have simulated the data and reconstructed the tomograms with three projections, one from each transducer. The reconstructed tomograms are shown in Fig. 16.

In order to better compare the results from this approach with the previous results which used nine projections, and to see the effect of utilizing multiple frequencies, we simulated the employment of three different frequencies for each transducer. The wavelengths used were 1, 1.1 and 1.2 times the original wavelength. The reconstructed images utilizing the nine projections we thus obtained are shown in Fig. 17. The results show that the images in Fig. 17 are of higher quality than those in Fig. 16 and are comparable to those in Figs. 6 and 11.

CONCLUSION

We have described three different schemes for data acquistion with STAM. Using computer simulations, we have shown that the two rotation schemes previously proposed are prone to introducing phase errors causing substantial degradation in the reconstructed tomograms. Random errors in positioning can make the images hard to recognize.

We have also shown that the third scheme, which uses multiple fixed transducers is capable of eliminating this type of error and of producing good images. Also, by using multiple frequencies to increase the totality of the projection data we should be able to further enhance the quality of the tomograms.

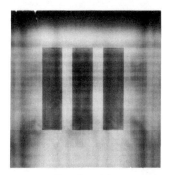

Fig. 16.   Simulated images obtained with the multi-transducer system with three projections.

Fig. 17.   Simulated images obtained with the multi-transducer system with nine projections.

REFERENCES

1. Z. C. Lin, H. Lee, and G. Wade, Scanning tomographic acoustic microscope: a review, IEEE Trans. Sonics Ultrason. SU-32:168 (1985).

2. Z. C. Lin, H. Lee, and G. Wade, Back-and-forth propagation for diffraction tomography, IEEE Trans. Sonics Ultrason. SU-31:626 (1984).

3. G. Wade, and A. Meyyappan, Scanning tomographic acoustic microscopy: principles and recent developments, in SPIE Proceedings 768 (1987) (accepted for publication).

4. Z. C. Lin, H. Lee, and G. Wade, M. G. Oravecz, and L. W. Kessler, Holographic image reconstruction in scanning laser acoustic microscopy, IEEE Trans. Ultrason. Ferroelec. Frequency Cont. UFFC-34:293 (1987).

5. H. Lee, and C. Ricci, Modification of the scanning laser acoustic microscope for holographic and tomographic imaging, Appl. Phys. Lett. 49:1336 (1986).

6. Z. C. Lin, H. Lee, and G. Wade, M. G. Oravecz, and L. W. Kessler, Data acquisition in tomographic acoustic microscopy, in Proc. IEEE Ultrason. Symp., Atlanta (1983).

7. H. Lee, C. F. Schueler, G. Flesher, and G. Wade, Ultrasound planar scanned tomography, in "Acoustical Imaging," vol. 11, J. Powers, ed., Plenum, New York (1982).

8. R. L. Rylander, "A laser scanned ultrasonic microscope incorporating a time-delay interferometric detector," Ph.D. dissertation, University of Minnesota, Minneapolis (1982).

METROLOGY POTENTIAL OF SCANNING LASER ACOUSTIC MICROSCOPES USING SURFACE

ACOUSTIC WAVES

W.P. Robbins, E.P. Rudd, and R.K. Mueller

Department of Electrical Engineering
University of Minnesota
123 Church Street
Minneapolis, MN 55455

ABSTRACT

The use of surface acoustic wave insonification in a scanning laser acoustic microscope (SLAM) for the quantitative characterization (metrology) of the mechanical properties of the surface/near-surface of a material or of a thin film deposited on a known substrate is discussed. Quantitative measurements of the mass-loading effects of a 5000 angstrom thick tungsten film on a lithium niobate surface wave delay line were obtained at 100 MHz which agreed with theoretical values. Measurements of the SAW velocity as a function of crystalline orientation on lithium niobate were also obtained. The potential of the SLAM for vertical (depth) profiling of mechanical properties using the dispersion of the surface wave velocity as the surface wave insonification frequency is varied is discussed. Methods of easily launching surface waves on nonpiezoelectric materials without requiring special sample preparation so that these potential capabilities can be applied to a wide range of samples is discussed.

I. INTRODUCTION

Both versions of the acoustic microscope, the scanning acoustic microscope (SAM) and the scanning laser acoustic microscope (SLAM), have demonstrated a wide range of imaging applications. Images have been obtained of silicon wafers, printed circuit boards, biological samples, composite materials (such as graphite fiber-epoxy materials), machine parts, etc. [1-3] Crisp acoustic images with submicron lateral spatial resolution have been demonstrated by the SAM and micron resolution by the SLAM. [1-3]

When used in the conventional manner, both acoustic microscopes produce acoustic images which are qualitative in nature and are ill-suited for quantitative characterization (for metrology) of the mechanical properties of a material. Yet there is increasing interest in such metrology applications especially of the surface/near-surface of a material or of thin films deposited on known substrates. In response to this interest, several investigations have shown that the SAM can be used to make such

553

measurements using the so-called acoustic material signature or V(Z) technique. [1,4-10] In this application of the SAM, surface waves are used (via mode conversion of bulk waves) in an indirect manner.

From a conceptual viewpoint, the scanning laser acoustic microscope would appear to be as well suited for using surface waves to make quantitative measurements as the SAM. Unfortunately the use of surface waves with a SLAM is virtually unexplored and so little is known about the metrology potential of a SLAM using surface waves. The purpose of this investigation is to demonstrate the operation of a SLAM using surface waves as the primary insonification source, to demonstrate the metrology potential of the SLAM for characterizing the surface/near-surface of a sample, and to consider and suggest possible methods of launching surface waves on nonpiezoelectric materials so that all types of materials can be examined in a SLAM using surface acoustic waves.

II. RATIONALE FOR USING SURFACE WAVES IN A SLAM

Intuitively it would appear that surface wave insonification would be preferable to use in a SLAM when the region of interest is the surface/near-surface of a sample. Surface waves propagate parallel to the surface rather than at some angle (usually near 90 degrees in practical situations) as do bulk waves and the energy in a surface wave is confined to roughly one wavelength of the surface. Thus for a given total insonification power level, the energy density in a surface wave is higher and more closely confined to the region of interest than is that of a bulk wave. Thus the signal-to-noise ratio will be higher with SAWs than with bulk waves and hence surface waves should be able to image smaller variations in acoustic properties than bulk waves. This, in fact, has been noted in images obtained with both the SAM and the SLAM. [1,11]

Another potential use of surface waves, which has been ignored, is for vertical (depth) profiling. Since the surface wave only penetrates to a depth of about one wavelength, changing the excitation frequency will change the sampling depth of the surface wave. If there is a change in acoustic properties with depth, then there will be a change in the SAW velocity with sampling depth. Hence a measurement of the SAW velocity dispersion with frequency should enable the nondestructive measurement of the variation in elastic properties with depth into the sample. This possibility has been demonstrated at low megahertz frequencies, [12,13] but not in conjunction with acoustic microscopes. The possibility of such vertical profiling with microscopic (micron) lateral spatial resolution is intriguing and should have a wide range of applications. There are many situations of technological importance where the mechanical properties of a material vary with depth because of some surface treatment or process to which they have been subjected. [12]

A SLAM using surface wave insonification may be able to detect acoustic features which are smaller than a SAW wavelength. In a SLAM the stimulus (insonification) to a sample is spatially separated from and based on a different physical mechanism than the detection (scanning laser beam) of the sample's elastic response. This means that the spatial resolution of the SLAM is affected as much by the characteristics of the laser beam as by the SAW wavelength. Stored elastic energy around small features gives rise to spatial frequencies in the elastic displacement that are larger than the SAW spatial frequency. Although these high spatial frequencies do not propagate, they will contribute to the surface displacement detected by the laser beam if the feature is within a SAW wavelength of the surface. Thus these high spatial frequencies (and hence the feature) will be detected by the laser as long as the laser spot size is smaller than the wavelength of these spatial frequencies.

III. EXPERIMENTS WITH THIN FILM TEST FEATURES

In order to examine the microscope's potential for metrology when using surface wave insonification, various thin metal film structure were deposited on the propagation surface of YZ lithium niobate SAW delay lines. The SAW delay line was chosen as the substrate in order to simplify the problem of generating surface waves on the substrate. A tungsten film, 5000 angstroms thick, was deposited on the lithium niobate and patterned using shadow masks during deposition. The entire surface of the delay line, exclusive of the interdigital transducers launching the surface waves, but including the tungsten features were then overcoated with about 1000 angstroms of aluminum. The purpose of the coating was two fold, first to provide a homogeneously reflecting surface for the scanning laser beam and secondly to eliminate the acoustic impedance discontinuity that would result from the shorting effect of the otherwise electrically isolated tungsten feature. With the aluminum coating, the only acoustic difference between the tungsten features and the rest of the sample surface would b e due to the mechanical pertubations (mass loading) of the tungsten film. The cross-section of the test structure is shown in Fig. 1.

A computer-controlled SLAM with digital data acquisition capability was used in this investigation rather than a conventional all-analog microscope because of the image/data processing needed for quantitative measurements. Detailed discussions of the system configuration and some of its capabilities have been presented elsewhere. [10,14]

A surface wave amplitude (magnitude) image of a feature in the shape of a finger is shown in Fig. 2. The field of view is about 1.5 mm square and the finger, with its rounded end, is barely discernible in the center of the image. The image data has been bandpass filtered in the spatial frequency domain with a gaussian-shaped filter (relative bandwidth or standard deviation of 4%) centered at the SAW spatial frequency. In addition the gray scale was windowed plus and minus 5% of the full gray scale range about the average gray scale value of the image data. The wide lines that criss-cross the image are thought to be due to diffraction in the incident SAW beam.

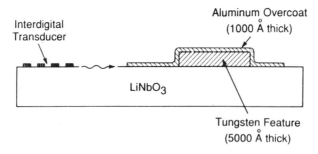

Figure 1. Cross-sectional view of SAW delay line with tungsten thin film test feature.

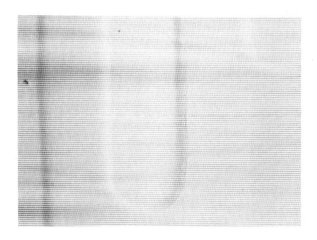

Figure 2. 100 MHz surface wave amplitude image
of tungsten thin film test feature.

Nonetheless, the image in Fig. 2 is at least recognizable. When bulk wave insonification is used to examine the same feature, nothing recognizable is produced even when spatial frequency filtering and gray scale windowing are used. The extreme thinness of the tungsten film (5000 angstroms) compared to the compressional wave wavelength in lithium niobate at 100 MHz (about 50 microns) is the basic reason why a compressional wave image of the finger cannot be obtained.

The SAW image shown in Fig. 2 is qualitative and does not give any quantitative information about the mechanical properties of the thin film. In order to obtain quantitative information, the raw data was processed by an algorithm that inverted the wave equation to produce an estimate of the wave number at each pixel. [11] The wave number estimates were then used for the gray scale of the reconstructed image, thus producing the wave number image shown in Fig. 3. The wave number data was bandpass filtered in the spatial frequency domain with a gaussian filter centered at the SAW spatial frequency and having a standard deviation of about 10% of the total spatial frequency range. Additionally the gray scale of the image in Fig. 3 was windowed plus and minus about 10% about the average value of the entire image's gray scale in order to enhance the contrast.

The relative brightness of the pixels at various locations in the image is a direct measure of the relative wave number values at those locations. Thus the ratio of the gray scale value in the center of the finger to a value outside the finger is the same as the ratio of the SAW velocities between the tungsten coated region and the aluminum-only coated region. For example a plot of the gray scale values as a function of position in the horizontal direction done in the upper half of the image of Fig. 3, a so-called horizontal line scan, is shown in Fig. 4. This is equivalent to a plot of relative SAW velocity change versus position along the scan direction. The measured velocity change between the tungsten coated region and the aluminized region is about 2%.

It is also possible to theoretically estimate what the velocity change caused by the tungsten film should be. Using pertubation theory [15] which assumes both the tungsten and the lithium niobate are elastically isotropic and using published values of the elastic constants [16], the velocity shift should be 2.2% which agrees well with the experimental value.

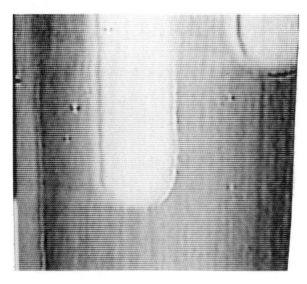

Figure 3. Surface wave wavenumber image of the
tungsten test feature.

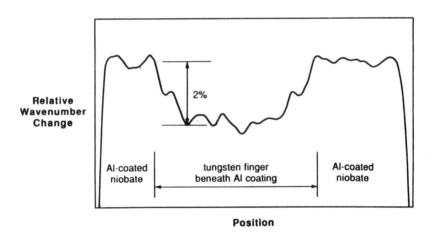

Figure 4. Relative SAW velocity versus position along
a horizonal scan of the wavenumber image of
Fig. 3.

IV. OTHER METROLOGY APPLICATIONS

The velocity measurements shown in the previous section are relative,
not absolute. They can be made absolute if the surface wave velocity is
either known apriori or can be measured with the microscope. One way of
measuring the velocity is to take the same raw data from which Figs. 2 and
3 were obtained and Fourier transform the data to the spatial frequency
domain. Such a transform is shown in Fig. 5 for surface waves propagating
in the Z direction on YZ lithium niobate at 100 MHz. A specific point
(pixel) in Fig. 5 corresponds to a specific acoustic wave with the radial
distance from the origin (center of the image) being equal to the number

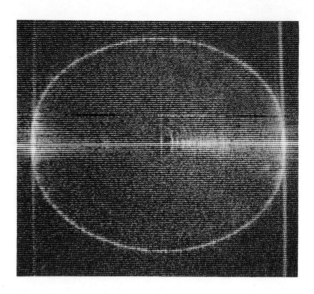

Figure 5. Experimental inverse surface wave velocity
or slowness curve for YZ lithium niobate
at 100 MHz.

of cycles of the wave contained in the spatial domain field of view.   The
angular position of the pixel with respect to a set of axes represents the
direction of propagation of the wave.   The scanning stage of the
microscope is very precise, so the absolute size of the spatial domain
field of view is accurately known.   This means that a given pixel in the
Fourier transform can be calibrated in terms of inverse wavelengths or
even inverse velocity if the incident insonification frequency is known.
A locus of such points is sometimes termed an inverse velocity or slowness
curve.

The image in Fig. 5 contains a large number of different points or
wavevectors because the incident surface waves have scattered off the
sides of the sample into all possible propagation directions.   Some of the
energy has also been mode-converted into bulk waves.   The outer oval-
shaped ring in Fig. 5 is easily identified as the locus of all possible
SAW wavevectors since the SAW wave length is shorter (and thus its
wavevector magnitude is larger) than for bulk waves.   The diffuse-
appearing points inside the ring correspond to bulk waves propagating at
shallow angles with respect to the surface and possibly to low amplitude
plate modes.

The inverse SAW velocity curve is much more useful than a casual
inspection might reveal.   By converting the data contained in the curve to
SAW velocity versus propagation direction (angle), we obtain the curve
shown in Fig. 6 which shows the anisotrophy in the SAW velocity in the XZ
plane of lithium niobate.   The theoretical curve [16] is also plotted in
Fig. 6 and it follows the experimental curve quite closely.   Of great
practical interest is the fact that the complete required only a single
imaging scan with only one fixed direction of incident insonifica-tion.
Moreover the measurements could have been made with the bulk wave
transducer that is commonly used in SLAMs because some of the bulk wave
energy would have mode converted into surface waves to produce the inverse
velocity curve. Previous measurements of SAW velocity anisotrophy required

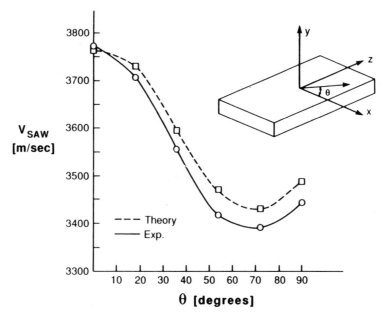

Figure 6. Comparison between experimental and theoretical curves of surface wave velocity vs
propagation direction on YZ lithium niobate.

tedious measurements with many different directions of incident
insonification. [10]

Vertical (depth) profiling could be accomplished by measuring the
average SAW velocity at several different frequencies. The various
frequencies, either uniformly or logorithmically distributed, should span
at least a 10:1 range [12] with the lowest frequency corresponding to a
wavelength equal to the maximum depth of interest. The broad frequency
range is needed in order to acquire sufficient data to accurately estimate
the variation in elastic parameters with depth. The need for the computer
which is an integral part of the SLAM for this application is clear. The
problem of how to provide a wide range of SAW frequencies and of properly
processing the data have yet to be addressed.

V. METHODS OF SURFACE WAVE TRANSDUCTION

It appears that a SLAM using surface waves could be useful for the
imaging and quantitative characterization (including depth profiling) of
the elastic properties of the surface or near surface of a sample.
However in order for these techniques to be generally useful on all
samples, that is nonpiezoelectric materials as well as piezoelectric, some
"practical" method must be developed for launching surface waves on such
samples.

Several different methods of launching SAWs on non piezoelectric
substrates have been demonstrated. A listing of these methods are given
below in Table I along with their principal strengths and weaknesses as we
perceive them for application as a surface wave insonification source in a
SLAM.

559

Method	Advantages	Disadvantages
Optical/thermo-elastic [17,18]	Non-contacting	Weak transduction
EMATs [19]	Non-contacting	Weak transduction, low frequencies
Edge-bonded transducers [20]	Strong transduction	Special sample preparation
ZnO thin film on sample [21,22]	Strong transduction	Involved sample preparation
Surface-to-surface transducer (STS) [23]	Strong transduction	Strong alignment sensitivity
Wedge transducers (liquid/solid) [24,25]	Strong transduction	Some alignment sensitivity
Mode conversion of backside bulk wave insonification [26,27]	Potential efficient transduction & little alignment sensitivity	Unproven concept with mechanically contacted grating

Table I. SAW Transduction Methods

The comments in Table I imply a set of desired characteristics for a SAW insonification source to be used with a SLAM. An explicit listing of the desired characteristics would include:

1. Efficient transduction.
2. Non-destructive.
3. No special sample preparation required.
4. Insensitive to minor misalignments.
5. Minimal number of additional components and utilize bulk wave transducer already in place, if possible.

Efficient transduction is an absolute requirement of any potential source because optical detection of acoustic energy is not a sensitive technique. Ultimately it is desired that the SLAM be capable of measuring small differences in acoustic properties, on the order of one part per thousand which will require signal-to-noise ratios greater than 60 db, with 80 db being desirable. S/N calculations indicate that SAW power levels of tens of milliwatts per centimeter of SAW beamwidth are needed for such S/N ratios. [18]

Insensitivity to minor misalignments of the components of the trans-duction scheme or between the transducer and the sample is another essential characteristic. Any source that requires delicate adjustments for efficient transduction would severely distract the user's attention away from his prime goal of using the SLAM to study the sample. In general, it is desirable that the SAW insonification source be as easy to use as the bulk wave sources currently used in SLAMs.

The need for the other characteristics is also clear. Any SAW insonification method that requires involved sample preparation, such as thin film deposition and photolithography as thin film ZnO transducers

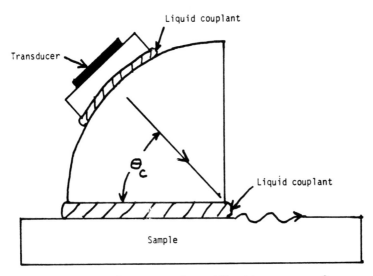

Figure 7. Surface wave insonification source for a
SLAM utilizing a wedge transducer.

need, would be too cumbersome and time-consuming to be practical. Any
insonification method that is potentially destructive to the sample must
be ruled out if there are nondestructive methods that could be used.

The wedge transducer is one of the two methods that appear the most
promising as SAW insonification sources for the SLAM. For this
application, we envision the wedge transducer as a compressional wave
transducer mounted on the sliding hemispherical structure shown in Fig. 7.
This structure will allow the angle at which the compressional wave is
incident on the sample to be varied so that samples having a wide range of
elastic constants (and hence critical angels, i.e. Rayleigh wave angles)
can be accommodated. The coupling medium between the wedge and the sample
could be either water or possibly a thin layer of RTV bonded to the bottom
side of the wedge. [25] Efficient generation of surface waves using
wedge transducers has been demonstrated at frequencies approaching 100
MHz. The wedge transducer is sensitive to misalignments, but not
inordinately so with misalignments of the angle of incidence of 2-3
degrees away from the critical angle not severely reducing the transduction efficiency. [25]

The other promising method is illustrated in Fig. 8. Bulk waves
incident on the sample surface from the backside of the sample are mode
converted into SAWs. The periodic grid shown in the figure must be in
contact with the sample surface in the area insonified by the incident
bulk waves or there will be little or no SAW generation. This method is
especially attractive because it makes use of the bulk wave transducer
already present on the microscope and requires a minimum of additional
components.

The basic principle of this method has already been demonstrated
using gratings etched into the sample of interest. [26,27] In the SLAM,
the required grating is etched into another substrate, probably a very
high impedance material such as tungsten,, and then put in mechanical
contact with the material of interest. The previous studies using etched
gratings [26,27] found the mode conversion to be very efficient, even at
high frequencies (880 MHz) and at normal (perpendicular) incidence. [27]
Further research will be required to see if this approach can be modified
to the arrangement shown in Fig. 7.

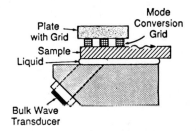

Figure 8. Surface wave insonification source for a
SLAM utilizing the mode conversion of back
side bulk waves via a grating in contact
with the sample surface.

VI.   CONCLUSIONS

It appears that a scanning laser acoustic microscope using surface
acoustic wave insonification has significant potential for the
quantitative characterization of the mechanical properties of materials.
This is especially true for the surface/near-surface of a sample or for
thin films deposited on a substrate. Quantitative information, including
vertical or depth profiling, can be obtained with surface waves that
cannot be measured with bulk waves. However, this potential will not be
realized until an efficient and easy to use method of surface acoustic
wave transduction onto any type of substrate (including nonpiezoelectric)
is developed. Several methods of transduction appear promising but
focused research efforts will be needed before any of them could be
considered "practical".

ACKNOWLEDGMENTS

This research was partially supported by the Microelectronics and
Information Sciences Center at the University of Minnesota.

REFERENCES

1.   "Acoustic Microscopy", Calvin F. Quate, Physics Today, p. 34, August,
      1985.

2.   "Acoustic Microscopy - 1979", L.W. Kessler and D.E. Yuhas, Proc.
      IEEE, 67, p. 526, (1979).

3.   "Instrumentation 'Seeing' Acoustically", R.K. Mueller and R.L.
      Rylander, IEEE Spectrum, p. 28, Feb., 1982.

4.   "Thin Film Characterization Using a Scanning Laser Acoustic
      Microscope with Surface Acoustic Waves", William P. Robbins, Rolf K.
      Mueller, and Eric Rudd, IEEE Trans. on Ultrasonics, Ferroelectrics,
      and Frequency Control, (accepted for publication).

5.   "Precision Measurement of Rayleigh Wave Velocity Pertubation", K.
      Liang, S.D. Bennett, B.T. Khuri-Yakub, and G.S. Kino, Appl. Phys.
      Lett., 41, p. 1124, (1982).

562

6. "Confocal Surface Acoustic Wave Microscoppy", I.R. Smith and H.K. Wickramasinghe, Appl. Phys. Lett., 42, p. 411, (1983)

7. "Effective Elastic Constants of Thin-Film Tungsten-Silicide from Surface Acoustic Wave Analysis", G.M. Crean, A. Golanski, and J.C. Oberlin, Appl. Phys. Lett., 50, p. 74, (1987)

8. "A New Focusing Method for Nondestructive Evaluation by Surface Acoustic Wave", B. Nongaillard, M. Ourak, J.M. Rouvaen, M. Houze, and E. Bridoux, J. Appl. Physics, 55, p. 75, (1984)

9. "Fourier Transform Approach to Materials Characterization with the Acoustic Microscope", J.A. Hildebrand, K. Liang, and S.D. Bennett, J. Appl. Physics, 54, p. 7016, (1983)

10. "Directional Acoustic Microscopy for Observation of Elastic Anisotropy", J.A. Hildebrand and L.K. Lam, Appl. Physics Lett., 42, p. 413, (1983)

11. "Scanning Laser Acoustic Microscope Using Surface Acoustic Waves", Rolf K. Mueller and William P. Robbins, 1984 IEEE Ultrasonics Symposium Proceedings, p. 561

12. "Characterization of Subsurface Anomalies by Elastic Surface Wave Dispersion", B. Tittmann, G.A. Alers, R.B. Thompson, and R.A. Young, 1974 IEEE Ultrasonics Symposium Proceedings, p. 561

13. "Nondestructive Subsurface Gradient Determination", Thomas L. Szabo, 1974 IEEE Ultrasonics Symposium Proceedings, p. 565

14. "Scanning Laser Acoustic Microscope with Digital Data Acquisition", E.P. Rudd, R.K. Mueller, W.P. Robbins, T. Skaar, and B. Soumekh, Rev. Sci. Instruments, 58, p. 46, (1987)

15. Surface Wave Filters, edited by Herbert Mathews, John Wiley & Sons, New York, (1977), Ch. 1

16. Microwave Acoustics Handbook, A.J. Slobodnik, E.D. Conway, and R.T. Delmonico, editors, Vol. 1A, Air Force Cambridge Research Laboratories

17. "Optical Generation of Continuous 76-MHz Surface Acoustic Waves on YZ-Lithium Niobate", G. Veith and M. Kowatsch, Appl. Phys. Lett., 40, p. 30, (1982)

18. "Theory of Laser Generation of Surface Waves Using Optically Adsorbing Coatings", R.E. Higashi, R.K. Mueller, and W.P. Robbins, 1983 IEEE Ultrasonics Symposium Proceedings, p. 357

19. "A Model for the Electromagnetic Generation of Rayleigh and Lamb Waves", R. Bruce Thompson, IEEE Trans. on Sonics and Ultrasonics, SU-20, p. 340, (1973)

20. "Surface Wave Edge Bonded Transducers and Applications", E. Lardat, 1974 IEEE Ultrasonics Symposium Proceedings, p. 433

21. "Surface-Elastic Wave Properties of DC-Triode Sputtered Zinc Oxide Films", F.S. Hickernell and J.W. Brewer, Appl. Phys. Lett., 21, p. 389, (1972)

22. "Theory of Interdigital Couplers on Nonpiezoelectric Substrates", G.S. Kino and R.S. Wagers, J. Appl. Physics, 44, p. 1480, (1973)

23. "A New Technique for Excitation of Surface and Shear Acoustic Waves on Nonpiezoelectric Materials", B.T. Khuri-Yakab and G.S. Kino, Appl. Phys. Lett., 32, p. 513, (1978)

24. "Characteristics of Wedge Transducers for Acoustic Surface Waves", Henry L. Bertoni and Theodor Tamir, IEEE Trans. on Sonics and Ultrasonics, SU-22, 415, (1975)

25. "The Design of Efficient Broadband Wedge Transducers", J. Fraser, B.T. Khuri-Yakab, and G.S. Kino, Appl. Phys. Lett., 32, p. 698, (1978)

26. "Acoustic Bulk-Surface-Wave Transducer", R.F. Humpryes and E.A. Ash, Elect. Lett., 5, p. 175, (1969)

27. "Generation on UHF Surface Wave by Transduction from Bulk Wave Using Fine Corrugation Grating on GaAs", M. Yamanishi, M. Ameda, K. Tsubouchi, T. Kawamura, and N. Mikoshiba, 1976 IEEE Ultrasonics Symposium Proceedings, p. 501

ANOTHER ANGLE ON ACOUSTIC MICROSCOPY

(OBLIQUE)

C. C. Cutler and H. R. Ransom

Stanford University, Stanford CA 94305

ABSTRACT

Our scanning, reflection acoustic microscope has 3 to 5 curved transducers which generate acoustic waves at around 600 MHz. The acoustic beams are projected through a single fused silica lens cavity in water and converge to a common focus. Previous observations have emphasized scattered intermodulation products at frequency (2f2-f1), which emerge in preferred directions. However, in this paper we concentrate on single frequency observations which illustrate the utility of oblique insonification. Oblique, specular and non-specular (dark field) scanned observations result in an interesting variety of images, emphasizing some details and subduing others. The anisotropic character of crystal domain structures, edge highlighting, and strong leaky Rayleigh interactions are captured in several of the acoustic images.

INTRODUCTION

Most mechanically scanned acoustic microscopes have used flat acoustical transducers which symmetrically insonify concave lens cavities at or near the Fresnel distance. In earlier papers (1-4), we have described an unusual configuration where several curved, off-axis transducers insonified a single hemispherical lens cavity asymmetrically. The unorthodox lens insonification provided an efficient means of generating two acoustic beams at different frequencies which overlap only near the focus. Our purpose in doing so was to observe intermodulation products generated in the object under investigation. Serendipitously, we observe some interesting advantages in our geometry with single frequency operation. The present configuration, shown in figure 1, is capable of insonifying an object at normal incidence, as well as obliquely from several directions. Clarity and increased contrast are gained in the observation of many objects and the resolution for our geometry is much better than one would expect.

Off-axis non-specular (i.e., dark field) insonification accentuates edges, subtle crystal domain interfaces, surface defects, and provides directional differentiation of Rayleigh wave interferences. The advantage of dark field viewing in emphasizing subtle detail is greater in type 2 microscopes than in type 1. This advantage is accentuated in our confocal lens system because the obliquity of the focused waves is not limited by the depth of focus.

Fused quartz is used as a substrate for our lens instead of sapphire because the converging beams could not easily be aligned in an anisotropic material. The frequencies of operation were a compromise between ease of construction, resolution, and laboratory convenience. The electronics is rather conventional except for the image acquisition circuitry which abets digital manipulation. The transducer and lens proportions were a compromise between the requirements for confocal operation, attenuation in the medium (water), the avoidance of false echoes, and physical space limitations.

Figure 1 illustrates the geometry of our lens system. We have used three or five transducers offset at various distances from the axis of symmetry. By properly proportioning and placing the transducers and lens, it is possible to focus several acoustic beams to a common point on the axis. However, in the observations to be described the foci are not quite coincident. They are adequately close for our oblique specular and oblique off-specular observations, but the present disparity in the focus limits the observation of intermodulation products.

Ours is not a simple single surface lens system. Because of the curvature of the acoustic sources, the lens is, in effect, a thick two surface focusing element. Numerical methods have been used to analyse the system and have determined the focal spot to be very nearly Gaussian, with the waist located by as much as four wavelengths beyond the axis crossing (3). There is sufficient overlap at focus, and sufficient intensity at the axis crossings to be useful. We expect to bring the foci and axis crossings more nearly into coincidence in later constructions;

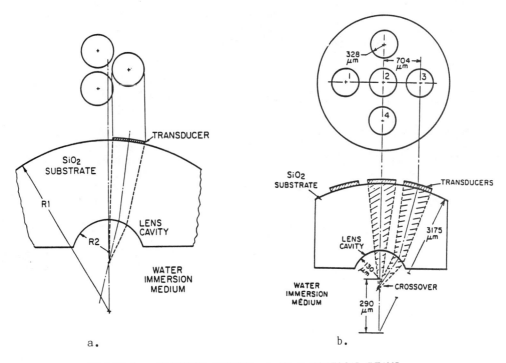

a.   b.

FIGURE 1: FOCUSING SYSTEM, 3 OR 5 CONFOCAL BEAMS

however, the transducer dimensions required for coincidence are too large to be accommodated in the present configuration. To obtain a common focus severely restricts the size of the transducer/lens.

In more conventional acoustic microscopes, the advantage of the high resolution that is characteristic of large numerical aperture lens systems (small F# used at short range) is sometimes not realized in images because of distortion caused by cross directional polarization (transverse particle motion) at the focus, or by beam convergence away from the focus. Essentially, one "sees around corners," and looses contrast at edges. The beam is already convergent at the lens cavity in our system, hence the numerical aperture is not reduced as much as one might expect; and as a result of the smaller numerical aperture, the image degradation is not as great. Because of obliquity, directional character is given to the image, with glints and shadows not seen when using normal incidence beams. In the dark-field modes of observation, i.e. the non-specular geometry, the modulation transfer function is positioned above zero spatial frequency and thus enhances details, resolution, and contrast. Also, in many cases the beams encompass the Rayleigh critical angle and excite surface waves in a direction given by the projection of the incident wave in the object plane, providing versatility in material characterization (5).

The three lens system of figure 1a has two linear modes of operation, bistatic and monostatic. Assuming reciprocity between transmitter and receiver, there are six conditions of operation: (three angles of insonification times the two modes). In the five lens system, there are five basic modes; and assuming reciprocity, 15 conditions of operation, (5 monostatic, 4 bistatic normal-oblique, 4 bistatic non-specular, and two bistatic specular). The modes are illustrated in figure 2. One of the modes is, of course, the conventional normal incidence mode used in conventional scanning acoustical microscopes. In nonlinear intermodulation observations, the 15 conditions are multiplied by 5, and the assumption of reciprocity is no longer valid. Frequently, the image character is a strong function of the axial focal position due to Rayleigh wave interactions. Also, if the transducers are not symmetrical about the axis of the system or if the lens is tilted, purposefully or not, the variety of observations is immense. This paper concentrates on observations using only a few linear modes on a single frequency.

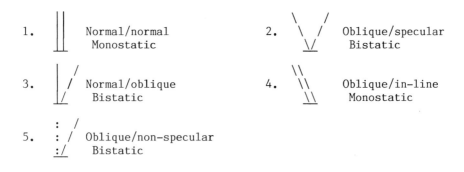

1. Normal/normal Monostatic

2. Oblique/specular Bistatic

3. Normal/oblique Bistatic

4. Oblique/in-line Monostatic

5. Oblique/non-specular Bistatic

FIGURE 2: MODES OF OPERATION

The directional, oblique-specular-bistatic mode differs from some directional microscopes because the highly concentrated acoustic waves are centered near the Rayleigh critical angle of many materials. Under these conditions, the reflection of the acoustic wave at the surface of a solid is very similar to the interaction described in the classical paper by Bertoni and Tamir. For an input wave incident on a solid at the Rayleigh wave critical angle, the input acoustic wave penetrates a short distance into the material, is shifted along the surface by a distance called the Schoch displacement, and then reradiated back into the liquid continuously as it propagates. The finite transmission of energy across the liquid-solid interface also may generate a (leaky) Rayleigh surface acoustic wave which reradiates energy back into the liquid. This wave component varies in phase with respect to other components in an acoustic microscope and, as a result, interference occurs between the components at the receiving transducer. When the surface of the object is located at focus, the specular wave components dominate. The effect of diffraction cancels most of the specular components when the object is located a few wavelengths in front of the focus. Here, interference effects become very pronounced and have a strong effect on imaging. The Rayleigh waves are sometimes reflected and/or scattered at discontinuities in the object, and result in striking interference patterns.

Our limited study of images does not begin to include all of the many possible modes of operation, but it does show considerable advantage in the use of several of the modes in characterizing a material or device, accentuating particular details. Computer manipulation and enhancement of images, of course, is also very helpful. We digitize and store 256 x 256 element eight bit pictures using a Computer Continuum Lab 40 acquisition board, and we display an adjusted six bit picture through Data Translation 2803 equipment. The frequency of operation is 640 MHz; the wavelength in water is near 2.4 microns; and the resolution is about 1.8 microns.

OBSERVATIONS

To study the modes of operation and evaluate the system, we have looked at a variety of common materials. First, we have a 1/8'th inch diameter steel ball bearing. Under small magnification, it is smooth and shiny, but with 100x optical magnification, imperfections are evident.

Using the normal/normal acoustical mode of observation in our microscope (plate 1, showing an area about 300 microns across) tiny pits are apparent, but the impression is still that it is a smooth object. The broad annular rings in the figure are caused by phase interference between the information signal and a reverberation in the delay medium and give a measure of depth, about 4 microns.

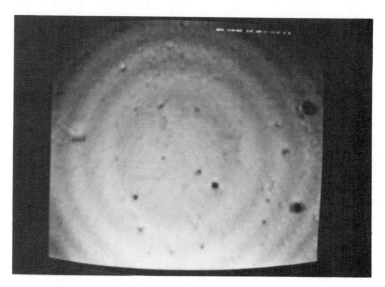

PLATE 1. STEEL BALL BEARING, NORMAL MODE

In plate 2, we show a section of the same image with four different contrasts. The contrast (photographic gamma) is doubled from one quadrant to the next. Clearly there is a lot of analytical advantage in simple contrast enhancement, made easy with digital image manipulation.

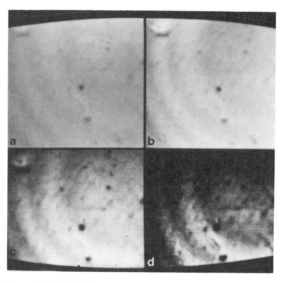

PLATE 2. STEEL BALL BEARING WITH CONTRAST ENHANCEMENT

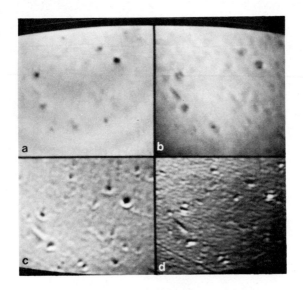

PLATE 3. STEEL BALL BEARING, 4 MODES

Plate 3, shows views of the same ball bearing observed
in four modes, each of the modes accentuating a different
character. They are, in order, a) normal, b) oblique
specular, c) oblique non-specular-bistatic, and c) oblique
monostatic. The oblique non-specular and the oblique
monostatic are much sharper and more detailed, and the contrast range and detail is greatly enhanced in the last image. Note the highlighting and shadow areas, the back-
slash' item just below center left and the scratches at the
lower left. There is no evidence of surface wave interactions in these images, and there is little change of image
character in moving out of focus in the Z direction.

Plate 4 is a randomly selected piece of granite, also about 300 microns across. Granite is a mixture of quartz, mica and feldspar. The most obvious feature of this picture is what appears like a lot of holes from a blast of 10 micron buckshot, which lie near crystal dislocations. We see at least three wavy patterns which parallel what we take to be crystal boundaries. There is a lot of character to this picture, and we guess that if one were studying the structure of granite, one could learn something from it. At least three of the wavy patterns we take to be interferences between direct surface reflections and signal returns which involve Rayleigh surface wave propagation, and reflection.

The surface acoustic wave velocity can be determined from the spacial period of the fringe pattern. The different periods (wave velocities) are evidence of the different acoustic properties of the various materials.

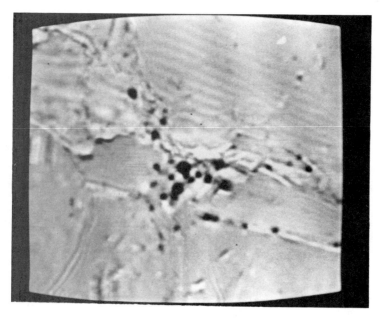

PLATE 4.   GRANITE,   NORMAL MODE

Plates 5 and 5' show four views taken from part of the same section of granite but using four other modes of observation. Image 5a is oblique specular, and gives a lot of contrast, highlights and shadows, mostly due to departure from surface flatness. (The surface was polished with 0.3 micron lapping discs, but clearly some of the constituents, were removed more effectively than others.) b) is normal-oblique. This geometry favors scattered waves, and the signal is strong and favors certain directions, especially from some of the crystal boundaries where the character is quite different than in the previous images. The oblique nonspecular bistatic image (c) is again different, with deep shadows and different highlights, while the oblique monostatic image (d) shows large effects due to interference between the Rayleigh wave reflections and backscattered radiation.

Plate 5' (oblique monostatic) is included to show the extent of the interference patterns produced by the reflection of Rayleigh waves at the crystal boundaries, and the variation with orientation by comparison with 5d. The strength of the interference is a function of the relative orientation of the acoustic beams to the boundaries; the direction is set by the boundaries; and the period is determined by the wave velocity.

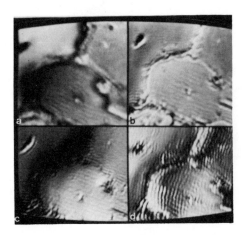 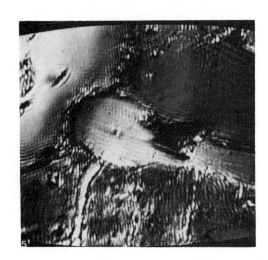

PLATES 5 AND 5', GRANITE, FOUR MODES

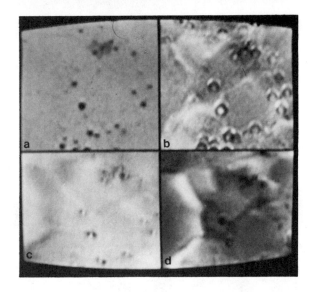

PLATE 6.   VIEWS OF POLYCRYSTALLINE FERRITE (INCONEL)

Plate 6 shows images from a polished polycrystalline piece of Ferrite. At normal incidence, (a) there is very little contrast, only a few pits at grain boundaries being in evidence. When the surface is brought inside the focus to emphasize the contribution of crystal boundaries, (b) (still at normal incidence) one sees a lot more of the character of the object. With this material, there is not much gain at focus using the oblique specular mode on the surface (c), but when focusing into the material, (d), the polycrystalline structure stands out much more sharply (6).

Plate 7, shows four images from the same object, all taken in the oblique specular mode, but with the object rotated around a normal to the surface in about 20 degree intervals. The returns from polycrystalline materials have very distinct directional signatures, the contrast between grains depending critically on orientation.

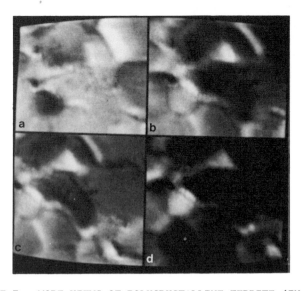

PLATE 7.   MORE VIEWS OF POLYCRYSTALLINE FERRITE (INCONEL)

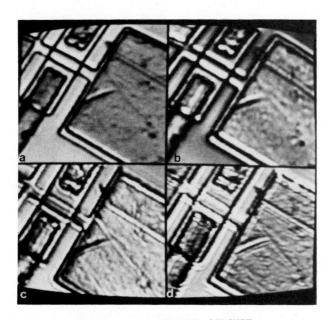

PLATE 8.   INTEGRATED CIRCUIT

Plate 8 shows images of a section of integrated circuit. The oblique-specular mode (a) shows some highlighting
not evident in the normal mode (b). (note the lines slanted
at '4 o'clock'). The oblique non-specular mode (c) shows a
little different detail, and the oblique monostatic mode (d)
accentuates detail in the flatter areas. (Note the detail
in the square pad area.)

Finally, in plate 9, we have images from a crystal of Neodymium Pentaphosphate, Nd2(PO3)5, a pure ferroelastic often referred to as NPP (7,8,9). The crystal belongs to the tetragonal class. Two orientation states are possible at room temperature and are separated by planar domain walls. Zigzag domain walls have been injected into the crystals by mechanical strains. Several images of the 010 plane of NPP are shown for different propagation directions with respect to the planar (horizontal) domain wall. The periodic (vertical) domain structure has a period of 51 microns. The domains were undetectable using a more conventional scanning acoustic microscope. The ferroelastic domains are not seen in our microscope using the normal mode focused at the surface, and are not apparent when the object is located at the focus in the oblique-specular mode. They become barely visible after translating the object towards the lens (image (a)). All of the images here were obtained with the object located 1.27 microns in front of the beam crossover and at an operating frequency of 760 MHz. The images (a) and (c) were obtained with the input acoustic wave directed along the A and C axes. Very little contrast is observed because surface acoustic wave velocities are almost the same along the principle axes on both domains. The only contrast in these images is due to surface acoustic waves scattering from the domain walls. Images (b) and (d) were obtained for propagation directions at 45 and 135 degrees. The periodic domain structure which is defined by the zigzag domain walls is visible right up to the planar domain walls in the acoustic images. This kind of detail is not seen in optical micrograms.

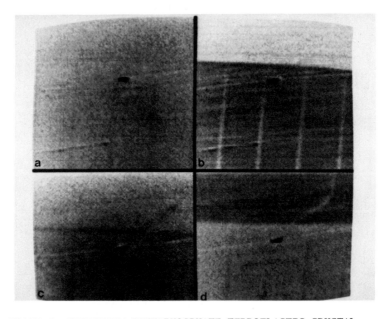

PLATE 9. NEODYMIUM PENTAPHOSPHATE FERROELASTIC CRYSTAL

CONCLUSIONS

The use of off-axis transducers, i.e., oblique illumination of objects, and non-specular observation enhances the characteristics of a scanning acoustic microscope by emphasizing the effect of scattered radiation, stronger Rayleigh wave interaction and interference effects, and by shifting the modulation transfer function to higher spacial frequencies. The incidence of the observing beams allows one to distinguish directional characteristics of anisotropic materials and surfaces, and enhances surface details by creating highlights and shadows. The directional character also gives clues to object structure and orientation. The three transducer configuration lends itself to higher contrast and resolution and a closer coincidence of foci, but the five transducer lens configuration described here makes comparison of conventional and oblique imaging more convenient.

ACKNOWLEDGEMENTS

Many past and present associates at Stanford contributed to the success of our work. Michael Tan initiated much of this work; Seyi Farotimi and Noa More contributed to the data acquisition programs. The assistance of Lance Goddard and Joseph Vhrel of the Ginzton laboratory technical staff for transducer fabrication and acoustic lens construction is acknowledged. This research is supported by the National Science Foundation contract ECS-8419202.

REFERENCES

1. M. R. Tan, H. L. Ransom, Jr., C. C. Cutler, and M. Chodorow, "Off-specular, linear and nonlinear observations with a scanning acoustic microscope" IEEE Symposium on Sonics and Ultrasonics, p. 598-603, 1984.

2. M. R. Tan, H. L. Ransom, Jr., C. C. Cutler, and M. Chodorow, "Oblique, off- specular, linear, and nonlinear observations with a scanning micron wavelength acoustic microscope", J. Appl. Phys. 57(11), p. 4931, 1 June 1985.

3. C. C. Cutler and H. L. Ransom, Jr., "Confocal beam formation for an oblique scanning acoustic microscope", Proc. IEEE Symposium on Sonics and Ultrasonics, p.740, 1985.

4. H. L. Ransom, Steven W. Meeks, and C. Chapin Cutler, "Linear and nonlinear imaging of ferroelastic domains in Neodymium Pentaphosphate," IEEE Symposium on Sonics and Ultrasonics, p731-734, 1986

5. N. Chubachi, "Ultrasonic micro-spectroscopy via Rayleigh waves" Proceedings of an International Symposium Organized by The Rank Prize Funds at The Royal Institute, 15-17 July, 1985 and published in Rayleigh Wave Theory and Application, ed. E. A. Ash and E. G. S. Paige p. 291-297, Springer-Verlag 1985.

6. M. G. Soumekh, G. A. D. Briggs, and C. Ilett, "The origins of grain contrast in the scanning acoustic microscope", Acoustic Imaging, ed. A. H. Ash, Vol. 13 , p. 107, October 1983.

7. S. W. Meeks and B. A. Auld, "Acoustic wave reflection from ferroelastic domain walls", Proc. IEEE Symposium on Sonics and Ultrasonics,
p.535, 1983.

8. S. Kojima and T. Suzuki, "Study of ferroelastic domains by acoustic
microscopy", Proc. 3rd Symp. Ultrasonics Electronics, Tokyo, Jpn. J.
Appl. Phys. 22, Suppl. 22-23, p. 81, 1983.

9. S. W. Meeks and B. A. Auld, "Periodic domain walls and ferroelastic
bubbles in neodymium pentaphosphate", Appl. Phys. Lett. 47(2), p. 102, 15
July 1985.

SURFACE SHAPES GIVING TRANSVERSE CUSP CATASTROPHES

IN ACOUSTIC OR SEISMIC ECHOES

Philip L. Marston

Department of Physics
Washington State University
Pullman, WA 99164-2814

INTRODUCTION

The reflection of sound from curved surfaces can give rise to cusp
caustics. This paper discusses the theory for locating the cusps and
describes the associated wavefields. The emphasis is on cases where a
cusp caustic opens up roughly transverse to the direction of propagation
of the reflected wave. The present work extends a previous discussion of
such transverse cusps[1] to allow for a wide range of point source and
receiver locations. The concepts of the caustic surface and the
coalescence of adjacent rays and echoes will also be discussed.

During the past decade, considerable advances were made in the
classification of caustics or foci in wavefields. These advances grew out
of the application of the mathematical theory of the singularities of
differential maps (sometimes known as "catastrophe theory") to the
specific mappings which describe rays. The relevant mapping becomes
singular where rays coalesce at caustics and the amplitude of the
wavefield is large near caustics.[2,3] This application of catastrophe
theory is often known as "catastrophe optics". Examples of nontrivial
caustics can easily be demonstrated in the reflection or refraction of
light at smooth surfaces as illustrated by Fig. 6 of the present paper.
The diffraction patterns associated with such caustics are often described
as "diffraction catastrophes" and the effects of diffraction are essential
in determining the local amplitude. Catastrophe optics has recently been
applied to various problems in the far-field scattering of light from
drops of water[4,5] as well as to the fluctuations or twinkling of the
intensity associated with random caustics[2,3].

The methods of catastrophe optics have been applied to various
problems in acoustic[1,6] and seismic[7,8] reflection and propagation.
Information concerning the location and classification of caustics is
useful for reconstructing the local shape of a reflecting surface[7]. The
emphasis of the present paper is not to solve a class of inverse problems
but rather to clarify certain geometric attributes of cusp caustics
associated with reflection. The coalescence and disappearing of rays or
glints as an aperture of an imaging system is scanned across a transverse
cusp curve is also explained. Applications include the reflection of
sound underwater from curved smooth surfaces, sensing of smooth surfaces
with airborne ultrasound, and seismic remote sensing.

LONGITUDINAL AND TRANSVERSE CUSP CAUSTICS

To explain the nature of transverse cusps it is appropriate to review
the better known case of a longitudinal cusp caustic, or "arête",

which opens up roughly along the direction of propagation[3,9]. Figure 1 shows how a curved cylindrical wavefront propagates to produce such a cusp. When point P is chosen (as shown) to lie within the cusp there are three rays (labeled A, B, and C) from points on the initial wavefront to P. As P is shifted down to lie on the adjacent cusp curve, rays A and B merge; if P is shifted to lie outside the cusp there is only one ray to P (ray C for P below the cusp.) If instead of the wavefront shown, we had considered the case of a wavefront corresponding to a sector of a perfect circular cylinder, there would be a perfect line focus at the center of the wavefront. It is therefore common to refer to the cusp focus or caustic, produced by the more typical cylindrical wavefront, as being a consequence of cylindrical aberration.[10] The wavefield near the cusped focus of a cylindrically aberrated wavefront is expressible for monochromatic waves in terms of a one-dimensional diffraction integral; the integral may be expressed in terms of a special function known as the Pearcey function,[10] defined below in Eq. (11).

In contrast to the longitudinal cusp, the transverse cusp caustic opens up roughly transverse to the propagation direction of the wavefront. The relevant geometry is illustrated in Fig. 2. It is supposed that a wave propagates from the exit plane so as to produce a transverse cusp and that the local normals of the initial wavefront are nearly parallel to the z axis. The (u,v) observation plane is taken to be perpendicular to the z axis. In that plane the caustic lies on a cubic cusp curve of the form

$$D(u - u_c)^3 = v^2 , \qquad\qquad (1)$$

where the parameters D and u_c may depend on the distance z from exit plane to the uv plane and u_c is the u coordinate of the cusp point (which is taken to lie on the u axis).

Evidence for the existence of optical transverse cusps from light scattering experiments[4,5] is reviewed in Ref. 1. Diffraction patterns were observed at large optical distances from water drops in planes perpendicular to the propagation direction of the outgoing light wave. These patterns were observed for oblate drops near the scattering angle of the primary rainbow. For a range of drop shapes, a portion of the diffraction pattern was clearly that of a cusp diffraction catastrophe (which is known to be describable by the Pearcey function.[2]) The conditions of the experiment indicate that a transverse cusp caustic can be established which retains its form even for observation planes infinitely distant from the drop.

GENERIC SHAPE FOR A WAVEFRONT WHICH PROPAGATES TO PRODUCE
TRANSVERSE CUSPS

This section reviews pertinent results of a two-dimensional propagation problem[1] giving transverse cusps in the near and far fields of the outgoing wave. The connection with other results of catastrophe theory will be noted. For the purposes of relating the shape of the outgoing wavefront to the location of the caustics, it is convenient to consider a monochromatic wave. In the exit plane (see Fig. 2), the pressure is given by the real part of $p(x,y)\exp(-i\omega t)$ where $p(x,y) = f(x,y) \exp[ikg(x,y)]$, $k = \omega/c > 0$, and c is the phase velocity which is assumed to be uniform throughout the region z > 0. Caustic locations which result from this analysis apply also to the problem of a pulsed wave having the same shape of wavefront, as specified by the slowly varying function $g(x,y)$. The analysis can also be used to anticipate caustics in certain situations where the medium is inhomogeneous. Let R denote the distance between representative points having coordinates (x,y) and (u,v) in the exit and observation planes respectively. The Fresnel approximation of the phase shift due to propagation between these points is

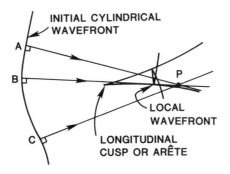

Fig. 1   Propagation of a cylindrical wavefront to form a longitudinal
          cusp caustic.

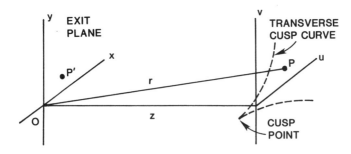

Fig. 2   A wave propagates from the exit plane toward the observation
          plane (u,v). For the problem considered the wavefront near the
          exit plane gives a cusp caustic in the uv plane.

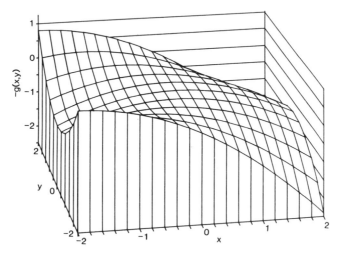

Fig. 3   Plot of $-g(x,y)$ from Eq. (8) with $a_3 = 0$ illustrates an example
          of a wavefront which gives a transverse cusp. The scales are
          in meters by taking $a_1 = 0.2$ m^{-1} and $a_2 = 0.2$m^{-2} or in cm by
          taking $a_1 = 0.2$ cm^{-1} and $a_2 = 0.2$ cm^{-2}.

$$kR \approx kz \left[1 + \frac{1}{2} (U^2 + V^2) \right] - k(xU + yV) + \frac{k(x^2 + y^2)}{2z}, \qquad (2)$$

where $U = u/z$ and $V = v/z$ are dimensionless. The Rayleigh-Sommerfeld formulation of scalar diffraction theory reduces to the following approximation for the complex pressure in the (u,v) plane[1]

$$p(u,v) \approx \frac{k}{2\pi i r} f(x=0, y=0) e^{ikr} F(u,v), \qquad (3)$$

$$F = \int\!\!\int_{-\infty}^{\infty} e^{ik\phi(u,v,x,y)} dxdy, \qquad \phi = g(x,y) - (xU + yV) + \frac{(x^2 + y^2)}{2z} \qquad (4,5)$$

where the amplitude $f(x,y)$ of the wave in the exit plane can be assumed to be sufficiently slowly varying that it may be well approximated by f evaluated at the origin O of the exit plane. In (3), r denotes the distance from O to (u,v,z) so that the first term in (2) has been approximated as kr. The phase $k\phi$ of the integrand in (4) is stationary where

$$\partial\phi/\partial x = 0 \quad \text{and} \quad \partial\phi/\partial y = 0. \qquad (6a,b)$$

If F is approximated using the method of stationary phase, the resulting approximation is equivalent to geometrical optics and gives $F \propto |H|^{-1/2}$ where H is the Hessian[2]

$$H = \frac{\partial^2 \phi}{\partial x^2} \frac{\partial^2 \phi}{\partial y^2} - \left(\frac{\partial^2 \phi}{\partial x \partial y} \right)^2, \qquad (7)$$

evaluated for a given (u,v) at the (x,y) which makes (6a) and (6b) true. The method fails at, and near, (u,v) for which $H = 0$. The locus of points giving $H = 0$ define a caustic in the (u,v) plane.

The following choice for $g(x,y)$ yields a transverse cusp specified by (1) with the caustic parameters u_c and D noted below[1]

$$g(x,y) = a_1 x^2 + a_2 y^2 x + a_3 y^2, \qquad (8)$$

$$u_c(z) = -2zb_1 b_3/a_2, \qquad D(z) = 4a_2/27 b_1^2 z, \qquad (9,10)$$

where $b_1 = a_1 + (2z)^{-1} \neq 0$ and $b_3 = a_3 + (2z)^{-1}$. The conditions on the constants a_j are that $a_2 \neq 0$ and that $-(2a_1)^{-1} \neq z$ for the selected observation plane. Since the parameter a_3 affects only the location u_c of the cusp point, the choice $a_3 = 0$ is allowed. Figure 2 illustrates the cusp's orientation for a case with $a_2 > 0$ and $u_c < 0$. Figure 3 shows a plot of $(-g)$ for representative parameters. In the paraxial approximation the wavefront near the exit plane is advanced along the z direction by a distance $\approx(-g)$ relative to the exit plane[1]. Hence Fig. 3 illustrates the local shape of an outgoing wavefront which propagates to produce a transverse cusp.

For g specified by (8), the diffraction integral $F(u,v)$ may be expressed using the generalized Pearcey function [1]

$$P_\pm(X,Y) = \int_{-\infty}^{\infty} \exp[\pm i(s^4 + s^2 X + sY)]ds, \qquad (11)$$

where X and Y are the flowing real-valued linear functions of U and V

$$X = (k/|b_1|)^{1\backslash 2} (U_c - U) \, \text{sgn}(a_2), \qquad Y = k^{3/4}|b_1|^{1/4}|2/a_2|^{1/2}v \, \text{sgn}(b_1) \qquad (12)$$

$U_c = u_c/z$, and $\text{sgn}(a_2) = 1$ if $a_2 > 0$ while $\text{sgn}(a_2) = -1$ if $a_2 < 0$. The analysis gives

$$F = (\pi/k|b_1|)^{1/2} e^{(\mp i\pi/4)} \exp(-ikU^2/4b_1) \ J(U,V), \tag{13}$$

$$J = (|b_1|/k)^{1/4} |2/a_2|^{1/2} P_{\pm}(X,Y) \tag{14}$$

where the upper (lower) sign is used if $b_1 < 0$ $(b_1 > 0)$; $|P_-(X,Y)| = |P_+(X,Y)|$ so that $|F|$ and the modulus of the diffracted wavefield which decorates the cusp may be obtained from plots[10] of $|P_+|$ irrespective of the sign of b_1. At the cusp point $X = 0$, $Y = 0$ so that $|p| \propto k^{1/4}$ and the amplitude diverges in the geometric optics limit, $k \rightarrow \infty$.

Discussion of Eq. (8) and relevant results of catastrophe theory is appropriate. For the elementary diffraction catastrophes, the form of the rapidly oscillating integrand of the canonical diffraction integrals is known as a consequence of Thom's theorem.[2,6,7] Unfortunately the shape of the wavefront which propagates to produce a given class of diffraction catastrophe is only specified up to smooth coordinate transformation and the construction of the transformation may not be a trivial task. For the case of a cusp diffraction catastrophe, it is well known that the Pearcey function is the appropriate integral so that for problems in which the diffraction integral is trivially one-dimensional, the shape of the required wavefront follows immediately by inspection of the form of the integrand of (11). For the transverse cusp diffraction catastrophe, reduction of the relevant diffraction integral to a one-dimensional integral is not as trivial so that the salient result of Ref. 1 is that $g(x,y)$ given by (8) is an appropriate two-dimensional (and non-cylindrical) shape. The actual wavefront shape which propagates to produce a transverse cusp is only specified by (8) up to a locally smooth transformation. For example, addition of linear terms $a_4x + a_5y$ to the right-hand side of (8) also yields a cubic cusp for the caustic with the cusp point shifted; the analysis given here is applicable with $U = (u/z) - a_4$ and $V = (v/z) - a_5$. It may be argued[1] that the form of (8) is consistent with various results of singularity theory and group theory.

TRANSVERSE CUSPS PRODUCED BY REFLECTION FROM SMOOTH CURVED SURFACES

Before considering the case of reflection of a wave radiated by a point source, it is appropriate to review results from Ref. 1 for the case of monochromatic plane waves incident on a surface whose height, relative to the xy plane, is

$$h(x,y) = c_1 x^2 + c_2 y^2 x + c_3 y^2 + c_4 x + c_5 y, \tag{15}$$

with $c_2 \neq 0$. Throughout this section the amplitude reflection coefficient ξ of the surface is assumed to be such a slowly varying function of the local angle of incidence that it may be well approximated as a constant. Consider first the case where a plane wave of amplitude p_o is directed vertically downward onto the surface. In the paraxial and Kirchhoff approximations, the reflected wave is equivalent to an upward directed wave in the exit plane having an amplitude $p_o \xi \exp[ikg(x,y)]$ where $g \approx -2h(x,y)$. Hence the analysis given in the previous section applies with $a_j \approx -2c_j$, $j = 1 - 5$, and the reflection will produce transverse cusps. In the case where the plane wave is directed downward at a small angle relative to the vertical, the approximations $a_4 \approx -2c_4$, and $a_5 \approx -2c_5$ need to be altered. Transverse cusps can also be produced by reflection from surfaces generated by smooth coordinate transformations of the form given in Eq. (15). This also applies to the case of a point source considered below.

The cusps and wavefields associated with reflection of a wave from a point source will now be analyzed. The geometry is illustrated in Fig. 4. The source is located a distance z_s above the xy plane with an offset from the z axis specified by u_s and v_s. The observation plane is allowed

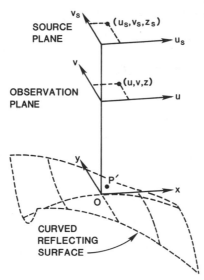

Fig. 4　For the problem considered, the wave from a point source
reflects to give a transverse cusp in the observation plane.
The reflecting surface illustrated has a height relative to the
xy plane of $h(x,y) = c_1x^2 + c_2y^2x$ with $c_1 < 0$ and $c_2 < 0$.

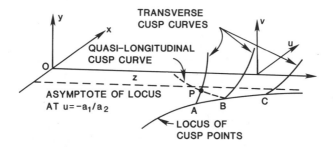

Fig. 5　The caustic surface sketched out from Eq. (1) by varying the distance
z to the observation plane in Eqs. (9) and (10). Point P on that
surface is at the intersection of cusp curves in orthogonal planes.

Fig. 6　Optical simulation of a transverse cusp for the reflection
geometry illustrated in Fig. 4. The screen on the left is the
observation plane and shows the cusp. The curved reflecting
surface is visible in the background.

to differ from that of the source. Discussion will be limited to surface profiles specified by Eq. (15) with $c_4 = c_5 = 0$ since, it may be shown that small linear terms here principally result in linear terms in $g(x,y)$ and a simple shift of the cusp point as noted above. As in the previous discussion, use will be made of a paraxial assumption that the incident and reflected rays are at small angles relative to the z axis. This assumption motivates the Fresnel approximation, as in Eq. (2), of propagation related phase shifts. In this approximation the distance R_s from the source to a given point in the xy plane becomes

$$R_s = (z_s^2 + s^2)^{1/2} \approx z_s(1 + s^2/2z_s^2),\tag{16}$$

where $s^2 = (x - u_s)^2 + (y - v_s)^2$. Let $r_s = (z_s^2 + u_s^2 + v_s^2)^{1/2}$ denote the distance of the source from the origin O of the xy plane. The amplitude of the upward directed wave in the exit plane, equivalent to the reflected wave may be approximated as

$$p(x,y) = (q\xi/r_s) \exp[ikz_s + ik(s^2/2z_s) - i2kh]\tag{17}$$

where q specifies the strength of the monopole source. Now the geometry for propagation to the uv plane is as shown in Fig. 2 and it follows that Eq. (3) is replaced by

$$p(u,v) = (kq\xi/2\pi i r_s r) \exp[ik(r_s + r)] F(u,v),\tag{18}$$

where in Eq. (4), Eq. (5) is replaced by

$$\phi = -2h(x,y) - (xU_e + yV_e) + [(x^2 + y^2)/2z_e],\tag{19}$$

The following effective parameters have been used: $U_e = (u/z) + (u_s/z_s)$, $V_e = (v/z) + (v_s/z_s)$, and $z_e = (z^{-1} + z_s^{-1})^{-1}$. Hence the phase $k\phi$ of the integrand of F has the form previously considered with a_j in Eq. (8) given by $-2c_j$. Inspection of the Eqs. (1), (9), and (10) shows that the caustic location is given by the condition

$$(4a_2/27b_1'^2)(U_e - U_{ec})^3 = V_e^2,\tag{20}$$

where here the dimensionless effective cusp point location is $U_{ec} = -2b_1'b_3'/a_2$ and $b_j' = -2c_j + (2z_e)^{-1}$. Equation (20) shows that a cusp curve is traced out by varying either (u,v) or (u_s, v_s) while holding the other point fixed. The wavefield is given by Eq. (18) where F is given by Eqs. (11)-(14) with U, U_c, V, and b_j replaced by U_e, U_{ec}, V_e, and b_j', respectively.

To understand the nature of the approximations used in deriving these results, it is helpful to consider the pathological case of reflections form a concave parabolic surface of revolution which corresponds to $c_2 = c_4 = c_5 = 0$ and $c_1 = c_3 > 0$. Inspection of (19) shows that all terms proportional to x^2 and y^2 vanish when the distances of the source and observation planes from the xy plane are such that $z^{-1} + z_s^{-1} = 4c_1$. The caustic degenerates to an image point which is located as per the usual rules of Gaussian or paraxial optics. In this level of approximation the image is point-like since both spherical aberration and the effects of finite aperture size were neglected. The Fresnel approximations, (2) and (16), were important to the derivation. These approximations were also used in the description of the cusps (and associated wavefields), even though the neglected phase terms are not necessarily small for all points on the reflecting surface. This is because it is only necessary for omitted phase terms to be negligible in the xy regions where the stationary phase condition (6) holds.[1] This also justifies other paraxial assumptions used in deriving (17) when the local surface slopes are small near the stationary phase points. Methods of calculating reflected waves more amenable to relaxation of the paraxial assumption and the analysis of transient signals are reviewed in Ref. 7.

THE CAUSTIC SURFACE

As the observation plane is moved, the caustic parameters change so that the cusp curve traces out a caustic surface. In the previous section, it was implicit that the reflected wave for the class of surface considered has a shape describable by Eq. (8), with the addition of linear terms. Consequently, the caustic surface of the reflection problem may be displayed by considering the caustic surfaces generated by a suitable class of wavefront shapes. The main features of the caustic surface associated cusps may be seen by omitting linear terms and taking $a_3 = 0$ in Eq. (8). We shall not concern ourselves here with wavefront shapes which cause either a birth or a joining of cusps known respectively as lips and beak-to-beak events.[2,7]

Figure 5 illustrates the main features of the caustic surface for $g = a_1x^2 + a_2y^2x$ with $a_1 > 0$ and $a_2 > 0$. The function $u_c(z)$ gives the locus of cusp points. Inspection of Eq. (9) with the present case of $b_3 = 1/2z$ shows that $u_c \rightarrow u_\infty \equiv -a_1/a_2$ as $z \rightarrow \infty$. Figure 5 shows three transverse cusps in uv planes having three different values of z and having cusp points at A, B, and C. The larger the cusp parameter D in Eq. (1), the faster the transverse cusp curve opens up. Inspection of Eq. (10) shows that for $z \gtrsim (a_1)^{-1}$, D decreases for increasing z. However, when the cusp curve is plotted in the dimensionless coordinates U and V, the corresponding cusp parameter is a constant as $z \rightarrow \infty$; see Eq. (20).

Consider now the intersection of the cusp surface with a plane of constant $u < u_\infty$. That plane intersects the locus of cusp points, say at point B as shown in Fig. 5. The intersection of the plane with the caustic surface traces out a quasi-longitudinal cusp curve, as may be seen from the following analysis. Inspection of Eqs. (1), (9), and (10) shows that the caustic curve in the plane of constant u is

$$-D_L(z - z_c)^3 = v^2, \qquad D_L = D/(2a_2zz_c)^3, \qquad (21,22)$$

where $z_c = [2a_2(u_\infty - u)]^{-1}$ is the value of z where the locus of cusp points intersects this plane. For z close to z_c, the z dependence of D_L is much slower than that of $(z - z_c)^3$ so that (21) is in essence a cubic cusp curve in the plane of constant u. The factor D_L is positive so the minus sign in (21) indicates that this quasi-cusp opens up in the negative z direction as shown in Fig. 5. Inspection of (22) shows that for $z - z_c > (2a_2)^{-1/2}$, $D_L < D$ so that the longitudinal caustic opens up slower than the transverse cusp having the same cusp point.

The salient differences of the quasi-longitudinal cusp with the longitudinal cusp illustrated in Fig. 1 can be seen by reviewing Fig. 1. By definition, a cylindrical wave is one for which the incident wave retains its shape as the plane of the figure is moved up or down parallel to itself. Furthermore the normal to the cylindrical surface is taken to lie parallel to the plane of Fig. 1. The u axis is taken as normal to this plane; it is therefore normal to the propagation direction. Neither the cusp point location nor the rate at which the cusp opens depend on u. For the quasi-longitudinal cusp, however, both z_c and D_L at $z = z_c$ depend strongly on the choice of the plane of constant u since $z_c \propto (u_\infty - u)^{-1}$ and for u close to u_∞, $D_L \propto (u_\infty - u)^7$.

OPTICAL SIMULATION OF ACOUSTICAL TRANSVERSE CUSPS

Some of the predictions of the previous sections were observed in the qualitative optical reflection experiment shown in Fig. 6. A diverging beam was produced by placing a short-focal-length lens in front of the He Ne laser visible on the right side of Fig. 6. Rays which diverge from this lens simulate those from the point source considered in Fig. 4. A polished mirror-like metal sheet ("Apollo metal") was bent into roughly the shape shown in Figs. 3 and 4 such that $c_1 < 0$ and $c_2 < 0$. This reflector was placed ~ 1 meter from the focus of the lens such that $(u_s, v_s, z_s) \approx (0.1m, 0, 1m)$. To make the transverse cusp visible, a ground glass screen was placed in the observation plane. Figure 6 was obtained by placing a camera ~ 2m behind the screen and setting the screen ~ 1m from the reflector ($z \approx 1m$). The orientation of the observed cusp agrees with the predictions of Eq. (20). (For this case $a_2 > 0$ and x and u increase toward the right side of Fig. 6.) Though there appears to be some diffraction related structure near the cusp point, the detailed features of a Pearcey pattern (see e.g. Ref. 1-3) were not resolved, evidently because of relative smallness of the wavelength, 633 nm.

Features evident in Fig. 5 were also observed. For example, quasi-longitudinal cusps and the cusp-point locus were made visible by placing a thin sheet of paper longitudinally in the field of the reflected light. It was found that the asymptote of the cusp-point locus was tilted with respect to the z axis because the source point was displaced from the z axis. Inspection of Eq. (20) shows this behavior is consistent with theory.

LOCATING CAUSTICS BY THE MERGING OF RAYS AND OF PULSED ACOUSTIC ECHOES

Recall from Fig. 1 that the rays from the initial wavefront to a given point P merge if P touches the cusp curve. This section describes rays directed to an observation point P for the case of a transverse cusp (see Fig. 2) and how these rays merge when P touches the cusp. The results are then used to predict the merging of travel times of reflected acoustic pulses. The salient difference between the three rays in Fig. 1 and those to an observer within a transverse cusp is that, in the latter case, the rays don't lie in a plane. This is evident from inspection of Fig. 6 where three glints are visible on the reflecting surface. These glints correspond to rays from the laser source which reflect off of the curved surface into the (narrow) lens aperture of the camera. The camera was purposely placed in the three-ray region such that none of the rays would be blocked by the ground glass screen used to view the cusp.

The procedure for calculating the sites of rays from the exit plane to an observer at P was previously described[1] for g given by Eq. (8). The equations are easily adapted to the geometry of Fig. 4, so only the method will be outlined here. In any problem of this type, the ray sites correspond to the simultaneous roots (x_i, y_i) of Eqs. (6a) and (6b) for an observer at (u, v, z). For the geometry of Fig. 4 the relevant $\phi(U_e, V_e, x, y)$ is given by Eq. (19). For P within the cusp, there are three roots ($i = 1, 2,$ and 3) which corresponds to intersections of a parabola specified by (6a) with a hyperbola specified by (6b). Let P′ in Fig. 4 denote a point on the reflector above some point (x, y) in the exit plane; to a distant observer the region around P′ appears bright as $(x, y) \rightarrow (x_i, y_i)$. The camera used for Fig. 6 had an aperture situated with $u/z \approx 0$. In this case the roots are such that $x_1 \approx x_3$, $y_1 \approx -y_3$, and $y_2 \approx 0$; the roots with $i = 1$ and 3 corresponds to the glints in the upper and lower left, respectively, and $i = 2$ corresponds to the central glint.

When the camera aperture is placed outside of the cusp, there is only single root and only one glint is visible. The way in which two of the roots disappear is analyzed in Ref. 1 and is consistent with the observed behavior of the glints summarized below. If the camera is moved vertically from its position for Fig. 6, the glints corresponding to $i = 1$ and 2 move together and merge when the aperture reaches the cusp curve. The merged glints (and corresponding root) disappears as the aperture crosses the cusp curve. Let $\hat{m}$ denote a unit vector in the xy plane parallel to the relative displacement of the merging roots as $(u, v) \rightarrow (u_{cc}, v_{cc})$ where (u_{cc}, v_{cc}) denotes a point on the cusp curve. Let $\hat{n}$ be tangent to the transverse cusp curve at (u_{cc}, v_{cc}). It may be shown that $\hat{m}$ is perpendicular to $\hat{n}$. The observations were in qualitative agreement with this prediction which suggests a novel method for inferring the orientation of the caustic.

The analysis of ray sites may be used to predict the sequences of pulses received at (u,v,z) for the reflection geometry illustrated in Fig. 4. (The direct pulse from the source is not included in this discussion.) For (u,v,z) in the three ray region, a pulse will be received for each ray, $i = 1, 2,$ and 3. Identification of the propagation related phase delays in Eqs. (18) and (19) shows that, from the source to the receiver, the travel time Δt_i for each pulse is such that

$$\Delta t_i\, c = r_s + r + \phi(U_e,\ V_e,\ x = x_i,\ y = y_i). \tag{23}$$

As the points (x_i, y_i) merge when (u,v,z) touches the cusp surface, so will the Δt_i of the corresponding echoes. Only a single echo remains as (u,v,z) is shifted into the one-ray region. The merging of echoes may be used to locate the cusp surface by moving either (u,v,z) or (u_s,v_s,z_s). These features are consistent with previous discussions of seismic echoes from a syncline.[7,11] The detailed impulse response may be inferred from Ref. 6.

ACKNOWLEDGEMENTS

This research was supported by O.N.R. I am grateful to W. P. Arnott, C. E. Dean, C. K. Frederickson, and S. G. Kargl for assistance.

REFERENCES

1. P. L. Marston, Transverse cusp diffraction catastrophes: Some pertinent wavefronts and a Pearcey approximation to the wavefield, J. Acoust. Soc. Am. 81:226 (1987). The sign of the right side of Eq. (23) there should be reversed.
2. M. V. Berry and C. Upstill, Catastrophe optics: Morphologies of caustics and their diffraction patterns, in: "Progress in Optics Vol. 18," E. Wolf ed., North Holland, Amsterdam, (1980).
3. M. V. Berry, Twinkling exponents in the catastrophe theory of random short waves, in: "Wave Propagation and Scattering," B. J. Uscinski, ed., University Press, Oxford, (1986).
4. P. L. Marston and E. H. Trinh, Hyperbolic umbilic diffraction catastrophe and rainbow scattering from spheroidal drops, Nature (London) 312:529 (1984).
5. P. L. Marston, Cusp diffraction catastrophe from spheroids:generalized rainbows and inverse scattering, Opt. Lett. 10:588 (1985).
6. M. G. Brown, The transient wavefields in the vicinity of the cuspoid caustics, J.Acoust.Soc.Am. 79:1367 (1986).
7. G. Dangelmayr and W. Güttinger, Topological approach to remote sensing, Geophys.J.R.Astr.Soc. 71:79 (1982).
8. J. F. Nye, Caustics in seismology, Geophys.J.R.Astr.Soc. 83:477 (1985).
9. A. D. Pierce, "Acoustics, An Introduction to its Physical Principles and Applications," McGraw-Hill, New York (1981).
10. S. Solimeno, B. Crosignani, and P. DiPorto, "Guiding, Diffraction, and Confinement of Optical Radiation," Academic, Orlando (1986).
11. F. J. Hilterman, Three-dimensional seismic modeling, Geophys. 35:1020 (1970).

588

ACOUSTIC IMAGERY OF THE SEA-BED

A. Farcy, M. Voisset, J.M. Augustin, and P. Arzelies

IFREMER - Centre de Toulon
B.P. 330
83507 La Seyne-Sur-Mer, Cédex, France

INTRODUCTION

The first equipment using acoustic wave reflection, following the side-scan sonar method, to create the image of the sea-bed, dates from about thirty years ago.

Their functioning principle can be compared to that of an optical camera which looks towards the horizon with the rising sun situated behind it for light. The objects or contours are made conspicuous essentially by the prolonged shadows they cast, this being due to the lighting angle. The image appears line by line, perpendicular to the displacement of the camera whose field of action is very narrow in the horizontal plane.

Various sonars of this type have been built with ranges varying from several hundred metres to tens of kilometres. The respective resolutions are from tens of centimetres to tens of metres, these parameters being directly linked to the frequency used.

The object of the SAR ("Systeme Acoustique Remorqué" or "Towed Acoustic System"), developped by IFREMER, was principally, to optimise the resolution-scope relationship in the range of resolutions of less than 1 metre and distances over 1 kilometre by introducing the principle of coherent processing of the signal received.

Besides the side-scan sonar, the SAR is equipped with a mud-penetrator which allows the various layers of sediment to be identified.

The system was conceived for working at depths of up to 6000 m.

The main components are : the acoustic aerials, signal-processing material, transmission, acquisition and presentation of the data live, but also the support vector whose characteristic stability during displacement is fundamental as for all photographic material.

The acoustic part with the transmission of the corresponding data was built by THOMSON-SINTRA.

The vector and transmission from navigation sensors and overall integration were produced by ECA.

In the panoply of IFREMER equipment for the exploration of the sea-bed, the SAR can be situated between the surface bathymetry probe ("Sea-Beam") which has a resolution of several tens of metres and optical systems used in manned and unmanned equipment whose resolution is about several centimetres.

DESCRIPTION OF THE SYSTEM (SYNOPTIC - FIGURE 1)

The two sonars (side-scan sonar and mud pentrator) are contained in a fish whose buoyancy is practically zero and which is towed at a low height above the sea-bed (on average 70 m) by means of a leash. This is attached to a depressor weighing about 2 tons which, itself, hangs on the end of an 8500 m cable. This cable carries out the double function of not only towing the fish but also supplying energy for the fish. The cable is controlled by a winch on the support ship.

This system was defined after much theoretical study and various trials using scale-models, were completed, in order to ensure the best possible stability of the fish in the roughest of seas. The yaw and pitch are the movements which disturb the image most.

The energy-conveying cable and the leash allow the energy to be supplied and the data to be recuperated live.

WHAT THE SYSTEM IS MADE UP OF

- Fish	: length	: 5 m
	diameter	: 1 m
	weight	: 2400 Kg
	buoyancy	: 50 Kg
- Leash	: length	: 40 m

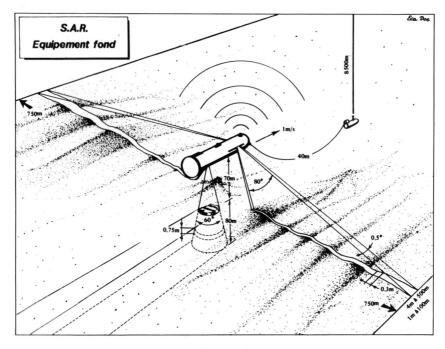

Figure 1

```
- Depressor          : length            : 1.6 m
                       diameter           : 0.5 m
                       weight             : 2000 Kg

- Energy-conveying cable :
                       length             : 8500 m
                       diameter           : 19.4 mm
                       breaking length    : 20 T
                       Both ends are equipped with slip-ring collectors.

- Transmission :
      electricity supply          : 1500 V, 1 A, 50 Hz
      digital multiplex capacity  : 300 Kbauds

- Side-scan sonar :
      frequencies                 : 170 and 190 KHz
      signal length               : 20 ms modulated on 2.5 KHz
      signal level                : 125 dB
      coherent processing gain    : 17 dB
      range                       : + 750 m
      longtitudinal directivity   : $\overline{2\theta_3}$ = 0.5°
      transversal directivity     : $2\theta_3$ = 80°

- Mud penetrator :
      frequency                   : 3.5 KHz
      signal length               : 7.5 ms modulated on 1 KHz
      signal level                : 117 dB
      coherent processing gain    : 8 dB
      directivity                 : 60° cone
      penetration                 : 50 to 80 m
      resolution                  : 0.75 m

- Miscellaneous sensors :
      immersion   : resolution $10^{-4}$ - precision $10^{-3}$
      heading     : resolution 0.5° - precision 1 %
      speed       : resolution 0.1 m/s
```

The height is determined by the side-scan sonar or the mud penetrator.

- Safety : the towing leash can be severed by acoustic remote-control which allows the fish to be recovered should the cable break.

- Data display (see Synoptic - Figure 2)
The image appears, line by line on a graphic recorder with various various scales possible :
. for the two side-scan sonar channels after slant correction and speed integration,
. for the mud penetrator after having re-established the defined reference level and integrated speed.

Several graphic recorders can be connected to restitute the image with maximum definition.

The data from the miscellaneous sensors : heading, speed, immersion and height, is displayed in front of the operator on a graphic screen together with the sea-bed bathymetry and vertical trajectory of the fish over periods of 20 minutes.

All the data is stored on magnetic tape to enable post-processing either by the system itself or on other computer facilities, thus allowing image-processing and reconstruction of the trajectory. The image-sampling produces 3000 samples per coded line (750 m), with each one on 256 levels.

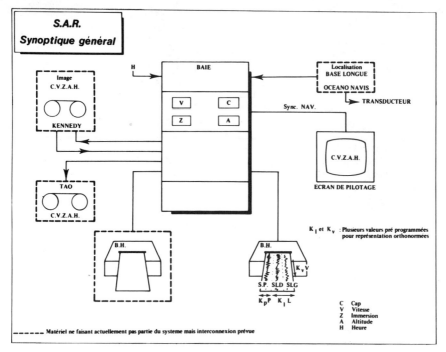

Figure 2

STEERING NEAR THE SEA-BED

The fish is raised or lowered by winch, with changes in speed which are always slow (maximum speed = 1 m/s).

The towing speed is 1-2 knots. This also affects the height of the fish but with much longer time-lags.

Simulation programmes allow one to calculate the length of cable to be released in relation to the immersion and the speed.

If a great length of cable is released, turning the boat, so as to pass methodically over a given piece of ground, can present some problems. Simulation programmes allow the turning procedures to be calculated so as to demand the shortest time possible.

The locating of the fish is carried out by acoustic systems called "Bases Longues" or "long base-line system" and "Bases Courtes" or "short base-line system".

ACOUSTIC SIDE-SCAN SONAR IMAGE-PROCESSING SOFTWARE

The live image obtained is not sufficient for the geologists to exploit. IFREMER has developped software equipment called TRIAS, which processes the data stored on magnetic tape.

This software equipment offers the various following possibilities :
- coding and demultiplexing numerical data
- geometrical adjustments
  . slant adjustment working on the assumption that the sea-bed is flat
  . slant adjustment working with an estimated sea-bed slope
  . slant adjustment when the bathymetry is known.

- speed adjustment
- yaw adjustment
- contrast enhancement
- elimination of halos due to non-uniform aerial diagrams
- image display using different scales
- navigation display
- navigation adjustment in accordance with image data
- plotting images on a mosaic mapping

TRIAS can process all sonar numerical data.

This software equipment has already been presented at different symposiums and so we would simply like to present here the automatic generation of mosaics and some results of processed images.

a) Automatic generation of mosaic images

TRIAS automatically generates the mosaic as a whole from all the images collected. It places the data from the sonar in its correct position on the map. The methodology is ilustrated in figure 3.

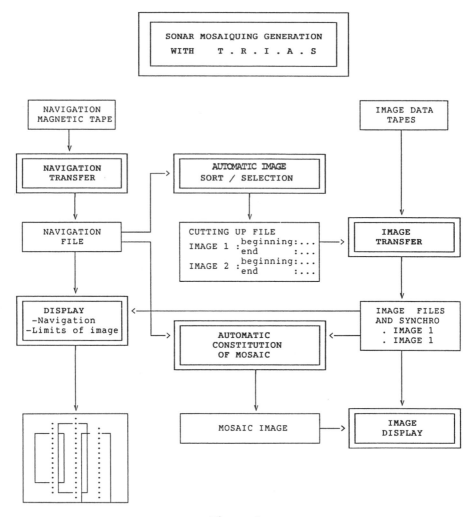

Figure 3

b) Illustration of parts of automatically generated mosaics
   (Figure 4)

 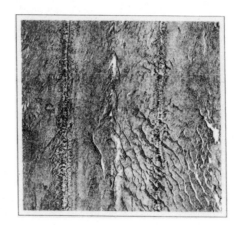

Figure 4

c) Various images after processing (figures 5, 6 and 7)

Figure 5

Figure 5 : TITANIC Zone - Depth : 3800 m. This image shows a field
of asymetric dunes. When you look at the downward slope, the white acoustic
shadows are cast by the dune crests. These structures are approximately
3 m high (estimated according to the shadow length) and 80 m in wavelength.

On the other side, the slopes reflect a powerful signal. The considerable acoustic reflectivity of the crests would lead one to suppose that
these structures are composed of coarse sediment such as gravel.

These dunes are the result of unusual flow conditions, perhaps strong
one-way sea-bottom currents.

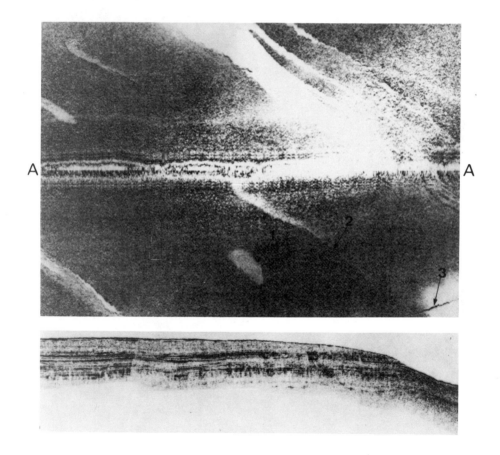

Figure 6

Zone with manganese nodules
Depth : 4900 m
Position : 131°W 14°N

. Side-scan sonar : scale 1 cm for 75 m

Arrow 1 : this white patch is characteristic of an area without nodules.
Arrow 2 : location of a small shelf 1 m high. The track (see arrow 3), of
          photographic apparatus towed by a guide-rope (a roll of metal
          1 m long and 0.5 m in diameter), goes through it.

Mud-penetrator : corresponding to the cross-section AA'
Scales : horizontal : 1 cm for 75 m
         vertical   : 1 cm for 50 m

This image completes that of the side-scan sonar for the geologist. Sedimentary deposit and the phenomenon of erosion are illustrated by the geometry of the acoustic reflectors. The excellent resolution of the mud-penetrator concords with that of the sonar.

Figure 7

PAPEETE - French Polynesia
Depth : 300 to 700 m
off Papeete harbour

The average pitch of the slope is 20 % (from up to down). The trace of the side-scan sonar being towed, spoils the quality of the image (from left to right) and the slope increases this effect.

Arrow 1 : an outcrop of basalt (2 m high and 5 m long) is described during a dive with a submersible. The contrast between the basalt (Arrow 2) and the sediment (Arrow 3) allows precise plotting on a map. The calibration of this image was realised by the French submersible CYANA.

IMAGES OF WRECKS

   a) Simulation

   A simulation programme on a computer allows one to evaluate the foresee
able image in function of the resolution. Figures 8 and 9 give an example of
the same wreck seen with different resolutions.

   b) Real image of a wreck

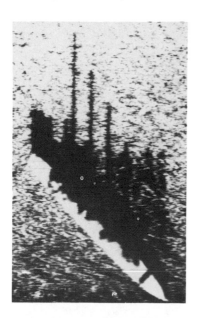

Figure 8

- Equivalent sonar : SAR
- Resolution :
   . lateral   : 2.6 to 300 m
   . in distance : 0.5 m

- Echo-reverberation contrast : 9 dB

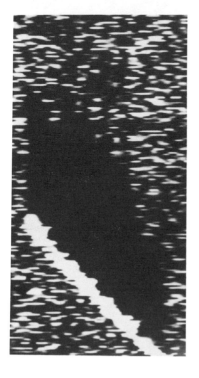

Figure 9

- Sonar with a resolution inferior to
  that of the SAR

- Resolution
   . lateral   : 8 to 300 m
   . in distance : 2.5 m

- Echo-reverberation contrast :   9 dB

Figures 10 and 11 respectively give the image with the shadow cast in black and the video inverse. They show details which stand out well more or less depending on the processing. So, in Figure 10, on the bridge, we can distinguish a circular opening with a diameter of about 10 m (see optic photograph, Figure 12). In Figure 11, it is easier to distinguish the black points which are amphoras and about 1 m high (see optic photograph, Figure 13). They probably come from a different wreck incidentally.

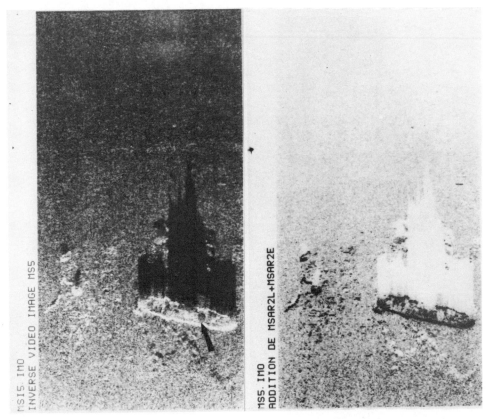

Figure 10   Figure 11

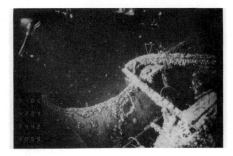   

Figure 12   Figure 13

CONCLUSION

Since its coming into service, at the begining of 1984, the SAR has completed many scientific and industrial missions, plus looking for wrecks, ... It has produced roughly 3000 Km2 of images of very different sea-beds in seas all over the world (such as the Mediterranean, Pacific, Atlantic and Indian Oceans, the China Sea, ...).

The running of the image live, in front of the operator is often very useful in leading to a successful mission. The quality of the image is generally good enough to allow a primary interpretation.

Played back, the software equipment (which processes and makes any adjustments necessary), allows the image to be improved and also makes it possible to make small details stand out but also to realise the complete image of a zone by constructing a mosaic.

The images thus obtained, in association with bathymetrical maps, will soon allow the systematical mapping of the sea-bed, as has already been done for the continents. There still remains about 360.000.000 Km2 to cover.

BIBLIOGRAPHY

AUGUSTIN J.M.   : Side Scan Acoustic Images Processing Software fourth
                       Symposium on Oceanographic date systems
                       SCRIPPS - La Jolla - Février 1986

AUGUSTIN J.M.   : Side Scan Acoustic Images Processing Software
                       Second Image Symposium CESTA
                       21/25 Avril 1986 - Nice/France

PERRON C.   : Rapport de stage IFREMER, Centre de Brest
                       Département Robotique - B.P. 337 - 29273 BREST Cédex

BERNE S. and   : Cartographie et Interprétation de la dynamique sédimen-
AUGUSTIN J.M.   taire des plateformes continentales : amélioration de la
                       technique d'observation par sonar latéral
                       Bull Soc. Géol. France 1986 (8) t II N° 3 p. 437-446

FARCY A. and   : Acoustic Imagery of sea floor
VOISSET M.   Ocean Engineering and the Environment
                       Conference record, SAN DIEGO CA, 1985 November 12/14

LE GUERCH E.   : The Deep towing of underwater fish behaviour patterns
                       during half-turn manoeuvres
                       Ocean Engineering, Vol. 14, N° 2, p. 145-162, 1987.

PARTIAL DISCRETE VECTOR SPACE MODEL FOR

HIGH-RESOLUTION BEAMFORMING

Hua Lee, Richard Chiao, and Douglas P. Sullivan

Department of Electrical and Computer Engineering
University of Illinois at Urbana-Champaign
Urbana, Illinois 61801 USA

Abstract- Beamforming has been an important branch of signal parameter estimation. It is widely used for direction finding in radar, sonar, infrared, and optical array systems. Because of the availability of the FFT, various versions of Fourier transformation are commonly used as the standard beamforming techniques. More recently, the ARMA method and Multiple Signal Characterization (often known as MUSIC) algorithm have also been applied to high-resolution beamforming.

In this paper, we will present a vector space approach to formulate direction finding as a partial discrete optimal inverse method. We model the source distribution as a continuous unknown variable, and the detected signals are in discrete form. Subsequently, we introduce two versions of the optimal beamforming algorithm and the subspace modification technique for resolution enhancement. The subspace modification is achieved by iteratively replacing the weighting of the scalar product of the inner product space with the previous estimate.

Introduction

Beamforming can be described as a typical linear parameter estimation problem in that we wish to identify the incident angle and the associated amplitude of each plane-wave component. The Fourier transform has been the most primitive and widely used beamforming algorithm due to its simplicity and the availability of the FFT. Recent efforts have been devoted to the development of high-resolution beamforming algorithms with extensive computational complexity[1,2].

Most linear beamforming methods are discrete over-deterministic approaches that assume a finite number of plane-wave components at fixed incident angles and the unknown parameters are the associated amplitude values. ARMA has also been applied to decompose the received wave-field into single-frequency components. However, the assumption of incident angles may not be feasible for most cases, and the filter size for ARMA approaches is often large and may not be realizable in practice.

In this paper, we introduce the partial discrete model to beamforming by using a continuous unknown for the representation of the source distribution and the estimation process is based on a finite number of discrete wave-field samples. This becomes an under-deterministic problem in that we estimate an infinite-dimensional continuous unknown with a finite number of discrete constraints[3,4]. We first present the general algorithm structure of the partial discrete model. Subsequently, we introduce two equivalent partial discrete beamforming algorithms, the direct form and the spectral estimation approach, as special cases of the model. The iterative subspace modification technique is also briefly discussed as a resolution enhancement method. Several simulations will be presented to demonstrate the resolving capability of the partial discrete model.

Partial Discrete Model

Denoting the source distribution as $A(x)$ and the received wave-field as $y(x)$, their linear relationship can be described by an integral

$$y(x) = \int_S A(\alpha)\, h(x, \alpha)\, d\alpha, \tag{1}$$

where $h(x)$ is the wave propagation kernel commonly known as Green's function, and S denotes the bounded source region. In the case of the partial discrete model, we assume the detected signal to be a finite number of discrete samples within the receiving aperture. Thus, the detected wave-field samples and the unknown source distribution are related by

$$y(n) \equiv y(x)\,|_{x=n\Delta} = \int_S A(\alpha)\, h(n\Delta, \alpha)\, d\alpha, \tag{2}$$

where Δ is the wave-field sample spacing.

We need to estimate the continuous infinite-dimensional source distribution based on a finite number of wave-field measurements. This is an under-deterministic linear estimation problem, and therefore, there exists an infinite number of solutions. In order to determine a unique solution, it is necessary to impose an additional optimization objective. The objective function commonly used is the norm of the feasible solutions. To formulate the structure of the optimal solution, we first define a space-domain inner product for two arbitrary continuous functions $a(x)$ and $b(x)$ within the source region

$$<a(x), b(x)> = \int_S a(x)\, b^*(x)\, w(x)\, dx, \tag{3}$$

where $w(x)$ is the weighting function. Then, we may rewrite Eq.(2) in terms of the inner product operation

$$y(n) = \int_S A(x)\, h(n\Delta, x)\, dx = \int_S [A(x)\, w^{-1}(x)]\, h(n\Delta, x)\, w(x)\, dx$$

$$= <A(x)w^{-1}(x), h^*(n\Delta, x)>. \tag{4}$$

The minimum norm solution of Eq.(4) is the orthogonal projection of the unknown source distribution into the finite-dimensional subspace spanned by the vectors {h*(nΔ,x)} and can be written as

$$\hat{A}(x) \, w^{-1}(x) = \sum_n z(n) \, h^*(n\Delta, x),$$

(5)

where $z(n)$ are the optimal coefficients of $\hat{A}(x)w^{-1}(x)$ in the subspace. It should be pointed out that proper selection of the weighting function can significantly improve the reconstruction, but without a priori knowledge of the source distribution, uniform weighting (unweighted case) is often used for the reconstruction of the source distribution.

With nonuniform weighting, the optimal estimate of the source distribution is given by

$$\hat{A}(x) = w(x) \sum_n z(n) \, h^*(n\Delta, x).$$

(6)

To linearly relate the optimal coefficients to the detected data samples, we utilize the measurement constraints

$$y(n) = \int_S \hat{A}(x) \, h(n\Delta, x) \, dx = \int_S [w(x) \sum_k z(k) \, h^*(k\Delta, x)] \, h(n\Delta, x) \, dx$$

$$= \sum_k z(k) \int_S w(x) h \, (n\Delta, x) \, h^*(k\Delta, x) \, dx.$$

(7)

If we define the elements of a square a matrix [H] as

$$H(n,k) = \int_S w(x) \, h(n\Delta, x) \, h^*(k\Delta, x) \, dx,$$

(8)

subsequently Eq.(7) can be rewritten as

$$y(n) = \sum_k H(n,k) \, z(k),$$

(9)

which in matrix form is

$$[y] = [H][z],$$

(10)

where [y] and [z] are the column vector representations of the data sequence $y(n)$ and the coefficients $z(n)$, respectively. Therefore, the optimal recombination coefficients can be computed by

$$[z] = [H]^{-1}[y].$$

(11)

It should be noted that the inverse of the [H] matrix is an enhancement operator which fully utilizes the characteristics of the kernel and the finite size of the source region. The matrix [H] is often ill-conditioned, and the inverse is commonly approximated by the singular value decomposition technique. The reconstruction algorithm for linear partial discrete data acquisition systems then can be summarized in the following steps:

Step 1: Compute $H(n,k)$

$$H(n,k) = \int_S w(x)\, h(n\Delta, x)\, h^*(k\Delta, x)\, dx$$

Step 2: Compute for the optimal coefficients $z(n)$

$$[z] = [H]^{-1}[y]$$

Step 3: Reconstruct the source distribution

$$\hat{A}(x) = w(x) \sum_n z(n)\, h^*(n\Delta, x)$$

Beamforming (Direct Form)

Consider a typical beamforming problem in which the wave-field at a linear aperture resulting from coherent point sources in the far field is given by

$$y(x) = \sum_k a_k \exp(j2\pi \frac{\sin\theta_k}{\lambda} x) + noise, \tag{12}$$

where λ is the wavelength of the sources, θ_k are the incident angles of the plane-waves, and a_k are the corresponding source amplitudes. Then, beamforming becomes a linear estimation process in that we estimate the unknown incident angles and associated amplitudes from the received wave-field distribution at the aperture. When the resultant wave-field is detected by a linear discrete array with sample spacing Δ, the relationship between the unknown parameters and the available data samples becomes

$$y(n) = y(x)|_{x=n\Delta} = \sum_k a_k \exp(j2\pi \frac{\sin\theta_k}{\lambda} n\Delta)$$

$$= \int_0^{2\pi} \sum_k a_k\, \delta(\theta-\theta_k)\, \exp(j2\pi \frac{\sin\theta}{\lambda} n\Delta)\, d\theta. \tag{13}$$

We then define the associated kernel as

$$h(n\Delta, \theta) = \exp(j2\pi \frac{\sin\theta}{\lambda} n\Delta), \tag{14}$$

604

and the source distribution as

$$A(\theta) = \sum_k a_k\ \delta(\theta-\theta_k).\qquad(15)$$

From Eq.(13) we realize that the beamforming problem is in the form of the partial discrete model, with the source region bounded between 0 and 2π. For the unweighted case, $w(\theta)=1$, the elements of the [H] matrix can be formulated in closed form

$$H(n,k) = \int_0^{2\pi} \exp[j2\pi\frac{\sin\theta}{\lambda}(n-k)\Delta]\ d\theta$$

$$= 2\pi\ J_0[\frac{2\pi(n-k)\Delta}{\lambda}].\qquad(16)$$

Therefore, the estimate of the source distribution can be written as

$$\hat{A}(\theta) = \sum_n z(n)\ \exp(-j2\pi\frac{\sin\theta}{\lambda}n\Delta),\qquad(17)$$

where the coefficients $z(n)$ can be computed by $[z] = [H]^{-1}[y]$. Figure (1) is the reconstruction of the source distribution consisting five point sources at the incident angles of 15, 30, 45, 60, and 75 degrees. This is reconstructed from 64 uniformly space wave-field samples with sample spacing 0.25λ.

Beamforming (Spectral Estimation Approach)

In the previous section, the direct form of beamforming is introduced as a special case of the partial discrete model. The solution of the source distribution is a function of the incident angle and the associated amplitude variation. The formation of the source distribution is often not computationally efficient because of the basis functions used in the recombination process given by Eq.(17). The beamforming problem has been commonly described as a spectral estimation process by using a nonlinear mapping to transform the incident angle to a spatial-frequency parameter. The mapping is also used in the conventional Fourier transform methods. This often improves the computation in the recombination process mainly due to the availability of the FFT algorithm. In this section, we introduce the spectral estimation approach for beamforming as an alternative format of the partial discrete model.

Consider the nonlinear mapping

$$f=\frac{\sin\theta}{\lambda}.\qquad(18)$$

Then the resultant wave-field can be written in the form of

$$y(x)=\sum_k a_k\ \exp(j2\pi f_k x) + noise.\qquad(19)$$

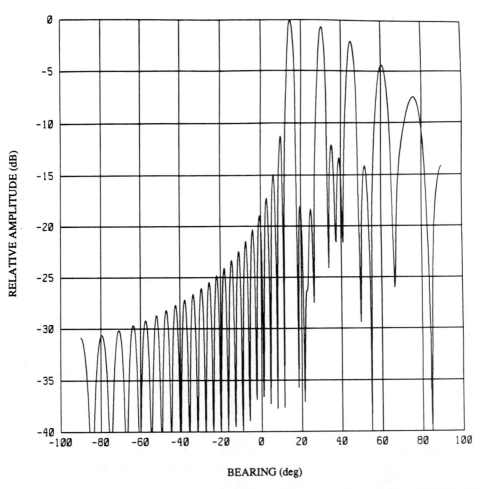

BEARING (deg)

Fig. 1. Direct method reconstruction of five point sources at incident angles of 15, 30, 45, 60, and 75 degrees from 64 unifromly spaced wave- field samples with sample spacing 0.25λ.

With this parameter transformation, the phase term of the kernel now becomes a linear function of the spatial-frequency parameter. Then the unknown spatial-frequency parameter is bounded within a source region by the cutoff frequency f_c.

Alternatively, we now define our source distribution as a function of spatial-frequency

$$A(f) = \sum_k a_k \, \delta(f-f_k), \qquad |f| \leq f_c. \tag{20}$$

Subsequently, we relate the unknown source distribution to the discrete data samples detected by the linear receiving array.

$$y(n) \equiv y(x) \,|_{x=n\Delta} = \sum_k a_k \exp(j2\pi f_k n\Delta)$$

$$= \int_{-f_c}^{f_c} A(f) \exp(j2\pi f n\Delta) \, df. \tag{21}$$

It can be seen that this relationship is again in the form of the linear partial discrete integral. For the unweighted case, $w(f)=1$, the elements of the [H] matrix now become

$$H(n,k) = \int_{-f_c}^{f_c} \exp[\,j2\pi f(n-k)\Delta\,] \, df$$

$$= 2f_c \, \text{sinc} \, [\, 2\pi f_c(n-k)\Delta \,], \tag{22}$$

with the Nyquist rate requirement

$$|f_c| < \frac{1}{2\Delta}, \tag{23}$$

which defines our maximum inter-sensor spacing

$$\Delta < \frac{\lambda}{2}. \tag{24}$$

Similar to the [H] matrix used in the previous section, the matrix we have for the spectral estimation approach is also symmetric, Toeplitz, and in closed form. As indicated in the previous section, the inverse of the [H] matrix is responsible for the enhancement of the resolution. When [H] becomes degenerate, this approach is then equivalent to the Fourier transform methods.

Even with these minor modifications such as the different formulation of the [H] matrix and the change of the basis functions for the recombination process, the overall structure of the algorithm remains identical. Fig.(2) shows the reconstructed source distribution using spectral estimation approach with the same set of data samples used for Fig.(1). The fundamental advantage is that now the recombination process can be performed by using the FFT algorithm which is more computationally efficient compared to the direct form. The optimal estimate of the wave-field spectrum can be

607

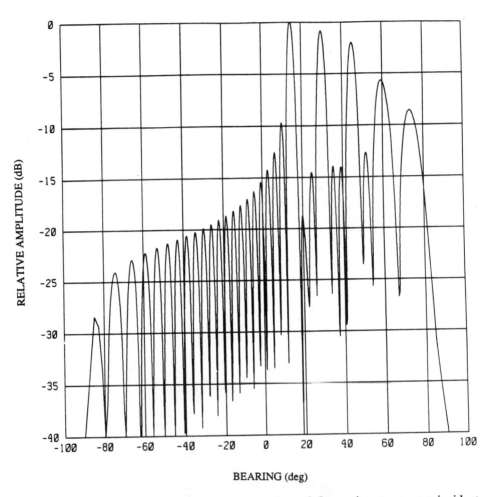

Fig. 2. Spectral estimation approach reconstruction of five point sources at incident angles of 15, 30, 45, 60, and 75 degrees from 64 unifromly spaced wave-field samples with sample spacing 0.25λ (uniform weighting).

obtained by taking the FFT of the coefficient sequence $z(n)$. Subsequently, we map the solution back to the original bearing domain with a nonlinear scaling factor which is the Jacobian associated with the change of the parameter between incident angle and spatial frequency

$$\hat{A}(\theta) = \frac{\cos\theta}{\lambda} \left. \hat{A}(f) \right|_{f = \frac{\sin\theta}{\lambda}} . \tag{25}$$

The nonlinear scaling can also be regarded as an additional lowpass windowing function applied to the source distribution.

Iterative Weighting Updating

It can be realized that the optimal solution is formed by a linear combination of a finite number of basis functions with the optimal coefficients $z(n)$. The estimate of the source distribution is a continuous function, however, the dimension is finite which is limited by the total number of detected data samples. Therefore, the resolution of the reconstruction of the source distribution is limited[5]. One approach to enhance the resolution is to utilize the subspace modification technique by changing the weighting of the inner product. This is to select a weighting to alter the structure of the subspace such that major components of the source distribution can be spanned by the subspace. We can initiate this iterative process with constant weighting. Then, we update the weighting recursively by replacing it with the previous estimate of the source distribution[6,7].

$$w_{k+1}(f) = \hat{A}_k(f) . \tag{26}$$

The formation of the optimal solution also becomes an iterative operation

$$w_k^{-1}(f) \hat{A}_k(f) = \sum_n z_k(n) \exp(-j2\pi f n\Delta) . \tag{27}$$

It can be seen that if the weighting is approaching the source distribution, the right hand side of the equation will be approaching constant which can then be completely spanned by the basis functions. This technique can considerably reduce the degradation of the reconstruction without substantial increase of computation. The resolution enhancement and convergence are illustrated by the results in Fig.(3).

Conclusion

The key objective of this paper is to apply the partial discrete model to high-resolution beamforming. The partial discrete model is suitable to imaging systems with discrete data acquisition. The main advantage of this approach is to keep the unknown source distribution continuous to avoid the artifacts due to discretization of the source. Therefore, the unknown source distribution of the beamforming problem is now a continuous function of the incident angle representing the direction of the point sources and the associated complex amplitude representing the strength of the sources. The estimation of this continuous unknown is based on a finite number of data samples detected by a linear discrete array.

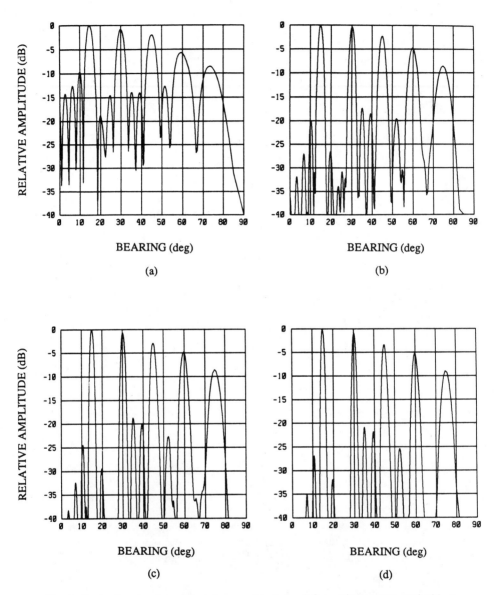

Fig. 3. Spectral estimation approach reconstruction of five point sources at incident angles of 15, 30, 45, 60, and 75 degrees from 64 unifromly spaces wave-field samples with sample spacing 0.25λ. (a) k=0, (b) k=1, (c) k=2, (d) k=3.

We first outlined the formulation of the generalized partial discrete model. Subsequently, we presented the beamforming problem as a special case in the form of this model. Then the direct form of the partial discrete beamforming algorithm is introduced according to this modeling technique. To simplify the computation of the image formation process, we also provided an alternative version of the algorithm by using a nonlinear scaling operation mapping the incident angle to the spatial-frequency domain. With this version, we can realize the similarity and difference of the algorithm structure with respect to the conventional Fourier transform approach. In addition. it significantly enhances the convergence of the iterative weighting method.

From the eigenvalue distribution of the [H] matrix, we will be able to outline the sensitivity and limitation of the data acquisition systems in the presence of noise. Coherent sources and one-dimensional discrete data acquisition arrays with uniform sample spacing are used in this paper for simplicity of the analysis. However, this algorithm can be extended to multi-dimensional beamforming for wide-band sources with nonuniform sample distribution with very minor modifications.

Acknowledgment

This research is partly supported by the Army Research Office under Contracts DAAL 03-86-K-0111 and DAAL 03-87-K-006.

References

[1]. Don H. Johnson, "The Application of Spectral Estimation Methods to Bearing Estimation Problems," *IEEE Proceedings,* vol. 70, no. 9, pp. 1018-1028, September 1982.

[2]. Don. H. Johnson and Stuart R. DeGraaf, "Improving the resolution of Bearing in Passive Sonar Arrays by Eigenvalue Analysis," *IEEE Transactions on Acoustics, Speech, and Signal Processing,* vol. ASSP-30, no. 4, pp. 638-647, August 1982.

[3]. Hua Lee, "Optimal Reconstruction Algorithm for Holographic Imaging of Finite Size Objects," *Journals of Acoustical Society of America,* 80(1), pp. 195-198, July 1986.

[4]. Hua Lee, "Alternative Approach for the Formulation of the Optimal Image Reconstruction Algorithm for Finite Size Objects," *Journals of Acoustical Society of America,* 81(4), pp. 1007-1008, April 1987.

[5]. Hua Lee, Zse-Cherng Lin, and Thomas S. Huang, "Performance and Limitation of Discrete Band-Limited Extrapolation Algorithms," *Proceedings of 1986 International Conference on Acoustics, Speech, and Signal Processing,* pp. 1645-1648, 1986.

[6]. Hua Lee, "Acoustical Image Reconstruction Algorithms: Structure, Performance, Sensitivity, and Limitation," *Proceedings of 1987 SPIE International Symposium on Pattern Recognition and Acoustical Imaging,"* 1987.

[7]. Hua Lee, Douglas P. Sullivan, and Thomas S. Huang, "Improvement of Discrete Band-Limited Signal Extrapolation by Iterative Subspace Modification," *Proceedings of 1987 IEEE International Conference on Acoustics, Speech, and Signal Processing,* vol. 3, pp. 1569-1572, 1987.

ACOUSTICAL RECOGNITION OF OBJECTS IN ROBOTICS*

II. DETERMINATION OF TYPE, POSE, POSITION, AND ORIENTATION

J.M. Richardson, K.A. Marsh, D. Gjellum, and M. Lasher

Rockwell International Science Center
Thousand Oaks, CA 91360

ABSTRACT

Optical recognition of objects in a robotics context encounters many sources of difficulty (e.g., polished objects, dark objects, extraneous light sources, etc.). However, acoustical recognition, i.e., recognition based upon the interpretation of scattered acoustical waves, has some intrinsic advantages, the most important of which is that in the absence of noise, the scattered waves depend only upon the external geometry and acoustical impedance of the object and the table that supports it. Specifically, our problem is to deduce the state of the object (i.e., type, pose, horizontal position, and azimuthal orientation) given a set of acoustical, pulse-echo, scattering data representing a sufficient diversity of directions and temporal frequencies. We use a decision-theoretic approach to the recognition problem, i.e., we determine the most probable state given the measured scattering data. A central element is a measurement model embracing certain statistical submodels and a representation of the scattering process. Here, we use an experimental approach to providing such a representation, thereby avoiding the errors in simple scattering theories (e.g., Kirchhoff). To illustrate the effectiveness of this approach, we have tested the recognition algorithm on real experimental scattering data obtained from a set of 4 objects with unknown states. The results were very satisfactory and constitute a proof of principle.

INTRODUCTION

Acoustical recognition differs from acoustical imaging in that in the former a finite list of possible objects is assumed while in the latter a practically infinite continuum of possibilities is assumed. Here we limit our attention to single objects in a robotics context, i.e., each object is isolated and rests upon a flat level surface. In this case the state‡ of the object is composed of the type, pose, position (horizontal), and orientation (azimuthal). The pose is a parameter labelling one of the stable orientations (other than azimuthal) of an object of a given type on the surface. We omit certain abnormal objects each of which has a continuum of poses

---

* This work was supported by the Independent Research and Development Funds of Rockwell International.
‡ In this treatment, we will use type as part of the definition of the state of a generic object, i.e., two objects of different types are regarded as a single object in two different states.

corresponding to a continuum of nonequivalent geometries. Here, the problem is to estimate* the state of an unknown object from a limited set of waveforms obtained from a set of broad-band pulse-echo measurements of the scattering of acoustical waves in air.

The concept of using acoustical scattering measurements for the recognition of objects in a robotics context is not new. For example, A. Wernerson (1984) has described a recognition system in which a small set of transducers produces scattered waveforms from an object that had been put into one of two standard azimuthal orientations by an ingenious presentation system. The type, pose, and one of the two possible orientations were determined by processing the waveforms with a suitable algorithm involving the analysis of the polar plots of amplitude vs phase of the signal in the temporal frequency representation. Ermert et al. (1986) are currently investigating the recognition of objects by acoustical scattering measurements using broad-band transducers but so far they have achieved only partial recognition as we define it.

Although the recognition of objects by optical means in the context of robotic acquisition has received a great deal of attention, many difficulties remain. Some of these are: (1) complexities in the interpretation of images of highly polished objects, (2) lack of sensed features of low reflectivity objects (except possibly for the silhouette), and (3) lack of 3-d input information in simple optical imaging systems. In contrast, the recognition of objects by acoustical means has the intrinsic advantage that the acoustical scattering data depends only upon the external geometry and acoustical impedance of the object and the surface upon which the body rests.

Our approach involves a straightforward decision-theoretic methodology using a measurement model in which the characterization of the scattering processes is represented by a set of experimental waveforms, thereby circumventing the errors associated with simple scattering theories. The decision-theoretic approach implies a probabilistic formulation in which the output of the recognition algorithm is the optimal estimate of the state of the object given the scattering data pertaining to an object in an unknown state. The general nature of the decision process is illustrated schematically in Fig. 1. The measurement model also contains a priori

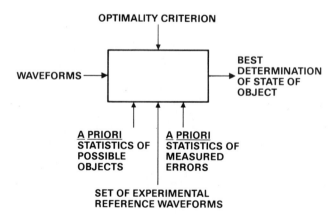

Fig. 1. Probabilistic methodology of acoustical object recognition.

---

* In the present discussions we generalize the word "estimate" to include discrete-valued variables as well as continuous-valued.

statistical information about the measurement error and the possible states of the object. The former is represented by an additive Gaussian random vector with zero mean and a specified covariance matrix. The latter (i.e., the possible states of the object) involves the assumption that the a priori probabilities of type and pose are known and that the a priori probability densities of position (horizontal) and orientation (azimuthal) are uniform over their domains of definition.

DEVELOPMENT OF THE RECOGNITION ALGORITHM

In this section we discuss the development of the recognition algorithm using the methodology mentioned in the last section. We will use an optimality criterion such that the optimal estimate of the state is the most probable value given the measured data. The definition of state will be made mathematically more precise later. The measurement system is composed of a set of transducers, as shown schematically in Fig. 2. The object is in an arbitrary position and orientation with its center within a specified square localization domain D_L in the xy-plane. The following additional assumptions are made: (1) each transducer is in the far field of the localization domain and vice versa, (2) each transducer makes a broad-band pulse-echo scattering measurement while the others are inactive, (3) each transducer is aimed at the origin, and (4) the scattered wave is defined as the perturbation (not necessarily small) of the pressure wave field due to the presence of the object.

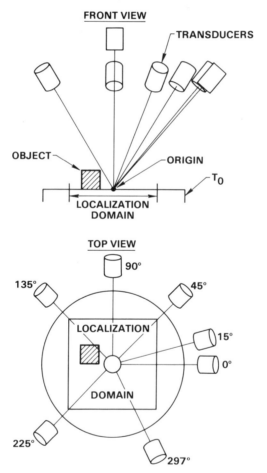

Fig. 2. Schematic representation of experimental setup.

The measurement model is given by the expression

$$f_n(t) = F(t-\tau_n, \phi_n-\alpha, \theta_n; \xi) + \nu_n(t) \tag{1}$$

where

$f_n(t)$ = Scattered waveform measured by transducer n in a pulse-echo mode at time t.

$\nu_n(t)$ = Measurement noise associated with the waveform $f_n(t)$.

$F(t,\phi,\theta,\xi)$ = Noiseless waveform obtained from a pulse-echo measurement by a transducer whose position relative to the origin has the polar angle θ and azimuthal angle ϕ when an object of type and pose labeled by the discrete-valued variable ξ is centered at the origin of xy-space and has a standard azimuthal orientation given by $\alpha = 0$.

α = Azimuthal orientation of preferred axes imbedded in the object of type and pose given by ξ. If the object has m-fold rotational symmetry we will require $0 \le \alpha < (2\pi/m)$, $m = m(\xi)$, and if it has ∞-fold symmetry we will require $\alpha = 0$.

Other forms of symmetry, if present, can be used to reduce further the range of α. For example, if an object with m-fold rotational symmetry has a plane of reflection symmetry passing through the axis then the range of α can be reduced by a factor of 2.

τ = $-2 c^{-1}(s_x \cos\phi + s_y \sin\phi)\sin\theta$ where $\underline{s} = \vec{e}_x s_x + \vec{e}_y s_y$ is the position of the center of the object in the xy-plane.

c = velocity of sound in air.

In the above expressions we will now assume for computational purposes that t is discrete-valued with a mesh size appropriate for the bandwidth. The other continuous-valued variables, i.e., $\underline{s}$ and α, will also be redefined on appropriate meshes. In the above definitions the unadorned variables ϕ and θ refer to possible transducer positions (suitably discretized) associated with the reference waveforms and the variables ϕ_n and θ_n refer to actual transducer positions used in the recognition process.

An important feature of our approach is the experimental determination of $F(t,\phi,\theta;\xi)$ thereby circumventing the use of approximate scattering theories. For a given value of ξ, this function is represented by a set of time-dependent waveforms corresponding to various transducer positions defined by ϕ and θ with the object placed in a standard position and orientation (i.e., $\underline{s} = 0$ and $\alpha = 0$). We will consider two values of the polar angle, i.e., $\theta = 0°, 45°$. Since, in the case of $\theta = 0°$, the azimuthal angle ϕ has no meaning, we consider a set of values of ϕ for $\theta = 45°$ only. The required spacing of these values depends upon size and complexity of the object, the shortest wavelength, etc.; however, the range of values is subject to the same considerations pertaining to the range of α.

To complete the specification of the measurement model we must specify the a priori statistical properties of the states of possible objects and the experimental errors. We assume that these two types of random variables are statistically independent of each other and that $\underline{s}$ is uniformly distributed in D_L, α is uniformly distributed in the interval $[0, 2\pi/m(\xi)]$, and ξ has an a priori probability $P(\xi)$. The experimental errors $\nu_n(t)$ are assumed to be Gaussian random variables with $E\nu_n(t) = 0$ and $E\nu_n(t)\nu_{n'}(t') = \delta_{nn'}\delta_{tt'} \sigma_\nu^2$.

The a posteriori probability of the object state clearly is given (aside from an additive constant) by

$$\log P(\underline{s},\alpha,\xi|f) =$$
$$-\frac{1}{2\sigma_\nu^2} \sum_{n,t} (f_n(t)-F(t-\tau_n,\phi_n-\alpha,\theta_n;\xi))^2 + \log P(\xi) \quad . \tag{2}$$

The optimal recognition procedure is simple to state, namely to determine the values of $\underline{s}$, α, and ξ that maximize the above expression within certain domain constraints. To obtain a practical approach to this maximization process we have adopted a two-step procedure: (1) obtain a preliminary estimate by determining α and ξ using a matching procedure based on a suitably chosen translationally invariant feature, followed by a determination of $\underline{s}$ by standard correlation techniques and (2) obtain a final estimate by applying a perturbation method in the original maximization problem with the state estimate in (1) as a starting point.

The translationally invariant feature chosen for step (1) was the temporal autocorrelation function of the scattered waveform, which for the measurements is given by:

$$g_n(\tau) = \sum_t f_n(t) f_n(t - \tau) \tag{3}$$

and for the hypothesized object is given by

$$G(\tau, \phi_n - \alpha, \phi_n; \xi) = \sum_t F(t - \tau_n, \phi_n - \alpha, \theta_n; \xi) F(t - \tau_n - \tau, \phi_n - \alpha, \theta_n; \xi) \tag{4}$$

The matching procedure in step (1) is based on minimizing a goodness-of-fit parameter $\epsilon(\alpha, \xi)$, defined by:

$$\epsilon(\alpha, \xi) = [\sum_{n, \tau} (g_n(\tau) - G(\tau, \phi_n - \alpha, \theta_n; \xi))^2]^{1/2} \tag{5}$$

where τ assumes the same set of values as t.

It should be noted that a 180° ambiguity may arise during the preliminary estimation process in the case of flat objects, due to the properties of the autocorrelation function. In such cases the ambiguity may be resolved in step (2) by comparing log $P(\underline{s}, \alpha, \xi | f)$ for the two possible orientations.

TEST OF ALGORITHM USING OBJECTS WITH UNKNOWN STATES

We consider four types of objects: a cube, a rectangular solid, a hemisphere, and a rectangular solid with a slot on one side. These types of objects are shown in Fig. 3. It is easy to see that the cube has 1 pose, the rectangular solid has 3 poses, the hemisphere has 2 poses, and, finally, the slotted rectangular solid has 4 poses. Although the hemisphere has 2 poses, we consider only 1 pose with the flat side down due to insufficient stability when the spherical side is down and resting on the rough surface of the turntable. Thus, we now have a total of $1 + 3 + 1 + 4 = 9$ types and poses. The test waveforms were obtained for each or the object types with the true positions (horizontal) and orientations (azimuthal) indicated in Figs. 4a-4d. In Fig. 4b we chose the true pose of the rectangular solid to correspond to a vertical orientation of the edges of intermediate length (pose #2). In Fig. 4d we chose the true pose of the slotted rectangular solid to correspond to the slotted side facing downwards (pose #2).

The objects discussed above are made of various solid materials whose acoustical impedances are so much larger than the acoustical impedance of air that they all behave very much like perfectly rigid objects. It is to be emphasized, however, that our methodology does not require this.

The function $F(t, \phi, \theta; \xi)$ was represented by a set of experimental waveforms for $\theta = 0$ and for $\theta = 45°$, $\phi = 0°$ to $360°/m(\xi)$ in 3° increments with each object of given type and pose in a standard position orientation. As discussed earlier, the integer $m(\xi)$ depends upon the symmetry of the object corresponding to type and pose labelled by ξ. The test data consisted of a set of experimental waveforms for $\theta = 0$ and for $\theta = 45°$, $\phi = 0°$, 15°, 90°, 135°, 225°, 297°. The sufficiency of this data set for the recognition of solid objects is supported by the fact that a smaller set was sufficient for the imaging of solid objects on a flat surface as was shown by

Richardson et al, (1984, 1985, 1986). To obtain satisfactory impedance matching with air we made use of high fidelity audio technology with a tweeter as the transmitter and a microphone as a receiver (both in sufficiently close proximity to approximate pulse-echo conditions to a satisfactory degree). The temporal frequency band extended from approximately 3 kHz to 30 kHz (at 20% of maximum amplitude). To minimize external noise and internal reverberations the experiments were performed in an anechoic chamber.

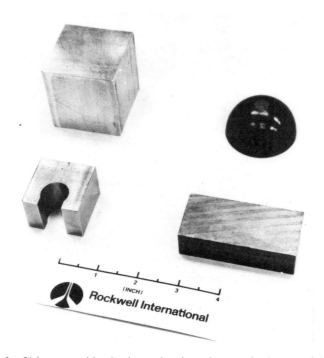

Fig. 3. Objects used in the investigation of acoustical recognition.

The test results were very satisfactory. The types and poses of the 4 test objects were determined perfectly. The positions and orientations were determined with adequate accuracy for robotic applications as is shown in Figs. 4a-4d. The levels of confidence to be associated with these results are appropriately represented by a sequence of goodness-of-fit tests. In Table 1 we show the results of such tests for the case in which the test object was a rectangular solid in pose #2 (i.e., the edge of intermediate length were vertical). In the left column, we list the 9 hypothetical types and poses. In the right column we list the corresponding values of the goodness-of-fit parameter $\varepsilon(\alpha, \xi)$ defined by (5). In each case the optimal orientation α of the hypothetical object has been determined. The reader will note that the $\varepsilon(\alpha, \xi)$ value corresponding to the correct hypothesis is significantly smaller than the $\varepsilon(\alpha, \xi)$ values corresponding to wrong hypotheses. The results of goodness-of-fit tests for other test objects were similar.

Table 1.   Goodness-of-Fit Tests for Various Hypotheses
(Test Object: Rectangle, Pose #2)

Object Type and Pose (hypoth.)	R.M.S. Residual $\varepsilon(\hat{\alpha}, \varepsilon)$
Cube	27.8
Rectangle, Pose 1	24.0
Rectangle, Pose 2	10.4
Rectangle, Pose 3	18.2
Hemisphere	24.4
Slotted Rectangle, Pose 1	22.1
Slotted Rectangle, Pose 2	21.8
Slotted Rectangle, Pose 3	21.5
Slotted Rectangle, Pose 4	21.9

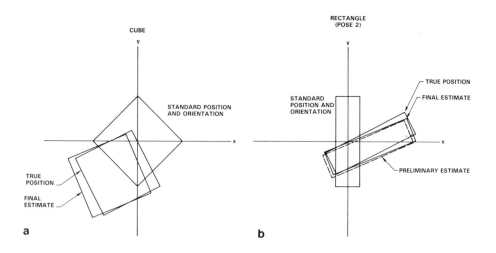

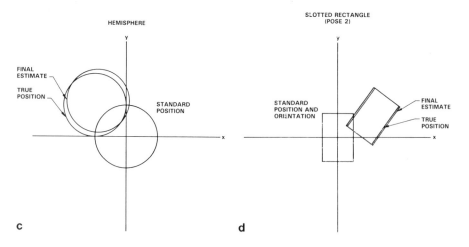

Fig. 4. (a)-(d) Estimation of position and orientation.

COMMENTS

This proof-of-principle investigation demonstrates the viability of the full version of acoustical recognition (i.e., the estimation of type, pose, orientation, and position). These results suggest the ultimate feasibility of acoustics for object recognition in a robotics context. However, there are a number of problems yet to be solved, some of which are:

1. Extension of the treatment to the cases of no objects and many objects.

2. Development of measures that would make the recognition process sufficiently insensitive to the level of noise (i.e., external noise, reverberations, etc.) existing in a typical work-cell environment.

3. Devising of real-time calibration techniques for correcting for variations of sound velocity and attenuation (due to variations of temperature and humidity).

REFERENCES

H. Ermert, J. Schmolke and G. Weth (1986), "An Adaptive Ultrasonic Sensor for Object Identification," Proc. 1986 IEEE Ultrasonics Symposium, pp. 555-558.

J.M. Richardson, K.A. Marsh, J.S. Schoenwald and J.F. Martin (1984), "Acoustic Imaging of Solid Objects in Air Using a Small Set of Transducers," Proc. 1984 IEEE Ultrasonics Symposium, p. 831.

J.M. Richardson, J.F. Martin, K.A. Marsh and J.S. Schoenwald (1985), "Acoustic Imaging for Robotic Acquisition," DARPA, Contract No. N00014-84-C-9985.

J.M. Richardson, K.A. Marsh, G. Rivera, M. Lasher and J.F. Martin (1986), "Acoustic Imaging of Solid Objects in Air Using a Small Set of Transducers: III. Experimental Demonstration," to appear in Acoustical Imaging, 15, ed. H.W. Jones, Plenum, New York.

A. Wernerson (1984), Private communication.

APPLICATIONS OF GENERALIZED RADON TRANSFORM

TO INVERSION PROBLEM OF GEOPHYSICS

Wan Suiren and Wei Yu

Dept. of Biomedical Engineering
Nanjing Institute of Technology
People's Republic of China

INTRODUCTION

This paper discusses the linearized multiparameter inversion problem with the S-P scattering system, which reconstructs shear modulus and velocity configuration from surface SH and P data. The generalized Radon transforms is used to complete the procedure.

Interaction between inhomogenous elastic medium gives rise to scattering P waves, and in the direction perpendicular to propagational direction of the original S wave the scattering P wave is due to the velocity perturbation $\delta v_s(\bar{r})$, so δv_s can be reconstructed by P wave equation whose source is proportional to δv_s. Therefore the parameter μ can be reconstructed by solving the inversion problem of S wave equation.

This approach, which only requires a complete set of receivers for each of one source location, represents a significant reduction in the data requirement from previous inversion methods and provides more exact solution.

RECONSTRUCTION OF THE VELOCITY PARAMETERS

By considering follow S wave equation

$$\nabla \cdot [\mu \nabla U] - \rho \frac{\partial^2 U}{\partial t^2} = -\rho_s S(t) \delta(\bar{r} - \bar{r}_s) \tag{1}$$

where μ is Lamè coefficient, ρ is medium density, ρ_s is density of source at $\bar{r}_s$. For convenience, suppose medium parameters

$$\mu = \mu_0[1 + a_1(r)]$$
$$\rho = \rho_0[1 + a_2(r)] \tag{2}$$

where μ_0 and ρ_0 are constants. Substituting Eq. (2) into Eq. (1) we obtain

$$[L_0 + L_1] \ U = -\frac{\rho_s}{\mu_0} S(\omega) \ (\bar{r} - \bar{r}_s) \tag{3}$$

621

where U and S(ω) are the Fourier transformation of U and S(t) in Eq. (1) and

$$L_0 = \nabla^2 + K^2 \quad ,$$

$$L_1 = \nabla \cdot (a_1 \nabla) + K^2 a_2 \quad .$$

$$K^2 = \frac{\omega^2}{V_o^2} = \frac{\omega^2 \rho_o}{\mu_o} \quad .$$

For Eq. (3), the solution can be written as

$$U = U_0 + U_1^S \tag{4}$$

where U_0 is original S wave, U_1^S is scattering S wave, and

$$L_0 U_0 = -\frac{P_s}{\mu_o} S(\omega) \delta(\vec{r} - \vec{r}_s) \quad ,$$

$$U_0 \gg U_1^S \quad .$$

substituting (4) into (3), we obtain

$$U_1^S \approx -L_0^{-1} L_1 U_0 \quad . \tag{5}$$

L_0^{-1} is defined in terms of the Green's function. Eq. (5) means Born approximation.

Since medium is inhomgenous, original S wave not only gives rise to wave U_1^S but also gives rise to scattering wave U_1^P. According to Scattering Theory in the elastic medium, scattering wave which is given rise by original wave does not vary with Lamè coefficient λ. There is a velocity disturbing scattering wave U_{1v}^P which is perpendicular to propagational direction of the original S wave. Following, we will omit subscripts v and seperate velocity disturbing parameter from operator L_1

$$L_1 U_0 = [\nabla a_1 \cdot \nabla + a_1 \nabla^2 + K^2 a_2] U_0$$

$$= [\nabla a_1 \cdot \nabla - K^2 a_1 + K^2 a_2] U_0 \quad . \tag{6}$$

Using $v_s^2 = \frac{\mu}{\rho} = v_o^2 \frac{1 + a_1}{1 + a_2}$, we obtain

$$a_1 - a_2 = \frac{2 \delta v_s}{v_o} \quad . \tag{7}$$

Let $\quad \delta v_s = \frac{v_o}{2} a_3 (\vec{r})$, then

$$L_1 U_0 = \nabla a_1 \cdot \nabla U_0 - K^2 a_3(\vec{r}) U_0 \quad . \tag{8}$$

Since the velocity disturbing a_3 gives rise to scattering wave U_1^P, it must be the solution of P wave equation

$$L_{op} U_1^P = -K^2 a_3(\vec{r}) U_0$$

where $L_{op} = \nabla^2 + K^2 n_o^2$, $n_o^2 = \frac{\mu_o}{2\mu_o + \lambda_o}$, λ_o is Lamè constant. Then we can obtain

$$U_1^P = -L_{op}^{-1} K^2 a_3(\vec{r}) U_0$$

$$= +K^2 \int_X G_{op}(\vec{r}, \vec{r}_\alpha) G_0(\vec{r}, \vec{r}_s) a_3(\vec{r}) d\vec{r} \tag{9}$$

where X is a region which is full of elastic medium, $\vec{r}_\alpha$ is the

622

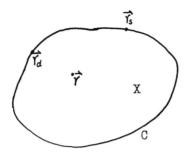

Fig. 1. Illustration of experiment.

position of receiver which is used to record scattering field U_1^P on the boundary C(see Fig. 1). G_{op} is P wave Green's function and G_o is S wave Green's function. Now we consider S-P scattering field

$$G_{op} = e^{-i\frac{\pi}{2}\cdot\frac{n+1}{2}} K^{\frac{n-3}{2}} A_{op}(\vec{r},\vec{r_d})e^{iK\phi_{op}(\vec{r},\vec{r_d})} \qquad (10)$$

where $\phi_{op}(\vec{r},\vec{r_d})$ satisfies the eikonal equation

$$(\nabla_{\vec{r}}\phi_{op})^2 = n_o^2 = \frac{\mu_o}{2\mu_o + \lambda_o} \qquad (11)$$

A_{op} satisfies the transport equation

$$A_{op}\nabla_{\vec{r}}^2\phi_{op} + 2\nabla_{\vec{r}}A_{op}\cdot\nabla_{\vec{r}}\phi_{op} = 0 \qquad (12)$$

∇_r represents ∇ operator with respect to $\vec{r}$. Similarly

$$G_o = e^{-i\frac{\pi}{2}\cdot\frac{n+1}{2}} K^{\frac{n-3}{2}} A_o(\vec{r},\vec{r_s})e^{iK\phi_o(\vec{r},\vec{r_s})} \qquad (13)$$

ϕ_o and A_o satisfy the eikonal equation and transport equation seperately. Measurement of U_1^P on the boundary is to reconstruct the parameter of velocity $a_3(\vec{r})$ by means of inversion of the causal generalized Radom transform.[2,3]

$$F(a_3)(\vec{r}) = R^*\begin{cases} \int [\frac{(-1)^{\frac{n-1}{2}}}{2(2\pi)^{n-1}}] U_P'(t,\vec{r_d},\vec{r_s}) & \begin{array}{l} \text{for } n=2m+1, \\ m=1,2,\cdots \end{array} \\ \frac{(-1)^{\frac{n}{2}}}{2(2\pi)^{n-1}}\cdot\frac{1}{\pi}\int_{-\infty}^{+\infty}\frac{U_P'(t',\vec{r_d},\vec{r_s})}{t-t'}dt' & \text{for } n=2m \end{cases} \qquad (14)$$

where F is Fourier integral operator, R^* is the dual transform of the generalized Radon transform.

For a fixed $\vec{r_s}$,

$$(R^*v)(\vec{r}) = \int v(t,\vec{r_d})\big|_{t=\phi(\vec{r},\vec{r_d},\vec{r_s})}b(\vec{r},\vec{r_d})d\vec{r_d} \qquad (15)$$

where
$$\phi(\vec{r},\vec{r_d},\vec{r_s}) = \phi_{op}(\vec{r},\vec{r_d})+\phi_o(\vec{r},\vec{r_s}) \qquad (16)$$

$$b(\vec{r},\vec{r_d}) = [h(\vec{r},\vec{r_d})/a(\vec{r},\vec{r_d})]\chi(\vec{r},\vec{r_d}) \qquad (17)$$

$$a(\vec{r},\vec{r_d}) = A_{op}(\vec{r},\vec{r_d})A_o(\vec{r}) \qquad (18)$$

623

$$h(\vec{r},\vec{r}_d) = \begin{bmatrix} \phi_{y_1} & \phi_{y_2} & \cdots & \phi_{y_n} \\ \hat{\phi}_{y_1\zeta_1} & \hat{\phi}_{y_2\zeta_1} & \cdots & \hat{\phi}_{y_n\zeta_1} \\ \vdots & \vdots & & \vdots \\ \hat{\phi}_{y_1\zeta_{n-1}} & \hat{\phi}_{y_2\zeta_{n-1}} & \cdots & \hat{\phi}_{y_n\zeta_{n-1}} \end{bmatrix} \qquad (19)$$

in Eq. (19), $\hat{\phi} = \phi_{op}$, y_i is the component of $\vec{r}$, ζ_j is the component of $\vec{r}_d$, $\chi(\vec{r},\vec{r}_d)$ is the cutoff function which cause $\chi(\vec{r},\vec{r}_d)$ $h(\vec{r},\vec{r}_d) \geqslant 0$.

For $\vec{r}_d(\vec{r}_s)$ case, in (15)

$$\phi = \phi(\vec{r},\vec{r}_d(\vec{r}_s),\vec{r}_s) \qquad (20)$$

$$b(\vec{r},\vec{r}_s) = [h(\vec{r},\vec{r}_s)/a(\vec{r},\vec{r}_d(\vec{r}_s),\vec{r}_s)]\chi(\vec{r},\vec{r}_s) \qquad (21)$$

$$h(\vec{r},\vec{r}_s) = \begin{bmatrix} \phi_{y_1} & \phi_{y_2} & \cdots & \phi_{y_n} \\ \vdots & \vdots & & \vdots \\ \phi_{y_1 n-1} & \phi_{y_2 n-1} & \cdots & \phi_{y_n n-1} \end{bmatrix} \qquad (22)$$

RECONSTRACTION OF THE LAME COEFFICIENT μ

Using Eq. (9) - (22) to obtain $a_3(\vec{r})$, then substitute it into Eq. (5). We have

$$U_1^S(\vec{r}) - L_o^{-1}K^2 a_3(\vec{r})U_o = -L_o^{-1}(\nabla a_r \nabla U_o) \qquad (23)$$

Let $\quad U_1^S(\vec{r}) - L_o^{-1}K^2 a_3(\vec{r})U_o = U_1^{S_2}(\vec{r})$, then

$$U_1^{S_2} = \int_X dr'[\nabla \cdot a_1(\vec{r}') \cdot \nabla_r U_o(\vec{r},\vec{r}_s)] \cdot G_o(\vec{r},\vec{r}') \qquad (24)$$

To reconstruct the paramenter, we can use the method of Esmersoy et al.[4] Let us introduce extrapolated field

$$P_e^*(\vec{r}) = \int_C dr' \, \hat{n} \cdot [G_o^*(\vec{r},\vec{r}')\nabla_r U_1^{S_2}(\vec{r}') - U_1^{S_2}(\vec{r}')\nabla_r G_o^*(\vec{r},\vec{r}') \qquad (25)$$

C is curved surface which surround X, $\hat{n}$ is outward normal unit vector for C, P_e satisfy

$$\begin{aligned} L_o P_e(\vec{r}) &= 0, & \vec{r} \in X \\ P_e(\vec{r}) &= U_1^{S_2^*}(\vec{r}), & \vec{r} \in C \end{aligned} \qquad (26)$$

$P_e|_C$ is measurable, since $U_1^{S_2}|_C$ is also measurable.

Applying Green's identity to Eq. (25), we find

$$P_e^*(\vec{r}) = 2i \int_X dr'[\nabla_r \cdot a_r \nabla_r G_o(\vec{r},\vec{r}_s)] \, I_m G_o(\vec{r},\vec{r}') \qquad (27)$$

Following EOL and HW methods,[4,5]

$$D(\vec{r},\vec{r}_s) = -8 \int_0^\infty dK \, \mathrm{Re}\frac{P_e^*(\vec{r})}{K} \qquad \vec{r} \in X$$

$$= \int_X d\sim' \left(\frac{1 - (\widehat{\vec{r}'-\vec{r}})\cdot(\widehat{\vec{r}'-\vec{r}_s})}{|\vec{r}' - \vec{r}_s|}\right) a_1 \delta(|\vec{r}-\vec{r}_s| - |\vec{r}-\vec{r}'|) \qquad (28)$$

where $\vec{r}$ is the unit vector in the $\vec{r}$ direction. Eq. (28) is in

624

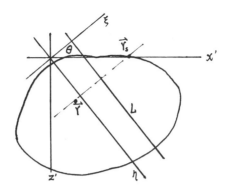

Fig. 2. Change of coordinates for a given source
location and a point in the medium.

the form of a generalized Radon transform. In order to transform Eq. (28) to a simpler form, we define a retated Cartesian
coordinate system ξ and η (see Fig. 2). In this coordinate
system, we find that

$$\overline{D}(\xi, \theta, \vec{r_s}) = |\vec{r} - \vec{r_s}| \, D(\vec{r}, \vec{r_s})$$

$$= \int_L d\eta' \, \frac{2(\xi - \xi_s)^2}{|\vec{r'} - \vec{r_s'}|^2} a_1(x', z') \qquad (29)$$

where
$$x' = \xi\cos\theta - \eta'\sin\theta$$
$$z' = \xi\sin\theta + \eta'\cos\theta \qquad . \qquad (30)$$

From Eq. (29) it follows that [6, 7]

$$a_1(\vec{r}) = |\vec{r} - \vec{r_s}|^2 \int_0^\pi \int_{-\infty}^{+\infty} \int_{-\infty}^{+\infty} \frac{1}{2} \overline{D}(\xi, \theta, \vec{r_s}) e^{i2\pi R(X\cos\theta + Z\sin\theta - \xi)} |R| \, d\xi \, dR \, d\theta \qquad (31)$$

Now, we have completed the process of using generalized
Radon transform to reconstract velocity parameter $a_3(\vec{r})$ and Lamè
coefficient parameter $a_1(\vec{r})$.

CONCLUSIONS

        We have presented a multiparameter inversion method with
the S-P system, which reconstructs shear modulus and velocity
configuration from surface SH and P data. This method represents
a significant reduction in the data requirments, from previous
multiparameter, multidimensional inversion methods.   It only
requires a complete set of receivers for each of one source
location.

REFERENCE

    1. KeiiTi Aki and P. G. Richards, " Quantitative Seis-
        mology Theory and Methods ", W. H. Freeman   and
        Company, 1980, Chap. 13;

    2. G. Beylkin, J. Math. Phys., 26(1), 1985, PP99-108;

    3. G. Beylkin, Common. Pure Appl. Math., 37,   1984,
        P 579;

4. Cengiz Esmersoy et al,   J.Acoust. Soc. Am. , 78(3),
    Sept. 1985, PP 1052-1057;

5. M. A. Hooshyar and A. B. Weglein,   J. Acoust. Soc.Am.
    79(5),   May 1986,   PP 1280-1283;

6. D. Ludwig, Commun. Pure Appl. Math.,XIX, 1966, P 49;

7. R. Lewitt, Proc. IEEE 71, 1983, P 390.

MINI-SPARKER AS A SOURCE IN SEISMIC MODELS

Ph. Pernod, B. Piwakowski[*], J.C. Tricot, and B. Delannoy

Laboratoire de Physique des Vibrations et d' Acoustique
(C.N.R.S. U.A. 832 - Valenciennes)
Institut Industriel du Nord, B.P. 48

INTRODUCTION

Physical models are commonly used to help in understanding the problems occurring in seismic acoustic imaging[1]. In the elaboration of the model, special attention should be paid to a correct choice of the source, to obtain the desired conformity between the real and modelled phenomena. Sources used in seismic models are commonly the piezoelectric transducers or small weight-drops.

In this paper we intend to present another type of source: a "mini-sparker" which we have succesfully applied as a source in the model measurements concerning shallow.seismic prospecting. Sparkers, well-known in marine[2] experiments, have found a recent and wider application to shallow seismics on land[3], but, in fact, there are very few studies concerning their behaviour and application to the models. After a short description of the source's construction, its pulse shape, frequency spectra and directivity patterns are shown and discussed, for free space and then for container-limited boundary conditions. Radiation in water and then in a plate of plexiglass are considered. Finally, as an example of application of the "mini-sparker", the records obtained on a two-layered seismic model are presented.

INSTRUMENTATION

The diagram of the "mini-sparker" is presented in Fig.1. It consists of a capacitor block charged by a power supply, a discharge gap and a triggering system. The conical electrodes of 2mm diameter are placed in a small cylindrical closed plastic container (1.5cmx1.5cm) filled with water, which assures the possibility of the free displacement of the source along the model. In the model presented, the distance between the electrodes is 0.25mm and the electrical energy is 0.4J, but these parameters can be easily modified.

---

* Permanent affiliation: Institut of Telecommunication, Technical University of Gdansk, POLAND

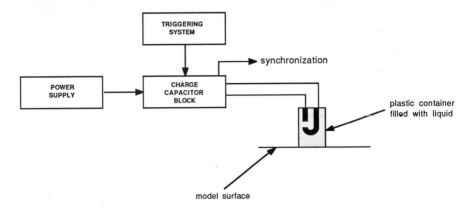

Fig. 1. Schematic diagram of a "mini-sparker" used for
model applications.

In order to characterize the properties of this source, the preliminary measurements were carried out for the following conditions :

1/ The source is applied in a semi-infinite water medium without plastic
container - this can be assumed as being free space conditions.

2/ The container is added to show its influence on the source behaviour.

3/ Finally the assembly (source within container) is applied to excite a
solid medium in similar conditions as shown in Fig.1; the plate of plexiglass is used as a solid.

In all cases, we present the acoustic signal generated, the associated
frequency spectrum and the directivity pattern of the source.

Fig.2 shows the experimental setup used for the first two cases: the
source emits in water, and a wide-band-receiving piezoelectric transducer
moves along a semi-cylindrical curve underneath. The receiver rotation
axis coincides with the source location. The complete signal, received in
a constant distance R from the source versus angular position θ, is recorded by a digital storage oscilloscope interfaced with a mainframe microcomputer. The electromagnetic signal associated to the spark discharge
was successfully applied to trigger the recording system.

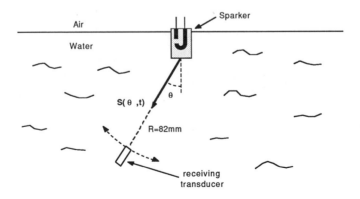

Fig.2 Experimental setup used for the measurements
of the source's radiation in water.

SOURCE PARAMETERS

Pressure waveforms and spectra

The direct arrival acoustic pressure signals generated by the mini-sparker and their frequency spectra are presented in Fig.3, for water conditions. The first two signals and spectra (Fig.3a,b) correspond to the conditions where source acts without the container, 10mm below the water surface, for two orientations of its electrodes: parallel and perpendicular with regard to the plane of receiver displacement. Fig.3c,d correspond to the conditions where the plastic container is applied and the source electrodes orientation is kept perpendicular: in section (c) the source is at the same position as in (b), in section (d) the source is placed precisely at the water surface.

The examination of the presented signals shows that the emitted wave is composed of three major parts, each one having its own principal frequency (200 kHz, 100 kHz and 50 kHz respectively). These parts might be attributed to three successive bubble collapses[4,5]. The relatively wide source signal spectrum should be noted. For the source power applied, the 20dB cut-off frequency is of 300kHz, but the signal seems to be exploitable even to 500kHz. Taking advantage of the - spectre/source power - dependence, the spectrum can be easily modified as needed[3].

The results, obtained under conditions where a solid is excited, are presented in Fig.4. Fig.4a shows the experimental setup applied. The source within container was coupled with the surface of solid by means of a silicon paste. The signal transmitted through a plate of plexiglass 12.3cm thick is shown in Fig.4b. Analysis of the complex section $S(\theta,t)$ allowed us to separate the P- and S-type arrivals - the corresponding spectra of these arrivals are presented in Fig.4c,d.

It is of interest to note the relatively simpler source signature when the solid is excited - the bubble effect is no longer observed. Finally the spectrum of P-type arrival is relatively smoother in regard with the spectra obtained in water and this allows us to consider our source as a genuine wide band transmitter. Comparison of the Fig.4c and 4d shows that S-type wave displays the narrower spectrum - this being typical for S-waves properties.

Finally the results obtained show that:

1/ There is no influence of the orientation of the source. Supposed masking action of the electrodes is not found (Fig.3a and 3b).

2/ The influence of the plastic container presence is generally not seen (Fig.3b and 3c). The supposed resonnance effects are not observed proving that, when correct choice of the container acoustic parameters is done, it can be considered as acoustically transparent.

3/ The comparison of the signal parameters from Fig.3c,d shows that the bottom radiation is essential in the source behaviour. This appears as an advantage when its application to the solid surface is considered.

4/ The source coupled with the solid displays the better radiation properties in regard with the water conditions: the signal signature is simpler and its spectrum is smoother.

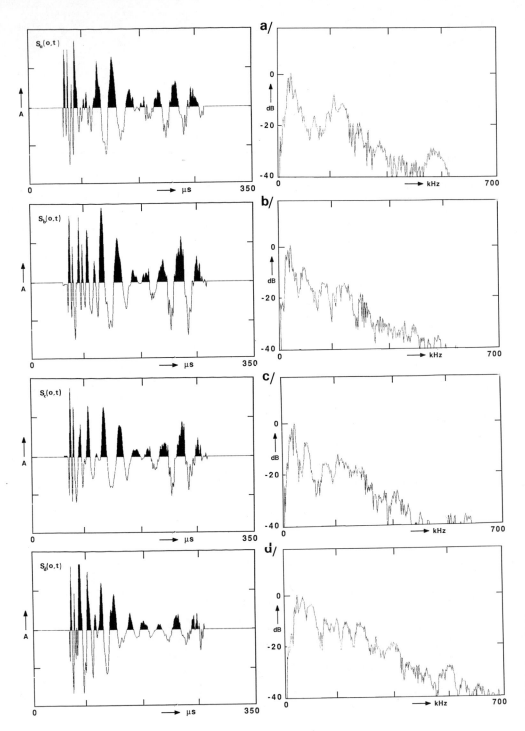

Fig.3 Acoustic pressure signals generated in water by the mini-
sparker, and associated frequency spectra. (a) Source without container, the electrodes are paralelle to the plane
of receiver displacement; (b) Same conditions as in section (a), but the electrodes are perpendicular to the plane;
(c) Same as section (b), but with a container; (d) Source
with container acting at the water surface.

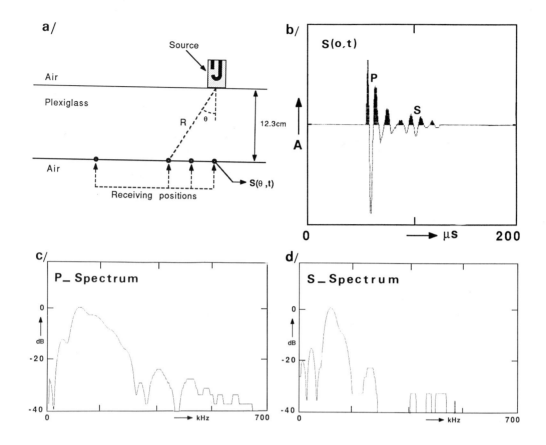

Fig.4 (a) Setup applied for the measurements of the source when
a solid layer (plexiglass) is excited; (b) Signal transmitted through the plate of plexiglass (θ=0), the P- and
S-type arrivals are seen; (c) Spectrum of the P-type arrival; (d) Spectrum of the S-type arrival.

Vertical directivity patterns

        The directivity patterns obtained for source radiation in the water
are presented in Fig.5a,b,c,d. Fig.5e shows the results for conditions as
applied in Fig.4 (solid). The experimental acoustic field produced by the
transducer was determined by measuring the amplitude of the first half-
cycle of the direct wave arrival. The source signal repetition was con-
troled by means of additionnal fixed transducer and we have verified that
source repeats very well. For the measurements in the solid (Fig.4a) where the value of the source-receiver distance R could not be kept constant, the corrections for spherical spreading (1/R) were applied. Moreover, we have considered the directivity patterns of the receiver, which
remains oriented normally to the bottom of the plate.

        The experiments made in water show that the source can generally be
assumed as non-directive; the drop observed in the directivity pattern
seen in Fig.5d for angle θ= +/-90° corresponds to the contribution of the
Rayleigh-Sommerfeld boundary conditions[6] (modification by means of multiplication by cosθ).

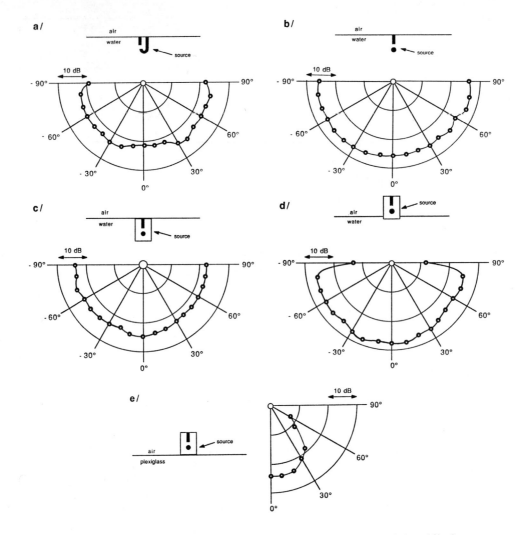

Fig.5 Directivity patterns of the source. (a), (b), (c), (d) Source is acting in water for the same cases as seen in Fig.3;
(e) Source is acting at the surface of a solid (plexiglass).

The directivity produced in the solid results from container aperture dimensions. Assuming the circular piston radiation model and, as before, the Rayleigh-Sommerfeld conditions, we obtain for container bottom
diameter 15mm and for arbitrary estimated central frequency of the signal
spectrum (Fig.4c) being of 200 kHz, that θ_{3b} is = 35°. This result is generally confirmed in Fig.5e. The modification of the source directivity
is then possible by changing of the container diameter. Based on data
from experiments made in water, we may suppose that this should not affect the source signature signal. We point out that it is difficult to
perform such a modification in the case of piezoelectric transducer (commonly applied to seismic models) where its dimensions directly affect the
frequency properties.

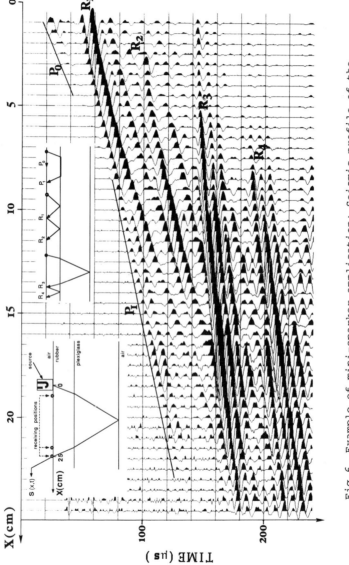

Fig.6 Example of mini-sparker application: Seismic profile of the two layered model; the signals are amplitude corrected only.

633

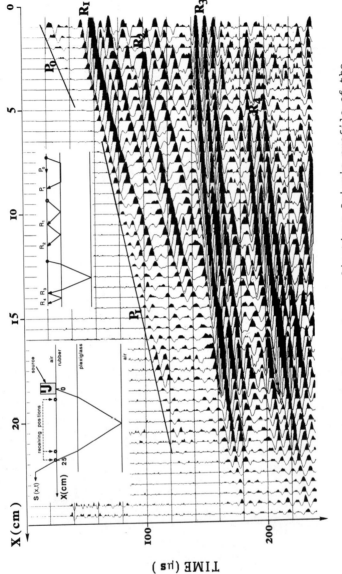

Fig.7 Example of mini-sparker application: Seismic profile of the two layered model, the same section as in Fig.6, but after automatic gain control application.

EXPERIMENTS MADE ON A TWO-LAYERED MODEL

The example of application of this source is shown in Figures 6,7 where a seismic profile was made on a two-layered model. The materials used for the model constitution were a plate of natural rubber 3 cm thick with acoustic P velocity $v_r = 1535ms^{-1}$ and a plate of plexiglass of thickness 12.3 cm and $v_p = 2758ms^{-1}$, coupled by means of special paste used in the medical applications. The source was applied on the rubber's surface, its position was x=0; space tracing was 0.5 cm. The seismic sections obtained are presented in Fig.6. The signals were amplitude corrected only in order to equilibrate their first arrivals amplitudes. In Fig.7 the same section is shown but after automatic gain control application. In both the figures, the common features characteristic of the two layered structure are observed: first layer direct P-arrival P_0 with velocity v_r, refraction on the layer's interface P_1 corresponding to the velocity v_p, P-type reflection in the first layer and its multiple (R_1 and R_2), reflection from the plexiglass bottom (R_3 and R_4). The surface and S-type events are not present on account of their weakening within rubber layer.

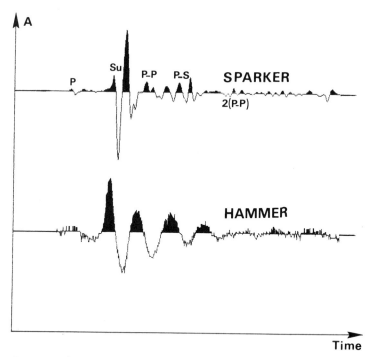

Fig.8 Comparison between a signal generated by a hammer shock with one generated by our mini-spaker, when a plate of plexiglass is excited.

CONCLUSIONS

1/ The "mini-sparker" presented combines three characteristics: poor directivity, wide band properties and sufficient output power to model applications . Such a combination is not available in the piezoelectric transducers where the transducer dimension should be enlarged to obtain the desired level of radiated signal. This generally involves the directivity increase. Finally, our source produces all the characteristics required to model a real seismic source. This is confirmed by the presented example (Fig.6,7).

2/ In comparison with the weight-drops which are also applied in seismic models, our source produces relatively more signals penetrating the model. In Fig.8 we have shown two signals recorded for conditions as applied in figures 6,7 but for plexiglass layer only, for source receiver distance 10cm. For a source the "mini-sparker" was used (Fig.8a) and then the hammer drop (Fig.8b). Observation of these records show that hammer radiation produces the S-type wave mainly and is poor in P-type signals which are essential in model exploration.

3/ The drawbacks are that this source produces a relatively high level of electromagnetic radiation, and it is necessary to take some precautions to protect the data-processing systems.

4/ In our opinion, such a type of source can be applied to the other numerous domains of acoustical imaging as well.

ACKNOWLEDGMENTS

The authors wish to thank M. Quayle for the language corrections.

REFERENCES

1. J.A. VOGEL, U. STELWAGEN, AND R. BREEUWER, 1985, Seismic analysis of thin beds aided by 3D physical model experiments, 14 th. symp. on Acoustical Imaging, Delft, The Netherlands.

2. WALTER C. BECKMANN, ARCHIE C. ROBERTS, AND BERNARD LUSKIN, Oct.1959, Sub-bottom depth recorder, Geophysics, vol.14, NO.4, pp.749-760.

3. G.B. CANNELI, E. D'OTTAVI AND S. SANTOBONI, 1985, Shallow propecting on land by means of a novel electroacoustical P-wave pulse generator, 14 th. symp. on Acoustical Imaging, Delft, The Netherlands.

4. J. CASSAND, M. LAVERGNE, 1970, Etinceleur multiélectrodes à haute résolution, Geophysical Prospecting, vol.18,NO.3,pp.0380-0388

5. A. SHIMA, K. TAKAYAMA, Y. TOMITA, AND N. MIURA, 1981, An experimental study on effects of a solid wall on the motion of bubbles and shock waves in bubble collapse, Acustica, vol.48, pp.293-301.

6. B. DELANNOY, H. LASOTA, C. BRUNNEL, R. TORGUET, AND E. BRIDOUX, The infinite planar baffles problem in acoustic radiation and its experimental verification,1979 , J. Appl. Phys., vol.50, pp.5189-5195.

SPATIAL FILTERING IN SEISMIC SHALLOW

PROSPECTING

B. Piwakowski*, J.C. Tricot, B. Delannoy, and PH. Pernod

Laboratoire de Physique de Vibrations et d'Acoustique
(C.N.R.S. U.A. 832 Valenciennes) - Institut Industriel du
Nord B.P. 48 - 59651 Villeneuve D'Ascq Cedex, France

INTRODUCTION

This paper concerns the problem of detection, by means of seismic reflection[1], of underground empty spaces appearing in the north of France,
which are the remains of the old chalk pits. They are situated in chalk at
a depth of 8-10 m and the terrain is covered by a clay layer. To explain
the problem let us consider the field situation (Fig.1.a.) where the source
is placed over the tunnel in position x = 0 and the signals s(x,t) are recorded along the x axis. The preliminary measurements and studies show[2]
that in these field conditions the P-type tunnel reflection r(x,t) is at
least 10-20 dB smaller than the coherent noise signal $C_n(x,t)$ observed at
the moment of reflection arrival. The most troublesome component of the coherent noise seems to be the refraction on top of the chalk, which has a
frequency spectrum ressembling the spectrum of reflection, and therefore
separating them by means of classic time-domain frequency filtering may
not, in fact, be effective. In Fig.2, we have shown the forms and the spectra of two signals : chosen as typical from the refraction, and recorded
at the tunnel vault which can be assumed as a potential tunnel reflection.
Both the signals have the similar spectrum up to 700 Hz, and apparent differences for greater frequencies seem to us to be inexploitable. Finally,
to detect the tunnel the reflection/coherent noise ratio (SNR) must be improved. The way to resolve this it is to suppress the nearly horizontally
running refraction by means of spatial filtering. To carry this out the
multi-channel acquisition system will be applied.

We present the following problem-solution approach : what is the proper trace spacing Δx to be applied and which part of the received signals
spectra should be exploited in order to maximise the SNR improvement for :
a given number of array elements N, a given spectrum characteristic for
the coherent noise signals $C_n(x,t)$ and a given spectrum and spatial characteristic for the enhanced signal r(x,t).

SPATIAL FILTER GAIN MEASURE

Let us define the tunnel reflection/coherent noise ration $SNR^{(1)}$ as the

* Permanent Affiliation : Institue of Telecommunication Technical
University of Gdansk, Poland.

637

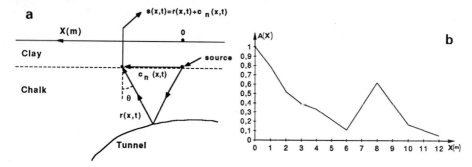

a

$s(x,t)=r(x,t)+c_n(x,t)$

X(m)

0

Clay

source

$c_n(x,t)$

Chalk

θ

r(x,t)

Tunnel

b

A(X)

1
0,9
0,8
0,7
0,6
0,5
0,4
0,3
0,2
0,1
0

0 1 2 3 4 5 6 7 8 9 10 11 12 X(m)

Fig. 1. (a) General field conditions for illustration of the considered
problem ; (b) distribution of the tunnel reflection amplitude as
a function of source-receiver offset.

quotient of energy of the enhanced and undesired signals in conditions where
only one (first) array channel is active (absence of spatial processing).
Based on the theorem of Parseval, such an expression can be determined in
the frequency domain. Designating $R(j\omega)$ and $CN(j\omega)$ as the Fourier transformations of $r(x,t)$ and $c(x,t)$ we have :

$$SNR^{(1)}(f_o,\Delta f) = 10\log \frac{\int_{f_o-\Delta f/2}^{f_o+\Delta f/2} R(\ ) \cdot R^*(\ )\ d\omega}{\int_{f_o-\Delta f/2}^{f_o+\Delta f/2} CN(\ ) \cdot CN^*(\ )\ d\omega} = 10 \log A/B \qquad /1/$$

where f_o and Δf are respectively the central frequency and the frequency
bandwidth of a rectangular filter applied to examine the $SNR^{(1)}$ behaviour in
various regions of the input signals spectrum. Let us introduce now the
coefficient $SNR^{(1)}$ as the quotient of the energies as before, but after the
application of spatial processing by means of the N-element array. We assume that the processing procedure will include the 2-D Fourier transformation multiplied by the 2-D filter and next the inverse transformation[3].
The $SNR^{(N)}$ can be found now in the ω-k domain, based on the 2-D version of
the theorem of Parseval :

$$SNR^{(N)}(f_o,\Delta f,\Delta x) = 10\log \frac{\int_0^{k_N}\int_{f_o-\Delta f/2}^{f_o+\Delta t/2} [R(,).F(,)][R(,).F(,)]^* \ dk\ d\omega}{\int_0^{k_N}\int_{f_o-\Delta f/2}^{f_o+\Delta f/2} [CN(,).F(,)][CN(,).F(,)]^* \ dk\ d\omega}$$

$$= 10 \log C/D \qquad /2/$$

where $R(j\omega,jk)$, $CN(j\omega,jk)$ are the 2-D transformations of the signals $r(x,
t)$ and $c(x,t)$, and $F(j\omega,jk)$ is the 2-D filter applied ; k_N is the spatial
sampling frequency.

Based on /1/ and /2/ we will now introduce the spatial filter gain
coeficient defined as :

$$G^{(N)}(f_o,\Delta f,\Delta x) = SNR^{(N)} - SNR^{(1)} = 10\log CB/AD \qquad /3/$$

This parameter compares the SNR ratio before (without) and after spatial
filtering and by these means expreses the real efficiency of the undesired
signal suppresion by means of the given 2-D filter, F. If the signals $r(x,
t)$ and $c_n(x,t)$ are taken separately then - due to the linearity of the
Fourier transformation - components A,B,C,D can be computed independently.

638

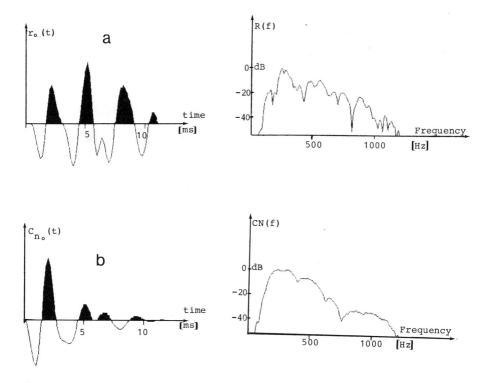

Fig. 2. Real signals recorded at the test site ; (a) Potential tunnel
reflection signal and its spectrum ; (b) example of a typical
refraction signal first arrival and its spectrum.

We shall take advantage of this in the following sections.

SYNTHETICS MODELLING

This problem will be discussed in light of various model $r(x,t)$ and
$c_n(x,t)$ signals by means of the analysis of the $G^{(N)}$ coefficient behaviour
as well as by means of comparing the filtered sections in the time domain.
We will use the real field parameters[2] : acoustic velocity of P waves in
chalk $c=1000$ m/s ; sampling frequency of signals $f_s = 5000$ Hz ; tunnel depth
$d=6$m. The array transducers are placed in line along the x axis (Fig.1.a.)
with the trace spacing Δx, at positions $x_i=(i-1)\Delta x$ $(i=1,2,,...N)$; array
length is $1 = (N-1)\Delta x$.

Case I - general considerations

We assume, for the sake of discussion, that :
- the refraction $c_n(x,t)$ represents the horizontally running signal with
the velocity c, with a form as shown in Fig 2.b ; $c_n(x,t) = c_{n_o}(t-x_i/c)=$
$c_{n_o}(x_i,t)$
- the tunnel refection $r(x,t)$ represents the vertically arriving signal
with a form as shown in Fig. 2.a ; $r(x,t) = r_o(x_i,t-t_o) = r_o(x_i,t)$, where
t_o represents the time of reflection arrival, $t_o \doteq 2d/c$.

The assumed model signals $r_o(x_i,t)$ and $c_{n_o}(x_i,t)$ are presented in
Fig. 3.a.b for $\Delta x=0.33$m and $N=16$. The spatial filter gain coefficient
$G^{(N)}(f_o,\Delta f,\Delta x)$ computed for these signals, based on definition /3/, is

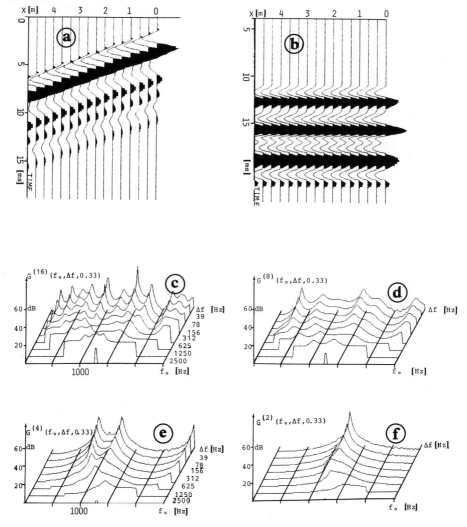

Fig. 3. (a) Synthetic model of assumed refraction $c_n(x_i,t)$; (b) synthetic model of assumed reflection $r_0(x_i,t)$; (c), (d), (e), (f) spatial filtrage gain characteristics as a function of filter central frequency f_0 and its bandwidth Δf, for array element number N=16,8, 4 and 2.

presented in Fig. 3.c,d,e,f, as a function of array element number N. In calculations for N<16 the signals from redundant array channels were set at zero. Each line of the $G^{(N)}$ curve plots represents its development as a function of the bandpass filter central frequency f_0, for a given filter bandwidth, Δf. Calculations were carried out for a frequency range of 0-$f_s/2$. For the 2-D ω-k filter F, we have applied the simplest solution - filtering the zero spatial frequency component only. Such a solution corresponds to the well known technique of the direct summation of the array outputs, and will be referred to as the F_0 filter.

Observation of the results indicates that when the total frequency band of the assumed signals is exploited (Δf=2500 Hz), the filtrage gain coefficient is $\cong$ 20dB for N=16. Its value decreases with the reduction of N and reaches 3 dB for N=2. Simultaneously, with the reduction of N, we

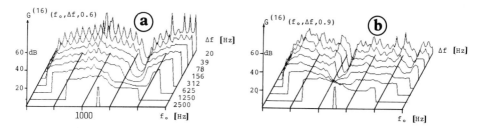

Fig. 4. Spatial filtrage gain characteristics (see Fig. 3.) for array
element number N=16 and for trace spacing : (a) Δx=0.6 m ;
(b) Δx=0.9 m.

observe the enhancement of the narrow-band properties of the array and for
N=2, an attenuation of the undesired signal is possible only for a single
frequency of f=c/2Δx=1515 Hz (a well known condition Δx=λ/2). The $G^{(N)}$ value observed for the narrow-band filter Δf=20 Hz represents the frequency
characteristic of the spatial filtering efficiency. It is seen that when
N increases this "filter" becomes wider. This observation is important in
our case because to obtain the better survey resolution, the spatial filtering efficiency should be such as to cover the total enhanced signal
spectrum (300-700 Hz for signal $r_o(t)$, Fig. 2.a.). The comparison of Fig.
3.c. and Fig.4.a.b. shows the evolution of the $G^{(N)}$ coefficient as a function of the trace spacing Δx : the array efficiency grows when its lengths
increases but the frequency properties are strongly modified. Appearence of
the minimum (at frequency f=c/Δx) limits the useful width of the enhanced
signal spectrum.

The computed filtrage gain values are confirmed in the results in the
time domain. Fig.5.a. shows the test input signal $s(x,t) = c_{n_o} (x_i,t)+0.1r_o$
(x_i,t), synthesized for Δx=0.6 m. Fig.5.c. shows this signal after spatial
processing for the following conditions : F_o 2-D filter, N=16 and Δf=2500
Hz (total spectrum of the signals). The apparent oscillations of the C_{n_o}
signal residues result from the sharp cut-off of the signal continuity at
both ends of the array. This involves the appearence of pseudo high frequen-
ces in the 2-D spectrum (Fig.5.e.). To overcome this, we have applied spatial tapering by means of Gaussian windowing (Fig.5.f.). Observation of the
spectra presented shows additionally that the enhanced signal spectrum oc-
cupes the larger region of the w-k plane than the F_o filter pass area.
Therefore, in the considered case, the better results are produced by the
application of the fan filter[4] designed to pass the signals arriving at the
angular θ range (Fig.1.a.) of $\pm$ 5 deg. The output signal processed for conditions including spatial tapering and fan filter application is shown in
Fig.5.d. The $G^{(16)}$ (1250,1250,0.6) value calculated in the time domain
(with input signals processed separately) produces the result of 25 dB -
this is 3 dB better than the reading from Fig.4.a. (22 dB). This difference
should result from fan filter application (spatial tappering insignifican-
tly affects the $G^{(N)}$ value). The residues of the c_{n_o} signal can be suppressed even better by means of post-filtering. Fig.5.b. represents the section from Fig. 5.d. filtered with a 300 - 1500 Hz filter, and the improvement of the $G^{(N)}$ value is remarkable. To compare the output section in the
simplified way, we have introduced the $G^{(N)}_c$ measure-calculated directly
from a comparison of the amplitudes of the filtered signals - before and
after processing. (This parameter ignores energy relations, and therefore,
in presented examples, reaches a value of several decibels greater than
$G^{(N)}$).

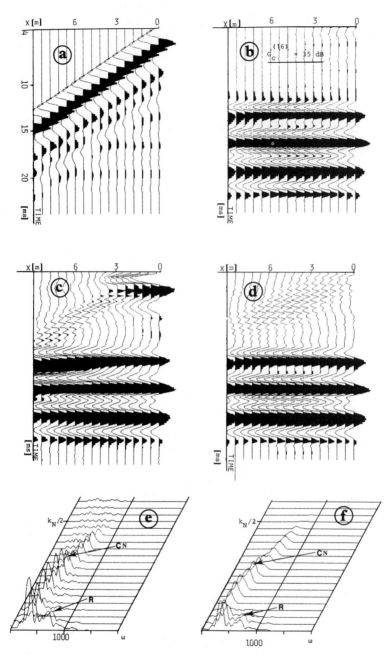

Fig. 5. (a) Input test signal $s(x,t)=c_n(x_i,t)+0.1r_o(x_i,t)$, for $\Delta x=0.6m$;
(b) final output version of processed input signal ; (c) signal
from section (a) processed with filter F_o ; (d) signal from section (a) spatially tapered and then processed with filter F_o ;
(e) 2-D spectrum of signal from section (a) (for coefficient 0.1
replaced by 1.0) ; (f) same as section (e) but with spatial tapering applied.

642

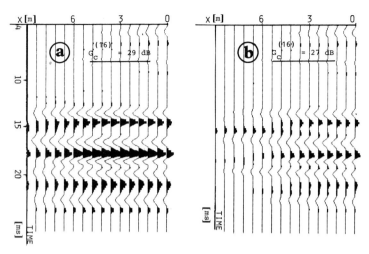

Fig. 6. The output signals processed for conditions as applied in Fig.5.b.
(a) for input test signal as considered in case II ; (b) for input
test signal as considered in case III.

Case II - spatial reflection distribution is considered

We assume now :
- a refraction signal the same as before,
- as reflection : the signal $r_1(x_i,t)=A(x_i)r_0(x_i,t)$
Function $A(x)$ (Fig.1.b.) represents the relative distribution of the suspected tunnel reflection amplitude as a function of transducer-receiver offset[2]
according to the conditions of our experiments. The calculation of the $G_c^{(N)}$
coefficient reveal now that its plots have the same form as before (Fig.3,
Fig. 4.) but the value is poorer, lowered by a value of $L(\Delta x)$ decibels.
This loss $L(\Delta x)$ can be estimated as :

$$L(\Delta x) = 20\log \left[\sum_{i=1}^{N} A(xi)/N \right] \qquad\qquad /4/$$

and results from the poorer total energy of the reflection signal computed
in the summation procedure of the 2-D transformation. The consequence of
this phenomenom is seen in Fig.6.a. where the output signal for a given
input $s(x,t) = c_{nse}(x_i,t)+0.1r_1(x_i,t)$ is shown - under the same processing
conditions as those applied in Fig. 5.b. Comparison of these results shows
a $\cong$ 7 dB loss of the enhanced signal level.

Case III - hyperbolic reflection distribution is considered

We assume, for the purpose of this model :
- a refraction signal again the same as before,
- a reflection signal $r_2(x_i,t)=r_1(x_i,t-t_i+t_o)$, where $t_i = \left[t_o^2 + (x_i/c)^2\right]^{\frac{1}{2}}$
The delay t_i represents the hyperbolic distribution[5] of the tunnel reflec-
tion. The model considered now regards all known apriori characteristics
of this reflection. In comparison with the preceeding case, the nonlinear
delays, t_i also have an unfavorable affect on the summation of the signals
in the 2-D transformation, which searches for the signals distributed in
line. The result of processing of the test signal $s(x,t) = c_{no_i}(x_i,t) +$
$0.1r_2(x_i,t)$ for the same conditions, as applied in Fig. 6.a., is presented
in Fig. 6.b. The relative $\cong$ 2 dB drop of the $G_c^{(16)}$ value results from the
considered phenomenom.

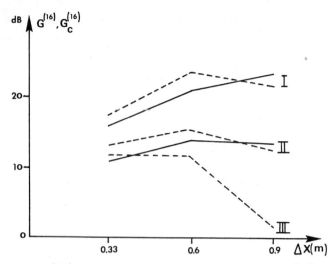

Fig. 7. Computed $G^{(16)}$ (continuous line) and calculated $G_c^{(16)}$ from filtered
sections (dotted line) spatial filter gain values for trace spacing
$\Delta x=0.33$, 0.6 and 0.9 m and for the three discussed cases I,II and
III.

Influence of the array length

For the considered reflection signal distribution A(x) (Fig.1.b.) the
loss $L(\Delta x)$ /4/ becomes smaller for short arrays of length $1<7$ m. At the
same time, the influence of the hyperbolic delay also becames negligible
because $t_i \cong t_0$ = const. But the potential $G^{(N)}$ value (case I) grows when the
array length increases. Finally the $G^{(N)}$ value computed for cases II and
III pass over the optimum point when array trace spacing increases. This
is illustrated in Fig.7. where computed $G^{(16)}$ and calculated $G_c^{(16)}$ values
are presented as a function of Δx, for cases I, II and III. It is seen that
both the computed and experimental data display a similar development
- this confirms the advantage of the introduced $G^{(N)}$ coefficient computation for the array design analysis.

Case IV - real refraction signal

We assume again :
- as refraction : the signal $c_{n_1}(x_i,t)$
The signal $c_{n_1}(x_i,t)$ (Fig.8.a.) was recorded with a trace spacing $\Delta x=0.66$m,
30m away from the tunnel under the same geological conditions as occur at
the tunnel position. We designed this signal as a representation of the coherent noise in our detection problem. Fig.8.b. shows the output signal
from the F_0 filter with an input test signal $s(x,t)=c_{n_1}(x_i,t)+r_0(x_i,t)$.
The obtained spatial filtering gain $G^{(16)}$ is 20 dB - when the attenuation
of the regular refractions first arrival is considered. The other parts
of the refraction signal correllated poorly with the neighbouring channels,
pass over the F_0 filter nearly without attenuation - because their spectra
are relatively wider in the spatial frequency domain. Finally, the presence
of these residues makes it impossible to distinguish the reflection signal -
when a more realistic reflection signal level is assumed. The way to improve
the signal spatial correlation is by means of the A.G.C. application. Fig.
8.d.c. shows the output signals for test inputs $s(x,t)=c_{n_1}(x_i,t)$ and $s(x,t)$
$=c_{n_1}(x_i,t)+0.1r_2(x_i,t)$. These signals were AGC controlled, bandpass filtered
and then AGC controlled and filtered once more, before spatial processing.
The comparison of these sections now shows signs of the enhanced r_2 signal

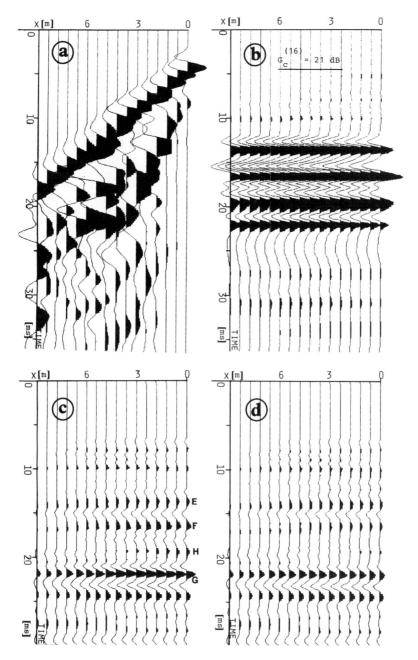

Fig. 8. (a) Real coherent noise signal $c_{n_1}(x_i,t)$ (amplitude and time corrected) ; (b) spatially processed with F_0 filter (+ spatial tapering) output section for test input $s(x,t)=c_{n_1}(x_i,t)+r_0(x_i,t)$;
(c) same as section (b) but for $s(x,t)=c_{n_1}(x_i,t)+0.1r_2(x_i,t)$ double
AGC controlled and filtered before spatial processing ; (d) same
as section (c) but for $s(x,t) = c_{n_1}(x_i,t)$ (reflection absence).

presence : an appearance of the event H and an amplification of events E, F,J. The supprising coinsidence of these events with the principal position of the synthetic reflection signal (seen in Fig. 8.b.) should be noted. This allows us to suspect the existence of a tunnel where, according to the available maps, no tunnels are present.

CONCLUSIONS

- The introduced spatial filtering gain coefficient seems to be useful for the array efficiency examination in the frequency domain. It enables the correct choice of field and processing parameters for a given input signal characteristic, especially when the relatively wider-band signal enhancement is required. This is often needed for shallow seismic prospection.

- The presence of both the enhanced and undesired signal, in all array channels, is essential for the spatial filtering efficiency. Therefore array length should be carefully chosen with regard to given field conditions. The optimum length should be considered as a compromise between the potential array gain and the presence of the well correlated signals. The AGC application is therefore desirable before spatial filtering.

- It seems justified to replace the spatial filtering, by means of CDP stacking[5] (direct summation of the array outputs), commonly applied in deep seismics, with the especially studied 2-D filter, in order to make the most of the enhanced signal energy. Spatial tapering of the array data seems to be indispensable.

ACKNOWLEDGMENTS

The authors wish to thank James J.Huerther for the language corrections and Annick Pennequin for her work in preparing the manuscript.

REFERENCES

1. J.C. TRICOT, B. DELANNOY, B. PIWAKOWSKI, Ph. PERNOD, Some problems and experimental results of seismic shallow propecting, 15 th Int. Symp. on Acoustical Imaging, Halifax, Canada.

2. B. PIWAKOWSKI, J.C. TRICOT, B. DELANNOY, Ph. PERNOD, Shallow seismic prospection of underground tunnels - evaluation of general detection conditions, deposed for acceptance for Geophysics in March 1987.

3. M. SUPRAJITNO, S.A. GREENHALGH, Separation of upgoing and downgoing waves in vertical seismic profiling by contour-slice filtering, Geophysics, 50, 6, 1985.

4. W.G. CLEMENT, Basic principles of two-dimensional digital filtering, Geophysical Prospecting 21, 1973.

5. M.B. DOBRIN, Introduction to geophysical prospecting, Tokyo Mc Graw-Hill, Inc.

PARTICIPANTS

Prof. R. Martin Arthur
Washington University
700 South Euclid Avenue
St. Louis, MO 63110
(314) 362-2135

Prof. A. R. Ashrafzadeh
Department of EE
University of Detroit
4001 McNichols Road
Detroit, MI 48221
(313) 927-1364

Prof. Mani Azimi
Electrical Engineering Dept.
Michigan State University
East Lansing, MI 48824
(517) 353-8980

Mingsian Robin Bai
Iowa State University
5175 Buchanan Hall
Ames, IA 50013
(515) 294-0075

Dr. H. Bartelt
Siemens AG, ZFE TPH 41
Paul-Gosseu-Str. 100
D-8520 Erlangen
Federal Republic of Germany
09131/720110

Michael Berggren
Dept. of Bioengineering
University of Utah
Room 2480 MEB
Salt Lake City, UT 84112
(801) 581-7064

Tim Blankenship
University of Missouri-Rolla
118 Elec. Engr. B ldg.
Rolla, MO 65401
(314) 341-4506

Carl Boehning
Allied-Signal Inc.
Bendix Kansas City Division
P. O. Box 619159, D/816, SA-1
Kansas City, MO 64141-6159
(816) 997-5266

M.S.S. Bolorforosh
King's College London
21-Firlodge, 3-Gipsy Lane
London 3W15
5.S.A., England
01-8764037

Steven R. Broadstone
Washington University
700 South Euclid Avenue
St. Louis, MO 63110
(314) 362-2135

Prof. Jim Burt
Physics Department
York University
4700 Keele Street
Downsview, Ontario M3J 1P3
Canada
(416) 623-2100 ext. 7722

Lawrence Busse
General Electric - AEBG
One Neumann Way, E-45
Cincinnati, OH 45215
(513) 552-4926

Didier Cassereau
Etudes et Productions Schlumberger
26 Rue de La Cavee
92120 Clamart, France
33-1-45-37-26-14

Thomas J. Cavicchi
University of Illinois
1406 West Green Street
Urbana, IL 61820
(217) 333-0188

Y. S. Cha
Argonne National Laboratory
9700 South Cass Avenue
Argonne, IL 60439
(312) 972-5899

Richard Y. Chiao
Department of ECE
University of Illinois
1406 West Green Street
Urbana, IL 61801
(217) 333-4465

Hajime Chiba
JEOL Trading Co., Ltd.
1-15-6 Ginza
Tokyo, Japan
(03) 567-4780

Al Chu
Mayo Clinic
200 First Street SW
Rochester, MN 55905
(507) 284-2115

Dr. Wei-Kom Chu
University of Nebraska Med. Ctr.
42nd and Dewey Avenue
Omaha, NE 68105
(402) 559-5275

Jen-Hui Chuang
Department of ECE
University of Illinois
1406 West Green Street
Urbana, IL 61801
(217) 333-6753

Prof. Noriyoshi Chubachi
Dept. of Electrical Engineering
Tohoku University
Sendai 980, Japan
(022) 222-1800 Ext. 4286

C. C. Cutler
Stanford University
Ginzton Lab
Stanford, CA 94305
(415) 723-0261

Patrick S. Davis
Hughes Aircraft Company
2000 East El Segundo Blvd.
BLD E-1, M/S L132
El Segundo, CA 90250
(213) 616-9990

Pr. J. F. de Belleval
Universite de Compiegne
BP233
60206 Compiegne, France
44-20-9960

Floyd Dunn
University of Illinois
1406 West Green Street
Urbana, IL 61801
(217) 333-3133

Andre Farcy
IFREMER
ZP de Bregaillom BP 330
83512 La Seyne Cedex
France
94 94 1836

Marc L. Feron
Department of ECE
Illinois Institute of Technology
3301 South Dearborn
Chicago, IL 60616
(312) 567-3400

Prof. Mathias Fink
University Paris VII
2 Place Jussieu
Paris 75005, France
(1) 4336-2525 ext. 4620

Dr. F. Stuart Foster
Physics Division
Ontario Cancer Institute
500 Sherbourne Street
Toronto, Ontario
Canada M4X 1K9
(416) 924-0671 ext. 5066

B. Froelich
Schlumberger
Old Quarry Road
Ridgefield, CT 06877
(203) 431-5331

Dr. M. J. Gebel
Div. Gastroenterol. and Hepatol
Medizinische Hochschule Hannover
3000 Hannover 61, POB. 610180
West Germany
511-5323305

Mark S. Geisler
Marquette University
8200 West Tower Avenue
Milwaukee, WI 53222
(414) 355-5000

648

Steve Gleason
Iowa State University
1300 Gateway, 301
Ames, IA 50010
(515) 292-3856

Prof. Sheryl Gracewski
Mechanical Engineering Dept.
University of Rochester
217 Hopeman
Rochester, NY 14627
(716) 275-7853

Philip S. Green
SRI International
333 Ravenswood Avenue
Meno Park, CA 94025
(415) 859-2882

Rolf Kahrs Hansen
Nutec
P.O. Box 6
N-5034 YTRE Laksevaag
Norway
47-5341600

John Harris
University of Illinois
216 Talbot Laboratory
104 South Wright
Urbana, IL 61820
(217) 333-7433

Jeff Hastings
ICFAR
611 North Capitol Avenue
Indianapolis, IN 46204
(317) 262-5071

Patrick B. Heffernan
Hewlett-Packard Labs
Building 286
1651 Page Mill Road
Palo Alto, CA 94306
(415) 857-7396

Gert Hetzel
Siemens
Henkestr. 127
Dept. STUE 12
8520 Erlangen
West Germany
9131-843488

Xie Huchen
University of Science & Technology
Hefei, Anhuei
China

Dr. Mehdi Vaez-Iravani
Philips Laboratories, NAPC
345 Scarborough Road
Briarcliff Manor, NY 10510
(914) 945-6429

Richard Johnson
Quantum Medical Systems
1040 12th Avenue N.W.
Issaquah, WA 98027
(206) 392-9180

Steven A. Johnson
Dept. of Bioengineering
University of Utah
Room 2480 MEB
Salt Lake City, UT 84112
(801) 581-8528

Dr. Patrick H. Johnston
NASA Langley Research Center
Mail Stop 231
Hampton, VA 23665
(804) 865-3036

Dr. Hugh W. Jones
Technical University of Nova Scotia
P.O. Box 1000
Halifax, Nova Scotia B3J 2X4
Canada
(902) 429-8300 ext. 2208

Martin P. Jones
Alcoa
ATC-B, EDD
Alcoa Center, PA 15069
(412) 337-2368

James Justice
Mobil R&D Corporation
2819 Carriage Lane
Carrollton, TX 75006
(214) 851-8249

Dr. Lawrence W. Kessler
Sonoscan, Inc.
530 East Green Street
Bensenville, IL 60106
(312) 766-8795

Sun I. Kim
Drexel University
Dept. of Biomedical
Philadelphia, PA 19104
(215) 895-1924

Gregory D. Lapin
Children's Memorial Hospital
Northwestern University
2300 Children's Plaza
Chicago, IL 60614
(312) 880-4730

Prof. Hua Lee
Electrical Engineering Dept.
University of Illinois
1406 West Green Street
Urbana, IL 61801
(217) 333-0167

Sidney Lees
Forsyth Dental Center
140 Fenway
Boston, MA 02115
(617) 262-5200

Dr. Robert M. Lerner
University of Rochester
Strong Memorial Hospital Diag. Rad.
601 Elmwood Avenue
Rochester, NY 14642
(716) 275-8365

D. Lesselier
CNRS-ESE
ESE Plateau du Moulon
91190 Gif-Sur-Yvette
France
(1) 69478040

Zse-Cherng Lin
SRI International
333 Ravenswood Avenue
Menlo Park, CA 94025
(415) 859-4904

Timothy Luce
Applied Research Lab
Penn State University
123 North Barnard
State College, PA 16801
(814) 238-4521

Prof. Philip L. Marston
Department of Physics
Washington State University
Pullman, WA 99164-2814
(509) 335-5343

Prof. J. F. McDonald
Rensselaer Polytechnic Institute
Center for Integrated Electronics
Troy, NY 12181
(518) 270-6033

Robert McWharf
General Electric Company
CSP-5 W2
Syracuse, NY 13146
(315) 456-1672

Serge Mensah
C.N.R.S.
31 Chemin de Joseph Aiguier
13402 Marseille Cedex 9
France
91-22-41-76

John W. Monzyk
Southern Illinois University
516 South Hays
Carbondale, IL 62901
(618) 549-2498

Dan Nagle
Illinois Institute of Technology
3366 South Michigan
Chicago, IL 60616
(312) 326-9830

M. Nikoonahad
Philips Laboratories
345 Scarborough Road
Briarcliff Manor, NY 10510
(914) 945-6312

Prof. Hiroji Ohigashi
Dept. of Polymer Materials Engrg.
Yamagata University
Jonan 4-3-16, Yonezawa, Yamagata
Yonezawa, 992, Japan
(0238) 22-5181

Michael G. Oravecz
Sonoscan, Inc.
530 East Green Street
Bensenville, IL 60106
(312) 766-8795

Harry Oung
Drexel University
32nd & Chestnut Streets
Philadelphia, PA 19104
(215) 895-1829

Prof. Song B. Park
Korea Advanced Inst. of Science & Tech
P.O. Box 150
Chongyangni, Seoul
Korea
967-8901   ext. 3711

Prof. Anna Pate
Dept. of Engineering Science & Mech
Iowa State University
Ames, IA 50011
(515) 294-0093

Bogdan Piwakowski
Institut Industriel du Nord
59 651 Villeneuve d'Ascq
B.P. 48 Cedex
France
20 91 01 15 ext. 323

Paul Reinholdtsen
Stanford University
Ginzton Lab
Stanford, CA 94305
(415) 723-0297

William D. Richard
Electrical Engineering Department
University of Missouri-Rolla
Rolla, MO 65401
(314) 364-6631

John M. Richardson
Rockwell International Science Center
1049 Camino Dos Rios
Thousand Oaks, CA 91360
(805) 373-4137

Karen L. Riney
Sonoscan, Inc.
530 East Green Street
Bensenville, IL 60106
(312) 766-8795

William P. Robbins
Department of Electronic Engineering
University of Minnesota
123 Church
Minneapolis, MN 55455
(612) 625-8014

Ronald A. Roberts
Argonne National Laboratory
9700 South Cass Avenue
Argonne, IL 60439
(312) 972-5198

Brent Robinson
Advanced Technology Labs, Inc.
22100 Bothell Hgwy, S.E.
P.O. Box 3003
Bothell, WA 98041-3003
(206) 487-7124

Prof. Mark S. Rondeau
Cornell Medical College
Dept. of Ophthomology
1300 York Avenue
New York, NY 10021
(212) 472-5616

Dr. Jafar Saniie
Electrical & Computer Engrg. Dept.
Illinois Institute of Technology
IIT Center, Dept. of ECE
Chicago, IL 60616
(312) 567-3400

Laura J. Santangelo
Sonoscan, Inc.
530 East Green Street
Bensenville, IL 60106
(312) 766-7088

Mark E. Schafer
Interspec
1100 East Hector Street
Conshohocken, PA 19428
(215) 834-1511

Jungen Schmolke
Uni-Erlangen, Hochfrequenztech.
Cauerstr. 9
D-8520 Erlangen
Germany
09131/85 72 27

Janet E. Semmens
Sonoscan, Inc.
530 East Green Street
Bensenville, IL 60106
(312) 766-8795

S. H. Sheen
Argonne National Laboratory
9700 South Cass Avenue
Argonne, IL 60439
(312) 972-7502

Michael Sherar
Physics Division
Ontario Cancer Institute
500 Sherbourne Street
Toronto, Ontario
Canada M4X 1K9
(416) 924-0671 ext. 5066

Akira Shiba
Fujitsu Laboratories Ltd.
1015, Kamiodanaka Nakahara-ku
Kawasaki 211
Japan
044-777-1111 ext. 2-6376

John Skelton
Albany International Rej Co.
1000 Providence Highway
Dedham, MA 02026
(617) 326-5500

Dr. Andrew M. Snoddy
ICFAR
611 North Capitol Avenue
Indianapolis, IN 46204
(317) 262-5080

Prof. Mehrdad Soumekh
Dept. of Electrical Engineering
State Univ. of New York
Buffalo, NY 14260
(716) 636-2425

Dr. Richard Stiffler
Alcoa
Alcoa Technical Center
Alcoa Center, PA 15069
(412) 337-5819

Wan Suiren
Nanjing Institute of Technology
Dept. of Biomedical Engineering
People's Republic of China
31700

Thomas L. Szabo
Hewlett Packard
3000 Minuteman Road
Andover, MA 01810
(617) 687-1501

Yasuhito Takeuchi
Yokogawa Medical Systems Ltd.
613 Sakae-cho
Tachikawa 190
Japan
81-425-36-8561

Lee Tatistcheff
Digital Equipment Corporation
30 Forbes Road, NR05/B4
Northboro, MA 01532
(617) 351-4875

Tat-Jin Teo
Biomedical Engineering Dept.
Drexel University
32nd & Chestnut Streets
Philadelphia, PA 19104
(215) 895-1829

Dr. Piero Tortoli
Dept. of Electronic Engineering
University of Florence
via Santa Marta, 3
Florence, Italy 50139
(055) 47951

Louis Tran Huu Huc
Laboratoire de Biophysique Medical
2 Bis Boulevard Tonnele
Tours 37000
France
47-37-66-69

W. Randal Vaughn
Independent Consultant & Publisher
One Holly Court Terrace
Lake Zurich, IL 60047
(312) 540-0618

Marija Vukovojac
Butterworth Scientific Limited
P.O. Box 63 Westbury House
Bury Street
Guildford, Surrey GU2 5BH
United Kingdom
0483-31261

Prof. Glen Wade
University of California
Department of ECE
Santa Barbara, CA 93106
(805) 961-2508

Tao Wang
Illinois Institute of Technology
3224 South Wallace Avenue
Chicago, IL 60616
(312) 842-4142

Zhao Qiu Wang
Laboratoire de Biophysique
Faculte de Medecine
2 Bd Tonnelle
Tours 37000
France
47-37-66-69

Richard J. Wombell
Physics Department
King's College London
Strand, London WC2R 2LS
United Kingdom
01-836-5454 ext. 2142

Debbie Yu
Sonoscan, Inc.
530 East Green Street
Bensenville, IL 60106
(312) 766-8795

AUTHOR INDEX